WHITE NOISE THEORY OF PREDICTION, FILTERING AND SMOOTHING

STOCHASTICS MONOGRAPHS

Theory and Applications of Stochastic Processes

A series of books edited by Mark Davis, Imperial College, London, UK

Volume 1
Contiguity and the Statistical Invariance Principle
P. E. Greenwood and A. N. Shiryayev

Volume 2
Malliavin Calculus for Processes with Jumps
K. Bichteler, J. B. Gravereaux and J. Jacod

Volume 3
White Noise Theory of Prediction, Filtering and Smoothing
G. Kallianpur and R. L. Karandikar

Additional volumes in preparation

This book is part of a series. The publisher will accept continuation orders which may be cancelled at any time and which provide for automatic billing and shipping of each title in the series upon publication. Please write for details.

WHITE NOISE THEORY OF PREDICTION, FILTERING AND SMOOTHING

By

G. Kallianpur

The University of North Carolina, USA

and

R. L. Karandikar

Indian Statistical Institute, Delhi, India

GORDON AND BREACH SCIENCE PUBLISHERS

New York London Paris Montreux Tokyo Melbourne

Gordon and Breach Science Publishers

Post Office Box 786
Cooper Station
New York, New York 10276
United States of America

Post Office Box 197
London WC2E 9PX
England

58, rue Lhomond
75005 Paris
France

Post Office Box 161
1820 Montreux 2
Switzerland

3-14-9, Okubo
Shinjuku-ku, Tokyo
Japan

Private Bag 8
Camberwell, Victoria 3124
Australia

Library of Congress Cataloging-in-Publication Data
Kallianpur, G.
White noise theory of prediction, filtering, and smoothing.

(Stochastics monographs, ISSN 0275-5785 ; v. 3)
Bibliography: p.
Includes index.
1. Gaussian processes. 2. Kalman filtering.
3. Prediction theory. I. Karandikar, Rajeeva L.,
1956- II. Title. III. Series.
QA274.4.K35 1988 519.2 88-7330
ISBN 2-88124-685-0

Dedicated to the memory
of Norbert Wiener and Andrei Nikolayevich Kolmogorov

Contents

Introduction to the Series

The journal *Stochastics* publishes research papers dealing with stochastic processes and their applications in the modelling, analysis and optimization of systems subject to random disturbances. Stochastic models are now widely used in engineering, the physical and life sciences, economics, operations research, and elsewhere. Moreover, these models are becoming increasingly sophisticated and often stretch the boundaries of the theory as it exists. A primary aim of *Stochastics* is to further the development of the field by promoting an awareness of the latest theoretical developments on the one hand and of all problems arising in applications on the other.

In association with *Stochastics*, we are now publishing *Stochastics Monographs*, a series of independently produced volumes with the same aims and scope as the journal. *Stochastics Monographs* will provide timely and authoritative coverage of areas of current research in a more extended and expository form than is possible within the confines of a journal article. The series will include extended research reports, material derived from lecture courses on advanced topics, and multi-author works with a unified theme based on conference or workshop presentations.

Mark Davis

Preface

This monograph is an expanded version of the work we have done over the past five years on the finitely additive white noise approach to filtering and prediction theory. Most of it has appeared in a series of papers and we have taken the opportunity to improve upon it and present it in as complete a form as possible.

We are indebted to Professor A. V. Balakrishnan whose conversations over the years with one of us aroused our interest in the subject and encouraged us to undertake a systematic development of the theory.

Our sincere thanks are due to Lee Trimble for her unfailing courtesy and the patience and care with which she has done the difficult job of typing.

G. Kallianpur
R. L. Karandikar

CHAPTER I

INTRODUCTION

The object of this monograph is to develop the theory of Gaussian white noise and apply it to a large class of non-linear filtering, prediction and interpolation problems. It should be made clear at the outset that we use the term 'white noise' to stand for the *finitely additive* Gaussian measure on (the field of cylinder sets of a separable) Hilbert space.

It is important first to explain the need for our use of finitely additive probabilities for it is the systematic use of the latter that sets the present work apart from the existing theory and, in our view at least, leads to a more natural and simplified presentation of the subject.

It is well known, of course, that the axiom of countable additivity is not directly linked to real phenomena. In his fundamental work [49], Kolmogorov remarks in connection with this axiom that countably additive probability spaces "occur only as idealized models of real random processes" and that we "*limit ourselves, arbitrarily to only those models which satisfy* Axiom VI [i.e., the axiom of countable additivity]." (Italics in the original.) Nevertheless, the inclusion of the axiom has enabled probability theory to exploit the richness of measure theory with such success that to question its validity (even in a field of application) might be considered by some to be a heresy.

There is one area of application in which finitely additive measure theory has been used to advantage. Finite additivity has been taken as a basic assumption in the systematic development of subjective theories of statistical inference and, more especially, in the study of optimal gambling sytems. We refer here to the books by Savage, and Dubins and Savage [69, 19]. However, the reasons motivating these authors seem to be quite different from ours.

Thoughout this work we will be concerned with models of the type

$$(1) \quad O_t = S_t + n_t \qquad (0 \leq t \leq T)$$

where information about a signal process S is obtained by observing the process O corrupted by Gaussian white noise n. The conventional method to deal rigorously with Gaussian white noise (GWN) is to replace (1) by the "integrated" version

$$(2) \quad Y_t = \int_0^t S_u du + W_t$$

or

$$(2)' \quad dY_t = S_t dt + dW_t.$$

Thus the Wiener process is used as a rigorous substitute for GWN. It is our view that the role of the Wiener process as a model for white noise should be separated from its other, extremely versatile uses in the theory of stochastic processes. We shall explain below why a finitely additive theory of GWN conforms to the intuitive meaning of "noise" and furnishes us with a more satisfying model for GWN at least for the application we have in mind.

Nonlinear filtering and prediction theory, which began with efforts to generalize the linear or Bucy-Kalman theory, has been developed using the techniques of the theory of continuous parameter semimartingales and Itô stochastic differential equations (SDE's). We will refer to this approach to the subject as the conventional or stochastic calculus theory. Despite the elegance and power of the mathematics involved and the enormous stimulus it has provided to the theory of SDE's (for instance, the SDE satisfied by the optimal nonlinear filter is one of the first "naturally" arising examples of an SDE governing an *infinite dimensional* process), the practical validity of the stochastic calculus theory is open to serious criticism.

A.V. Balakrishnan was the first to express the view (shared by some engineers) in a series of papers that the conventional approach based on SDE's is not suitable for applications since the results obtained cannot be instrumented. He was also the first to advocate and initiate the use of finitely additive white noise in filtering problems.

A main objective of this monograph is to develop rigorously the theory of Gaussian white noise measures on Hilbert spaces and apply them to give a comprehensive account of nonlinear filtering theory including prediction and smoothing.

Two points are worthy of mention in connection with the theory which we propose to present in this book:

(1) The natural space of observations and of the noise is a Hilbert space (or some subspace of it) and zero Wiener measure. An insistence on the use of countably additive probabilistic techniques would require an enlargement of this space either to a Wiener space of continuous functions $C([0,T];\mathbb{R}^m)$ or to an even larger space such as $\mathcal{S}^*(\mathbb{R}^d)$, the Schwarz space of tempered distributions. (The latter interesting possibility has not yet been fully explored in the

stochastic calculus set up). The alternative is to resort to a finitely additive measure on Hilbert space to describe the statistics of white noise.

(2) A central question that has to be settled first is the search for an alternative to the Wiener process as a model for white noise. It is our view that this aspect of the Wiener process has not received sufficient attention in the vast literature devoted to the study of Brownian motion. Wiener himself was well aware (in a 1955 conversation with one of the authors at the Indian Statistical Institute) that the Wiener process W defines a finitely additive white noise $n(f) := \int_0^1 f(t)dW_t$ over the Hilbert space $L^2[0,1]$. The latter is indeed an example of a cylinder measure or weak distribution used by I.E. Segal at about the same time and later by L. Gross. But all of this work has been almost exclusively devoted to linear problems. In Chapter 3 and subsequent chapters we will develop white noise calculus for application to nonlinear filtering theory. As a preliminary to it we will, in this section, begin by making more specific comments to throw more light on points (1) and (2) mentioned above.

It should be mentioned at once that the problem arises only in the case of continuous parameter processes. In discrete time problems, as for instance, in problems of regression analysis is statistics, noise is customarily modeled by a sequence of mutually independent (or uncorrelated) random variables (sometimes assumed to be Gaussian).

In the remainder of this chapter we give a brief account of the probabilistic concepts essential for the development of white noise calculus. The additive nonlinear filtering model will involve a stochastic process (usually a

Markov process) defined on a σ-additive probability space and a Gaussian white noise that is not, and cannot be defined on a countably additive probability space natural to the problem.

The stochastic calculus techniques used in conventional filtering theory require a fairly comprehensive background of stochastic integrals and semimartingale theory which it will take us too far from our present purpose to discuss here. Besides, exhaustive treatments of these topics are available in several books devoted to filtering theory. Our concern in this chapter will be to concentrate on those areas of probability theory and stochastic processes which will contribute to a better understanding of the finitely additive white noise theory developed in the later chapters of the book.

Although portions of this monograph can serve as an introduction to the study of finitely additive (f.a.) probability measures on Hilbert spaces and could also be used for certain problems of Quantum Physics, we have included it pri marily as necessary background material (most of it new and hardly any of it available in books) for analysis of the "noise" part of the rigorous filtering model introduced in Chapter 7. However, before developing the white noise filtering theory proper, we have to consider the other ingredient in the problem--the so called signal or system process X, the process of statistical interest. Since a crucial requirement of the theory is to derive recursive estimates for the optimal filter or predictor in the form of solutions to partial differential (or other evolution) equations, we will be assuming, for the most part, that X is a Markov process. Hence we begin, in the next chapter, with a discussion of Markov processes and their generators. Chapters 3 through

6 will be devoted to f.a. white noise calculus. We will start with filtering theory from Chapter 7 onwards with the introduction of the abstract filtering (as well as prediction and smoothing) model and the Bayes formula.

1. PROBABILISTIC PRELIMINARIES

$(\Omega,\mathcal{A},P)$ is a complete probability space if (i) Ω is an arbitrary set, (ii) $\mathcal{A}$ is a σ-field of subsets of Ω and (iii) P is a (countably additive) probability measure on $\mathcal{A}$. It is further assumed that P is complete with respect to $\mathcal{A}$, i.e., if $N \in \mathcal{A}$ with $P(N) = 0$ then any subset $N' \subset N$ belongs to $\mathcal{A}$.

A function $f: \Omega \to \mathbb{R}$ which is measurable with respect to $\mathcal{A}$ is called a random variable. It will be necessary in our work to consider random mapping taking values in more general topological spaces. A complete, separable metric space S is called a Polish space. The σ-field $\mathcal{B}(S)$ of topological Borel sets of S is defined to be the σ-field generated by all the open sets (equivalently, all the closed sets) of S.

A map X from a complete probability space $(\Omega,\mathcal{A},P)$ into a Polish space S is called an *S-valued r.v.* if it is $\mathcal{B}(S)/\mathcal{A}$-measurable, i.e., $X^{-1}(B) \in \mathcal{A}$ for every $B \in \mathcal{B}(S)$. The term 'random variable' without qualification will always mean a real valued r.v.

A sequence (X_n) of S-valued r.v.'s *converges in probability* to an S-valued r.v. X if $P\{\omega: \rho(X_n(\omega),X(\omega)) > \epsilon\} \to \infty$ as $n \to \infty$, for every $\epsilon > 0$, where ρ is a metric on S. $X_n \to X$ P-almost surely (P-a.s.) if and only if

$$P\{\omega: \lim_{n\to\infty} \rho(X_n(\omega),X(\omega)) = 0\} = 1$$

Similar definitions apply to a continuous parameter family of r.v.'s (X_t).

Linear spaces of random variables

Let $\mathscr{L} := \mathscr{L}(\Omega,\mathscr{A},P)$ be the linear space of real random variables on $(\Omega,\mathscr{A},P)$. We introduce some well known spaces of r.v.'s. More precisely, they will be spaces of equivalence classes of r.v.'s where f and g belong to the same equivalence class if and only if $f = g$ P-a.s. For p, $(1 \le p < \infty)$ define $\mathscr{L}^p := \mathscr{L}^p(\Omega,\mathscr{F},P)$ to be the space of f's defined on $(\Omega,\mathscr{F},P)$ and such that $E|f|^p < \infty$. Then, for each p, $\mathscr{L}^p$ is a Banach space. Of particular interest is the Hilbert space $\mathscr{L}^2$.

The space $\mathscr{L}_0$

For $f,g \in \mathscr{L}$ define

$$d(f,g) = \int_\Omega (|f(\omega)-g(\omega)| \wedge 1)dP.$$

It is easy to check that d is a metric on $\mathscr{L}$. For any $\epsilon > 0$,

$$\begin{aligned} d(f,g) &= \int_{|f-g|\le\epsilon} (|f-g| \wedge 1)dP + \int_{|f-g|>\epsilon} (|f-g| \wedge 1)dP \\ &\le \epsilon + P(|f-g| > \epsilon). \end{aligned} \tag{1.1}$$

Also

$$d(f,g) \ge (\epsilon \wedge 1)P(|f-g| > \epsilon). \tag{1.2}$$

From (1.1) and (1.2) it follows that $d(f_n,f) \to 0$ if and only if $f_n \to f$ in P-probability. Moreover, a sequence (f_n) is d-Cauchy if and only if (f_n) is Cauchy in P-probability. Hence, denoting the space of equivalence classes also by $\mathscr{L}$ it follows that $(\mathscr{L},d)$ is a complete metric space. In general, however, it is not a Polish space because $(\mathscr{L},d)$ may fail to be separable. If $\mathscr{A}$ is countably generated, i.e., if $\mathscr{A}$ is the σ-field generated by a countable family of sets, then it can be shown that $(\mathscr{L},d)$ is separable. The space $(\mathscr{L},d)$ is usually denoted by $\mathscr{L}_0$.

Conditional expectation

Let $\mathscr{F}$ be a sub σ-field of $\mathscr{A}$. Suppose that f is a r.v. on $(\Omega,\mathscr{A},P)$ with finite expectation, i.e., such that $E|f| < \infty$. Define

$$Q(f) := \int_F f dP \quad \text{for all } F \in \mathscr{F}.$$

Clearly Q is a finite signed measure on $\mathscr{F}$ with $Q(\Omega) = E(f)$, and such that Q is absolutely continuous with respect to the restriction of P to $\mathscr{F}$, denoted by $P^{\mathscr{F}}: Q << P^{\mathscr{F}}$. By the Radon-Nikodym theorem, there exists a ($P^{\mathscr{F}}$-a.s. unique) $\mathscr{F}$-measurable r.v. $\frac{dQ}{dP^{\mathscr{F}}}$ such that for all $F \in \mathscr{F}$,

$$\int_F f dP = \int_F \frac{dQ}{dP^{\mathscr{F}}} dP.$$

The conditional expectation $E(f|\mathscr{F})$ of f given $\mathscr{F}$ is defined as

$$E(f|\mathscr{F}) := \frac{dQ}{dP^{\mathscr{F}}} .$$

Some of the basic properties of conditional expectation are listed below. Throughout, f is an integrable random variable.

(1) $f \geq 0$ implies $E(f|\mathscr{F}) \geq 0$. a.s.

(2) If $\mathscr{F}$ and $\mathscr{G}$ are sub σ-fields and $\mathscr{F} \subset \mathscr{G}$,

$$E\{E(f|\mathscr{G})|\mathscr{F}\} = E(f|\mathscr{F}) \quad \text{a.s.}$$

(3) (a) Let $\mathcal{F}_n \subseteq \mathcal{F}_{n+1}$ $(n \geq 1)$ be an increasing family of sub σ-fields of $\mathcal{A}$. Then

$$\lim_{n\to\infty} E(f|\mathcal{F}_n) = E(f|\bigvee_{n=1}^{\infty} \mathcal{F}_n) \quad \text{a.s.,}$$

where $\bigvee_{n=1}^{\infty} \mathcal{F}_n$ is the smallest σ-field containing all the $\mathcal{F}_n$'s.

(b) Let $\mathcal{F}_n \supseteq \mathcal{F}_{n+1}$ $(n \geq 1)$ be a decreasing sequence of sub σ-fields of $\mathcal{A}$. Then

$$\lim_{n\to\infty} E(f|\mathcal{F}_n) = E(f|\bigcap_{n=1}^{\infty} \mathcal{F}_n) \quad \text{a.s.}$$

3(a) and 3(b) are consequences of well known martingale convergence theorems.

Now consider r.v.'s with finite second moments, i.e. such that $E|f|^2 < \infty$. We can regard equivalence classes of such r.v.'s as elements of the Hilbert space $H := L^2(\Omega,\mathcal{A},P)$. If $\mathcal{F} \subseteq \mathcal{A}$, $M := L^2(\Omega,\mathcal{F},P)$ is a closed linear subspace of H. In this set up, the conditional expectation with respect to $\mathcal{F}$ is identifiable with the orthogonal projection operator in H with range M which we shall denote here by Proj_M. A basic and important property of orthogonal projections says that $f - \text{Proj}_M f$ is orthogonal to g for all $g \in M$ and

$$\|f - \text{Proj}_M f\|^2 = \min_{g\in M} \|f-g\|^2. \tag{1.4}$$

Translated into the language of conditional expectations (1.4) can be rewritten as

$$E[f - E(f|\mathcal{F})]^2 = \min_{g \in L^2(\Omega,\mathcal{F},P)} E[f-g]^2. \tag{1.5}$$

When the σ-field $\mathcal{F} = \sigma(X)$, the σ-field generated by an S-valued r.v. X, the conditional expectation $E(f|\mathcal{F}) = \phi(X)$ where ϕ is the unique (up to $P \circ X^{-1}$ a.s. equivalence) $\mathcal{B}(S)$-measurable function on S to $\mathbb{R}$ such that

$$\int f \cdot 1_{[X \in B]} dP = \int_B \phi dP \circ X^{-1}.$$

The above characterization of $E(f|\mathcal{F})$ follows from the fact that

$$\int_B \phi dP \circ X^{-1} = \int \phi(X) 1_{[X \in B]} dP$$

and that every $\sigma(X)$-measurable function g is of the form $g(\omega) = g_1(X(\omega))$ for some $\mathcal{B}(S)$-measurable g_1.

Let X defined on the probability space $(\Omega,\mathcal{F},P)$ be an S-valued map (or random variable). Then the induced probability measure $P \circ X^{-1}$ is defined by

$$P \circ X^{-1}(B) = P[X^{-1}(B)] \quad \text{for all } B \in \mathcal{B}(S).$$

In general we cannot define EX for an S-valued r.v. X. However, if S is in addition a linear space such as a Banach or Hilbert space, EX may be defined as follows: Let S = B, a separable Banach space and X a B-valued r.v. Then $y[X(\omega)]$ is a real r.v. for every $y \in B^*$. Assume the following:

(i) $E|y[X(\omega)]| < \infty$ for every $y \in B^*$;

(ii) There exists $J \in B$ such that for every $y \in B^*$,

$$y[J] = E(y[X]).$$

Then J is defined to be the expected value of X, i.e.

$J = E(X)$.

2. PROBABILITY MEASURES IN POLISH SPACES

Let S be a Polish space and $\mathcal{B}(S)$ the σ-field of topological Borel sets in S. If P is a probability measure on $(S,\mathcal{B}(S))$ then P concentrates most of its mass on compact subsets of S. More precisely, for any $\epsilon > 0$ there is a compact K_ϵ in S such that $P(K_\epsilon) > 1-\epsilon$. Denote by $\mathcal{M}_0(S)$ the family of all probability measures on $\mathcal{B}(S)$.

Definition: A sequence (P_n) $(n \geq 1)$ in $\mathcal{M}_0(S)$ converges weakly to $P \in \mathcal{M}_0(S)$ written as $P_n \xrightarrow{d} P$ if and only if

$$(1) \qquad \lim_{n\to\infty} \int_S f(x)P_n(dx) = \int_S f(x)P(dx)$$

for all $f \in C_b(S)$. (Here $C_b(S)$ denotes the family of all bounded, real continuous functions on S.) The following two conditions are equivalent to (1):

$$(1a) \qquad \liminf_{n\to\infty} P_n(G) \geq P(G) \quad \text{for all open sets } G,$$

$$(1b) \qquad \limsup_{n\to\infty} P_n(F) \leq P(F) \quad \text{for all closed sets } F.$$

A metric ρ can be defined on $\mathcal{M}_0(S)$ under which $(\mathcal{M}_0(S),\rho)$ becomes a Polish space and $\rho(P_n,P) \to 0$ if and only if P_n converges weakly to P. We give two examples of metrics on $\mathcal{M}_0(S)$ for the topology of weak convergence.

Let $C_b(S)$ be the space of bounded continuous functions f from S into $\mathbb{R}$ with the norm $\|f\| = \sup_x |f(x)|$. When S is compact, $C_b(S)$ is separable and if $\{f_j: j\geq 1\}$ is a countable dense subset of $C_b(S)$, then

$$d_0(P_1,P_2) := \sum_{j=1}^{\infty} 2^{-j}\{(\int f_j dP_1 - \int f_j dP_2) \wedge 1\}$$

defines a matric on $\mathcal{M}_0(S)$ which gives rise to the weak convergence topology.

When S is not compact, we can choose a metric d' on S such that d'-completion of S is compact. Then the space $U_{d'}(S)$ of bounded d'-uniformly continuous functions is separable under uniform norm and if $\{f_j\}$ is a dense subset of $U_{d'}(S)$, then d_0 defined above is a metric on $\mathcal{M}_0(S)$ such that d_0 convergence is the same as weak convergence. However, $\mathcal{M}_0(S)$ is not complete under d_0. A second example is that of the Prokhorov metric ρ under which $\mathcal{M}_0(S)$ becomes a Polish space.

A sequence $\{P_n\}$ in $\mathcal{M}_0(S)$ is relatively compact if every subsequence of $\{P_n\}$ contains a further subsequence which is weakly convergent. The sequence $\{P_n\}$ is relatively compact if and only if it is tight, i.e., if for every $\epsilon > 0$ there is a compact $K_\epsilon \subset S$ such that $P_n(K_\epsilon) > 1-\epsilon$ for all $n = 1,2,\ldots$. Hence to show the weak convergence of the sequence $\{P_n\}$, one has to show (i) that $\{P_n\}$ is tight and (ii) that all its limit (cluster) points coincide.

For weak convergence of probability measures on C[0,1] we have the following well known result.

<u>Proposition</u>: A sequence $\{P_n\}$ of probability measures on $(C,\mathcal{B}(C))$ converges weakly to a probability measure P on the same space if the finite dimensional distribution of P_n converge weakly to those of P and if $\{P_n\}$ is tight.

The other important Polish spaces that occur in connection with the weak convergence of measures induced by Markov processes are the Skorokhod spaces D([0,T];S) or

$D([0,\infty);S)$, the space of paths $t \to x(t) \in S$ (S, a Polish space) which are right continuous and have left limits at every point. (Right continuity and left limit have to be understood suitably at the end points 0 and T.) A metric can be defined under which $D([0,T];S)$ or $D([0,\infty);S)$ becomes a Polish space.

For criteria for tightness for probability measures in these spaces as well as on other Polish spaces, the reader is referred to monographs on the subject.

When S is a Banach or Hilbert space it is possible to define the characteristic functional of P so that techniques of harmonic analysis are available. Let P be a probability measure on the Borel sets of a separable Banach space B. Then the function

$$\theta_P(f) = \int_B e^{if[x]} P(dx), \qquad f \in B^*$$

is called the characteristic functional of P. When B = H, a Hilbert space, we write (identifying H^* with H),

$$\theta_P(f) = \int_H e^{i(f,x)} P(dx) \quad \text{for } f \in H.$$

The use of the characteristic functional as a tool in the study of weak convergence is complicated by the fact that while $P_n \xrightarrow{d} P$ implies $\theta_{P_n}(f) \to \theta_P(f)$ for each f, the converse is false as is seen by the following simple example: Take $H = \ell_2$, and let $e_j = (0,\ldots, 1,0,\ldots)$ (1 in the jth place, 0's elsewhere). For each n,

$$e^{-\frac{1}{2}\sum_{j=1}^{n}(f,e_j)^2}$$

is obviously the characteristic functional of a Gaussian measure P_n on ℓ_2 that, in fact, concentrates its mass on an n-dimensional subspace of ℓ_2. Then clearly

$$\lim_{n\to\infty} \theta_{P_n}(f) = e^{-\frac{1}{2}|f|^2} \quad \text{for all } f \in \ell_2.$$

It is easy to see (and will be shown later) that there is no countably additive probability measure P on ℓ_2 such that $e^{-\frac{1}{2}|f|^2} = \theta_P(f)$. Thus, $e^{-\frac{1}{2}|f|^2}$ is clearly not the characteristic functional of a probability measure on $(H,\mathcal{B}(H))$. The reason why the theorem which is true for finite dimensional linear spaces breaks down for an infinite dimensional Hilbert (or Banach) space is that, in the latter case, the given topology is too strong, in other words, the continuity of θ_P with respect to the given topology is too weak a condition. For Hilbert spaces, the "correct" topology known as the Sazonov (or the Sazonov-Gross) topology τ_S is generated by the complete system of neighborhoods of the origin, of the form $\{f \in H: (Sf,f) < 1\}$ where S ranges over the class of positive, self-adjoint, trace class (or nuclear) operators S on H. We state the result for future reference:

<u>Theorem</u>: Let θ: $H \to \mathbb{C}$ satisfy the following conditions

(i) $\theta(-f) = \overline{\theta(f)}$, (i.e., θ is Hermitian);

(ii) For any $n \geq 1$, arbitrary $f_j,\ldots,f_n \in H$ and complex numbers $\alpha_1,\ldots,\alpha_n$,

$$\sum_{i,j=1}^{n} \theta(f_i-f_j)\alpha_i\overline{\alpha_j} \geq 0, \quad (\theta \text{ is nonnegative definite});$$

(iii) $\theta(0) = 1$.

Then θ is the characteristic functional of a probability measure on $(H,\mathcal{B}(H))$ if and only if θ is continuous in the Sazanov topology of H.

We will now apply Sazonov's theorem to show the following: Let $R = (R(t,s))$, $(t,s) \in [a,b]^2$ be a continuous covariance function and let $\mathcal{H}(R)$ be the corresponding reproducing kernel Hilbert space (RKHS). It is a well known fact shown by Kolmogorov's theorem that there is a unique Gaussian probability measure P defined on the Kolmogorov σ-field $\mathcal{A}$ of $\mathbb{R}^{[a,b]}$ such that the coordinate process $\{X(t)\}$, $X \in \mathbb{R}^{[a,b]}$ is Gaussian with zero mean and covariance function $R(t,s)$. Now $\mathcal{H}(R)$ is a subset of $\mathbb{R}^{[a,b]}$ and is, of course, a Hilbert space of functions on $[a,b]$. An easy application of Sazonov's theorem shows that $\mathcal{H}(R)$ cannot carry the centered Gaussian measure with covariance function R. More precisely, suppose that μ is a probability measure on the Borel sets of $\mathcal{H}(R)$ such that for all $t_1,\dots,t_n$, $(n \geq 1)$, $(x_{t_1},\dots,x_{t_n})$ has a centered Gaussian distribution under μ with $Ex_{t_1}x_{t_j} = R(t_i,t_j)$, where $x \in \mathcal{H}(R)$.

Denote by $\mathcal{S}$ the class of all elements f of the form $f = \Sigma_{i=1}^{k} a_i R(\cdot,t_i)$ ($t_i \in [a,b]$ and a_i's real numbers). Then $\mathcal{S}$ is a dense subset of $\mathcal{H}(R)$. By assumption, it follows easily that for $f \in \mathcal{S}$, $\langle x,f\rangle$ is a Gaussian random variable under μ, with characteristic function $E_\mu e^{i\langle x,f\rangle} = e^{-\frac{1}{2}\|f\|^2}$. Let $f \in \mathcal{H}(R)$ and $\{f_n\} \subset \mathcal{S}$ such that $\|f_n - f\| \to 0$.

An easy verification shows that

$$E_\mu e^{i\langle x,f\rangle} = \lim_{n\to\infty} Ee^{i\langle x,f_n\rangle} = e^{-\frac{1}{2}\|f\|^2}.$$

Thus the characteristic functional of μ is clearly <u>not</u> continuous with respect to the Sazonov topology of $\mathcal{H}(R)$, thus contradicting Sazonov's theorem.

3. PROBABILITY MEASURES IN FUNCTION SPACES

The problem of defining probability measures in function spaces (or more general, infinite dimensional spaces) occurs in connection with stochastic processes. Suppose $\{X_t(\omega), t\in J\}$ (where J may be $(-\infty,\infty)$, $[0,\infty)$ or a finite interval $[a,b]$) is a family of real r.v.'s on a probability space $(\Omega,\mathcal{F},P)$.

Under the conditions of a particular problem it may be reasonable to expect that the ensemble of functions or paths: $\{\omega: t \to X_t(\omega)\}$ is a space useful for further analysis; for instance, a space of continuous functions, or of functions which are right continuous with left hand limits. In fact, $(\Omega,\mathcal{F},P)$ is seldom given a priori in most physical problems. In such situations, the choice of a suitable function space on which P can be defined is a problem of the first importance. Before going into such questions, we begin with the basic result due to Kolmogorov.

Let $\Omega = \mathbb{R}^{\mathbb{T}}$, the family (linear space) of all real valued functions $\omega(\cdot)$ defined on $\mathbb{T}$. Let $\mathcal{H}(\Omega)$ be the σ-field generated by all finite dimensional Borel cylinder subsets of Ω, i.e., of the form $\{\omega \in \Omega: [\omega(t_1),\dots,\omega(t_n)] \in B\}$ where $\{t_i, i=1,\dots,n\}$ is an arbitrary subset of $\mathbb{T}$ and B is any Borel set in $\mathbb{R}^n$. We will also assume that $\mathbb{T}$ has arbitrary cardinality. Let $\mathbb{F} = \{F_{t_1,\ldots,t_n}\}$ be a family of distribution functions in $\mathbb{R}^n$ where $\{t_1,\ldots,t_n\}$ ranges over all finite subsets of $\mathbb{T}$. $\mathbb{F}$ is a consistent family if the following conditions are satisfied:

(a) For any $\{t_1,\ldots,t_n\}$, $\{x_1,\ldots,x_n) \in \mathbb{R}^n$ and any permutation $\pi = (\pi_1,\ldots,\pi_n)$ of $(1,\ldots,n)$ we have

$$F_{t_1,\ldots,t_n}(x_1,\ldots,x_n) = F_{t_{\pi_1},\ldots,t_{\pi_n}}(x_{\pi_1},\ldots,x_{\pi_n}).$$

(b) $$\lim_{x_j \to +\infty} F_{t_1,\ldots,t_n}(x_1,\ldots,x_n)$$

$$= F_{t_1,\ldots,t_{j-1},t_{j+1},\ldots,t_n}(x_1,\ldots,x_{j-1},x_{j+1},\ldots,x_n)$$

for any j, $1 \leq j \leq n$.

<u>Theorem</u> (Kolmogorov): A consistent family $\mathbb{F}$ defines a unique probability measure P on $(\Omega,\mathcal{B}(\Omega))$ under which the coordinate stochastic process $\tilde{X}_t(\omega) := \omega(t)$ has $\mathbb{F}$ for its family of finite dimensional distributions, i.e.,

$$P\{\omega \in \Omega: \tilde{X}_{t_i}(\omega) \leq x_i,\ i=1,\ldots,n\} = F_{t_1,\ldots,t_n}(x_1,\ldots,x_n)$$

Let $\mathcal{B}'(\Omega)$ be the completion of $\mathcal{B}(\Omega)$ with respect to P. It is customary to refer to $\tilde{X} := (\tilde{X}_t)$ defined on $(\Omega,\mathcal{B}'(\Omega),P)$ as the canonical process.

While the above result is fundamental, its importance is more of an existential nature and it is of limited practical value as a function space model for stochastic processes (as was noted by Kolmogorov himself) when $\mathbb{T}$ is nondenumerable. The following remarks will help to clarify the point.

Let $\mathbb{T} = [0,1]$ or $[0,\infty)$.

1(a). Let $\Omega_c = \{\omega \in \Omega\colon t \to \omega(t) \text{ continuous}\}$. Then $\Omega_c \notin \mathcal{K}'(\Omega)$. Furthermore, let P_* be the inner measure determined by P. Then $P_*(\Omega_c) = 0$.

(b). Let $\Omega_m = \{\omega \in \Omega\colon t \to \omega(t)$ is Lebesgue measurable$\}$. Then, again $P_*(\Omega_m) = 0$.

1(a) and (b) show that the σ-field $\mathcal{K}'(\Omega)$ is too small. In fact, it can be shown that if $\Lambda \in \mathcal{K}(\Omega)$, there exists a denumerable subset $\{t_1, t_2, \ldots\}$ of $\mathbb{T}$ and $A \in \mathcal{B}(\mathbb{R}^\infty)$, the Borel σ-field in $\mathbb{R}^\infty$ the countable product of real lines such that

$$\Lambda = \{\omega \in \Omega\colon [\omega(t_1), \omega(t_2), \ldots] \in A\}.$$

These examples show that if we are to construct a suitable function space probability model for a stochastic process (given initially only by its family of finite dimensional distributions) we cannot use the Kolmogorov theorem directly. Nevertheless, it furnishes a method for such a construction.

Let M be a subset of $\mathbb{R}^{\mathbb{T}}$ (in applications M is usually a linear space) and let $\mathcal{K}(M)$ be the σ-field of subsets of the form $M \cap A$ where $A \in \mathcal{K}(\Omega)$. $\mathcal{K}(M)$ will be called the Kolmogorov σ-field on M.

<u>Definition</u>: A stochastic process $X = (X_t)$ has a *realization* in M if there exists a probability measure μ on $(M, \mathcal{K}(M))$ such that the coordinate process $\tilde{X}$ defined by $\tilde{X}_t(\omega) = \omega(t)$ for $\omega \in M$ has the same finite dimensional distributions as X.

It can be shown that the process $\tilde{X}$ has a realization in M if and only if $P^*(M) = 1$ where P^* is the outer measure determined by the probability measure P of Kolmogorov's theorem.

Definition: The process $X = (X_t)$ where $t \in [0,T]$ is a *measurable* process (or sometimes referred to as (t,ω) measurable) if the function $(t,\omega) \to X_t(\omega)$ is $\mathcal{B}[0,T] \times \mathcal{F}$-measurable where $\mathcal{B}[0,T]$ is the σ-field of Borel sets in $[0,T]$.

Definition: The process $X = (X_t)$ has a *measurable version* $\tilde{X} = (\tilde{X}_t)$ if $\tilde{X}$ is a measurable process defined on some probability space $(\tilde{\Omega},\tilde{\mathcal{F}},\tilde{P})$ and $\tilde{X}$ has the same family of finite dimensional distributions as X.

We shall not go into the question of the existence of measurable versions. In most practical situations it is possible to choose a suitable measurable version directly.

Definition: Wiener Process: A real valued process $W = (W_t)$, $t \in [0,T]$ defined on a complete probability space $(\Omega,\mathcal{F},P)$ is called a Wiener process (or Brownian motion) with variance parameter σ^2 if it has the following properties:

(1) $W_0(\omega) = 0$ P-a.s..

(2) For each s and t $(s \le t)$, $W_t - W_s$ has a Gaussian distribution with mean zero and variance $\sigma(t-s)$.

(3) For $0 \le t_1 \le t_2 \le \dots \le t_n \le T$, the increment random variables $W_{t_2} - W_{t_1}$, $W_{t_3} - W_{t_2}$, ..., $W_{t_n} - W_{t_{n-1}}$ are independent.

(4) For P-almost all ω, the paths $t \to W_t(\omega)$ are continuous.

The above conditions are equivalent to the following definition: $W = (W_t)$ is a Wiener process if W is a sample continuous Gaussian process with $EW_t = 0$ for all t and covariance function $E(W_tW_s) = \sigma^2\min(t,s)$. If $\sigma^2 = 1$, W is called a standard Wiener process.

Properties (1) - (3) pertain only to finite dimensional distributions and are sufficient (via an application of Kolmogorov's theorem) to determine a "Wiener" measure on $\mathbb{R}^{[0,T]}$. However, condition (4) is an essential part of our definition and a construction of a process satisfying (1) - (4) can be given in many different ways. Perhaps the simplest and the most appealing is the one stated in the following result given here without proof.

<u>Theorem</u>: Let $\eta_j(\omega)$ be independent N(0,1) random variables defined on $(\Omega,\mathcal{F},P)$. [We may take $\Omega = \mathbb{R}^\infty$, $\mathcal{F} = \mathcal{B}(\mathbb{R}^\infty)$ and $P = N^\infty$, the infinite product of N(0,1)-measures]. Let (ϕ_j) be an arbitrary CONS in $L^2[0,T]$. Then the series

$$\sum_{j=1}^{\infty} \eta_j(\omega)\int_0^t\phi_j(u)du \quad \text{converges uniformly in t, P-a.s.,}$$

say, for $\omega \in \Omega_0$. Defining $W_t(\omega)$ to be the sum of this series for $\omega \in \Omega_0$ and to be zero for all t when $\omega \notin \Omega_0$, it is seen that $W = (W_t)$ satisfies conditions (1)-(4), and so defines a Wiener process on $(\Omega,\mathcal{F},P)$.

The proof of the above becomes particularly simple if we choose for (ϕ_j), the Haar system given by $\{g_{n,j}$, $n = 0$, $j = 0$; $n \geq 1$, $j = 0,1,..,2^{n-1}-1\}$ where

$$g_{00} = 1,\ g_{nj}(u) = \begin{cases} 2^{(n-1)/2} & \text{if } u \left[\frac{j}{2^{n-1}}, \frac{j+\frac{1}{2}}{2^{n-1}}\right], \\ -2^{(n-1)/2} & \text{if } u \left[\frac{j+\frac{1}{2}}{2^{n-1}}, \frac{j+1}{2^{n-1}}\right], \\ 0 & \text{otherwise.} \end{cases}$$

N-dimensional Wiener Process.

$W_t = (W_t^1, \ldots, W_t^N)$ is a standard Wiener process in $\mathbb{R}^N$, defined on $(\Omega, \mathcal{F}, P)$ if

(i) each component W_t^j is a real valued Wiener process as defined above, and

(ii) the components $W^j = (W_t^j)$ are mutually independent.

Brownian motion in several parameters plays an important role in problems involving random fields. Here we restrict our attention to the two-parameter Brownian motion also known as the Yeh-Wiener process or the Brownian sheet.

Definition: A sample continuous Gaussian process $W = (W_{ts})$, $(0 \le t \le T,\ 0 \le s \le T)$ is a Yeh-Wiener process if $EW_{ts} = 0$ for all t and s and

$$EW_{ts}W_{t's'} = (t \wedge t')(s \wedge s').$$

The existence of a Yeh-Wiener process is proved exactly as for a one parameter Wiener process with the help of a CONS in $L^2([0,T] \times [0,T])$.

The N-dimensional Wiener process W_t induces a probability measure μ on $(C_N, \mathcal{K}(C_N))$ where $C_N = C([0,T], \mathbb{R}^N)$ and $\mathcal{K}(C_N)$ is the Kolmogorov σ-field. It can be shown that $\mathcal{K}(C_N)$

coincides with $\mathcal{B}(C_N)$, the Borel σ-field on C_N. Hence μ, known as the Wiener measure, is an example of a probability measure on the Polish space $(C_N, \mathcal{B}(C_N))$.

The equality $\mathcal{K}(C_N) = \mathcal{B}(C_N)$ is a special case of the following more general result.

Proposition: Let B be an arbitrary separable Banach space and B^* its strong dual. For $y \in B^*$ and $x \in B$ denote $y[x]$ the evaluation of y at x. Let $\mathcal{K}(B)$ be the σ-field generated by sets $[x: (y_1[x], \ldots, y_n[x]) \in A_n]$ where $y_1, \ldots, y_n \in B^*$, $A_n \in \mathcal{B}(\mathbb{R}^n)$, $n \geq 1$. Finally let $\mathcal{B}(B)$ be the σ-field of topological Borel sets in B. Then $\mathcal{K}(B) = \mathcal{B}(B)$.

We shall briefly comment on the relationship of the various Wiener measures and Wiener spaces introduced above to the concept of abstract Wiener spaces. To fix the ideas we consider only an N-dimensional Wiener process of a single parameter.

Let μ be the standard Wiener measure on $C([0,T], \mathbb{R}^N)$ the Banach space of $\mathbb{R}^N$-valued continuus functions with the sup norm. Define

$$\mathcal{H} := \{f \in C: f(t) = \int_0^t f'(u)du \quad \text{with } f' \in L^2([0,T], \mathbb{R}^N)\}.$$

$\mathcal{H}$ is a subset of C and is also a Hilbert space under the norm $\|f\| = \|f'\|_{L^2}$. $\mathcal{H}$ is called the reproducing kernel Hilbert space (RKHS) of the Wiener process. It is easy to show that $\mu(\mathcal{H}) = 0$. Despite this fact, $\mathcal{H}$ plays a significant role in the study of the Wiener process. $\mathcal{H}$ and its extensions are discussed in greater detail in Chapter 3 in connection with the finitely additive white noise theory. As will be seen in some of the examples in Chapter 3 the relation between $\mathscr{H}$ and μ is subsumed in the geneal theory of abstract Wiener spaces.

We shall also have occasion in Chapter 3 in the context of representations of finitely additive Gauss measures and in applications to filtering problems involving random fields, to consider Wiener processes (of a single parameter) taking values in infinite dimensional linear spaces such as Banach spaces and the space $\mathscr{S}^*(\mathbb{R}^N)$ of Schwartz distributions.

4. GAUSSIAN WHITE NOISE

We now take up the question of defining "white noise" (WN) rigorously, keeping in mind that such a definition should prove useful in application.

Serious difficulties arise when one tries to attribute properties of WN to a continuous parameter family of random variables. An obvious choice would be to adopt the same definition as in the discrete parameter case, i.e., define $X = (X_t)$ to be a WN process if X_t and X_s are independent and have the same distribution whenever $t \neq s$. We can apply Kolmogorov's theorem to obtain a stochastic process (viz. the canonical or coordinate process) on $\Omega = \mathbb{R}^{[0,T]}$ which has the requisite property of independence at every point. However, such a process cannot give rise to a useful theory since it has no regularity properties that would render it amenable to the usual analytical operations. The following two remarks illustrate the point.

Remark 1: The WN process X does not have a realization in C[0,1]. (T is taken to be 1 for simplicity.)

Proof: If not, let μ be the probability measure on C[0,1] induced by X. Writing $A = \{x \in C: x(1) > a\}$ we choose a such that $\mu(A) = 1-\alpha$ $(0 < \alpha < 1)$. Let $A_k = \{x \in C: x(a - \frac{1}{n}) > a, n = k,k+1,\ldots,2k\}$. Then $\mu(A_k) = (1-\alpha)^k$ so that $\Sigma_{k=1}^{\infty}\mu(A_k) < \infty$. The Borel-Cantelli lemma then implies $\mu(\lim\sup A_k) = 0$. Hence $\mu(A^c) = 1$ which is a contradiction since $\mu(A) > 0$.

Even more remarkable is the nonexistence of a measurable realization which we show under the added assumptions that $EX_t^2 = \sigma^2$, $(\sigma^2 > 0)$ and $EX_t = 0$ for all t.

Remark 2: The WN X under the above condition does not have a measurable realization.

Proof: Suppose that X is a measurable realization of X on some probability space $(\tilde{\Omega},\tilde{\mathcal{F}},\tilde{P})$. Since $E\tilde{X}_t^2 < \infty$ for each t,

$$\int_{\Omega}\int_I\int_I |\tilde{X}_t(\tilde{\omega})\tilde{X}_s(\tilde{\omega})|\, dsdt\tilde{P}(d\tilde{\omega}) < \infty.$$

Using Fubini's theorem,

$$E(\int_I\tilde{X}_t(\tilde{\omega})dt)^2 = E\int_I\int_I\tilde{X}_t(\tilde{\omega})\tilde{X}_s(\tilde{\omega})dsdt = \int_I\int_I E(\tilde{X}_t\tilde{X}_s)dsdt = 0.$$

Hence

$$\int_I\tilde{X}_t(\tilde{\omega})dt = 0.$$

for $\tilde{\omega} \in N_I$ where $\tilde{P}(N_I) = 0$. Letting I range over all sub-intervals of [0,1] with rational endpoints and setting $N = \cup_I N_I$ we have $\int_a^b\tilde{X}_t(\tilde{\omega})dt = 0$ for all subintervals [a,b] of

$[0,1]$ if $\tilde{\omega} \notin N$. Hence, if $\tilde{\omega} \in N^c$, $\tilde{X}_t(\tilde{\omega}) = 0$ for all t in $[0,1]$ except possibly for a set of Lebesgue measure zero. Hence by Fubini's theorem, $\int_{\Omega}\int_0^1 \tilde{X}_t^2(\tilde{\omega})\tilde{P}(d\tilde{\omega})dt = 0$. This is impossible since the integral on the left hand side equals $\int_0^1 E(\tilde{X}_t^2)dt = \sigma^2 > 0$.

Abandoning this naive approach we next look for a more acceptable definition of Gaussian white noise (GWN) based on the requirements of engineering problems. Suppose $X = (X_t)$ is a stationary Gaussian process with mean zero and covariance function $R(\tau) := E(X_{t+\tau}X_t)$. Then by Bochner's theorem

$$R(\tau) = \int_{-\infty}^{\infty} e^{i\tau\lambda} dF(\lambda)$$

where F is the spectral distribution function and $f(\lambda) := F'(\lambda)$ (if it exists) is called the power spectral density and is then given by the inverse Fourier transform

$$f(\lambda) = \frac{1}{2\pi} \int_{-\infty}^{\infty} e^{-i\tau\lambda} R(\tau) d\tau.$$

Idealized GWN is that process X which has constant power spectral density. From the above expression for $f(\lambda)$ it will be seen that this can only happen if $R(\tau)$ is the Dirac δ-function which cannot be the covariance function of a stochastic process X consisting of real valued random variables. Among the commonly used approximations to GWN is the following based on the Ornstein-Uhlenbeck (OU) process

$$X_t^{\alpha} = e^{-\alpha t} X_0^{\alpha} + e^{-\alpha t} \int_0^t e^{\alpha s} dW_s$$

where $\alpha > 0$, W is a standard Wiener process and X_0^α is a $N(o,\sigma_0^2\alpha)$ r.v. independent of W. Then $EX_t^\alpha = 0$ for all t and the covariance $R_\alpha(\tau) = \frac{\sigma_0^2}{2} e^{-\alpha|\tau|}$, ($\sigma_0^2$ is a positive constant). As $\alpha \to \infty$, $R_\alpha(\tau) \to \frac{\sigma_0^2}{2}\delta(0)$. The power spectral density of $R_\alpha(\tau)$ is given by $f_\alpha(\lambda) = \frac{\sigma_0^2}{1 + \frac{\lambda^2}{\alpha^2}}$ which converges as $\alpha \to \infty$ to $f_\infty(\lambda) = \sigma_0^2$ for all λ. It is customary, there-fore, to define white noise as that stationary process which has constant spectral density. As we have indicated only the "generalized" function δ qualifies and in practice it is usual to employ approximations such as band-limited noise, i.e., where the spectral density $f(\lambda)$ is constant over $[-\lambda_0,\lambda_0]$ where λ_0 is large and falls off sharply to zero for $|\lambda| > \lambda_0$.

Suppose $\mathcal{D}[0,1]$ is the subspace of $L^2[0,1]$ consisting of C^∞-functions vanishing at 0 and 1 and endowed with Schwarz's topology. For all $\phi \in \mathcal{D}[0,1]$, defining $\xi^\alpha[\phi] = \int_0^1\xi_t^\alpha\phi(t)dt$, we have $E \exp i\xi^\alpha[\phi] \to \exp -\frac{1}{2}\|\phi\|_{L^2}^2$. It can then be shown that the limit is the characteristic functional of a countably additive measure of GWN on the dual space $\mathcal{D}^*[0,1]$ of Schwarz distributions.

Let us now come back to the definition of the Wiener process using the infinite series method discussed earlier. For each n define

$$W_t^n := \sum_{j=1}^{n} \eta_j \int_0^t \phi_j(s)ds$$

where (ϕ_j) is a CONS in L^2. Hence (Φ_j) where $\Phi_j(t) = \int_0^t \phi_j(s)ds$ is a CONS in the RKHS $\mathcal{H}$. Hence $W_\cdot^n \in \mathcal{H}$ for each $n \geq 1$. Assuming that $\phi_j(s)$ is continuous, we have the derivative

$$\dot{W}_t^n = \sum_{j=1}^{n} \eta_j \phi_j(t) \in L^2 \quad \text{for all } n.$$

$(\dot{W}^n)$ does not converge in L^2.

Let us suppose that λ_n is an increasing sequence of numbers such that $0 < \lambda_1$ and $\lambda_n \to \infty$, $\Sigma_{j=1}^{\infty} \lambda_j^{-2r}$ for $r > \alpha > 0$, but $\Sigma_{j=1}^{\infty} \lambda_j^{-2\alpha} = \infty$. Define

$$\Phi = \{\phi \in L^2 : \sum_{j=1}^{\infty} \lambda_j^{2r} (\phi, \phi_j)^2 < \infty \quad \text{for all } r\}$$

where (ϕ_j) is a CONS in L^2. On Φ define the family of Hilbertian seminorms $\|\cdot\|_r$ by

$$\|\phi\|_r^2 = \sum_{j=1}^{\infty} \lambda_j^{2r} (\phi, \phi_j)^2.$$

Let Φ_r be the Hilbert space obtained by completing Φ with respect to $\|\cdot\|_r$. Then $\Phi = \cap_r \Phi_r$ and Φ is a countably Hilbertian space whose topology is determined by the family $\{\|\cdot\|_r, r \geq 0\}$. The strong dual of Φ can be shown to be $\Phi' = \bigcup_{r>0} \Phi_{-r}$, Φ_{-r} being the dual of the Hilbert space Φ_r. Note that $\Phi_0 = L^2$. In view of the condition on the λ_j's (which has the effect of making the topology of Φ nuclear) we have, noting that (ϕ_j) is a common orthogonal system for all Φ_r,

$$\|\sum_{j=m+1}^{n} \eta_j(\omega)\phi_j\|_{-r}^2 = \sum_{j=m+1}^{n} \eta_j(\omega)\|\phi_j\|_{-r}^2$$

$$= \sum_{j=m+1}^{n} \eta_j(\omega)\lambda_j^{-2r} \to 0 \quad \text{a.s. if } r > \alpha.$$

Hence $\|\dot{W}^m_{\bullet} - \dot{W}^n_{\bullet}\|_{-r} \to 0$ a.s., so that there exists an element $\dot{W}$, say, belonging (a.s.) to Φ_{-r} such that $\dot{W}^n \to \dot{W}$ in Φ_{-r}, $\dot{W}$ being given by

$$\dot{W} = \sum_{j=1}^{\infty} \eta_j\phi_j \quad \text{a.s. (in } \Phi_{-r}).$$

The map $\omega \to \dot{W}(\omega)$ from Ω into Φ_{-r} has all the properties of Gaussian white noise (= derivative of Wiener process) and the induced measure $\mu' := P\dot{W}^{-1}$ on the Borel sets of Φ_{-r} is a (countably additive) GWN measure. However, two things may be noted concerning this definition:

(a) $\Phi_{-r} \underset{\neq}{\supset} \Phi_0$ $(= L^2)$, and

(b) There is no "smallest" space Φ_{-r} on which a GWN measure can be defined since the above construction is valid for any $r > \alpha$.

Thus the model given above for a countably additive GWN measure is subject to the same criticism as the construction of a model for the Wiener process as the a.s. limit in uniform norm of the sequence $\{W^n\}$ belonging to $\mathcal{H}$ – namely that in both cases we have to extend the space from $\mathcal{H}$ (and L^2) to $C[0,1]$ (and Φ_{-r}), the extended spaces being measure theoretically very "large." Furthermore, there does not exist a smallest such extension. (These remarks apply also to the general construction of abstract Wiener measures.)

The implication of all this is that if the natural space of observations in an experiment is the Hilbert space $\mathcal{H}$ or L^2, one would have to extend these spaces in order to obtain a countably additive model of Gaussian noise (whether one models noise by a Wiener process or its derivative). The extensions, moreover, are such that $\mu(\mathcal{H}) = \mu'(L^2) = 0$ where μ and μ' are, respectively, Wiener measure on $C[0,1]$ and the GWN measure on Φ_{-r}.

It is clear from the above discussion that for each $f \in L^2$, $\int_0^1 f(s)\dot{W}_s^n ds \to \zeta(f)$ a.s. where $\zeta(f)$ is a centered Gaussian r.v. with variance equal to $|f|^2_{L^2}$; moreover, for f_1 and f_2 belonging to L^2 and any real numbers c_1 and c_2, we have

$$\zeta(c_1 f_1 + c_2 f_2) = c_1(f_1) + c_2(f_2) \quad \text{a.s.}$$

Thus $\zeta(f)$ may be interpreted as the (Wiener) integral $\int_0^1 f(s)dW_s$ or, equivalently, as a cylinder measure or weak distribution in the Hilbert space L^2.

The example at the end of Section 2 shows that the Gaussian family of finite dimensional distributions defined there does not extend to a countably additive probability measure on the Borel sets of $\mathcal{H}(R)$ but determines a finitely additive Gauss measure (called the canonical Gauss measure in Chapter III) on the field of finite dimensional cylinder sets. In Chapter VII and in the nonlinear filtering models considered in this book, Gaussian white noise is defined in relation to a particular Hilbert space, viz., $L^2([0,T],\mathbb{R}^N)$. In view of the example referred to above, it will be seen that Gaussian white noise could also be defined with respect to $\mathcal{H}(R)$ where R is any covariance, not necessarily that of a Wiener process.

CHAPTER II

MARKOV PROCESSES

In the first two sections of this chapter, we present a brief account of the properties of the semigroup associated with a Markov process and its generator. In Section 3, we define a diffusion process and state results on existence, uniqueness and convergence of diffusion processes. The last section is devoted to a comprehensive discussion of the Feynman-Kac formula.

For definitions and elementary properties of martingales, local martingales, stopping times, etc., we refer the reader to standard books on the subject.

1. DEFINITIONS

Let $(\Omega,\mathcal{A},\Pi)$ be a countably additive probability space and $(S,\mathcal{S})$ a measurable space. $J(S,\mathcal{S})$ will denote the class of all real valued $\mathcal{S}$-measurable bounded functions on S.

Let $\{\mathcal{A}_t\}$ be an increasing family of sub σ-fields of $\mathcal{A}$ such that $\mathcal{A}_0$ contains all Π-null sets.

Definition 1.1: An S-valued process (X_t) is said to be $\{\mathcal{A}_t\}$ adapted if for each t, X_t is $\mathcal{A}_t/\mathcal{S}$ measurable.

Definition 1.2: An S-valued $\{\mathcal{A}_t\}$ adapted process (X_t) is said to be a *Markov process with respect to the family* $\{\mathcal{A}_t\}$ if for all $0 \leq s < t$, $B \in \mathcal{S}$, we have

$$E_{\Pi}[1_B(X_t)|\mathcal{A}_s] = E_{\Pi}[1_B(X_t)|\sigma(X_s)]. \tag{1.1}$$

The condition that (X_t) is $\{\mathcal{A}_t\}$ adapted means that $\sigma(X_u: u \leq s) \subseteq \mathcal{A}_s$. Thus if (X_t) is a Markov process with respect to $\{\mathcal{A}_t\}$, then it is also a Markov process with respect to $\mathcal{F}_t^X$. Here and in the sequel, for a process (Z_t), $\mathcal{F}_t^Z$ will denote the smallest σ-field with respect to which the family $\{Z_u: 0 \leq u \leq t\}$ is measurable and contains all Π-null sets. When $\mathcal{A}_t = \mathcal{F}_t^X$ in Definition 1.2, we will refer to (X_t) simply as a Markov process.

The Markov property (1.1) has several equivalent versions which are stated below.

For all $0 \leq s < t$ and $f \in J(S,\mathcal{S})$

$$E_\Pi[f(X_t)|\mathcal{A}_s] = E_\Pi[f(X_t)|\sigma(X_s)]; \tag{1.2}$$

for all $0 \leq s < t$ and $f \in J(S,\mathcal{S})$, there exists a $g \in J(S,\mathcal{S})$ such that

$$E_\Pi[f(X_t)|\mathcal{A}_s] = g(X_s). \tag{1.3}$$

Let (X_t) be a Markov process with respect to $\mathcal{A}_t$. Suppose that there exists a function P: $\{(s,x,t,B)\colon s,t \in [0,\infty),\ x \in S,\ B \in \mathcal{S}\}$ into $[0,1]$ such that

(i) for all $s,t \in [0,\infty)$, $x \in S$; $P(s,x,t,\cdot)$ is a probability measure on $(S,\mathcal{S})$;

(ii) for all $t \in [0,\infty)$, $B \in \mathcal{S}$; $(s,x) \to P(s,x,t,B)$ is $\mathcal{B}([0,\infty))\otimes\mathcal{S}$ measurable;

(iii) for all $s,t_1,t_2 \in [0,\infty)$, $x \in S$, $B \in \mathcal{S}$,

$$P(s,x,t_1+t_2,B) = \int P(s,x,t_1,dz)P(s+t_1,z,t_2,B); \tag{1.4}$$

and

(iv) for all $s,t \in [0,\infty)$, $B \in \mathcal{S}$,

$$E_{\Pi}[1_B(X_{s+t})|\mathscr{A}_s] = P(s,X_s,t,B) \quad \text{a.s.}\Pi. \tag{1.5}$$

Then P is called the *transition probability function* of the Markov process (X_t), for in view of (1.5), P(s,x,t,B) represents the conditional probability that $X_{s+t} \in B$ given that $X_s = x$. It should be noted that the Markov property (1.1) implies the existence of P satisfying (1.5), but in general, it may not be possible to choose a function P satisfying (i) - (iv). If such a P exists, (X_t) is said to admit a transition probability function.

The Wiener process (W_t), which is also a Markov process, admits a transition probability function P given by

$$P(s,x,t,B) = \frac{1}{\sqrt{2\pi t}} \int_B \exp\{-\frac{1}{2t}(y-x)^2\}dy. \tag{1.6}$$

In general, given a Markov process (X_t), it may be difficult to decide whether or not it admits a transition probability function. On the other hand given a function P satisfying (i) - (iv) above, one can construct a process (X_t) (on some probability space) such that (X_t) is a Markov process with P as its transition probability function and X_0 as some specified distribution, say μ — known as the initial distribution. This can be done because μ and P determine the finite dimensional distributions of (X_t). In this sense, the transition probability function contains all the information on the evolution law of the process (X_t).

We want to draw the attention of the reader to the fact that a Markov process (X_t) on a finite interval [0,T] can be defined exactly as in the case of $[0,\infty)$; we require that (1.1) holds for all $0 \leq s < t \leq T$. Such a process is said to admit a transition probability function P if P satisfies (i) - (iv) above (for $s,t,t_1,t_2 \geq 0$ such that $s+t \leq T$,

$s+t_1+t_2 \leq T$).

Let $(X_t)_{o\leq t\leq T}$ be an S-valued Markov process with respect to $\{\mathscr{A}_t\}$ and admitting a transition probability function P. Define

$$X_t = X_T \quad \text{and} \quad \mathscr{A}_t = \mathscr{A}_T \quad \text{for } t \geq T$$

and for all $x \in S$, $B \in \mathscr{S}$, $P(s,x,t,B)$ for $s+t > T$ by

$$\begin{aligned} P(s,x,t,B) &= P(s,x,T-s,B) && \text{if } s \leq T,\ s+t \geq T \\ &= 1_B(x) && \text{if } s \geq T,\ t \geq 0. \end{aligned}$$

Then it is easy to see that $(X_t)_{0\leq t<\infty}$ is an S-valued Markov process with respect to the family $\{\mathscr{A}_t\}$ and that P is its transition probability function. Thus all the definitions and results for a Markov process on $[0,\infty)$ also apply to a Markov process on $[0,T]$.

2. THE ASSOCIATED SEMIGROUP AND ITS GENERATOR

In this section (X_t) will denote an S-valued Markov process with respect to $\{\mathscr{A}_t\}$. We will assume that it admits a transition probability function P.

It is well known that if (X_t) is time homogeneous, i.e., when $P(s,x,t,B) = P'(x,t,B)$, then $T'_t\colon J(S,\mathscr{S}) \to J(S,\mathscr{S})$ defined by

$$(T'_t f)(x) = \int f(y)P'(x,t,dy)$$

is a one parameter semigroup, i.e., for $t_1, t_2 \geq 0$, $T'_{t_1+t_2} = T'_{t_1} \circ T'_{t_2}$.

When (X_t) is not necessarily time homogeneous, a two parameter semigroup $\{V^s_t\colon s \leq t\}$ can be associated with P as follows. For $f \in J(S,\mathcal{S})$, $0 \leq s \leq t < \infty$,

$$(V_t^s f)(x) = \int f(y)P(s,x,t-s,dy). \tag{2.1}$$

The family $\{V_t^s: s \leq t\}$ satisfies

$$V_r^s \circ V_t^r = V_t^s, \qquad 0 \leq s \leq r \leq t, \tag{2.2}$$

and hence is called a two parameter semigroup.

The generator of the one parameter semigroup associated with a time homogeneous Markov process plays an important role in the study of the process. Thus, it would be useful to imbed $\{V_t^s\}$ into a one parameter semigroup. This is achieved by considering the process $\hat{X}_t = (t,X_t)$ which becomes a time homogeneous Markov process. For this purpose, let $\hat{S} = [0,\infty)\times S$, $\hat{\mathcal{S}} = \mathcal{B}([0,\infty))\otimes\mathcal{S}$. For $t \geq 0$, $f \in J(\hat{S},\hat{\mathcal{S}})$, let $(T_t f): \hat{S} \to \mathbb{R}$ be defined by

$$(T_t f)(s,x) = \int f(s+t,z)P(s,x,t,dz). \tag{2.3}$$

The measurability properties of P imply that $T_t f \in J(\hat{S},\hat{\mathcal{S}})$ and (1.4) yields the semigroup property, namely

$$T_{t_1+t_2} f = T_{t_1}(T_{t_2} f), \quad t_1,t_2 \geq 0. \tag{2.4}$$

The relation (1.4) gives, for $0 \leq s < t$, $f \in J(\hat{S},\hat{\mathcal{S}})$,

$$E_\Pi[f(t,X_t)|\mathcal{A}_s] = (T_{t-s} f)(s,X_s). \tag{2.5}$$

Taking f to be an indicator function, it follows that T_t determines P and thus T_t contains all the information about the evolution of (X_t).

From the knowledge of $\{T_t: 0 \le t \le \epsilon\}$ for any $\epsilon > 0$ one can recover T_t using the semigroup property: choose $k \ge 1$ such that $\frac{t}{k} < \epsilon$ and then $T_t = (T_{t/k})^k$. This suggests that perhaps

$$Lf = \lim_{t \downarrow 0} \frac{T_t f - f}{t} \tag{2.6}$$

will also determine $\{T_t\}$. The limit in (2.6) is to be interpreted suitably. When the limit is taken in the uniform norm on $J(\hat{S},\hat{\mathcal{S}})$, it is known as the strong generator and the class of functions f for which the limit in (2.6) exists in uniform norm is the domain of the strong generator. It is useful to consider convergence weaker than uniform convergence and define a corresponding weak generator.

For $f \in J(\hat{S},\hat{\mathcal{S}})$, let $\|f\| = \sup\{|f(s,x)|: (s,x) \in \hat{S}\}$. A sequence $\{f_k\} \subseteq J(\hat{S},\hat{\mathcal{S}})$ will be said to converge weakly to f, written as $w - \lim_{k\to\infty} f_k = f$ if $f_k(s,x)$ converges to $f(s,x)$ for each $(s,x) \in \hat{S}$ and $\|f_k\|$ is bounded. Similarly, for $\{f_t: t \in (0,\infty)\} \subseteq J(\hat{S},\hat{\mathcal{S}})$, we will say that $w - \lim_{t\downarrow 0} f_t = f$ if (i) $f_t(s,x) \to f(s,x)$ as $t \downarrow 0$ for each $(s,x) \in \hat{S}$ and (ii) for some $\epsilon > 0$, $\sup\{\|f_t\|: 0 < t < \epsilon\} < \infty$. It can be checked that $w - \lim_{t\downarrow 0} f_t = f$ if and only if f_{t_k} converges weakly to f for every sequence $t_k \downarrow 0$.

Let $J_0 = \{f \in J(\hat{S},\hat{\mathcal{S}}): w - \lim_{t\downarrow 0} (T_t f) = f\}$. Let $\mathcal{D}$ be the class of functions $f \in J_0$ such that the limit

$$w - \lim_{t\downarrow 0} \frac{T_t f - f}{t} = g \tag{2.7}$$

exists and $g \in J_0$. For $f \in \mathcal{D}$, define $Lf = g$, where g is given by (2.7). L will be called the *weak generator* and $\mathcal{D}$ will be called the *domain of the weak generator*. L is also known as the weak infinitesimal operator.

Example 1: First consider the one dimensional deterministic motion with velocity v. It corresponds to the time homogeneous probability transition function

$$P(s,x,t,B) = 1_B(x+tv),$$

$B \in \mathcal{B}(\mathbb{R})$, $x \in \mathbb{R}$. The operators $\{T_t\}$ are given by

$$(T_t f)(s,x) = f(s+t,\ x+tv).$$

Thus $C_b([0,\infty)\times\mathbb{R}) \subseteq J_0$. The class $C_b^1(\mathbb{R})$ of bounded functions f with bounded continuous derivatives $\frac{\partial f}{\partial t}$ and $\frac{\partial f}{\partial x}$ is contained in $\mathcal{D}$ and for $f \in C_b^1(\mathbb{R})$,

$$Lf = \frac{\partial f}{\partial t} + v\cdot\frac{\partial f}{\partial x}\ .$$

Example 2: Let us return to the example of the Wiener process (W_t). Its transition function is given by (1.10). The corresponding semigroup is defined by

$$(T_t f)(s,x) = \int f(s+t,y)\ \frac{1}{\sqrt{2\pi t}}\ \exp\{-\ \frac{1}{2t}(y-x)^2\}dy.$$

It can be proved that for this semigroup $C_b([0,\infty)\times\mathbb{R}) \subseteq J_0$ and the class $C_b^{1,2}([0,\infty)\times\mathbb{R})$ of bounded functions f for which the derivatives $\frac{\partial f}{\partial t}$, $\frac{\partial f}{\partial x}$ and $\frac{\partial^2 f}{\partial x^2}$ exist and are bounded continuous is contained in the domain $\mathcal{D}$ and for $f \in C_b^{1,2}$, Lf is given by

$$Lf = \frac{\partial f}{\partial t} + \frac{1}{2}\frac{\partial^2 f}{\partial x^2}.$$

The Wiener process is a time homogeneous Markov process and so we could have defined a semigroup T'_t as in the first paragraph of this section. The weak generator L' of T'_t is given by

$$L'g = \frac{1}{2}\frac{\partial^2 g}{\partial x^2} \qquad \text{for } g \in C_b^2(\mathbb{R}).$$

Example 3: Let a(t), b(t) be bounded continuous functions on $[0,\infty)$. For $0 \leq s, t < \infty$, let

$$\mu(s,t) = \int_s^{s+t} b(u)du \quad \text{and} \quad \sigma^2(s,t) = \int_s^{s+t} a^2(u)du.$$

For $s \geq 0$, $t > 0$, $x \in \mathbb{R}$, $B \in \mathcal{B}(\mathbb{R})$, let

$$P(s,x,t,B) = \frac{1}{\sqrt{2\pi\sigma^2(s,t)}} \int_B \exp\{-\frac{1}{2\sigma^2(s,t)}(y-x-\mu(s,t))^2\}dy$$

and

$$P(s,x,0,B) = 1_B(x).$$

It can be checked that P so defined is a transition probability function. Also in this example as well we have

$$C_b([0,\infty)\times\mathbb{R}) \subseteq J_0,$$

$$C_b^{1,2}([0,\infty)\times\mathbb{R}) \subseteq \mathcal{D}$$

and for $f \in C_b^{1,2}$,

$$(Lf)(t,x) = \frac{\partial}{\partial t} f(t,x) + b(t)\frac{\partial}{\partial x} f(t,x) + \tfrac{1}{2}a^2(t)\frac{\partial^2 f}{\partial x^2}(t,x).$$

As in the above examples, the form of the generator L is usually calculated for a class of smooth functions in the domain $\mathcal{D}$. It is seldom that one can or needs to find Lf for all $f \in \mathcal{D}$. In what follows we refer, somewhat loosely, to L defined over a subclass of $\mathcal{D}$ as the generator.

We now return to the general setup and state some properties of the semigroup $\{T_t\}$ and its generator.

For all $f \in \mathcal{D}$, $t \geq 0$,

$$T_t f = f + \int_0^t T_u(Lf)du. \tag{2.8}$$

Indeed, $f \in \mathcal{D}$ if and only if there exists $g = Lf \in J_0$ satisfying (2.8).

Suppose P' is another probability transition function and T_t', J_0', $\mathcal{D}'$, L' are the semigroup etc. corresponding to P'. Then $\mathcal{D} = \mathcal{D}'$ and $Lf = L'f$ for all $f \in \mathcal{D}$ implies $J_0 = J_0'$ and $T_t f = T_t' f$ for all $f \in J_0$.

Recall that $\mathcal{G} \subseteq J(\hat{S},\hat{\mathcal{S}})$ is called a measure determining class on $(\hat{S},\hat{\mathcal{S}})$ if for finite measures μ_1,μ_2 on $(\hat{S},\hat{\mathcal{S}})$,

$$\int f d\mu_1 = \int f d\mu_2 \quad \text{for all } f \in \mathcal{G} \Rightarrow \mu_1 = \mu_2. \tag{2.9}$$

If J_0 is a measure determining class on $(\hat{S},\hat{\mathcal{S}})$, then $\mathcal{D} = \mathcal{D}'$ and $Lf = L'f$ for all $f \in \mathcal{D}$ implies $P = P'$. In this sense the weak generator determines the transition probability function.

The resolvent operator R_λ for $\lambda > 0$ is defined by

$$(R_\lambda f)(s,x) = \int_0^\infty \exp(-\lambda t)(T_t f)(s,x)dt.$$

Then for $f \in J_0$,

$$R_\lambda f \in \mathcal{D}, \quad L(R_\lambda f) = \lambda R_\lambda f - f \tag{2.10}$$

and

$$w - \lim_{\lambda\to\infty} \lambda R_\lambda f = f. \tag{2.11}$$

Remark 2.1: Let $P(s,x,t,B)$ be a transition function for which $P(s,x,t,\cdot)$ may not be a probability measure, but is a positive finite measure, and let T_t be the corresponding semigroup. Suppose $\|T_t f\| \leq \|f\| \exp(tM)$ for some $M < \infty$. Then the definition and properties of the weak generator described above continue to be valid.

The following result will be used in Chapter VIII.

Theorem 2.1: Suppose J_0 is a measure determining class. Then $\mathcal{D}$ and $\mathcal{D}^{(2)} := \{f \in \mathcal{D}: Lf \in \mathcal{D}\}$ are both measure determining classes.

Proof: Let μ_1, μ_2 be finite measures on $(\hat{S}, \hat{\mathcal{S}})$. Let

$$J' = \{f \in J(\hat{S}, \hat{\mathcal{S}}): \int f d\mu_1 = \int f d\mu_2\}.$$

Clearly, if $f_k \in J'$ and $w - \lim_{k\to\infty} f_k = f$, then $f \in J'$.

To prove that $\mathcal{D}$ is a measure determining class, we need to prove that $\mathcal{D} \subseteq J'$ implies $\mu_1 = \mu_2$.

Given $f \in J_0$, set $f_k = kR_k f$. By (2.10), $f_k \in \mathcal{D}$ and by (2.11), $w - \lim_{k\to\infty} f_k = f$. Since $\mathcal{D} \subseteq J'$, $J_0 \subseteq J'$ and hence $\mu_1 = \mu_2$ because J_0 is a measure determining class.

For the second part, observe that if $f \in \mathcal{D}$ then f_k defined above belongs to $\mathcal{D}^{(2)}$ and from the argument given

above, $\mathscr{D}^{(2)} \subseteq J' \Rightarrow \mathscr{D} \subseteq J' \Rightarrow J_0 \subseteq J'$ and hence $\mu_1 = \mu_2$.
□

Extended Generator

Suppose that $f \notin \mathscr{D}$, but there exists a $g \in J(\hat{S},\hat{\mathscr{S}})$ such that the pair (f,g) satisfies

$$T_t f = f + \int_0^t T_u g\,du. \tag{2.12}$$

Since $f \notin \mathscr{D}$, g cannot belong to J_0, but we can express the relation (2.12) as 'Lf = g' in an extended sense. While doing so, we would also like to include unbounded functions f in the domain of the extended generator.

For a measurable function f on $(\hat{S},\hat{\mathscr{S}})$, let $T_t|f|$ be defined by (2.3) and if $T_t|f| < \infty$, then $T_t f$ also be defined by (2.3). Let

$$\hat{J} = \{f: \hat{S} \to \mathbb{R}: \hat{\mathscr{S}}\text{-measurable}, T_t|f| < \infty,$$
$$\int_0^t T_u|f|du < \infty \text{ for all } t\}.$$

We will say that $f \in \hat{J}$ belongs to the *domain* $\mathscr{D}^{(e)}$ of the *extended generator* $L^{(e)}$ (of $\{T_t\}$) if there exists a $g \in \hat{J}$ such that (2.12) holds and, in that case, define $L^{(e)}f = g$.

It is possible that for a given f, (2.12) holds for more than one g. Thus $L^{(e)}$ is not defined uniquely. However, if g_1, g_2 both satisfy (2.12) for a given f, then we have

$$\int_0^t T_u g_1 du = \int_0^t T_u g_2 du. \tag{2.13}$$

If we define an equivalence relation ~ on $\hat{J}$ by $g_1 \sim g_2$ if

(2.13) holds, then $L^{(e)}f$ is uniquely defined upto ~.

The term extended generator is generally used to denote the operator $L^{(e)}f = g$, where f,g are related by (2.12). But there is no standard convention about its domain $\mathscr{D}^{(e)}$: some authors allow unbounded functions whereas some do not. It follows from (2.8) that if f is bounded and $L^{(e)} = g \in J_0$, then $f \in \mathscr{D}$ and $Lf = L^{(e)}f$. In this sense, $L^{(e)}$ is an extension of the weak generator L.

<u>Example 4</u>: In Example 3, suppose that a(t), b(t) are no longer continuous, but are measurable functions such that

$$\int_0^t |a^2(s)|ds < \infty, \quad \int_0^t |b(s)|ds < \infty \quad \text{for all } t.$$

Then it can be verified that $C_b^{1,2}([0,\infty)\times\mathbb{R}) \subseteq \mathscr{D}^{(e)}$ and for $f \in C_b^{1,2}([0,\infty)\times\mathbb{R})$,

$$(L^{(e)}f)(t,x) = \frac{\partial f}{\partial t}(t,x) + b(t)\frac{\partial f}{\partial x}(t,x) + \frac{1}{2}\, a^2(t)\frac{\partial^2 f}{\partial x^2}(t,x). \tag{2.14}$$

Here, if a(t), b(t) are not continuous, $(L^{(e)}f)(t,x)$ may not be continuous or bounded and hence may not belong to J_0.

The generator L of a Markov process is related to the process (X_t) via a martingale property. For this we need to assume that (X_t) is $\mathscr{A}_t$-progressively measurable, i.e., for all t, the mapping $(u,\omega) \to X_u(\omega)$ from $[0,t]\times\Omega$ into S is $\mathscr{B}([0,t] \otimes \mathscr{A}_t/\mathscr{S}$ measurable.

<u>Theorem 2.2</u>: Let (X_t) be an S-valued $\{\mathscr{A}_t\}$-Markov process which is $\{\mathscr{A}_t\}$-progressively measurable. Let L be its weak generator and $\mathscr{D}$ be the domain of L. Then for all $f \in \mathscr{D}$,

$s \geq 0$,

$$M_t^f = f(s+t, X_{s+t}) - f(s,X_s) - \int_s^{t+s}(Lf)(u,X_u)du \qquad (2.15)$$

is an $\{\mathcal{G}_t\}$-martingale, where $\mathcal{G}_t = \mathcal{A}_{s+t}$.

Proof: Progressive measurability of (X_t) implies that M_t^f is $\mathcal{G}_t$-adapted. Also, f, $Lf \in J_0$ implies that M_t^f is bounded and hence integrable.

Note that for any $g \in J_0$, $0 \leq r \leq t$ and $A \in \mathcal{G}_r$, we have

$$E_\Pi[1_A f(s+t, X_{s+t})] = E_\Pi[1_A(T_{t-r}f)(s+r, X_{s+r})]. \qquad (2.16)$$

Using (2.16), it can be verified that for all $A \in \mathcal{G}_r$ $(0 \leq r \leq t)$

$$E_\Pi[1_A(M_t^f - M_r^f)] \qquad (2.16)$$

$$= E_\Pi[1_A\{T_{t-r}f - f - \int_0^{t-r}(T_uLf)du\}(s+r,X_{s+r})] = 0$$

since f, Lf satisfy (2.8). It follows that

$$E_\Pi[M_t^f|\mathcal{G}_r] = M_r^f,$$

i.e., (M_t^f) is an $\{\mathcal{G}_t\}$-martingale. □

Remark 2.2: The same proof shows that if $f \in \mathcal{D}^{(e)}$ instead of $\mathcal{D}$ in Theorem 2.2, then M_t^f defined by (2.15) with $L^{(e)}$ instead of L is still a $\{\mathcal{G}_t\}$-martingale.

3. DIFFUSION PROCESSES

An $\mathbb{R}^d$-valued Markov process with continuous paths is said to be a diffusion process if its transition probability function P satisfies, for every bounded open set U containing x,

$$\lim_{t\downarrow 0} \frac{1}{t} [1 - \int_U P(s,x,t,dz)] = 0 \tag{3.1a}$$

$$\lim_{t\downarrow 0} \frac{1}{t} \int_U (z^i - x^i) P(s,x,t,dz) = b_i(s,x) \tag{3.1b}$$

$$\lim_{t\downarrow 0} \frac{1}{t} \int_U (z^i - x^i)(z^j - x^j) P(s,x,t,dz) = a_{ij}(s,x). \tag{3.1c}$$

for suitable functions a,b. When a,b are smooth, the process (X_t) can be constructed as a solution to an appropriate stochastic differential equation. It can be shown that the (weak or strong) generator of a process satisfying (3.1) must be a second order differential operator. We take the latter property as our definition of a diffusion process.

For $s < T$, $C^{1,2}([s,T]\times\mathbb{R}^d)$ will denote the class of functions $f: [s,T]\times\mathbb{R}^d \to \mathbb{R}$ for which the derivatives $\frac{\partial f}{\partial t}$, $\frac{\partial f}{\partial x^i}$, $\frac{\partial f}{\partial x^i \partial x^j}$ exist and are continuous in $(t,x) \in [s,T]\times\mathbb{R}^d$. $\frac{\partial f}{\partial t}$ at $t = s$ $(t = T)$ is to be understood as right (respectively left) derivative. $C_0^{1,2}([s,T]\times\mathbb{R}^d)$ will denote the class of functions in $C^{1,2}([s,T]\times\mathbb{R}^d)$ which have compact support.

The classes $C^{1,2}((s,T]\times\mathbb{R}^d)$ and $C^{1,2}([s,T)\times\mathbb{R}^d)$ are defined similarly except that the derivatives are assumed to exist and be continuous in $(t,x) \in (s,T]\times\mathbb{R}^d$ and $(t,x) \in [s,T)\times\mathbb{R}^d$ respectively.

Suppose that functions a_{ij}, b_i from $[0,T]\times\mathbb{R}^d$ into $\mathbb{R}$ are given, for $1 \leq i,j \leq d$ satisfying the following conditions:

a_{ij}, b_i are measurable functions and are bounded (3.2)
on each compact subset of $[0,T]\times\mathbb{R}^d$;

for all (t,x), the matrix $\{a_{ij}(t,x)\}$ is (3.3)
symmetric and nonnegative definite..

For $g \in C^2(\mathbb{R}^d)$, let $L_t g$ be defined by

$$(L_t g)(x) := \tfrac{1}{2} \sum_{i,j=1}^{d} a_{ij}(t,x)\frac{\partial^2}{\partial x^i \partial x^j} f(x) + \sum_{i=1}^{d} b_i(t,x)\frac{\partial f}{\partial x^i}(x). \tag{3.4}$$

Definition 3.1: An $\mathbb{R}^d$-valued process (X_t) is said to be a *diffusion process* if

(i) (X_t) is a Markov process (with respect to the σ-fields $\mathscr{F}_t^X$) and admits a transition probability function P;

(ii) (X_t) is a continuous process; and

(iii) there exist functions $a = (a_{ij})$, $b = (b_i)$ satisfying (3.2), (3.3) (called the diffusion and drift coefficients of (X_t) respectively) such that for $f \in C_0^{1,2}([0,T]\times\mathbb{R}^d)$, we have

$$(T_t f)(s,x) = f(s,x) + \int_0^t [T_u(\frac{\partial}{\partial u} + L_u)f](s,x)du \tag{3.5}$$

for all $(s,x) \in [0,T]\times\mathbb{R}^d$, $0 \le t \le T-s$, where $\{T_t\}$ is the semigroup associated with P.

Remark 3.1: In (3.5), f is assumed to have compact support and hence in view of (3.2), $g(t,x) = (\frac{\partial}{\partial t} + L_t)f(t,x)$ is a bounded function. Thus, (3.5) can be expressed as: $f \in \mathscr{D}^{(e)}$ and $L^{(e)}f = (\frac{\partial}{\partial t} + L_t)f$. When a,b are continuous, the condition (iii) above is equivalent to $C_0^{1,2}([0,T]\times\mathbb{R}^d) \subseteq \mathscr{D}$ and

$(Lf)(t,x) = (\frac{\partial}{\partial t} + L_t)f(t,x)$, $f \in C_0^{1,2}([0,T]\times\mathbb{R}^d)$.

All the examples of Markov processes considered in Section 2 are diffusion processes as well. Before we proceed, we record a property connecting the process $\{X_t\}$ with (a,b) that follows from Theorem 2.2 and Remark 2.2.

<u>Theorem 3.1</u>: Let (X_t) be an $\mathbb{R}^d$-valued diffusion with diffusion and drift coefficients a,b. Let $f \in C_0^{1,2}([0,T]\times\mathbb{R}^d)$. Then for all $s \in [0,T]$,

$$M_t^f = f(s+t,\ X_{s+t}) - f(s,X_s) - \int_s^{s+t}[(\frac{\partial}{\partial u} + L_u)f](u,X_u)du \tag{3.6}$$

is an $\mathcal{G}_t = \mathcal{F}_{s+t}^X$-martingale.

<u>Corollary 3.2</u>: In Theorem 3.1, let $f \in C^{1,2}([0,T]\times\mathbb{R}^d)$. Then M_t^f is an $\mathcal{G}_t$-local martingale.

The following result will be used only in the proof of Theorem 4.2.

<u>Theorem 3.3</u>: Let (X_t) be an $\mathbb{R}^d$-valued diffusion process with diffusion and drift coefficients a,b respectively. Suppose that for some constants C_1, C_2, C_3, C_4, we have

$$|a_{ij}(t,x)| \le C_1, \qquad |b_i(t,x)| \le C_2(1+|x|) \text{ and}$$

$$E \exp(C_3|X_0|^2) \le C_4.$$

Then there exist constants C_5, C_6 depending only on C_1, C_2, C_3, C_4, T and d such that

$$E \exp(C_5 \sup_{0\le t\le T} |X_t|^2) \le C_6. \tag{3.7}$$

The above theorem is an easy extension of the result proved in [34] where (X_t) is taken to be the solution of an SDE with smooth coefficients a,b. The part played by the smoothness assumption there was to ensure existence of a solution to the SDE.

For later application we require two results due to Stroock and Varadhan on the existence, uniqueness and convergence of diffusion processes. It is beyond the scope of this monograph to include an outline of the proofs of these results. To state them we need the following notation. Fix $T > 0$ and let $\Omega_d = C([0,T],\mathbb{R}^d)$. An element of Ω_d will be denoted by ω'. Let $Z_t(\omega') = \omega'(t)$ be the co-ordinate mappings on Ω_d. For $0 \le s \le t \le T$, let $\mathcal{G}_t^s = \sigma(Z_u: \ s \le u \le t)$ and $\mathcal{G} = \mathcal{G}_T^0$. We improve the following conditions on a,b.

a_{ij}, b_j are continuous functions from $[0,T]$ into $\mathbb{R}$ (3.8)

into $\mathbb{R}$ and for some constants $C_T < \infty$, $\delta > 0$, we have

$$\sup_{0\le t\le T} |a_{ij}(t,x)| \le C_T(1+|x|^2) \tag{3.9}$$

$$\sup_{0\le t\le T} |b_j(t,x)| \le C_T(1+|x|) \tag{3.10}$$

$$\sum_{i,j=1}^{d} a_{ij}(t,x)z_i z_j \ge \delta \sum_{i=1}^{d} (z_i)^2 \tag{3.11}$$

for all $x \in \mathbb{R}^d$, $t \in [0,T]$.

Theorem 3.4: Suppose a,b satisfy (3.8) - (3.11) and assume the following conditions to hold:

(i) For each $(s,x) \in [0,T)\times\mathbb{R}^d$, there exists a unique countably additive probability measure $Q_{s,x}$ on $(\Omega_d, \mathcal{G}_T^s)$ such that

$$Q_{s,x}(Z_s = x) = 1 \tag{3.12}$$

and for all $f \in C_0^{1,2}([s,T]\times\mathbb{R}^d)$, M_t^f: $0 \leq t \leq T-s$ defined by

$$M_t^f = f(s+t,\ Z_{s+t}) - f(s,Z_s) - \int_s^{s+t} (\frac{\partial}{\partial u} + L_u) f(u,Z_u)du \tag{3.13}$$

is a $Q_{s,x}$ martingale.

(ii) Given $\mu \in \mathcal{M}_0(\mathbb{R}^d)$, let Π_μ be the probability measure on $(\Omega_d, \mathcal{G})$ defined by

$$\Pi_\mu(B) = \int_{\mathbb{R}^d} Q_{0,x}(B)d\mu(x),\ B \in \mathcal{G}. \tag{3.14}$$

Then (Z_t) is a diffusion process (on $(\Omega_d, \mathcal{G}, \Pi_\mu)$) with drift coefficient b, diffusion coefficient a and transition probability function $P(s,x,t,B)$ given by

$$P(s,x,t,B) = Q_{s,x}(Z_{s+t} \in B),\ B \in \mathcal{B}(\mathbb{R}^d). \tag{3.15}$$

The measure $Q_{s,x}$ is referred to as the solution to the martingale problem for (a,b) starting from (s,x). It can be proved that if a,b satisfy the conditions of Theorem 3.4 and (X_t) is a diffusion process with coefficients (a,b) on some probability space $(\Omega, \mathcal{A}, \Pi)$, then the distribution $\Pi\circ(X)^{-1}$ of the process on Ω_d is Π_μ, where μ is $\Pi\circ X_0^{-1}$ and Π_μ is defined by (3.14). In this sense, a,b and the initial distribution μ uniquely determine the distribution of the corresponding diffusion process (X_t).

For later convenience, we extend the measure $Q_{s,x}$ to all of $\mathcal{G}$ by defining $Q_{s,x}(Z_u = x: 0 \leq u \leq s) = 1$.

Theorem 3.5: Suppose that (a,b) and (a_k,b_k), $(k \geq 1)$ satisfy (3.8) - (3.11), and that the constant C_T appearing in (3.9), (3.10) can be chosen independent of k.

Further, suppose that for each R, $1 \leq i,j \leq d$,

$$\lim_{k\to\infty} \int_0^T \Big[\sup_{|x|\leq R} \{ |a^k_{ij}(t,x) - a_{ij}(t,x)| \qquad (3.16)$$

$$+ |b^k_j(t,x) - b_j(t,x)| \} \Big] dt = 0.$$

Let $Q_{s,x}$, $Q^k_{s,x}$ be, respectively, the solution to the martingale problem for (a,b) and (a^k,b^k) starting from (s,x). Then for $s_k \to s$, $x_k \to x$, we have

$$Q^k_{s_k,x_k} \xrightarrow{d} Q_{s,x}. \qquad (3.17)$$

The convergence above is to be understood as weak convergence.

By taking $(a^k,b^k) \equiv (a,b)$, it follows that the family $\{Q_{s,x}\}$ obtained in Theorem 3.4 is continuous in (s,x).

4. THE FEYNMAN-KAC FORMULA

Let (W_t) be a one dimensional Wiener process. In an effort to evaluation the distribution

$$F(y) = \text{Prob}(\int_0^T c(W(t)+x)dt \leq y)$$

of the Wiener functional $Z = \int_0^T c(W(t)+x)dt$ for a given function c (say bounded, continuous) on $\mathbb{R}$, Kac discovered a relation between F(y) and the fundamental solution $\Gamma_\beta(s,x,t,y)$ to the partial differential equation

$$\frac{\partial v(t,x)}{\partial t} + \tfrac{1}{2}\frac{\partial^2}{\partial x^2} v(t,x) + \beta c(x)v(t,x) = 0 \tag{4.1}$$

in the domain $[0,T]\times\mathbb{R}$. He showed that for suitable functions g, v_β defined by

$$v_\beta(s,x) = \int \Gamma_\beta(s,x,T,y)g(y)dy \tag{4.2}$$

which is a solution to (4.1) with boundary condition v(T,y) = g(y) has a representation

$$v_\beta(0,x) = E[g(W_T)\cdot\exp(\beta\int_0^T c(W(t)+x)dt)]. \tag{4.3}$$

Taking g = 1 in (4.3), we get an expression for the Laplace transform, i.e., the moment generating function of Z and hence can determine F. Kac used this approach to compute distributions of several Wiener functionals. In his work, Kac was influenced by the ideas of Feynman on representing the solution of the Schrödinger equation as an "integral" over path space. Thus the representation (4.3) has been called the Feynman-Kac formula.

It was observed that the differential operator appearing in (4.1), namely $\frac{1}{2}(d^2/dx^2)$, is the generator of the Wiener process, and thus (4.3) was generalized by replacing (W(t)+x) by a diffusion (X_t) and $\frac{1}{2}(d^2/dx^2)$ in (4.1) by its generator, namely L_t.

We will obtain this representation via martingale theory. For the rest of the section, it is assumed that (X_t) is a diffusion with drift and diffusion coefficients a,b. Suppose a,b satisfy the conditions of Theorem 3.4 and let $Q_{s,x}$ be the solution to the martingale problem starting at (s,x), $(s,x) \in [0,T]\times\mathbb{R}^d$.

<u>Theorem 4.1</u>: Suppose that c: $[0,T]\times\mathbb{R}^d$ into $\mathbb{R}$ is a continuous function. Suppose $v \in C^{1,2}([0,T)\times\mathbb{R}^d) \cap C([0,T]\times\mathbb{R}^d)$ is a solution to the PDE

$$\left(\frac{\partial v}{\partial t} + L_t v\right)(t,x) + c(t,x)v(t,x) = 0 \tag{4.4}$$

for $(t,x) \in [0,T)\times\mathbb{R}^d$ and

$$\text{c is bounded above and v is bounded.} \tag{4.5}$$

Then, for all $(s,x) \in [0,T)\times\mathbb{R}^d$,

$$v(s,x) = E_{Q_{s,x}}[v(T,Z_T)\ \exp(\textstyle\int_s^T c(u,Z_u)du)] \tag{4.6}$$

and

$$v(s,X_s) = E_{\Pi}[v(T,X_T)\ \exp(\textstyle\int_s^T c(u,X_u)du)\,|\sigma(X_s)]. \tag{4.7}$$

<u>Proof</u>: Fix $(s,x) \in [0,T)\times\mathbb{R}^d$ and consider the measure $Q_{s,x}$ on $(\Omega_d,\mathcal{G})$. Let $s \le t_0 < T$. Then $v \in C^{1,2}([s,t_0]\times\mathbb{R}^d)$ and hence

$$M_t := v(s+t,\ Z_{s+t}) - v(s,Z_s) \tag{4.8}$$
$$- \int_s^{s+t}\left(\frac{\partial}{\partial u} + L_u\right)v(u,Z_u)du,$$

c

$0 \leq t \leq t_0-s$, is a local martingale (with respect to $\mathcal{A}_t = \mathcal{G}^0_{s+t}$ for the measure $Q_{s,x}$). Let

$$N_t = \exp(\int_s^{s+t} c(u,Z_u)du) - 1. \tag{4.9}$$

Then by an integration by parts formula or, alternatively, using Itô's formula, it follows that

$$M'_t := M_t N_t - \int_0^t M_s dN_s \tag{4.10}$$

is also a local martingale. Then $\overline{M}_t := M_t + M'_t$ is a local martingale under $Q_{s,x}$. Using the fact that v satisfies (4.4), it is easily verified that

$$\overline{M}_t = v(s+t, Z_{s+t}) \exp(\int_s^{s+t} c(u,Z_u)du) - v(s,Z_s). \tag{4.11}$$

The assumption (4.5) on c and v implies that $\overline{M}_t$ is bounded so that $\overline{M}_t$ is a bounded local martingale. Thus $(\overline{M}_t,\mathcal{A}_t)$, $0 \leq t \leq t_0-s$, is a martingale under $Q_{s,x}$ for every $t_0 < T$ and further, since $\overline{M}_t$ is bounded and continuous in $t \in [0,T-s]$, we have

$$E_{Q_{s,x}}[\overline{M}_{T-s}] = E_{Q_{s,x}}[\overline{M}_0]. \tag{4.12}$$

Equality (4.12) is the same as (4.6). The other relation (4.7) can be proved by proceeding as above, taking $s = 0$, $Q_{s,x} = \Pi$ and $Z_t = X_t$. □

Let us note that the condition (4.5) on c,v was used only to deduce that the local martingale $\overline{M}_t$ is a martingale. Instead of (4.5), if we have some other condition which implies that the local martingale $\overline{M}_t$ is martingale, then we would still have the representations (4.6) and (4.7). The next result is a variant of Theorem 4.1.

<u>Theorem 4.2</u>: In the setup of Theorem 4.1, suppose that instead of (4.5), we have for constants $K_1 < \infty$, $K_2 < \infty$, $\epsilon > 0$,

$$|v(t,x)| \leq \exp(K_1(1+|x|^2)^{1-\epsilon}) \tag{4.13}$$

and

$$c(t,x) \leq K_2(1+|x|^2)^{1-\epsilon}. \tag{4.14}$$

Further, suppose that the diffusion coefficient a is bounded. Under these conditions v is given by the formulas (4.6) and (4.7).

<u>Proof</u>: As noted earlier, $(\overline{M}_t)$ defined by (4.11) is a $Q_{s,x}$ local martingale. Theorem 3.3 implies that for some $\delta > 0$,

$$E_{Q_{s,x}} \exp(\delta \sup_{s \leq t \leq T} |Z_t|^2) < \infty. \tag{4.15}$$

From conditions (4.13) and (4.14), we have

$$\sup_{0 \leq t \leq T-s} |\overline{M}_t| \leq \exp(K_3 \sup_{s \leq t \leq T} (1+|Z_t|)^{2-\epsilon}) \tag{4.16}$$

for some constant $K_3 < \infty$. Hence

$$E_{Q_{s,x}} (\sup_{0 \leq t \leq T-s} |\overline{M}_t|) < \infty \tag{4.17}$$

follows from (4.15) and proves that $\overline{M}_t$ is a martingale. The remaining part of the proof is the same as in Theorem 4.1.

□

The proof of these two versions of the Feynman-Kac formula suggests that the above result might continue to hold even if $\frac{\partial v}{\partial t}$ and $L_t v$ are understood to exist in a generalized sense. In Theorem 4.4 we give another set of conditions under which the representations (4.6) and (4.7) are valid. The result will be used in Chapter VII.

Suppose that $g_0, g_1, \ldots, g_N \in C^{1,2}([0,T)\times\mathbb{R}^d) \cap C([0,T]\times\mathbb{R}^d)$ and $f_1, f_2, \ldots, f_N \in L^1([0,T])$. Let $v\colon [0,T]\times\mathbb{R}^d \to \mathbb{R}$ be defined by

$$v(t,x) := g_0(t,x) \exp\{\sum_{j=1}^{N} g_j(t,x)\int_0^t f_j(s)ds\}. \tag{4.18}$$

Then $F_j(t) := \int_0^t f_j(s)ds$ is differentiable for almost all t. Let D_j be the set of t where F_j is not differentiable and let $D = \cup_j D_j$. Then $\frac{\partial}{\partial t}v(t,x)$ exists for $t \notin D$. For $t \in D$, set $\frac{\partial}{\partial t}v(t,x) = 0$. Note that for each t, $v(t,\cdot) \in C^2(\mathbb{R}^d)$ and hence $L_t v(t,\cdot)$ is well defined.

<u>Lemma 4.3</u>: Let v be defined by (4.18). Then $(M_t)_{0\le t\le t_0-s}$ given by (4.8) is a local martingale.

<u>Proof</u>: For each j, let $(h_{jm} : m \ge 1)$ be a sequence of continuous functions on [0,T] such that

$$\lim_{m\to\infty} \int_0^T |f_j(t) - f_{jm}(t)|dt = 0 \tag{4.19}$$

and let

$$v_m(t,x) = g_0(t,x) \exp(\sum_{j=1}^{N} g_j(t,x)\int_0^t f_{jm}(s)ds). \tag{4.20}$$

Then $v_m \in C^{1,2}([0,T)\times\mathbb{R}^d) \cap C([0,T]\times\mathbb{R}^d)$ and it can be easily seen that for each R,

$$\lim_{m\to\infty}\left[\sup_{|x|\leq R}\{|v_m(t,x)-v(t,x)| + |L_t v_m(t,x)-L_t v(t,x)|\}\right] = 0 \tag{4.21}$$

and

$$\lim_{m\to\infty}\left[\sup_{|x|\leq R}\int_0^T\left|\frac{\partial}{\partial t}v_m(t,x) - \frac{\partial}{\partial t}v(t,x)\right|dt\right] = 0. \tag{4.22}$$

Define the stopping times

$$\tau_R(\omega') = \inf\{t \geq 0: |Z_{s+t}(\omega')| \geq R\} \wedge (t_0-s)$$

and let M_t^m be the local martingale given by (4.8), with v_m in place of v. Then $M^m_{t\wedge\tau_R}$ is a bounded local martingale and hence a martingale for each m and R. Conditions (4.21), (4.22) imply that

$$E_{Q_{s,x}}|M^m_{t\wedge\tau_R} - M_{t\wedge\tau_R}| \longrightarrow 0 \text{ as } m\to\infty \tag{4.23}$$

and hence $(M^m_{t\wedge\tau_R})$ is a martingale, i.e., (M_t) is a local martingale. □

<u>Theorem 4.4</u>: Let a be bounded and (a,b) satisfy conditions of Theorem 3.4. Let c: $[0,T]\times\mathbb{R}^d \to \mathbb{R}$ be a measurable function such that

$$c(t,x) \leq \alpha(t)(1+|x|^2)^{1-\epsilon} \tag{4.24}$$

where $\int_0^T|\alpha(t)|dt < \infty$. Let $f_j\in L^1([0,T])$, $g_j\in C^{1,2}([0,T)\times\mathbb{R}^d) \cap C([0,T]\times\mathbb{R}^d)$, $0 \leq j \leq N$ and v be defined by (4.18). Further assume that v satisfies the growth condition (4.13) and satisfies the PDE (4.4) for all x, for almost all t, i.e., the set

$$D_1 = \{t: \frac{\partial v}{\partial t}(t,x) + L_t v(t,x) \qquad (4.25)$$
$$+ c(t,x)v(t,x) \neq 0 \text{ for some } x\}$$

has Lebesgue measure zero. Then v is given by formulas (4.6) and (4.7).

Proof: We will use the notation established in the proof of Theorem 4.1. By Lemma 4.3, (M_t) is a local martingale and hence as in the proof of Theorem 4.1, it follows that M_t' is a local martingale. Since by assumption the Lebesgue measure of D_1 is zero, it can be checked that

$$\int_s^{s+t}[(\frac{\partial}{\partial u} + L_u)v(u,Z_u) + c(u,Z_u)v(u,Z_u)]du = 0.$$

Hence as in the proof of Theorem 4.1, it follows that

$$\overline{M}_t = v(s+t,Z_{s+t}) \exp(\int_s^{s+t} c(u,Z_u)du) - v(s,Z_s)$$

is a $Q_{s,x}$ local martingale. The bounds on a,c,v now ensure that $\overline{M}_t$ is a martingale. This part is the same as in Theorem 4.2. Formulas (4.6) and (4.7) now follow as in Theorem 4.1.

□

Remark 4.1: In this section, we have concentrated on diffusion processes. The analogue of (4.7) is easy to prove when (X_t) is any Markov process, $v \in \mathcal{D}$ and

$$Lv + cv = 0. \qquad (4.26)$$

CHAPTER III

CYLINDER PROBABILITIES

1. CYLINDER PROBABILITIES

Let H be a real separable Hilbert space. The inner product in H will be denoted by $(\cdot,\cdot)$ and the norm by $|\cdot|$. We will assume throughout that H is infinite dimensional.

Let $\mathscr{P} = \mathscr{P}(H)$ be the class of orthogonal projections on H with finite dimensional range. For $P \in \mathscr{P}$, let

$$\mathscr{C}_P := \{P^{-1}B\colon B \text{ Borel in Range } P\}.$$

Sets in $\mathscr{C}_P$ are called cylinder sets with base P. Clearly $\mathscr{C}_P$ is a σ-field. Let $\mathscr{C}$ be the class of all cylinder sets, i.e.,

$$\mathscr{C} = \bigcup_{P\in\mathscr{P}} \mathscr{C}_P.$$

It should be noted that $\mathscr{C}$ is the union of $\{\mathscr{C}_P\colon P \in \mathscr{P}\}$ and not the σ-field generated by the union. $\mathscr{C}$ is not a σ-field. Sets is $\mathscr{C}$ can be alternatively described as follows.

<u>Proposition 1.1</u>:

(i) For $P \in \mathscr{P}$, $\mathscr{C}_P$ consists of sets of the form

$$C = \{h\colon ((h,h_1),\ldots,(h,h_k)) \in B_1\} \tag{1.1}$$

where $h_1,\ldots,h_k \in PH$, $B_1 \in \mathscr{B}(\mathbb{R}^k)$, $k \geq 1$. Further, $h_1,h_2,\ldots,h_k$ can be taken to be an orthonormal basis of Range P.

(ii) $\mathscr{C}$ consists of sets of the form (1.1) with

$$h_1,\ldots,h_k \in H,\ B_1 \in \mathscr{B}(\mathbb{R}^k),\ k \geq 1.$$

Proof: (i) Let B be a Borel set in Range P and let $C = P^{-1}B$. We first show that C can be represented in the form (1.1). Let $e_1,e_2,\ldots,e_j$ be an orthonormal basis (ONB) in Range P. The map Φ given by

$$\Phi(h) = ((h,e_1),\ldots,(h,e_j)) \tag{1.2}$$

is a topological homeomorphism from PH onto $\mathbb{R}^j$ and hence Φ is a Borel isomorphism. Let $B_1 = \{\Phi(h)\colon h \in B\}$. Then $B_1 \in \mathscr{B}(\mathbb{R}^j)$. Now

$$\begin{aligned} h \in P^{-1}B = C \quad &\text{if and only if} \quad Ph \in B \\ &\text{if and only if} \quad \Phi(Ph) \in B_1 \\ &\text{if and only if} \quad ((Ph,e_1),\ldots,(Ph,e_j)) \in B_1 \\ &\text{if and only if} \quad ((h,e_1),\ldots,(h,e_j)) \in B_1. \end{aligned}$$

Thus C can be expressed in the form (1.1).

Conversely, let C be of the form (1.1). First observe that $C \in \mathscr{B}(H)$ as the mapping $h \to ((h,h_1),\ldots,(h,h_j))$ is continuous and hence Borel measurable. Let $B = C \cap (PH)$. Then $B \in \mathscr{B}(PH)$. Now $h \in C$ if and only if $Ph \in C$, and hence if and only if $Ph \in B$, because $(h,h_i) = (Ph,h_i)$ for $h_i \in PH$ and thus $C = P^{-1}B$.

(ii) is an immediate consequence of (i). □

Definition 1.1: A *cylinder probability* n on H is a finitely additive non-negative set function on $\mathscr{C}$ with $n(H) = 1$ such that for all P in $\mathscr{P}(H)$, the restriction n_P of n to $\mathscr{C}_P$ is countably additive.

A trivial example of a cylinder probability on H is the restriction to $\mathscr{C}$ of any countably additive probability measure on $(H,\mathscr{B}(H))$. However, there exist cylinder probabilities on $(H,\mathscr{C})$ that do not admit any countably additive extension to $(H,\mathscr{B}(H))$ as will be seen later.

Cylinder Characteristic Functional

Definition 1.2: Let n be a cylinder probability on $(H,\mathscr{C})$. The function ψ: $H \to C$ defined by

$$\psi(h) := \int_H \exp(i(h,h_1))dn(h_1) \tag{1.3}$$

is called the *cylinder characteristic functional* of n.

The integral appearing in the definition of $\psi(h)$ is well defined as the integrand is $\mathscr{C}_P$ measurable, for any $P \in \mathscr{P}$ such that $h \in PH$ and n is countably additive on $\mathscr{C}_P$. As in the finite dimensional case, there is one-to-one correspondence between cylinder probabilities and their characteristic functionals as the following theorem shows.

Theorem 1.2: (a) Let n be a cylinder probability on $(H,\mathscr{C})$ and let ψ be the cylinder characteristic functional of n. Then

(i) $\psi(-h) = \overline{\psi(h)}$ and $\psi(0) = 1$.

(ii) ψ is positive definite, namely

$$\sum_{i,j=1}^{k} a_i \bar{a}_j \psi(h_i - h_j) \geq 0$$

for $h_1,\ldots,h_k \in H$, $a_i,\ldots,a_k \in \mathbb{C}$ and $k \geq 1$. Here $\overline{a_j}$ is the complex conjugate of a_j.

(iii) The restriction of ψ to any finite dimensional subspace H_1 of H is continuous (in the relative topology on H_1).

(b) Conversely, if ψ is any function from $H \to \mathbb{C}$ satisfying (i), (ii), (iii) above, then there exists a unique cylinder probability n on $(H,\mathscr{C})$ such that ψ is its cylinder characteristic functional.

<u>Proof</u>: (a) The properties (i), (ii) of ψ are proved exactly as in the finite dimensional case. For (iii), let $P \in \mathscr{P}$ be the orthogonal projection onto H_1 and let $\psi_1 = \psi|_{H_1}$. Let $n_1 = n|_{\mathscr{C}_P}$ and $n_2 = n_1 \circ P^{-1}$. Then for $h \in H_1$

$$\psi_1(h) = \psi(h) = \int_H \exp(i(h,h_1))dn(h_1)$$

$$= \int_H \exp(i(h,h_1))dn_1(h_1)$$

as the integrand is $\mathscr{C}_P$-measurable

$$= \int_H \exp(i(h,Ph_1))dn_1(h_1)$$

as $h \in H_1 = PH$

$$= \int_{H_1} \exp(i(h,h'))dn_2(h')$$

by substituting $Ph_1 = h'$.

Since n_1 is countably additive by the definition of the cylinder measure n, n_2 is countably additive and thus ψ_1 is the characteristic functional of a countably additive probability on a finite dimensional space H_1. Thus ψ_1 is continuous. This proves (a).

For (b), fix $P \in \mathscr{P}$ and let $H_0 = PH$ and ψ_0 be the restriction of ψ to H_0. Then by Bochner's theorem, there exists a unique countably additive probability measure, say

n'_P such that ψ_0 is the characteristic functional of n'_P. Let n_P be defined on $\mathscr{C}_P$ by

$$n_P(P^{-1}B) = n'_P(B), \quad B \in \mathscr{B}(H_1).$$

We now claim that $\{n_P\colon P \in \mathscr{P}\}$ is a consistent family, i.e., if range $P \subset$ range P_1, then

$$n_P = n_{P_1}\big|_{\mathscr{C}_P}.$$

This follows from the uniqueness part of Bochner's theorem. Indeed, let $n_1 = n_{P_1}\big|_{\mathscr{C}_P}$ and $n_2 = n_1 \circ P^{-1}$. Then it can be easily checked that ψ_0 is the characteristic functional of n_2 and hence $n_2 = n'_P$ which implies $n_1 = n_P$. Thus $\{n_P\colon P \in \mathscr{P}\}$ is consistent. Thus for $C \in \mathscr{C}$,

$$n(C) := n_P(C) \quad \text{if } C \in \mathscr{C}_P$$

properly defines a set function on $(H,\mathscr{C})$, which by its construction is a cylinder probability and ψ is its cylinder characteristic functional. The uniqueness of n follows from uniqueness in the finite dimensional case. □

Remark 1.1: The Sazonov Theorem and Theorem 1.2 above give us the following result. A cylinder probability n on $(H,\mathscr{C})$ has a countably additive extension to $(H,\mathscr{B}(H))$ if and only if its cylinder characteristic functional is continuous in the $\mathscr{S}$-topology on H.

The Canonical Gauss Measure

Example 1.1: Let $\psi(h) = e^{-\frac{1}{2}|h|^2}$, $h \in H$. It is very easy to see that ψ satisfies the conditions of Theorem 1.2 and hence is the characteristic functional of a cylinder probability on $(H,\mathcal{C})$.

Definition 1.3: The unique cylinder probability m on $(H,\mathcal{C})$ whose cylinder characteristic functional is $e^{-\frac{1}{2}|h|^2}$ is called the *canonical Gauss measure* on H.

The canonical Gauss measure is the infinite dimensional analogue of the standard normal distribution. The next result gives some properties of m.

Theorem 1.3: For $h_0 \in H$, let $f_{h_0} : H \to \mathbb{R}$ be defined by $f_{h_0}(h) = (h_0,h)$, $h \in H$.

(a) For $h_1,h_2,\ldots,h_k \in H$, the joint distribution of $(f_{h_1},\ldots,f_{h_k})$ under m is multivariate normal (or Gaussian) with mean zero and variance-covariance matrix $\{(h_i,h_j)\}$, $1 \le i, j \le k$.

(b) For $h_1,h_2,\ldots,h_k \in H$ such that $(h_i,h_j) = 0$, $1 \le i \le k$, $1 \le j \le k$, $i \neq j$, the random variables $f_{h_1},f_{h_2},\ldots,f_{h_k}$ are independent (under m).

(c) m does not admit a countably additive extension to $(H,\mathcal{B}(H))$.

Proof: (a) If $P \in \mathcal{P}$ is such that $h_1,h_2,\ldots,h_k \in$ Range P, then $(f_{h_1},\ldots,f_{h_k})$ is $\mathcal{C}_P$ measurable and hence

$$\mu = m\circ(f_{h_1},\ldots,f_{h_k})^{-1}$$

is a countably additive probability measure on $(\mathbb{R}^k,\mathcal{B}(\mathbb{R}^k))$. For $t_1,\ldots,t_k \in \mathbb{R}$,

$$\int_{\mathbb{R}^k} \exp(i \sum_{j=1}^{k} t_j x_j) d\mu(x) = \int_H \exp(i \sum_{j=1}^{k} t_j f_{h_j}(h)) dm(h) \qquad (1.4)$$

$$= \int_H \exp(i \sum_{j=1}^{k} t_j (h_j, h)) dm(h) = \int_H \exp(i(\sum_{j=1}^{k} t_j h_j, h)) dm(h)$$

$$= \exp(-\tfrac{1}{2} |\sum_{j=1}^{k} t_j h_j|^2) = \exp(-\tfrac{1}{2} \sum_{j=1}^{k} \sum_{r=1}^{k} t_j t_r (h_j, h_r)).$$

The relation (1.4) implies that μ is normal with mean zero, variance-covariance matrix $\{(h_j, h_r)\}$.

(b) Follows from (a) and properties of normal random variables.

(c) Suppose m admits a countably additive extension $\hat{m}$ to $\sigma(\mathscr{C}) = \mathscr{B}(H)$. Let $\{e_j\}$ be a CONS in H. Then by (a), (b), $\{f_{e_j}\}$ is a sequence of independent random variables on $(H, \mathscr{B}(H), \hat{m})$, the distribution of each of them is normal with mean zero, variance 1. For any positive integers N and M, we have

$$\hat{m}\{h: \sum_{j=1}^{N} (h, e_j)^2 \leq M\} = \hat{m}\{h: \sum_{j=1}^{N} (f_{e_j}(h))^2 \leq M\} \qquad (1.5)$$

$$\leq \hat{m}\{h: |f_{e_j}(h)| \leq M^{\frac{1}{2}}: 1 \leq j \leq N\}$$

$$= \prod_{j=1}^{N} \hat{m}\{h: |f_{e_j}(h)| \leq M^{\frac{1}{2}}\}$$

by independence of $\{f_{e_j}\}$

$$= [\hat{m}\{h: |f_{e_1}(h)| \leq M^{\frac{1}{2}}\}]^N$$

$$= \left[\frac{1}{\sqrt{2\pi}} \int_{-M^{\frac{1}{2}}}^{M^{\frac{1}{2}}} e^{-\frac{1}{2}x^2} dx\right]^N.$$

Since for all $M > 0$,

$$\left|\frac{1}{\sqrt{2\pi}} \int_{-M^{\frac{1}{2}}}^{M^{\frac{1}{2}}} e^{-\frac{1}{2}x^2} dx\right| < 1$$

the relation (1.5) implies

$$\lim_{N\to\infty} \hat{m}\{h: \sum_{j=1}^{N} (h,e_j)^2 \leq M\} = 0. \tag{1.6}$$

Since

$$\bigcap_{=1}^{\infty} \{h: \sum_{j=1}^{N} (h,e_j)^2 \leq M\} = \{h: \sum_{j=1}^{\infty} (h,e_j)^2 \leq M\}$$

$$= \{h: |h|^2 \leq M\}$$

and

$$\{h: \sum_{j=1}^{N+1} (h,e_j)^2 \leq M\} \subseteq \{h: \sum_{j=1}^{N} (h,e_j)^2 \leq M\},$$

the countable additivity of $\hat{m}$ and (1.6) implies

$$\hat{m}\{h: |h|^2 \leq M\} = 0. \tag{1.7}$$

The sets $\{h: |h|^2 \leq M\}$ increase with M and their union is H, thus (1.7) implies

$$\hat{m}(H) = 0 \tag{1.8}$$

which is a contradiction as $\hat{m}(H) = m(H) = 1$.

Thus m cannot be extended to a countably additive probability on $(H,\mathscr{B}(H))$. □

Corollary 1.4: The canonical Gauss measure can be described as follows. Let $C \in \mathscr{C}$ be given by

$$C = \{h: ((h,e_1),\ldots,(h,e_k)) \in B\} \tag{1.9}$$

where $k \geq 1$, $B \in \mathscr{B}(\mathbb{R}^k)$, $\{e_1,\ldots,e_k\}$ is an orthonormal set in H. Then

$$m(C) = \left(\frac{1}{\sqrt{2\pi}}\right)^k \int_B \exp\left(-\tfrac{1}{2}\sum_{i=1}^{k} x_i^2\right) d\lambda(x_1,x_2,\ldots,x_k) \tag{1.10}$$

where λ is the Lebesgue measure on $(\mathbb{R}^k,\mathscr{B}(\mathbb{R}^k))$.

Proof: This follows from part (a) of Theorem 1.3. □

Other examples of cylinder probabilities which do not admit a countably additive extension can be obtained as follows.

Example 1.2: Let $(\Omega_1,\mathscr{A}_1,\Pi_1)$ be a probability space and let $\{X_j\}$ be a sequence of real valued independent, identically distributed random variables on $(\Omega_1,\mathscr{A}_1,\Pi_1)$. Suppose $EX_1 = 0$, $EX_1^2 = \sigma_2$, $0 < \sigma^2 < \infty$. By a classical result, if $\{a_j\}$ is a sequence of real numbers with $\sum_{j=1}^{\infty} a_j^2 < \infty$, then the series $\sum_{j=1}^{\infty} a_j X_j(\omega)$ converges a.s. Π_1.

Now let $\{e_j\}$ be a CONS in H. For $h \in H$, define

$$L(h)(\omega) = \sum_{j=1}^{\infty} (h,e_j)X_j(\omega) \tag{1.11}$$

if the series converges, and 0 otherwise.

Since the series in (1.11) converges a.s., it is easy

to see that L is linear, i.e.,

$$L(a_1h_1 + a_2h_2) = a_1L(h_1) + a_2L(h_2) \quad \text{a.s. } \Pi_0 \tag{1.12}$$

for $h_1, h_2 \in H$, $a_1, a_2 \in \mathbb{R}$.

Let ϕ: $H \to \mathbb{C}$ be defined by

$$\phi(h) = E_{\Pi_1}(e^{iL(h)}). \tag{1.13}$$

Using (1.12) it is easy to see that ϕ defined by (1.13) satisfies the conditions (i), (ii) of Theorem 1.2. (Same computations as in proving these properties for a characteristic functional). We will prove that (1.12) implies the condition (iii) in Theorem 1.2 as well. Let H_0 be a finite dimensional subspace of H and let $\{e_1, \ldots, e_k\}$ be its basis. Let ϕ_1 be the restriction of ϕ to H_0. Then

$$\phi_1\left(\sum_{j=1}^{k} x_j e_j\right) = E_{\Pi_1}\left(\exp\left(iL\left(\sum_{j=1}^{k} x_j e_j\right)\right)\right) \tag{1.14}$$

$$= E_{\Pi_1}\left(\exp\left(i \sum_{j=1}^{k} x_j L(e_j)\right)\right) \quad \text{by (1.12).}$$

Thus $(x_1, \ldots, x_k) \to \phi_1(\Sigma_{j=1}^{k} x_j e_j)$ is a continuous function from $\mathbb{R}^k$ to $\mathbb{R}$. This means ϕ_1 is continuous.

Thus by Theorem 1.2, there exists a unique cylinder probability n on $(H, \mathscr{C})$ such that ϕ is its cylinder characeristic functional. We will prove that n can not be extended to $\mathscr{B}(H)$ as a countably additive probability measure.

Suppose n has an extension μ to $(H, \mathscr{B}(H))$, where μ is countably additive. Let Y_j: $H \to \mathbb{R}$ be defined by

$$Y_j(h) = (h, e_j).$$

Then

$$E_\mu \exp(i \sum_{j=1}^{k} x_j Y_j) = \int \exp(i \sum_{j=1}^{k} x_j (h, e_j)) dn(h) = \phi(\sum_{j=1}^{k} x_j e_j)$$

$$= E_{\Pi_1}(\exp(i \sum_{j=1}^{k} x_j L(e_j))) = E_{\Pi_1} \exp(i \sum_{j=1}^{k} x_j X_j).$$

Thus the finite dimensional distributions of $\{X_j\}$ under Π_1 and $\{Y_j\}$ under μ are identical. Since Π_1 and μ are countably additive, this means that the distributions of $\{X_j\}$ under Π_1 and $\{Y_j\}$ under μ are identical.

Now, by the strong law of large numbers

$$\frac{1}{k} \sum_{j=1}^{k} X_j^2(\omega) \to 1 \qquad \text{a.s. } \Pi_1.$$

But, $\Sigma_{j=1}^{\infty} Y_j^2(h) = \Sigma_{j=1}^{\infty} (h, e_j)^2 = |h|^2 < \infty$, so that $Y_j^2(h) \to 0$, for all h; hence,

$$\frac{1}{k} \sum_{j=1}^{k} Y_j^2(h) \to 0 \qquad \text{a.s. } \mu \text{ (for all } h \in H),$$

a contradiction. Hence n cannot be extended as a countably additive probability to $(H, \mathcal{B}(H))$.

Example 1.3: In Example 1.2 above, if instead, $\{X_j\}$ is a sequence of independent normal random variables with $EX_j = 0$, $EX_j^2 = \sigma_j^2$, then it can be shown that the cylinder measure n (determined by (1.11), (1.12), (1.13) and the condition that ϕ be its cylinder characteristic functional) admits a countably additive extension to $(H, \mathcal{B}(H))$ if and only if $\Sigma_{j=1}^{\infty} \sigma_j^2 < \infty$.

Example 1.4: Let Q be a symmetric, positive definite bilinear form on H, i.e., $Q: H \times H \to \mathbb{R}$ is such that

(i) $Q(h_1, h_2) = Q(h_2, h_1)$

(ii) $Q(a_1 h_1 + a_2 h_2, h_0) = a_1 Q(h_1, h_0) + a_2 Q(h_2, h_0)$

(iii) $\Sigma_{i=1}^{k} \Sigma_{j=1}^{k} a_i a_j Q(h_i, h_j) \geq 0$

for all $a_j \in \mathbb{R}$, $h_j \in H$, $0 \leq j \leq k$. Let

$$\psi(h) = \exp(-\tfrac{1}{2} Q(h_1, h_2)).$$

Using (i), (ii), (iii) above, it is easy to see that ψ satisfies the condition of Theorem 1.2 and is hence a cylinder characteristic functional of a cylinder probability n on $(H, \mathcal{C})$. If f_{h_0} is as in Theorem 1.3, then the distribution of $(f_{h_1}, \ldots, f_{h_k})$ under n is multivariate normal with mean 0 and variance-covariance matrix $(Q(h_i, h_j))$. The cylinder probability n is called the Gaussian cylinder probability with covariance form Q.

2. INTEGRATION WITH RESPECT TO CYLINDER PROBABILITIES

A cylinder probability n on $(H, \mathcal{C})$ is, in particular, a finitely additive probability measure and thus we do have a theory of integration with respect to n. [See Dunford-Schwartz [20] and Section III.5.] However, the class of integrable functions for this integral is not large enough and moreover does not contain some functions which arise naturally in the context of filtering theory, likelihood ratios, etc.

In this section, we will present a theory of integration with respect to n which takes into account the fact that underlying space is a Hilbert space and the field is the cylinder field. This follows closely the integral defined by Gross [27] for the Gauss measure m. The integral with respect to n is the same as the one defined in our earlier papers [35, 38, 40]. This will be established in the next section.

Cylinder Functions

Definition 2.1: A function $f: H \to \mathbb{R}$ is called a *cylinder function* if f is $\mathcal{C}_P$-measurable for some $P \in \mathcal{P}$.

The following proposition describes the class of cylinder functions.

Proposition 2.1: A function $f: H \to \mathbb{R}$ is a cylinder function if and only if f can be expressed as

$$f(h) = f_1((h,h_1),\dots,(h,h_k)) \tag{2.1}$$

where $k \geq 1$, $h_1,\dots,h_k \in H$ and $f_1: \mathbb{R}^k \to \mathbb{R}$ is Borel measurable.

Proof: Let f be given by (2.1) and let H_1 be the (finite dimensional) subspace generated by $\{h_1,\dots,h_k\}$. Let $P \in \mathcal{P}$ be the orthogonal projection onto H_1. Then the mappings $h \to (h,h_j)$, $1 \leq j \leq k$ are $\mathcal{C}_P$-measurable and hence f is $\mathcal{C}_P$-measurable.

Conversely, let f be $\mathcal{C}_P$-measurable for some $P \in \mathcal{P}$. Then

$$f(h) = f(Ph) \quad \text{for all } h \in H, \tag{2.2}$$

for if $f(h_0) \neq f(Ph_0)$ for some $h_0 \in H$ and $A = \{h: f(h) = f(h_0)\}$, we see that $h_0 \in A$ but $Ph_0 \notin A$, so that $A \notin \mathcal{C}_P$, which contradicts the assumption that f is $\mathcal{C}_P$-measurable.

Hence (2.2) holds.

Now, let $h_1,h_2,\ldots,h_k$ be an orthonormal basis in PH. Let $f_1: \mathbb{R}^k \to \mathbb{R}$ be defined by

$$f_1(x_1,x_2,\ldots,x_k) = f(\sum_{j=1}^{k} x_j h_j). \tag{2.3}$$

Since the mapping $(x_1,x_2,\ldots,x_k) \to \Sigma_{j=1}^{k} x_j h_j$ is continuous, it is Borel measurable and hence f_1 is Borel measurable. Now

$$f_1((h,h_1),\ldots,(h,h_k)) = f(\sum_{j=1}^{k} (h,h_j)h_j) = f(Ph) = f(h)$$

by (2.2) since $\{h_1,h_2,\ldots,h_k\}$ is an ONB for PH. □

<u>Remark 2.1</u>: If f is Borel measurable and $P \in \mathcal{P}$, it can be seen as in the above proof that $f \circ P$ is a $\mathcal{C}_P$-measurable cylinder function.

Let f be a real valued $\mathcal{C}_P$-measurable function. Since $n_P = n|_{\mathcal{C}_P}$ is countably additive, the probabilities of events like $\{|f| \geq \alpha\}$, $\{f > \alpha\}$, $f^{-1}B$ (α real and $B \in \mathcal{B}(\mathbb{R})$) are defined and if f is bounded or positive, the Lebesgue integral $\int f dn$ is also defined. We will extend this integral to a larger class of functions. The first step is to introduce the class of 'measurable functions.' For this purpose, let us define a partial ordering $\leq$ on $\mathcal{P}$ by $P_1 \leq P_2$ if and only if range $P_1 \subseteq$ range P_2.

<u>Definition 2.2</u>: Let $\mathcal{L}(H,\mathcal{C},n)$ be the class of Borel measurable functions $f: H \to \mathbb{R}$ such that for all $\epsilon > 0$, $\delta > 0$, $\exists$ $P_0 \in \mathcal{P}$ such that $P_1, P_2 \in \mathcal{P}$, $P_0 \leq P_i$, $i = 1,2$ implies

$$n(h: |f \circ P_1(h) - f \circ P_2(h)| > \delta) < \epsilon. \tag{2.4}$$

Remark 2.2: For real valued cylinder functions g, g', let

$$d(g,g') = \int [\,|g(h) - g'(h)| \wedge 1]dn(h).$$

Then it can be shown that for cylinder functions g_k, g, $d(g_k, g)$ converges to zero as $k \to \infty$ if and only if g_k converges to g in n-probability, i.e., for all $\epsilon > 0$,

$$\lim_{k\to\infty} n(|g_k - g| > \epsilon) = 0.$$

With this notation, $\mathcal{L}(H,\mathcal{C},n)$ is the class of real valued Borel measurable functions f on H such that the net $\{f\circ P: P \in \mathcal{P}\}$ of cylinder functions is Cauchy in the metric d.

Real valued cylinder functions trivially belong to $\mathcal{L}(H,\mathcal{C},n)$. The following result shows, in particular, that the class $\mathcal{L}(H,\mathcal{C},n)$ is closed under additions, product, maximum of two functions.

Theorem 2.2: Let $k \geq 1$, $f_1,\dots,f_k \in \mathcal{L}(H,\mathcal{C},n)$ and let $g: \mathbb{R}^k \to \mathbb{R}$ be a continuous function. Then

$$g(f_1,f_2,\dots,f_k) \in \mathcal{L}(H,\mathcal{C},n). \tag{2.5}$$

Proof: Let $\epsilon > 0$, $\delta > 0$ be given and let $P_1 \in \mathcal{P}$ be such that $P_1 \leq P'$, $P_1 \leq P''$ implies

$$n(|f_i\circ P' - f_i\circ P''| > 1) < \epsilon/8k, \quad 1 \leq i \leq k. \tag{2.6}$$

Let M be such that

$$n(|f_i\circ P_1| > M-1) < \epsilon/8k, \quad 1 \leq i \leq k. \tag{2.7}$$

Combining (2.6), (2.7), we get that for $P_1 \leq P'$,

$$n(|f_i\circ P'| > M, \text{ for some } i,\ 1 \leq i \leq k) < \epsilon/4. \tag{2.8}$$

Let $K = \{\underline{x} = (x_1,\dots,x_k) \in \mathbb{R}^k: \ |x_i| \le M, \ 1 \le i \le k\}$. Then K is compact and hence g is uniformly continuous on K. Thus, there exists $\delta_1 > 0$ such that $\underline{x} = (x_1,x_2,\dots,x_k) \in K$, $\underline{x}' = (x_1',\dots,x_k') \in K$, $|x_i-x_i'| \le \delta_1$, $1 \le i \le k$ implies $|g(\underline{x}) - g(\underline{x}')| \le \delta$. Thus

$$\begin{aligned}(|g(f_1\circ P',\dots,f_k\circ P') - g(f_1\circ P'',\dots,f_k\circ P'')| > \delta) \quad &(2.9)\\ \subseteq ((f_1\circ P',\dots,f_k\circ P') \notin K) \cup ((f_1\circ P'',\dots,f_k\circ P'') \notin K)&\\ \cup \sum_{i=1}^{k} (|f_i\circ P' - f_i\circ P''| > \delta_1).&\end{aligned}$$

Now, let P_2, $P_1 \le P_2$ be such that $P_2 \le P'$, $P_2 \le P''$ implies

$$n(|f_i\circ P' - f_i\circ P''| > \delta_1) < \frac{\epsilon}{2k}, \quad 1 \le i \le k. \qquad (2.10)$$

Thus for $P_2 \le P'$, $P_2 \le P''$, from (2.8), (2.9), (2.10) we get

$$\begin{aligned}&n(|g(f_1\circ P',\dots,f_k\circ P') - g(f_1\circ P'',\dots,f_k\circ P'')| > \delta)\\ &< \frac{\epsilon}{4} + \frac{\epsilon}{4} + \sum_{i=1}^{k} \frac{\epsilon}{2k} = \epsilon.\end{aligned}$$

This implies (2.5). □

Let $f_1,f_2 \in \mathcal{L}(H,\mathcal{C},n)$. By taking $k = 2$ and $g(x_1,x_2) = x_1 + x_2$, $g(x_1,x_2) = x_1\cdot x_2$, $g(x_1,x_2) = x_1 \vee x_2$ respectively in Theorem 2.1, it follows that $f_1 + f_2$, $f_1\cdot f_2$ and $f_1 \vee f_2 \in \mathcal{L}(H,\mathcal{C},n)$.

Let $f \in \mathcal{L}(H,\mathcal{C},n)$ be bounded, say by M. Given $\epsilon > 0$, let $P_0 \in \mathcal{P}$ be such that (2.4) holds for $\delta = \epsilon$. Then for $P_1,P_2 \in \mathcal{P}$, $P_0 \le P_1$, $P_0 \le P_2$,

$$\int_H |f\circ P_1(h) - f\circ P_2(h)|dn(h)$$

$$\leq \epsilon + \int |f\circ P_1(h) - f\circ P_2(h)| 1_{\{|f\circ P_1 - f\circ P_2| > \epsilon\}} dn(h)$$

$$\leq \epsilon + 2M\epsilon$$

and hence the net of real numbers $\{\int f\circ P(h)dn(h): P \in \mathcal{P}\}$ is Cauchy. In view of this, we make the following definition.

Definition 2.3: Let $f \in \mathcal{L}(H,\mathcal{C},n)$ be bounded. Then the integral of f with respect to n is defined by

$$\int f dn := \lim_{P\in\mathcal{P}} \int f\circ P dn.$$

We now extend the integral to positive functions first and later to real valued functions.

Definition 2.4: Let $f \in \mathcal{L}(H,\mathcal{C},n)$ be non-negative. Define

$$\int f dn := \lim_{k\to\infty} \int (f\wedge k)dn.$$

For $f \in \mathcal{L}(H,\mathcal{C},n)$, $f \geq 0$, $\int f dn$ may be finite or $+\infty$. Let

$$\mathcal{L}^1(H,\mathcal{C},n) := \{f \in \mathcal{L}(H,\mathcal{C},n): \int |f| dn < \infty\}.$$

Definition 2.5: Let $f \in \mathcal{L}^1(H,\mathcal{C},n)$. Define

$$\int f dn := \int f^+ dn - \int f^- dn$$

where $f^+ = f\vee 0$, $f^- = f\wedge 0$.

By arguments similar to those in countably additive measure theory, we can prove that the mapping $f \to \int f dn$ from $\mathcal{L}^1(H,\mathcal{C},n)$ into $\mathbb{R}$ is linear and monotone, i.e., for $f_1, f_2 \in \mathcal{L}^1(H,\mathcal{C},n)$, $a_1, a_2 \in \mathbb{R}$, $a_1f_1 + a_2f_2 \in \mathcal{L}^1(H,\mathcal{C},n)$ and

$$\int(a_1 f_1 + a_2 f_2)dn = a_1\int f_1 dn + a_2\int f_2 dn \tag{2.11}$$

and if $f_1, f_2 \in \mathcal{L}^1(H,\mathcal{C},n)$, $f_1 \leq f_2$, then

$$\int f_1 dn \leq \int f_2 dn. \tag{2.12}$$

We will now give an example to show that the classes $\mathcal{L}(H,\mathcal{C},n)$ and $\mathcal{L}^1(H,\mathcal{C},n)$ are strictly larger than the class of cylinder functions.

Example 2.1: Let A be a self-adjoint Hilbert-Schmidt (H-S) operator on H and let

$$f(h) = \|Ah\|^2.$$

Suppose that Range A has infinite dimension. Let $\{\alpha_j\}$ be the eigenvalues of A and $\{e_j\}$ be the corresponding eigenvectors so that

$$Ah = \sum_{j=1}^{\infty} \alpha_j (h,e_j)e_j, \quad h \in H$$

and thus

$$f(h) = \sum_{j=1}^{\infty} \alpha_j^2 (h,e_j)^2, \quad h \in H.$$

A is Hilbert-Schmidt implies that $\|A\|^2_{H.S.} = \Sigma_{j=1}^{\infty}\alpha_j^2 < \infty$.

For each k, let $P_k \in \mathcal{P}$ be given by

$$P_k h = \sum_{j=1}^{k} (h,e_j)e_j.$$

Then for $P' \in \mathcal{P}$, $P_k \leq P'$, we have

$$|f(P'h) - f(P_k h)| = |\sum_{j=1}^{\infty} \alpha_j^2 (P'h,e_j)^2 - \sum_{j=1}^{\infty} \alpha_j^2 (P_k h,e_j)^2| \tag{2.13}$$

$$= \sum_{j=k+1}^{\infty} \alpha_j^2 (P'h,e_j)^2$$

as $(P_k h,e_j) = (h,e_j) = (P'h,e_j)$ for $j \leq k$ and $(P_k h,e_j) = 0$, $j > k$. Fix $P' \in \mathscr{P}$. Let $g_k(h) = \Sigma_{j=k+1}^{\infty} \alpha_j^2 (P'h,e_j)^2$. Now $g_k(h) = g_k(P'h)$ is a $\mathscr{C}_{P'}$-measurable positive cylinder function. Thus (recall that $m_{P'}$ is a countably additive probability measure) for any $\epsilon' > 0$, we have

$$m(h\colon g_k(h) > \epsilon') \leq \frac{1}{\epsilon'} \int g_k(h) dm_{P'}(h) \tag{2.14}$$

$$= \frac{1}{\epsilon'} \sum_{j=k+1}^{\infty} \alpha_j^2 \int (P'h,e_j)^2 dm_{P'}(h)$$

$$= \frac{1}{\epsilon'} \sum_{j=k+1}^{\infty} \alpha_j^2 \int (h,P'e_j)^2 dm_{P'}(h)$$

$$= \frac{1}{\epsilon'} \sum_{j=k+1}^{\infty} \alpha_j^2 |P'e_j|^2 \leq \frac{1}{\epsilon'} \sum_{j=k+1}^{\infty} \alpha_j^2 .$$

Given $\epsilon > 0$, $\delta > 0$, choose k_0 such that $\sum_{j=k_0+1}^{\infty} \alpha_j^2 < \frac{\epsilon\delta}{4}$ (can be done as $\Sigma_{j=1}^{\infty} \alpha_j^2 < \infty$). Then for $P', P'' \in \mathscr{P}$, $P_{k_0} \leq P'$, $P_{k_0} \leq P''$, we have from (2.13), (2.14)

$$m(h\colon |f(P'h)-f(P''h)| > \delta) \leq m(h\colon |f(P'h)-f(P_{k_0}h)| > \tfrac{\delta}{2})$$

$$+ m(h\colon |f(P''h)-f(P_{k_0}h)| > \tfrac{\delta}{2}) \tag{2.15}$$

$$\leq \frac{2}{\delta} \sum_{j=k_0+1}^{\infty} \alpha_j^2 + \frac{2}{\delta} \sum_{j=k_0+1}^{\infty} \alpha_j^2 < \epsilon$$

by the choice of k_0. Hence $f \in \mathscr{L}(H,\mathscr{C},m)$.

For any $T > 0$,

$$\int (f \wedge T)dm = \lim_{P\in\mathscr{P}} \int (f\circ P)\wedge T \; dm = \lim_{k\to\infty} \int (f\circ P_k)\wedge T \; dm \tag{2.16}$$

[follows from (2.15)]. But

$$\int (f\circ P_k)(h)dm = \int \sum_{j=1}^{k} \alpha_j^2 (h,e_j)^2 = \sum_{j=1}^{k} \alpha_j^2 \leq \sum_{j=1}^{\infty} \alpha_j^2 = \|A\|_{H.S.}^2$$

Thus by (2.16)

$$\int (f\wedge T)dm \leq \|A\|_{H.S.}^2$$

so that by the definition of $\int f dm$ for $f \geq 0$,

$$\int f dm = \lim_{T\to\infty} \int (f\wedge T)dm \leq \|A\|_{H.S.}^2 < \infty.$$

Hence $f \in \mathscr{L}^1(H,\mathscr{C},m)$.

Remark 2.3: In the next section, we will develop other techniques which will make it easier to show that a given function belongs to $\mathscr{L}(H,\mathscr{C},m)$ and give a method of computing the integrals. In the above example, it can be shown that $\int f dm = \|A\|_{H.S.}^2$.

We have remarked that elements of $\mathscr{L}(H,\mathscr{C},n)$ are real valued 'measurable' functions on $(H,\mathscr{C},n)$ and we have defined integral for a subclass of these functions. For each 'measurable' function f, we now introduce the notion of its distribution $n\circ f^{-1}$ in a natural way. The latter is a countably additive probability measure which has the property that f is integrable if and only if $\int|x|dn\circ f^{-1}(x) < \infty$. More generally, for a bounded continuous function $g: \mathbb{R} \to \mathbb{R}$

$$\int g\circ f\, dn = \int g(x)dn\circ f^{-1}(x).$$

Proposition 2.3: Let $f \in \mathscr{L}(H,\mathscr{C},n)$. Then the net (of countably additive probability measures) $\{n\circ[f\circ P]^{-1}: P \in \mathscr{P}\}$ converges in the sense of weak convergence of probability measures to a countably additive probability measure λ on $(\mathbb{R},\mathscr{B}(\mathbb{R}))$. The measure λ will be denoted by $n\circ f^{-1}$, and will be called the distribution of f under n. Further

$$f \in \mathscr{L}^1(H,\mathscr{C},n) \text{ if and only if } \int|x|dn\circ f^{-1}(x) < \infty \tag{2.17}$$

and then

$$\int f dn = \int x dn\circ f^{-1}(x). \tag{2.18}$$

Also, for $g \in C_b(\mathbb{R})$,

$$\int g(f)dn = \int g(x)dn\circ f^{-1}(x).$$

Proof: For $P \in \mathscr{P}$, let $\lambda_P = n\circ[f\circ P]^{-1}$ and let

$$\phi_P(t) = \int e^{itx}d\lambda_P(x) = \int e^{it(f\circ P)(h)}dn(h).$$

Then for any $\epsilon > 0$

$$|\phi_P(t)-\phi_{P'}(t)| \le \int|e^{itf\circ P(h)} - e^{itf\circ P'(h)}|dn(h)$$

$$\le \sup_{|\theta|\le\epsilon} |1-e^{it\theta}| + 2n(|f\circ P-f\circ P'| > \epsilon).$$

Thus

$$\sup_{|t|\leq T} |\phi_P(t)-\phi_{P'}(t)| \leq \sup_{|y|\leq T\epsilon} |1-e^{iy}| + 2n(|f\circ P-f\circ P'| > \epsilon). \tag{2.19}$$

Since $f \in \mathscr{L}(H,\mathscr{C},n)$, (2.19) implies that the net of functions $\{\phi_P(t): P \in \mathscr{P}\}$ converges uniformly on compact subsets of $\mathbb{R}$ to $\phi(t)$. From Levy's continuity theorem for characteristic functions, it follows that ϕ is the characteristic function of a countably additive probability measure λ on $(\mathbb{R},\mathscr{B}(\mathbb{R}))$ and that λ_P converges weakly to λ, i.e., for all $g \in C_b(\mathbb{R})$,

$$\int g(x)d\lambda_P(x) \to \int g(x)d\lambda(x)$$

or

$$\int g(f\circ P)(h)dn(h) \to \int g(x)d\lambda(x). \tag{2.20}$$

Since for $g \in C_b(\mathbb{R})$, $g(f)$ is bounded, we get (by the definition of the integral $\int g(f)dn$)

$$\int g(f)dn = \int g(x)d\lambda(x). \tag{2.21}$$

In particular, we get for all $T > 0$

$$\int f^+\wedge T dn = \int x\wedge T 1_{(x>0)}d\lambda(x)$$

which implies (by definition of $\int f^+ dn$) that

$$\int f^+ dn = \int x1_{(x>0)}d\lambda(x). \tag{2.22}$$

Similarly

$$\int f^- dn = -\int x1_{(x<0)}d\lambda(x). \tag{2.23}$$

The relations (2.22) and (2.23) together imply (2.17) and (2.18). The last assertion in the theorem has already been proved - (2.21). □

In the case of random variables on a countably additive probability space, we do not distinguish between random variables which are equal to each other outside a set of probability zero. We introduce a similar concept on a cylinder probability space.

Definition 2.6: Let $f, g \in \mathcal{L}(H, \mathcal{C}, n)$. We say that f is equal to g modulo n, written as $f \equiv g$ mod[n] if for all $\epsilon > 0$, $\exists$ $P_0 \in \mathcal{P}$ such that for all $P \in \mathcal{P}$, $P_0 \leq P$,

$$n(H: |f \circ P(h) - g \circ P(h)| > \epsilon) < \epsilon. \tag{2.24}$$

From the definition, it is clear that 'equal to mod[n]' is an equivalence relation on $\mathcal{L}(H, \mathcal{C}, n)$. We may think of elements in $\mathcal{L}(H, \mathcal{C}, n)$ as equivalence classes under this equivalence relation. The following theorem gives some properties of this relation.

Theorem 2.4: (a) $f \equiv g$ mod[n] if and only if $f - g \equiv 0$ mod[n]. (b) $f \equiv 0$ mod[n] if and only if $n \circ f^{-1} = \delta_{\{0\}}$ where $\delta_{\{0\}}$ is the unit mass at 0.

Proof: (a) follows from definition 2.6. For (b), observe that (2.24) (for f) would imply that

$$n \circ (f \circ P)^{-1}((-\infty, -\epsilon) \cup (\epsilon, \infty)) < \epsilon$$

for all $P \in \mathcal{P}$, $P_0 \leq P$. Since $n \circ (f \circ P)^{-1} \to n \circ f^{-1}$ in the sense of weak convergence, this implies that

$$n \circ f^{-1}((-\infty, -\epsilon) \cup (\epsilon, \infty)) \leq \epsilon$$

for all $\epsilon > 0$, so that $n \circ f^{-1}(\{0\}) = 1$. □

Remark 2.4: We will later prove (in the next section) that if $f \in \mathcal{L}^1(H,\mathcal{C},n)$ and $\int_C f dn = 0$ for all $C \in \mathcal{C}$, then $f \equiv 0$ mod[n]. This will show that 'equal to mod[n]' is the right analogue of 'equal to a.e.' in our set.up. However, to prove this statement, we need some more machinery, which will be developed in the next section. This statement is proved in Theorem 3.9.

3. REPRESENTATION AND LIFTING MAPS

In the previous section, we have proved that for all $f \in \mathcal{L}(H,\mathcal{C},n)$, there exists a unique countably additive probability measure -- denoted by $n \circ f^{-1}$ and called the distribution of f under n -- on $(\mathbb{R},\mathcal{B}(\mathbb{R}))$ such that for all $g \in C_b(\mathbb{R})$

$$\int g \circ f \, dn = \int g(x) d(n \circ f^{-1})(x).$$

We will show that for each $f \in \mathcal{L}(H,\mathcal{C},n)$ the measure $n \circ f^{-1}$ can be represented by a random variable $R_n(f)$ on some fixed countably additive probability space $(\Omega_0,\mathcal{A}_0,\Pi_0)$ (depending on H and n) such that

(i) $\Pi_0 \circ [R_n(f)]^{-1} = n \circ f^{-1}$

(ii) the map $f \to R_n(f)$ is linear and multiplicative.

This map R_n will be called an n-lifting. This is a very useful computational tool.

Integration with respect to the Gauss measure m was defined by Gross [27] via the lifting map (he had not used the term lifting and he denoted $R_n(f)$ by $\tilde{f}$). In our earlier work, we followed Gross's approach to introduce integration with respect to a general cylinder probability via the lifting map [35, 38, 40]. We will show in this section that the definition via the lifting map is equivalent to the direct definition given in Section III.2.

Weak Distribution

The notion of weak distribution is due to Segal [70]. Consider the class of mappings

$$L\colon H \to \mathscr{L}(\Omega_0, \mathscr{A}_0, \Pi_0)$$

where $(\Omega_0, \mathscr{A}_0, \Pi_0)$ is some countably additive probability space, satisfying

$$L(a_1h_1+a_2h_2) = a_1L(h_1) + a_2L(h_2) \quad \text{a.s. } \Pi_0 \tag{3.1}$$

for all $h_1, h_2 \in H$, $a_1, a_2 \in \mathbb{R}$.

Two such mappings L, L' $(L'\colon H \to \mathscr{L}(\Omega_1, \mathscr{A}_1, \Omega_1))$ are said to be equivalent if for all $k \geq 1$, $h_1, h_2, \ldots, h_k \in H$,

$$\Pi_0 \circ [L(h_1), \ldots, L(h_k)]^{-1} = \Pi_1 \circ [L'(h_1), \ldots, L'(h_k)]^{-1}. \tag{3.2}$$

It is easy to see that this is an equivalence relation. The equivalence classes under this relation are called weak distributions and a mapping L satisfying (3.1) is called a representation of the weak distribution [L], which is the equivalence class determined by L.

In view of the linearity (3.1), it is easy to see that L and L' are equivalent, i.e., (3.2) holds, if and only if

$$E_{\Pi_0} e^{iL(h)} = E_{\Pi_1} e^{iL'(h)} \quad \text{for all } h \in H.$$

Let [L] be a weak distribution and let $\phi\colon H \to \mathbb{C}$ be defined by

$$\phi(h) = E_{\Pi_0} e^{iL(h)}. \tag{3.3}$$

It was observed in Example 1.2 that ϕ satisfies the conditions of Theorem 1.2 and hence determines a unique cylinder probability n whose cylinder characteristic functional is ϕ. This correspondence [L] → n can be described by the relation

$$E_{\Pi_0} e^{iL(h)} = \int e^{i(h,h')} dn(h'). \tag{3.4}$$

This is indeed a one-to-one correspondence as the next theorem states.

<u>Theorem 3.1</u>: There is a one-to-one correspondence between weak distribution and cylinder probabilities given by relation (3.4).

<u>Proof</u>: We have already proved the existence of n satisfying (3.4) given a weak distribution [L] (Example 1.2).

Conversely, let n be a cylinder probability. Let $\Omega_0 = \mathbb{R}^H$, $\mathcal{A}_0 = \underset{H}{\otimes}\mathcal{B}(\mathbb{R})$ and let L(h) be the coordinate mappings on $\mathbb{R}^H$, i.e., $L(h)(\omega) = \omega(h)$, $\omega \in \Omega_0$, $h \in H$. Let $\mathcal{A}_1$ be the field of finite dimensional cylinder sets in $\mathcal{A}_0$, namely

$$\mathcal{A}_1 = \{\{\omega: (L(h_1)(\omega),\ldots,L(h_k)(\omega)) \in B\}: h_i \in H,\ B \in \mathcal{B}(\mathbb{R}^k),\ k \geq 1\}.$$

Define a set function Π_0 on $\mathcal{A}_1$ by

$$\Pi_0(\omega:\ (L(h_1)(\omega),\ldots,L(h_k)(\omega)) \in B) \tag{3.5}$$
$$= n(h:\ ((h,h_1),\ldots,(h,h_k)) \in B)$$

$h_i \in H$, $B \in \mathcal{B}(\mathbb{R}^k)$, $k \geq 1$. By the Kolmogorov consistency theorem, Π_0 extends uniquely to $\mathcal{A}_0$ as a countably additive probability, also denoted by Π_0. Of course, (3.5) continues to hold. We will show that the mapping $h \to L(h)$ is linear, i.e., (3.1) holds. This will complete the proof, as the weak distribution [L] is the one we are looking for -- since (3.5) implies (3.4).

Let $a_1,a_2 \in \mathbb{R}$ and $h_1,h_2 \in H$ be given. Let $h_3 = a_1h_1 + a_2h_2$ and let $B \in \mathcal{B}(\mathbb{R}^3)$ be given by

$$B = \{(x_1,x_2,x_3) \in \mathbb{R}^3:\ a_1x_1 + a_2x_2 = x_3\}.$$

By (3.5) (for $k = 3$), we get

$$n(h:\ ((h,h_1),(h,h_2),(h,h_3)) \in B) \tag{3.6}$$
$$= \Pi_0(\omega:\ (L(h_1)(\omega),L(h_2)(\omega),L(h_3)(\omega)) \in B).$$

Since $h_3 = a_1h_1 + a_2h_2$,

$$((h,h_1),(h,h_2),(h,h_3)) \in B \text{ for all } h \in H$$

and hence by (3.6)

$$\Pi_0(\omega:\ (L(h_1)(\omega),L(h_2)(\omega),L(h_3)(\omega)) \in B) = n(H) = 1,$$

i.e.,

$$\Pi_0(\omega:\ L(h_3)(\omega) = a_1L(h_1)(\omega) + a_2L(h_2)(\omega)) = 1.$$

This shows that L satisfies (3.1). As observed earlier, this completes the proof. □

Definition 3.1: A pair (L,Π_0) satisfying (3.5) will be called a *representation* of n.

The term representation is more convenient to use than 'representative of the weak distribution corresponding to n.' Theorem 3.1 implies that a representation of n always exists.

For the remaining part of this section, we fix a representation (L,Π_0) of n.

We will first define the lifting $R_n(f)$ for cylinder functions f and then extend it to the class $\mathcal{L}(H,\mathcal{C},n)$. We need the following lemma for this.

Lemma 3.2: Let f be a real valued cylinder function and suppose

$$f(h) = f_1((h,h_1),\dots,(h,h_k)) = f_2((h,h_1'),\dots,(h,h_j')) \quad (3.7)$$

where $h_1,\dots,h_k; h_1',\dots,h_j' \in H$, $f_1\colon \mathbb{R}^k \to \mathbb{R}$ and $f_2\colon \mathbb{R}^j \to \mathbb{R}$ are Borel measurable functions. Then

$$f_1(L(h_1),\dots,L(h_k)) = f_2(L(h_1'),\dots,L(h_1')) \quad \text{a.s. } \Pi_0. \quad (3.8)$$

Proof: Let $B \in \mathcal{B}(\mathbb{R}^{k+j})$ be given by

$$B := \{(x_1,x_2,\dots,x_{k+j}) \in \mathbb{R}^{k+j}\colon$$
$$f_1(x_1,\dots,x_k) \neq f_2(x_{k+1},\dots x_{k+j})\}.$$

Then (3.7) implies

$$A := \{h\colon ((h,h_1),\dots,(h,h_k),\ (h,h_1'),\dots,(h,h_j')) \in B\} = \phi.$$

Hence by (3.5)

$$\Pi_0\{\omega\colon (L(h_1)(\omega),\ldots,L(h_k)(\omega),\ L(h_1')(\omega),\ldots,L(h_j')(\omega)) \in B\}$$
$$= n(A) = 0$$

and (3.8) is proved. □

Definition 3.2: Let f be a real valued cylinder function given by

$$f(h) = f_1((h,h_1),\ldots,(h,h_k)) \tag{3.9}$$

with $h_i \in H$, $f_1\colon \mathbb{R}^k \to \mathbb{R}$ Borel measurable. Then the *n-lifting* of f is defined

$$R_n(f) := f_1(L(h_1),\ldots,L(h_k)). \tag{3.10}$$

In view of Lemma 3.2, the relation (3.10) defines $R_n(f)$ up to Π_0-null sets, i.e., $R_n(f)$ is uniquely determined as an element of $\mathcal{L}(\Omega_0,\mathcal{A}_0,\Pi_0)$. $R_n(f)$ of course depends on the representation (L,Π_0) of n.

Lemma 3.3: Let $f\colon H \to \mathbb{R}$ be a cylinder function. Then

$$n\circ f^{-1} = \Pi_0\circ[R_n(f)]^{-1}. \tag{3.11}$$

Proof: Let f be given by (3.9) so that $R_n(f)$ is given by (3.10). Let $\theta\colon H \to \mathbb{R}^k$ be defined by

$$\theta(h) = ((h,h_1),\ldots,(h,h_k)).$$

From (3.5) we have

$$n\circ\theta^{-1} = \Pi_0\circ[L(h_1),\ldots,L(h_k)]^{-1}. \tag{3.12}$$

Now $f = f_1 \circ \theta$ so that

$$n \circ f^{-1} = [n \circ \theta^{-1}] \circ f_1^{-1} \tag{3.13}$$

$$= [\Pi_0 \circ (L(h_1), \ldots, L(h_k))]^{-1} \circ f_1^{-1} \quad \text{by (3.12)}$$

$$= \Pi_0 \circ [R_n(f)]^{-1} \quad \text{by (3.10).}$$

□

Proceeding exactly as in the proof of Lemma 3.3, it can be proved that if $f_1, f_2, \ldots, f_j$ are real valued cylinder functions, then

$$n \circ (f_1, f_2, \ldots, f_j)^{-1} = \Pi_0 \circ (R_n(f_1), \ldots, R_n(f_j))^{-1}. \tag{3.14}$$

The relation (3.14) implies that for cylinder functions f_1, f_2 and real numbers a_1, a_2, we have

$$R_n(a_1 f_1 + a_2 f_2) = a_1 R_n(f_1) + a_2 R_n(f_2) \quad \text{a.e. } \Pi_0 \tag{3.15}$$

and

$$R_n(f_1 \cdot f_2) = R_n(f_1) \cdot R_n(f_2) \quad \text{a.e. } \Pi_0. \tag{3.16}$$

To see that (3.15) holds, let $B \in \mathscr{B}(\mathbb{R}^3)$ be as in the proof of Theorem 3.1. Then by (3.14), writing $f_3 = a_1 f_1 + a_2 f_2$, we have

$$\Pi_0((R_n(f_1), R_n(f_2), R_n(f_3)) \in B) = n \circ (f_1, f_2, f_3)^{-1}(B).$$

But $(f_1(h), f_2(h), f_3(h)) \in B$ for all h by the choice of f_3, B. Thus $n \circ (f_1, f_2, f_3)^{-1}(B) = 1$ and hence

$$\Pi_0((R_n(f_1), R_n(f_2), R_n(f_3)) \in B) = 1,$$

i.e.,

$$\Pi_0(R_n(f_3) = a_1R_n(f_1) + a_2R_n(f_2)) = 1.$$

The relation (3.16) can be proved similarly by taking

$$B = \{(x_1,x_2,x_3) \in \mathbb{R}^3: x_3 = x_1 \cdot x_2\}.$$

<u>Theorem 3.4</u>: For every $f \in \mathcal{L}(H,\mathcal{C},n)$, there exists a random variable $X_f \in \mathcal{L}(\Omega_0,\mathcal{A}_0,\Pi_0)$ such that the net $\{R_n(f\circ P): P \in \mathcal{P}\}$ converges to X_f in Π_0-probability, i.e., for all $\epsilon > 0$, $\exists\, P_1 \in \mathcal{P}$ such that $P \in \mathcal{P}$, $P_1 \leq P$ implies

$$\Pi_0(|R_n(f\circ P) - X_f| > \epsilon) < \epsilon. \tag{3.17}$$

<u>Proof</u>: First observe that given $\epsilon_1 > 0$, there exists a $P_0 \in \mathcal{P}$ such that whenever $P_0 \leq P'$ and $P_0 \leq P''$ $(P',P'' \in \mathcal{P})$

$$n(H: |f\circ P'(h) - f\circ P''(h)| > \epsilon_1) < \epsilon_1, \tag{3.18}$$

which in view of the relation (3.5) and the fact that $R_n(f\circ P'-f\circ P'') = R_n(f\circ P') - R_n(f\circ P'')$ (see (3.15)) can be written as

$$\Pi_0(|R_n(f\circ P') - R_n(f\circ P'')| > \epsilon_1) < \epsilon_1. \tag{3.19}$$

For each $k \geq 1$, let $P_k \in \mathcal{P}$ be such that $P_1',P_2' \in \mathcal{P}$, $P_k \leq P_1'$, $P_k \leq P_2'$ implies

$$\Pi_0(|R_n(f\circ P_1') - R_n(f\circ P_2')| > \frac{1}{2^k}) < \frac{1}{2^k}. \tag{3.20}$$

Without loss of generality, we can assume that $P_k \leq P_{k+1}$. (If not, inductively get P_{k+1}' such that $P_{k+1} \leq P_{k+1}'$ and $P_k' \leq P_{k+1}'$). Then we have

$$\Pi_0(|R_n(f\circ P_k) - R_n(f\circ P_{k+1})| > \frac{1}{2^k}) < \frac{1}{2^k}.$$

From the Borel-Cantelli lemma it now follows that

$$\Pi_0(|R_n(f\circ P_k) - R_n(f\circ P_{k+1})| > \frac{1}{2^k} \text{ infinitely often}) = 0. \tag{3.21}$$

The relation (3.21) implies that the sequence of random variables $\{R_n(f\circ P_k): k \geq 1\}$ converges a.s. to X_f, say. Now

$$\begin{aligned} &\Pi_0(|R_n(f\circ P_{k+1})-X_f| > \frac{1}{2^k}) \\ &\leq \sum_{j=k+1}^{\infty} \Pi_0(|R_n(f\circ P_j)-R_n(f\circ P_{j+1})| > \frac{1}{2^j}) < \sum_{j=k+1}^{\infty} \frac{1}{2^j} \\ &= \frac{1}{2^k} . \end{aligned} \tag{3.22}$$

The relation (3.20) and (3.22) imply that for any $P_1' \in \mathcal{P}$, $P_k \leq P_1'$, we have (since $P_k \leq P_{k+1}$),

$$\begin{aligned} \Pi_0(|R_n(f\circ P_1')-X_f| > \frac{1}{2^{k-1}}) &\leq \Pi_0(|R_n(f\circ P_1')-R_n(f\circ P_{k+1})| > \frac{1}{2^k}) \\ &\quad + \Pi_0(|R_n(f\circ P_{k+1})-X_f| > \frac{1}{2^k}) \\ &< \frac{1}{2^k} + \frac{1}{2^k} = \frac{1}{2^{k-1}} . \end{aligned} \tag{3.23}$$

Thus the net $\{R_n(f\circ P): P \in \mathcal{P}\}$ converges in probability to X_f. □

Compare the above theorem with Remark 2.2.

<u>Definition 3.3</u>: For $f \in \mathcal{L}(H,\mathcal{C},n)$, define the *n-lifting of f* $R_n(f)$ by

$$R_n(f) := \lim_{P \in \mathcal{P}} \text{ in } \Pi_0\text{-probability } R_n(f \circ P) \tag{3.24}$$

The limit in (3.24) exists by Theorem 3.4. It is simple to check that if f is a cylinder function, then $R_n(f)$ as defined earlier satisfies (3.24) so that the definition given above is an extension of the earlier one for cylinder functions. Indeed, if f is $\mathcal{C}_{P_0}$-measurable, then $f \circ P = f$ for all $P \in \mathcal{P}$, $P_0 \leq P$ which implies $R_n(f) = R_n(f \circ P)$ for all $P \in \mathcal{P}$ such that $P_0 \leq P$.

The n-lifting R_n is a mapping from $\mathcal{L}(H,\mathcal{C},n)$ into $\mathcal{L}(\Omega_0,\mathcal{A}_0,\Pi_0)$. It depends upon the underlying representation (L,Π_0) of n. When we wish to emphasize this dependence, or when there is more than one representation under consideration, we will call R_n as defined above the n-lifting corresponding to the representation (L,Π_0) of n and $(\Omega_0,\mathcal{A}_0)$ will be called the underlying representation space.

The following lemma connects the class $\mathcal{L}^1(H,\mathcal{C},n)$ and the lifting map. It also shows that the integral $\int f dn$ defined in Section III.2 is the same as the one defined in our earlier papers [35, 38, 40].

<u>Theorem 3.5</u>: Let $f \in \mathcal{L}(H,\mathcal{C},n)$. Then

(i) $$n \circ f^{-1} = \Pi_0 \circ [R_n(f)]^{-1} \tag{3.25}$$

(See Proposition 2.3 for the definition of $n \circ f^{-1}$).

(ii) $f \in \mathcal{L}^1(H,\mathcal{C},n)$ if and only if $\int |R_n(f)| d\Pi_0 < \infty$

and for $f \in \mathcal{L}^1(H,\mathcal{C},n)$

$$\int f dn = \int R_n(f) d\Pi_0. \tag{3.26}$$

Proof: (i) Since convergence in probability implies convergence in distribution, (3.24) implies that the net

$$\Pi_0 \circ [R_n(f \circ P)]^{-1} \to \Pi_0 \circ [R_n(f)]^{-1} \tag{3.27}$$

in the sense of weak convergence. By Proposition (2.1),

$$n \circ (f \circ P)^{-1} \to n \circ f^{-1} \tag{3.28}$$

in the sense of weak convergence. Now (3.26), (3.28) and the fact that for all $P \in \mathcal{P}$

$$n \circ (f \circ P)^{-1} = \Pi_0 \circ [R_n(f \circ P)]^{-1}$$

(as $f \circ P$ is a cylinder function) implies (3.25).

Part (ii) follows from (3.25) and relations (2.17), (2.18) in Proposition 2.3. □

The class $\mathcal{L}(H, \mathcal{C}, n)$ and the lifting map R_n have been defined in terms of convergence of nets. It is easier to work with sequences (and convergence almost everywhere in place of convergence in probability). The following theorem gives a characterization of $\mathcal{L}(H, \mathcal{C}, n)$ and R_n using sequential convergence.

For $\{P_i\} \subseteq \mathcal{P}$, say that $P_i \uparrow I$ if $P_i \leq P_{i+1}$ for all i and $|P_i h - h| \to 0$ for all $h \in H$.

Theorem 3.6: Let $f\colon H \to \mathbb{R}$ be a Borel measurable function. Then the following are equivalent.

(a) $f \in \mathcal{L}(H, \mathcal{C}, n)$.

(b) There exists $\{P_k\} \subseteq \mathcal{P}$, $P_k \uparrow I$ such that for all $\{P'_k\} \subseteq \mathcal{P}$, $P_k \leq P'_k$, $R_n(f \circ P'_k)$ converges a.e. Π_0.

(c) There exists $\{P_k\} \subseteq \mathcal{P}$, $P_k \uparrow I$ such that for all

$\{P'_k\} \subseteq \mathcal{P}$, $P_k \leq P'_k$, $R_n(f \circ P'_k)$ converges in Π_0-probability.

Further, if $f \in \mathcal{L}(H, \mathcal{C}, n)$, then for $\{P'_k\}$ as in (b), [as in (c)] $R_n(f \circ P'_k)$ converges to $R_n(f)$ a.s. Π_0 [in Π_0-probability].

Proof: Suppose (a) holds. Then for each $k \geq 1$, let $\tilde{P}_k \in \mathcal{P}$ be such that $P', P'' \in \mathcal{P}$, $\tilde{P}_k \leq P'$, $\tilde{P}_k \leq P''$ imply

$$n(h: |f \circ P'(h) - f \circ P''(h)| > \frac{1}{2^k}) < \frac{1}{2^k} . \tag{3.29}$$

Choose any sequence $\{P_k\} \subseteq \mathcal{P}$, $P_k \uparrow I$ such that $\tilde{P}_k \leq P_k$ for all k. Then for all $P', P'' \in \mathcal{P}$, $P_k \leq P'$, $P_k \leq P''$, (3.29) holds, which in view of (3.14) may be written as

$$\Pi_0(|R_n(f \circ P') - R_n(f \circ P'')| > \frac{1}{2^k}) < \frac{1}{2^k} . \tag{3.30}$$

Thus if $\{P'_k\} \subseteq \mathcal{P}$, $P_k \leq P'_k$, then for all $k \geq 1$

$$\Pi_0(|R_n(f \circ P'_{k+1}) - R_n(f \circ P'_k)| > \frac{1}{2^k}) < \frac{1}{2^k} . \tag{3.31}$$

By the Borel-Cantelli lemma,

$$\Pi_0(|R_n(f \circ P'_{k+1}) - R_n(f \circ P'_k)| > \frac{1}{2^k} \text{ infinitely often}) = 0, \tag{3.32}$$

and hence

$R_n(f \circ P'_k)$ converges a.s. Π_0.

This proves (a) => (b).

Since convergence a.s. Π_0 implies convergence in Π_0-probability for a sequence of random variables on $(\Omega_0, \mathcal{A}_0, \Pi_0)$, it is easy to see that (b) => (c).

We will now prove that (c) => (a). Let (c) hold and let $\{P_k\} \subseteq \mathcal{P}$ be as in (c). The first thing to note is that for all $\{P'_k\} \subseteq \mathcal{P}$, $P_k \leq P'_k$, $R_n(f \circ P'_k)$ converges in Π_0-probability to the same limit. To see this, let $P'_k = P_k$, so that $R_n(f \circ P_k)$ converges in Π_0-probability. Let

$$X := \lim_{k \to \infty} \text{ in } \Pi_0\text{-probability } R_n(f \circ P_k)$$

Then for all $\{P'_k\} \subseteq \mathcal{P}$, $P_k \leq P'_k$, we have

$$R_n(f \circ P'_k) \to X \quad \text{in } \Pi_0\text{-probability}. \tag{3.33}$$

For if (3.33) does not hold, and if $R_n(f \circ P'_k) \to Y$ in Π_0-probability, $\Pi_0(Y \neq X) > 0$, then taking $\{\tilde{P}_k\} \subseteq$ such that

$$\tilde{P}_{2k-1} = P_k, \quad \tilde{P}_{2k} = P'_k, \quad k \geq 1$$

we have that

$$R_n(f \circ \tilde{P}_{2k}) \to Y \quad \text{in } \Pi_0\text{-probability}$$

$$R_n(f \circ \tilde{P}_{2k-1}) \to X \quad \text{in } \Pi_0\text{-probability}$$

and $\Pi_0(Y \neq X) > 0$, so that $R_n(f \circ \tilde{P}_k)$ cannot converge in Π_0-probability, which contradicts (c). This proves (3.33).

We now claim that

$$\lim_{P \in \mathcal{P}} \text{ in } \Pi_0\text{-probability } R_n(f \circ P)$$

exists and is equal to X. If not, there exists $\epsilon > 0$, $\delta > 0$

such that for each $P_0 \in \mathcal{P}$, $\exists$ $P_0' \in \mathcal{P}$, $P_0 \leq P_0'$ with

$$\Pi_0(|R_n(f\circ P_0') - X| > \delta) \geq \epsilon.$$

Using this for $P_0 = P_k$, choose P_k', $P_k \leq P_k'$ such that

$$\Pi_0(|R_n(f\circ P_k') - X| > \delta) \geq \epsilon. \tag{3.34}$$

Now, (3.34) contradicts (3.33). Hence the net $\{R_n(f\circ P): P \in \mathcal{P}\}$ converges. Thus $f \in \mathcal{L}(H,\mathcal{C},n)$ and $R_n(f) = X$. This proves (c) => (a) and also the remaining assertions.

□

We now list some properties of the lifting map R_n. These could be proved directly without invoking Theorem 3.6 and using convergence of nets, but Theorem 3.6 makes these properties almost obvious. Also, Theorem 2.1 is a consequence of the following result and the proof here is much simpler.

<u>Theorem 3.7</u>: Let $f_1, f_2, \ldots, f_k \in \mathcal{L}(H,\mathcal{C},n)$, let $a_1, a_2 \in \mathbb{R}$ and let $g: \mathbb{R}^k \to \mathbb{R}$ be a continuous function, $k \geq 1$, Then

(i) $a_1f_2 + a_2f_2 \in \mathcal{L}(H,\mathcal{C},n)$ and

$$R_n(a_1f_1 + a_2f_2) = a_1R_n(f_1) + a_2R_n(f_2) \quad \text{a.s. } \Pi_0. \tag{3.35}$$

(ii) $f_1\cdot f_2 \in \mathcal{L}(H,\mathcal{C},n)$ and

$$R_n(f_1\cdot f_2) = R_n(f_1)\cdot R_n(f_2) \quad \text{a.s. } \Pi_0. \tag{3.36}$$

(iii) $g(f_1,f_2,\ldots,f_k) \in \mathcal{L}(H,\mathcal{C},n)$ and

$$R_n(g(f_1,f_2,\ldots,f_k)) = g(R_n(f_1),\ldots,R_n(f_k)), \text{ a.s. } \Pi_0. \tag{3.37}$$

(iv) If $f_1 > 0$ and $\Pi_0(R_n(f_1) > 0) = 1$, then

$$\frac{1}{f_1} \in \mathcal{L}(H,\mathcal{C},n) \quad \text{and} \quad R_n(\frac{1}{f_1}) = \frac{1}{R_n(f_1)} \quad \text{a.s. } \Pi_0. \tag{3.38}$$

(v) If $f_1 \leq f_2$, then

$$R_n(f_1) \leq R_n(f_2) \quad \text{a.s. } \Pi_0. \tag{3.39}$$

Proof: For $1 \leq j \leq k$, let $\{P_i^j\} \subseteq \mathcal{P}$ be such that (b) in Theorem 3.6 holds for f_j. Let $\{P_i\} \subseteq \mathcal{P}$, $P_i \uparrow I$ be such that $P_i^j \leq P_i$ for all i.j. Then (b) holds for this $\{P_i\} \subseteq \mathcal{P}$, for each of $f_1, f_2, \ldots, f_j$. Thus for any $\{P_i'\} \subseteq \mathcal{P}$, $P_i \leq P_i'$, we have, for $1 \leq j \leq k$

$$R_n(f_j \circ P_i') \to R_n(f_j) \quad \text{a.s. } \Pi_0. \tag{3.40}$$

We will prove (iii), (iv), (v). (i) and (ii) are special cases of (iii): Take $k = 2$, $g(x_1,x_2) = a_1x_1 + a_2x_2$ and $g(x_1,x_2) = x_1 \cdot x_2$ respectively. By the definition of the n-lifting for cylinder functions, it can be checked that

$$R_n((g(f_1,f_2,\ldots,f_k))\circ P_i') = R_n(g(f_1\circ P_i',\ldots,f_k\circ P_i')) \tag{3.41}$$

$$= g(R_n(f_1\circ P_i',\ldots,f_k\circ P_i')).$$

Hence by (3.40), (3.41), for all $\{P_i') \subseteq \mathcal{P}$, $P_i \leq P_i'$, we have

$$R_n((g(f_1,f_2,\ldots,f_k))\circ P_i') \to g(R_n(f_1),\ldots,R_n(f_k)) \quad \text{a.s. } \Pi_0.$$

By Theorem (3.6), this implies $g(f_1,\ldots,f_k) \in \mathcal{L}(H,\mathcal{C},n)$ and that (3.37) holds. This proves (iii) and hence (i), (ii) as observed. For (v), observe that if $f_1 \leq f_2$, $f_1 \circ P_i' \leq f_2 P_i'$ and hence by (3.14)

$$R_n(f_1 \circ P_i') \leq R_n(f_2 \circ P_i'). \tag{3.42}$$

Now (3.41) and (3.42) imply (3.39). For (iv), the relations $f_1 > 0$, $\Pi_0(R_n(f_1) > 0) = 1$ and (3.40) give us

$$R_n\left[\frac{1}{f_1} \circ P_i'\right] = R_n\left[\frac{1}{f_1 \circ P_i'}\right] = \frac{1}{R_n(f \circ P_i')} \to \frac{1}{R_n(f)} \text{ a.s.} \Pi_0 \tag{3.43}$$

Since (3.43) holds for all $\{P_i'\} \subseteq \mathcal{P}$, $P_i \leq P_i'$, by Theorem 3.6, we get (3.38). □

We are now going to introduce subclasses of $\mathcal{L}(H,\mathcal{C},n)$ and $\mathcal{L}^1(H,\mathcal{C},n)$. Recall that a sequence $\{P_i\} \subseteq \mathcal{P}$ is said to converge strongly to I, i.e., $P_i \xrightarrow{s} I$, if for all $h \in H$, $|P_i h - h| \to 0$ as $i \to \infty$. Thus, $P_i \uparrow I$ if and only if $P_i \leq P_{i+1}$ and $P_i \xrightarrow{s} I$. In our next result we will need to use the fact that if $P_i \xrightarrow{s} I$, and $\{P_i'\} \subseteq \mathcal{P}$ is such that $P_i \leq P_i'$ for all i, then $P_i' \xrightarrow{s} I$. This follows from the inequality $|P_i' h - h| \leq |P_i h - h|$ which is a consequence of $P_i \leq P_i'$.

<u>Definition 3.4</u>: Let $\mathcal{L}^*(H,\mathcal{C},n)$ consist of Borel measurable functions $f: H \to \mathbb{R}$ such that for all $P_i \xrightarrow{s} I$, $R_n(f \circ P_i)$ converges in Π_0-probability.

<u>Definition 3.5</u>: Let $\mathcal{L}^{1*}(H,\mathcal{C},n)$ consist of Borel measurable functions $f: H \to \mathbb{R}$ such that for all $P_i \xrightarrow{s} I$, $R_n(f \circ P_i)$ converges in $\mathcal{L}^1(\Omega_0, \mathcal{A}_0, \Pi_0)$.

Clearly $\mathcal{L}^{1*}(H,\mathcal{C},n) \subseteq \mathcal{L}^*(H,\mathcal{C},n)$. We also have:

Theorem 3.8:

(i) $\mathcal{L}^*(H,\mathcal{C},n) \subseteq \mathcal{L}(H,\mathcal{C},n)$ and for $f \in \mathcal{L}^*(H,\mathcal{C},n)$

$$R_n(f \circ P_i) \to R_n(f) \quad \text{in } \Pi_0 \text{ probability}$$

for all $P_i \xrightarrow{s} I$.

(ii) $\mathcal{L}^{1*}(H,\mathcal{C},n) \subseteq \mathcal{L}^1(H,\mathcal{C},n)$ and for $f \in \mathcal{L}^{1*}(H,\mathcal{C},n)$, $P_i \xrightarrow{s} I$,

$$R_n(f \circ P_i) \to R_n(f) \quad \text{in } \mathcal{L}^1(\Omega_0,\mathcal{A}_0,\Pi_0).$$

Proof: (i) Let $f \in \mathcal{L}^*(H,\mathcal{C},n)$. Fix a sequence $\{P_i\} \subseteq \mathcal{P}$, $P_i \uparrow I$. Then if $\{P_i'\} \subset \mathcal{P}$, $P_i \leq P_i'$, then $P_i' \xrightarrow{s} I$ as observed earlier. Thus (c) of Theorem 3.6 holds for this $\{P_i\}$ and so $f \in \mathcal{L}(H,\mathcal{C},n)$ and further the common limit in Π_0-probability is $R_n(f)$.

(ii) follows from (i). □

Example 3.1: We now return to Example 2.1. Let $f: H \to \mathbb{R}$ be given by

$$f(h) = \sum_{j=1}^{\infty} \alpha_j^2 (h,e_j)^2$$

where $\sum_{j=1}^{\infty} \alpha_j^2 < \infty$ and $\{e_j\}$ is an orthonormal sequence in H. We will show that $f \in \mathcal{L}^{1*}(H,\mathcal{C},m)$.

Let (L,Π_0) be a representation of m--(the canonical Gauss measure on H). Since $|e_j| = 1$, $(e_j,e_{j'}) = 0$ $(j \neq j')$, it follows that $\{L(e_j): j \geq 1\}$ is a sequence of i.i.d. normal random variables with mean 0 and variance 1. (See Theorem 1.3 and the relation (3.5)).

Thus

$$\sum_{j=1}^{\infty} \alpha_j^2 (L(e_j)(\omega))^2 < \infty \quad \text{a.s. } \Pi_0. \tag{3.44}$$

Let $X(\omega)$ denote the infinite sum appearing in (3.44) when it is finite and zero otherwise.

Let $P_k \overset{s}{\to} I$, $P_k \in \mathcal{P}$. We will prove that

$$R_m(f \circ P_k) \to X \qquad \text{in } \mathcal{L}^1(\Omega_0, \mathcal{A}_0, \Pi_0) \tag{3.45}$$

where R_m is the m-lifting corresponding to (L, Π_0). Observe that

$$f \circ P_k(h) = \sum_{j=1}^{\infty} \alpha_j^2 (P_k h, e_j)^2 = \sum_{j=1}^{\infty} \alpha_j^2 (h, P_k e_j)^2.$$

Fix $k \geq 1$ and let $\phi_1, \ldots, \phi_{d_k}$ be an ONB in Range P_k. Then

$$P_k e_j = \sum_{i=1}^{d_k} (\phi_i, e_j) \phi_i$$

and thus

$$\begin{aligned} f \circ P_k(h) &= \sum_{j=1}^{\infty} \alpha_j^2 [\sum_{i=1}^{d_k} (\phi_i, e_j) L(\phi_i)]^2 \qquad (3.46) \\ &= \sum_{j=1}^{\infty} \alpha_j^2 [L(\sum_{i=1}^{d_k} (\phi_i, e_j) \phi_i)]^2 \qquad \text{by } (3.1) \\ &= \sum_{j=1}^{\infty} \alpha_j^2 (L(P_k e_j))^2. \end{aligned}$$

Thus

$$\int |R_m(f \circ P_k) - X| d\Pi_0 = \int | \sum_{j=1}^{\infty} \alpha_j^2 [L(P_k e_j)]^2 - \sum_{j=1}^{\infty} \alpha_j^2 [L(e_j)]^2 | d\Pi_0 \tag{3.47}$$

$$\leq \sum_{j=1}^{\infty} \alpha_j^2 \int |(L(P_k e_j))^2 - (L(e_j))^2| d\Pi_0$$

$$= \sum_{j=1}^{\infty} \alpha_j^2 \beta_j^k$$

where

$$\beta_j^k = \int |(L(P_k e_j))^2 - (L(e_j))^2| d\Pi_0$$

$$\leq \int |(L(P_k e_j))^2 + (L(e_j)^2)| d\Pi_0$$

$$= |P_k e_j|^2 + |e_j|^2 = 2.$$

The distribution of $(L(P_k e_j), L(e_j))$ is bivariate normal with mean 0, $\mathrm{Var}(L(P_k e_j)) = |P_k e_j|^2$, $\mathrm{Var}(L(e_j)) = 1$, $\mathrm{Cov}(L(P_k e_j), L(e_j)) = (P_k e_j, e_j)$. Since $|P_k e_j|^2 \to |e_j|^2 = 1$ and $(P_k e_j, e_j) \to |e_j|^2 = 1$ as $k \to \infty$, it follows that

$$\lim_{k\to\infty} \beta_j^k = 0. \tag{3.48}$$

Now (3.48), $\beta_j^k \leq 2$ and the fact that $\Sigma_{j=1}^{\infty} \alpha_j^2 < \infty$ implies

$$\lim_{k\to\infty} \sum_{j=1}^{\infty} \alpha_j^2 \beta_j^k = 0.$$

Hence

$$R_m(f \circ P_k) \to X \quad \text{in } \mathcal{L}^1(\Omega_0, \mathcal{A}_0, \Pi_0)$$

for all $\{P_k\} \subset \mathcal{P}$, $P_k \xrightarrow{s} I$. Thus

$$f \in \mathcal{L}^{1*}(H, \mathcal{C}, n)$$

and

$$R_m(f) = X.$$

Further

$$\int f dm = \int X d\Pi_0 = \int \sum_{j=1}^{\infty} \alpha_j^2 (L(e_j))^2 d\Pi_0 = \sum_{j=1}^{\infty} \alpha_j^2.$$

If A is as in Example 2.1 in Section 2, so that $f(h) = |Ah|^2$, then $\|A\|^2_{H.S.} = \Sigma_{j=1}\alpha_j^2$ and we obtain from above that

$$\int |Ah|^2 dm(h) = \|A\|^2_{H.S.} \qquad \square$$

We now come to the proof of the statement made at the end of the last section about 'null functions'.

<u>Theorem 3.9</u>:

(a) Let $f \in \mathscr{L}(H,\mathscr{C},n)$. Then

$$f \equiv 0 \text{ mod}[n] \quad \text{if and only if } R_n(f) = 0 \text{ a.s. } \Pi_0$$

(b) Let $f \in \mathscr{L}^1(H,\mathscr{C},n)$. Then

$$\int_C f dn = 0 \quad \text{for all } C \in \mathscr{C} \text{ implies } f \equiv \text{mod}[n].$$

<u>Proof</u>: (a) By Theorem 2.3, $f \equiv 0$ mod[n] if and only if $n \circ f^{-1} = \delta_{\{0\}}$. Now $\Pi_0 \circ [R_n(f)]^{-1} = n \circ f^{-1}$ and $\Pi_0 \circ [R_n(f)]^{-1} = \delta_{\{0\}}$ if and only if $R_n(f) = 0$ a.s. Π_0. This proves (a).

(b) Let $\mathscr{D}_0 = \{B \in \mathscr{A}_0 : 1_B = R_n(1_C) \text{ for some } C \in \mathscr{C}\}$. Using Theorem 3.7, it is easy to verify that $\mathscr{D}_0$ is a field. Let $\mathscr{D} = \sigma(\mathscr{D}_0)$ and $\overline{\mathscr{D}}$ be the Π_0-completion of $\mathscr{D}$. It follows from the definition of the lifting for cylinder functions that if f is a cylinder function, then $R_n(f)$ is $\overline{\mathscr{D}}$-measurable. Now if $f \in \mathscr{L}(H,\mathscr{C},n)$ is arbitrary, get $P_k \uparrow I$ (Theorem 3.6) such that $R_n(f\circ P_k) \to R_n(f)$ a.s. Π_0. Since $R_n(f\circ P_k)$ is $\overline{\mathscr{D}}$ measurable as observed earlier, this implies that $R_n(f)$ is $\overline{\mathscr{D}}$ measurable. Now

$$
\begin{aligned}
\int_C f dn = 0 \text{ for all } C \in \mathscr{C} &\Rightarrow \int R_n(1_C f) d\Pi_0 = 0 \text{ for all } C \in \mathscr{C} \\
&\Rightarrow \int R_n(1_C) R_n(f) d\Pi_0 = 0 \quad \text{for all } C \in \mathscr{C} \\
&\Rightarrow \int 1_B R_n(f) d\Pi_0 = 0 \qquad \text{for all } B \in \mathscr{D}_0 \\
&\Rightarrow \int 1_B R_n(f) d\Pi_0 = 0 \qquad \text{for all } B \in \mathscr{D}
\end{aligned}
$$

as $\mathscr{D} = \sigma(\mathscr{D}_0)$, $\mathscr{D}_0$ is a field. Since $R_n(f)$ is $\overline{\mathscr{D}}$-measurable, this last statement implies $R_n(f) = 0$ a.s. Π_0 which is the same as $f \equiv 0 \bmod[n]$. □

4. EXAMPLES OF REPRESENTATIONS OF THE CANONICAL GAUSS MEASURE

In this section we consider specific examples of H of importance in the applications. As before, m denotes the canonical Gauss measure on H. Some of the examples are chosen because of their connection with the filtering problem when the state space of the signal process entering the observation model is an infinite dimensional Hilbert space. In each case we produce a representation space $(\mathscr{X}\ \mathscr{B}(\mathscr{X}),\mu)$ where $\mathscr{X}$ is a suitable topological space, $\mathscr{B}(\mathscr{X})$, the topological Borel σ-field and μ is Wiener measure (to be appropriately defined in each case). Moreover, $\mathscr{X}$ is an *enlargement* of H in the sense that the latter can be regarded as imbedded as a dense subset of $\mathcal{X}$. It is also known (although not proved here) that $\mu(H) = 0$.

The spaces which feature in the construction of $\mathcal{X}$ are, respectively, an abstract Wiener space (γ, H, B) and the Schwartz space $\mathcal{S}(\mathbb{R}^d)^*$, the strong dual of $\mathcal{S}(\mathbb{R}^d)$. We now turn to the definition and basic properties of abstract Wiener spaces. For details, we refer the reader to Kuo [56].

Abstract Wiener Space

Let H be a separable real Hilbert space.

Definition 4.1: A norm or a seminorm $\|\cdot\|_1$ on H is said to be measurable if for every $\epsilon > 0$, there exists $P_\epsilon \in \mathcal{P}$ such that

$$m(h: \|Ph\|_1 > \epsilon) < \epsilon \quad \text{for all } P \perp P_\epsilon,\ P \in \mathcal{P}. \tag{4.1}$$

Remark 4.1: Let $f: H \to \mathbb{R}^+$ be given by $f(h) := \|h\|_1$. In the terminology of previous sections, if $\|\cdot\|_1$ is a measurable norm (or seminorm) then $f \in \mathcal{L}(H, \mathcal{C}, m)$. To see this note that (4.1) can be rewritten as

$$m(h: \|P'h - P_\epsilon h\|_1 > \epsilon) < \epsilon \quad \text{for all } P' \in \mathcal{P},\ P_\epsilon \leq P'. \tag{4.2}$$

Since $\big| \|P'h\|_1 - \|P_\epsilon h\|_1 \big| \leq \|P'h - P_\epsilon h\|_1$ by the triangle inequality for $\|\cdot\|_1$, (4.2) implies

$$m(h: |f(P'h) - f(P_\epsilon h)| > \epsilon) < \epsilon \quad \text{for all } P' \in \mathcal{P},\ P_\epsilon \leq P'. \tag{4.3}$$

Thus $f \in \mathcal{L}(H, \mathcal{C}, m)$.

Example 4.1: Let A be a self-adjoint Hilbert-Schmidt operator on H and $\|h\|_1 = |Ah|$. Then $\|\cdot\|_1$ is a measurable seminorm. In Example 2.1, it was shown that $f(h) := |Ah|^2 \in \mathscr{L}(H,\mathscr{C},m)$. We will use the notation in Example 2.1. Thus, $\{\alpha_j\}$ are the eigenvalues of A and $\{e_j\}$ are the corresponding eigenvectors of A and P_k is the orthogonal projection onto span$\{e_1,\dots,e_k\}$. Now, for $P' \perp P_k$, $P' \in \mathscr{P}$, we have

$$\|P'h\|_1^2 = |AP'h|^2 = \sum_{j=1}^{\infty} \alpha_j^2 (P'h,e_j)^2 \tag{4.4}$$

$$= \sum_{j=k+1}^{\infty} \alpha_j^2 (P'h,e_j)^2 \qquad (\text{as } P' \perp P_k).$$

Thus, for $P' \perp P_k$, $P' \in \mathscr{P}$, for any $\epsilon > 0$,

$$m(h\colon \|P'h\|_1 > \epsilon) = m(h\colon \|P'h\|_1^2 > \epsilon^2) \tag{4.5}$$

$$= m(h\colon g_k(h) > \epsilon^2) \le \frac{1}{\epsilon^2} \sum_{j=k+1}^{\infty} \alpha_j^2$$

where $g_k(h)$ is as in Example 2.1 (it denotes the right hand side of (4.4)) and we have used (2.14) for the last inequality. Given $\epsilon > 0$, if we choose k such that $\Sigma_{j=k+1}^{\infty}\alpha_j^2 < \epsilon^3$, then (4.5) implies that for $P' \perp P_k$, $P' \in \mathscr{P}$, we have

$$m(h\colon \|P'h\|_1 > \epsilon) < \epsilon.$$

Thus $\|\cdot\|_1$ is measurable.

The concept of a measurable norm as well as the following basic result (Theorem 4.1) are due to Gross.

A measurable seminorm on H is necessarily weaker than the given Hilbertian norm $|\cdot|$. Indeed, if $\|\cdot\|_1$ is a measurable seminorm, then there exists a constrant c such that $\|h\|_1 \le c|h|$ for all $h \in H$. It can be shown that a measurable seminorm $\|\cdot\|_1$ is strictly weaker than $|\cdot|$ and that H cannot be complete under $\|\cdot\|_1$.

Let B denote the Banach space which is the $\|\cdot\|_1$-completion of H. Let γ: H $\rightarrow$ B denote the injection. Then γ is continuous and $\gamma(H)$ is dense in B. The adjoint γ^* of γ maps the dual B^* of B continuously into the dual H^* of H and again $\gamma^*(B^*)$ is dense in H^*. Since H is a Hilbert space, H^* can be identified with H. Thus we have

$$B^* \underset{\gamma^*}{\longrightarrow} H^* = H \underset{\gamma}{\longrightarrow} B.$$

Let $\mathscr{B}_0$ denote the cylinder sets in B, i.e., sets of the form

$$\{x: (f_1[x],\ldots,f_k[x]) \in A\} \tag{4.6}$$

for $k \geq 1$, $f_1 \in B^*$, $A \in \mathscr{B}(\mathbb{R}^k)$. We will use the notation f[x] to denote the action of an element $f \in B^*$ on an element $x \in B$. By the definition of adjoint, we have

$$f[\gamma(h)] = (\gamma^*(f),h) \tag{4.7}$$

Using (4.7), it can be checked that

$$\gamma^{-1}(\mathscr{B}_0) \subseteq \mathscr{C} \tag{4.8}$$

and hence $m\circ\gamma^{-1}$ is a finitely additive probability measure on $(B,\mathscr{B}_0)$. For $f_1,f_2,\ldots,f_k \in B^*$ fixed, if $\mathscr{B}_1$ denotes the collection of sets of the form (4.6) for some $A \in \mathscr{B}(\mathbb{R}^k)$, then it is easy to see that $m\circ\gamma^{-1}$ is countably additive on

$\mathcal{B}_1$ and for $t_1, t_2, \ldots, t_k \in \mathbb{R}$,

$$\int_B e^{i \sum_{j=1}^{k} t_j f_j[x]} \, dm\circ\gamma^{-1}(x) = \int_H \exp\{i \sum_{j=1}^{k} t_j(\gamma^*(f_j), h\} dm(h)$$
$$= \exp\{-\tfrac{1}{2}|\sum_{j=1}^{k} t_j\gamma^*(f_j)|^2\}. \tag{4.9}$$

Theorem 4.1: The finitely additive measure $m\circ\gamma^{-1}$ on $\mathcal{B}_0$ admits a countably additive extension μ to $\mathcal{B}(B)$, the σ-field of Borel sets in B.

Proof: Let (L, Π_0) be a fixed (but arbitrary) representation of m, with representation space $(\Omega_0, \mathcal{A}_0)$. For each $r \geq 1$, taking $\epsilon = 2^{-r}$ in (4.1), choose $P_r \in \mathcal{P}$ such that

$$m(h: \|Ph\|_1 > 2^{-r}) < 2^{-r}, \text{ for all } P \perp P_r,\ P \in \mathcal{P}. \tag{4.10}$$

Without loss of generality, we can assume that $P_r \uparrow I$. Let $Q_r = P_{r+1} - P_r$. Then $Q_r \perp P_r$ and hence from (4.10)

$$m(h: \|Q_r h\|_1 > 2^{-r}) < 2^{-r}. \tag{4.11}$$

If $g(h) = \|h\|_1$, then from the definition of m-lifting R_m, we have

$$\Pi_0(\omega: R_m(g\circ Q_r)(\omega) > 2^{-r}) < 2^{-r}. \tag{4.12}$$

Let $k_r = \dim(P_r H)$. Then $\dim Q_r H = k_{r+1} - k_r$. For each r, let $\{e_j: j = k_r+1, k_r+2, \ldots, k_{r+1}\}$ be an ONB in Q_r H. Then it is easy to see that $\{e_j: 1 \leq j \leq k_r\}$ is an ONB in $P_r H$. Since $P_r \uparrow I$, $\{e_j: 1 \leq j < \infty\}$ is a CONS in H. For each $r \geq 1$, let

$$Z_r(\omega) = \sum_{j=1}^{k_r} L(e_j)(\omega)\gamma(e_j) \ : \omega \in \Omega_0.$$

Clearly, $Z_r\colon \Omega_0 \to B$ is a measurable mapping. From the definition of m-lifting for cylinder functions, it follows that

$$R_m(g\circ Q_r) = R_m(\| \sum_{j=k_r+1}^{k_{r+1}} (\cdot, e_j)e_j \|_1) \tag{4.13}$$

$$= \| \sum_{j=k_r+1}^{k_{r+1}} L(e_j)e_j \|_1 = \|Z_{r+1} - Z_r\|_1.$$

Hence by (4.12)

$$\Pi_0(\omega\colon \|Z_{r+1}(\omega) - Z_r(\omega)\|_1 > 2^{-r}) < 2^{-r}. \tag{4.14}$$

By the Borel-Cantelli Lemma, we have

$$\Pi_0(\omega\colon \|Z_{r+1}(\omega) - Z_r(\omega)\|_1 > 2^{-r} \text{ infinitely often}) = 0 \tag{4.15}$$

so that

$$\sum_{r=1}^{\infty} \|Z_{r+1}(\omega) - Z_r(\omega)\|_1 < \infty \quad \text{a.s. } \Pi_0. \tag{4.16}$$

Thus

$$\sup_{\substack{j_1\geq k\\ j_2\geq k}} \|Z_{j_1}(\omega) - Z_{j_2}(\omega)\|_1 \leq \sum_{r=k}^{\infty} \|Z_{r+1}(\omega) - Z_r(\omega)\|_1 \tag{4.17}$$

$$\to 0 \text{ as } k \to \infty \text{ a.s. } \Pi_0 \text{ (by (4.16))}$$

and hence for almost all $\omega \in \Omega$, the sequence $\{Z_r(\omega)\}$ of elements of B is Cauchy in $\|\cdot\|_1$ norm. Since B is complete under

$\|\cdot\|_1$, this implies the a.s. convergence of Z_r. Let

$$Z(\omega) = \lim_r Z_r(\omega) \quad \text{if limit exists in } \|\cdot\|_1 \tag{4.18}$$
$$= 0 \quad \text{otherwise.}$$

Then $Z_r \to Z$ a.s. Π_0. Let $\mu = \Pi_0 \circ Z^{-1}$. Clearly μ is a countably additive probability measure on $(B, \mathscr{B}(B))$. To see that μ is an extension of $m \circ \gamma^{-1}$, let us calculate the characteristic functional of μ. For $f \in B^*$, we have

$$\int_B \exp\{if[x]\} d\mu(x) = \int_{\Omega_0} \exp\{if[Z(\omega)]\} d\Pi_0(\omega) \tag{4.19}$$
$$= \lim_{r\to\infty} \int_{\Omega_0} \exp\{if[Z_r(\omega)]\} d\Pi_0(\omega)$$
$$= \lim_{r\to\infty} \int_{\Omega_0} \exp\{i \sum_{j=1}^{k_r} L(e_j)(\omega) f[\gamma(e_j)]\} d\Pi_0(\omega)$$
$$= \lim_{r\to\infty} \exp\{-\tfrac{1}{2} \sum_{j=1}^{k_r} (f[\gamma(e_j)])^2\}$$

(as $(L, \Pi_0$ is a representation of m)

$$= \exp\{-\tfrac{1}{2} \sum_{j=1}^{\infty} (f[\gamma(e_j)])^2\}.$$

Now $f[\gamma(e_j)] = (\gamma^*(f), e_j)$ as observed earlier, so that

$$\sum_{j=1}^{\infty} (f[\gamma(e_j)])^2 = \sum_{j=1}^{\infty} (\gamma^*(f), e_j)^2 = |\gamma^*(f)|^2 \tag{4.20}$$

since $\{e_j\}$ is a CONS in H. Thus,

$$\int_B \exp(if[x])d\mu(x) = \exp\{-\tfrac{1}{2}|\gamma^*(f)|^2\}. \tag{4.21}$$

The relations (4.9) and (4.21) imply that the restrictions of μ to $\mathscr{B}_0$ is $m\circ\gamma^{-1}$. Thus μ is an extension of $m\circ\gamma^{-1}$.

□

Each element $f \in B^*$ is a continuous real valued function on B and hence is Borel measurable. Thus $f \in B^*$ can be regarded as a random variable on the probability space $(B,\mathscr{B}(B),\mu)$. The relation (4.21) gives us the following result.

Corollary 4.2:

(i) Let $(f_1,\ldots,f_k) \in B^*$. The joint distribution of $(f_1,\ldots,f_k)$ under μ is multivariate normal or Gaussian with mean vector zero and variance-covariance matrix given by $(\gamma^*(f_i),\gamma^*(f_j))$, $1 \le i, j \le k$.

(ii) Let $\{f_k\} \subseteq B^*$ be such that $\gamma^*(f_k)$ is an ortho-normal set in H. Then $\{f_k\}$ is a sequence of independent identically distributed random variables on $(B,\mathscr{B}(B),\mu)$ with common Gaussian or normal distribution with mean zero and unit variance.

The triple (γ,H,B) is called an *abstract Wiener space* and the measure μ is called an *abstract Wiener measure*.

A Representation of m With $(B,\mathscr{B}(B))$ as the Representation Space

Let (γ,H,B) be an abstract Wiener space and let μ be the abstract Wiener measure on $(B,\mathscr{B}(B))$. A representation of m with $(B,\mathscr{B}(B))$ as the representation space can be obtained as follows.

Recall that $\gamma^*(B^*)$ is dense in $H^* = H$, so that we can choose a CONS $\{\phi_j\}$ of H such that $\{\phi_j\} \subseteq \gamma^*(B^*)$. Let $\{f_j\} \subseteq B^*$ be such that $\phi_j = \gamma^*(f_j)$. Define L_1: $H \to \mathcal{L}(B,\mathcal{B}(B),\mu)$ by, for $h \in H$, $x \in B$

$$L_1(h)(x) = \sum_{j=1}^{\infty} (h,\phi_j)f_j[x] \quad \text{if the series converges} \tag{4.22}$$

$$= 0 \quad \text{otherwise.}$$

Theorem 4.8: (L_1,μ) is a representation of the canonical Gauss measure m with representation space $(B,\mathcal{B}(B))$.

Proof: By choice of $\{f_j\}$, $\phi_j = \gamma^*(f_j)$ is a CONS in H and hence as seen in Corollary 4.2, $\{f_j\}$ is a sequence of independent identically distributed random variables on $(B,\mathcal{B}(B),u)$ with common normal distribution with $E_\mu f_1 = 0$, $E_\mu f_1^2 = 1$. Hence the series in (4.22) converges a.s. μ_0. Thus L_1 is linear, i.e., it satisfies (3.1). Also, we have

$$\int_B \exp(iL_1(h)(x))d\mu(x) = \lim_{k\to\infty} \int_B \exp(i \sum_{j=1}^{k} (h,\phi_j)f_j[x])d\mu(x)$$

$$= \lim_{k\to\infty} \exp(-\tfrac{1}{2} \sum_{j=1}^{k} (h,\phi_j)^2)$$

$$= \exp(-\tfrac{1}{2}|h|^2).$$

This proves the assertion that (L_1,μ) is a representation of m. □

Example 4.2: An important example of an abstract Wiener space is provided by taking $H = \mathcal{H}$, where

$$\mathcal{H} = \{\eta\colon [0,T] \to \mathbb{R}^N : \eta_t^j = \int_0^t D\eta_s^j ds, \tag{4.23}$$

$$\int_0^T (D\eta_s^j)^2 ds < \infty,\ 1 \le j \le N\}.$$

Recall our convention that for an $\mathbb{R}^N$-valued function η, η^j, $1 \le j \le N$ denote its components, i.e., $\eta = (\eta^1, \eta^2, \dots, \eta^N)$. In (4.23), $D = \frac{d}{ds}$ and $D\eta_s^j$ is assumed to exist a.e.

$\mathcal{H}$ is a Hilbert space with inner product

$$(\eta, \hat{\eta}) := \sum_{j=1}^N \int_0^T (D\eta_s^j)\cdot(D\hat{\eta}_s^j) ds, \quad \eta, \hat{\eta} \in \mathcal{H}. \tag{4.24}$$

We will denote the norm on $\mathcal{H}$ by $\|\cdot\|$ so that $\|\eta\|^2 = (\eta, \eta)$.

The Hilbert space $\mathcal{H}$ is known in the theory of Gaussian processes and is the reproducing kernel Hilbert space (RKHS) of the N-dimensional standard Wiener measure. Let $\|\cdot\|_1$: $\mathcal{H} \to \mathbb{R}^+$ be defined by

$$\|\eta\|_1 := \sup_{o \le t \le T} \left(\sum_{j=1}^N (\eta_t^j)^2 \right)^{\frac{1}{2}},\ \eta \in \mathcal{H}. \tag{4.25}$$

It is easy to see that the $\|\cdot\|_1$ completion of $\mathcal{H}$ is $\mathcal{X} = C_0([0,T], \mathbb{R}^N)$, the Banach space of continuous functions from $[0,T]$ into $\mathbb{R}^N$, which vanish at 0.

It can be shown that $\|\cdot\|_1$ defined by (4.25) is a measurable norm on $\mathcal{H}$. (See [56]). Thus, if γ denotes the injection of $\mathcal{H}$ into $\mathcal{X}$ (so that $\gamma(\eta) = \eta$), then $(\gamma, \mathcal{H}, \mathcal{X})$ is an abstract Wiener space. Let the corresponding abstract Wiener measure be μ. We will show that $\mu = \mu_w$ -- the standard Wiener measure on $(\mathcal{X}, \mathcal{B}(\mathcal{X}))$.

It is well known that the dual $\mathcal{X}^*$ of $\mathcal{X}$ is given by

$$\mathcal{X}^* = \{\theta = (\theta^1,\dots,\theta^N): \theta^i \text{ a signed measure on } (0,T],\ 1 \le i \le N\}$$

and the action $\theta[x]$ of $\theta \in \mathcal{X}^*$ on $x \in \mathcal{X}$ is given by

$$\theta[x] = \sum_{j=1}^N \int_0^T x^j(u)d\theta^j(u)$$

By the definition of the adjoint map, we have for $\theta \in \mathcal{X}^*$, $\eta \in \mathcal{H}$

$$(\gamma^*(\theta),\eta) = \theta[\gamma(\eta)] = \sum_{j=1}^N \int_0^T \eta_u^j d\theta^j(u) \tag{4.26}$$

$$= \sum_{j=1}^N \int_0^T\int_0^T D\eta_s^j 1_{[0,u)}(s)dsd\theta^j(u)$$

$$= \sum_{j=1}^N \int_0^T\int_0^T 1_{(s,T]}(u)d\theta^j(u)D\eta_s^j ds$$

(By Fubini's Theorem)

$$= \sum_{j=1}^N \int_0^T \theta^j((s,T])\cdot D\eta_s^j ds.$$

Since (4.26) stands for all $\eta \in \mathcal{H}$, it follows that

$$\gamma^*(\theta)(s) = \left(\int_0^s \theta^j((t,T])dt\right)_{1\le j\le N}, \quad 0 \le s \le T. \tag{4.27}$$

For $1 \le j \le N$, let the signed measure θ^j be given by

$$\theta^j(A) := \sum_{r=1}^{d} b_r^j 1_A(t_{r+1}), \quad A \in \mathscr{B}([0,T]) \tag{4.28}$$

where $b_r \in \mathbb{R}^N$, $r = 1,2,\ldots,d$ and $0 = t_1 < t_2 < \ldots < t_{d+1} = T$. Then for $t_i < t \leq t_{i+1}$, we have

$$\theta^j((t,T]) = \sum_{r=i}^{d} b_r^j := a_i^j \quad \text{(say)} \tag{4.29}$$

so that

$$\gamma^*(\theta)(s) = \int_0^s \sum_{r=1}^{d} a_r 1_{(t_r,t_{r+1}]}(t) dt. \tag{4.30}$$

and

$$\|\gamma^*(\theta)\|^2 = \sum_{j=1}^{N} \int_0^s \Big(\sum_{r=1}^{d} a_r^j 1_{(t_r,t_{r+1}]}(t)\Big)^2 dt \tag{4.31}$$

$$= \sum_{j=1}^{N} \sum_{r=1}^{d} (a_r^j)^2 (t_{r+1}-t_r) = \sum_{r=1}^{d} \sum_{j=1}^{N} (a_r^j)^2 (t_{r+1}-t_r).$$

Further, for θ given by (4.28),

$$\theta[x] = \sum_{j=1}^{N} \sum_{r=1}^{d} b_r^j x^j(t_{r+1}) = \sum_{j=1}^{N} \sum_{r=1}^{d} (a_r^j - a_{r+1}^j) x^j(t_{r+1})$$

with $a_{d+1} \equiv 0$, since $a_r - a_{r+1} = b_r$. Since $a_{d+1} = 0$, $x^j(t_1) = x_j(0) = 0$,

$$\sum_{r=1}^{d} (a_r^j - a_{r+1}^j) x^j(t_{r+1}) = \sum_{r=1}^{d} a_r^j (x^j(t_{r+1}) - x^j(t_r)) \tag{4.32}$$

and thus

$$\theta[x] = \sum_{j=1}^{N} \sum_{r=1}^{d} a_r^j (x^j(t_{r+1}) - x^j(t_r)). \tag{4.33}$$

Now, from the relation (4.21), we have

$$\int_{\mathcal{X}} \exp\{i\theta[x]\} d\mu(x) = \exp(-\tfrac{1}{2}\|\gamma^*(\theta)\|^2)$$

which in view of (4.33) and (4.31) gives

$$\int_{\mathcal{X}} \exp\{i \sum_{j=1}^{N} \sum_{r=1}^{d} a_r^j (x^j(t_{r+1}) - x^j(t_r))\} d\mu(x) \tag{4.34}$$

$$= \exp\{-\tfrac{1}{2} \sum_{j=1}^{N} \sum_{r=1}^{d} (a_r^j)^2 (t_{r+1} - t_r)\}.$$

This implies that the abstract Wiener measure μ is equal to μ_w -- the standard Wiener measure on $\mathcal{X} = C_0([0,T],\mathbb{R}^N)$.

Thus, in view of Theorem 4.3, there exists a representation (L_1, μ_w) of the canonical Gauss measure m on $\mathcal{H}$, with the representation space $(\mathcal{X}, \mathcal{B}(\mathcal{X}))$, $\mathcal{X} = C_0([0,T],\mathbb{R}^N)$.

The Wiener Integral

The representation (L_1, μ_w) of m in Example 4.2 above can be expressed as a Wiener integral, which we now describe.

Let $H_0 = L^2([0,T],\mathbb{R}^N)$ with the usual L^2-norm denoted by $|\cdot|$. It is easy to see that simple (i.e., step) functions of the form

$$\phi(u) = \sum_{r=1}^{d} a_r 1_{(t_r, t_{r+1}]}(u) \tag{4.35}$$

where $a_r \in \mathbb{R}^N$, $1 \leq r \leq d$ and $0 = t_1 < \ldots < t_{d+1} = T$, are dense in H_0. Let H_1 be the collection of such simple functions. Let $(\mathscr{X},\mathscr{B}(\mathscr{X}),\mu_w)$ be as in Example 4.2. For $\phi \in H_1$ given by (4.35), define

$$I(\phi)(x) := \sum_{r=1}^{d} \sum_{j=1}^{N} a_r^j(x^j(t_{r+1})-x^j(t_r)). \tag{4.36}$$

It is usual to denote $I(\phi)$ by $\int_0^T \phi \cdot dx := \sum_{j=1}^{N} \int_0^T \phi^j dx^j$ for obvious reasons. It should be noted that $x \in \mathscr{X}$ does not imply that x^j is of bounded variation and hence $\int_0^T \phi^j \cdot dx^j$ cannot be defined as a Rieman-Steljes integral. But for $\phi \in H_1$, $\int_0^T \phi \cdot dx$ has an obvious meaning. From the properties of μ_w (see also Example 4.2)

$$E_{\mu_w} \exp(iI(\phi)) = \exp(-\tfrac{1}{2} \sum_{r=1}^{d} \sum_{j=1}^{d} (a_r^j)^2(t_{r+1}-t_r)) \tag{4.37}$$

$$= \exp(-\tfrac{1}{2}|\phi|_{H_0}^2).$$

In particular, for $\phi \in H_1$,

$$E_{\mu_w}(I(\phi))^2 = |\phi|_{H_0}^2.$$

Clearly, $I: H_1 \to \mathscr{L}^2(\mathscr{X}.\mathscr{B}(\mathscr{X}),\mu_w)$ is a linear mapping. In view of (4.38), it can be extended as a linear mapping, again denoted by I, from H_0 into $\mathscr{L}^2(\mathscr{X}.\mathscr{B}(\mathscr{X}),\mu)$ such that (4.38) holds for all $\phi \in H_1$. It can be checked that

$$E_{\mu_w} \exp(iI(\phi)) = \exp(-\tfrac{1}{2}|\phi|_{H_0}^2). \tag{4.39}$$

<u>Definition 4.2</u>: For $\phi \in H_0$, $I(\phi) \in \mathscr{L}^2(\mathscr{X}.\mathscr{B}(\mathscr{X}),\mu_w)$ is called

the Wiener integral of ϕ with respect to the Wiener process $(x(t))$ and is denoted by

$$I(\phi) = \int_0^T \phi \cdot dx. \tag{4.40}$$

Let $L_0: H_0 \to \mathcal{L}(\mathcal{X}.\mathcal{B}(\mathcal{X}),\mu_w)$ be defined by

$$L_0(\phi)(x) = \int_0^T \phi \cdot dx. \tag{4.41}$$

Then in view of (4.39) and (4.40), (L_0,μ_w) is a representation of the canonical Gauss measure m on H_0.

<u>Example 4.2 (continued)</u>: Let $J: \mathcal{H} \to H_0$ be defined by

$$J(\eta) = D\eta \tag{4.42}$$

where $D = (D\eta^1,\ldots,D\eta^N)$, $D = \frac{d}{ds}$. Then from the definition of the respective norms, it follows that J is an isometry. Let $L_0': \mathcal{H} \to \mathcal{L}(\mathcal{X}.\mathcal{B}\mathcal{X},\mu_w)$ be defined by

$$L_0'(\eta) = L_0(D\eta) = \int_0^T (D\eta) \cdot dx \tag{4.43}$$

where L_0 is as in (4.41). Then L_0' is linear, and

$$E_{\mu_w} \exp(iL_0'(\eta)) = \exp(-\tfrac{1}{2}|D\eta|^2_{H_0}) = \exp(-\tfrac{1}{2}\|\eta\|)$$

from (4.39) and the fact that J is an isometry and thus (L_0',μ_w) is a representation of the canonical Gauss measure m on $\mathcal{H}$.

Now we will show that for a proper choice of a CONS for $\mathcal{H}$, the representation (L_1,μ_w) of m given by (4.22) is equal to (L_0',μ_w) given above.

For this, let θ be given by (4.28). Then from (4.30)

$$J(\gamma^*(\theta)) = \phi$$

where ϕ is given by (4.35). Now, from (4.33) and (4.36)

$$\theta[x] = \int_0^T \phi \cdot dx$$

and hence

$$\theta[x] = L_0(\phi)(x) = L_0(J(\gamma^*(\theta))(x) = L_0'(\gamma^*(\theta))(x). \quad (4.44)$$

Since H_1 is dense in H_0 and J is an isometry, we can choose a CONS $\gamma^*(\theta_k)$, where each θ_k^j is a discrete measure of the form (4.29). For this $\{\theta_k\}$, let L_1 be given by (4.22) so that

$$L_1(\eta)(x) = \sum_{k=1}^{\infty} (\eta, \gamma^*(\theta_k))\theta_k[x] \qquad \text{a.s. } \mu_W \quad (4.45)$$

Thus, denoting $\gamma^*(\theta_k)$ by $\eta_{(k)}$, (4.44) and (4.45) give

$$L_1(\eta)(x) = \sum_{k=1}^{\infty} (\eta, \eta_{(k)})L_0'(\eta_{(k)})(x) \qquad \text{a.s. } \mu_W$$

$$= \lim_{r\to\infty} \sum_{k=1}^{r} (\eta, \eta_{(k)})L_0'(\eta_{(k)})(x) \qquad \text{a.s. } \mu_W$$

$$= \lim_{r\to\infty} L_0'\Big(\sum_{k=1}^{r} (\eta, \eta_{(k)})\eta_{(k)}\Big)(x) \qquad \text{a.s. } \mu_W.$$

As $r \to \infty$, $\sum_{i=1}^{r} (\eta, \eta_{(k)})\eta_{(k)} \to \eta$ in $\mathcal{H}$ as $\{\eta_{(k)}\}$ is a CONS. Hence

$$\lim_{r\to\infty} L_0'\Big(\sum_{k=1}^{r} (\eta, \eta_{(k)})\eta_{(k)}\Big) \to L_0'(\eta)$$

in $\mathscr{L}^2(\mathfrak{X}.\mathscr{B}(\mathfrak{X}),\mu_w)$ as $L_0': \mathscr{H} \to \mathscr{L}^2(\mathfrak{X}.\mathscr{B}(\mathfrak{X}),\mu_w)$ preserves the norm. Thus

$$L_1(\eta)(x) = L_0'(\eta)(x).$$

So, we have shown that for a proper choice of the CONS in $\mathscr{H}$, the general representation (L_1,μ_w) of m given by (4.2) actually reduces to

$$L_1(\eta)(x) = \int_0^T(D\eta)\cdot dx. \tag{4.46}$$

<u>Example 4.3</u>: We now give an example of a representation space which is not a Banach space but is the Schwartz space $\mathscr{S}(\mathbb{R}^d)^*$ of distributions. The latter is the strong dual of $\mathscr{S}(\mathbb{R}^d)$ the space of rapidly decreasing functions on $\mathbb{R}^d$ given by an increasing sequence of Hilbertian seminorms $|\cdot|_p$, $(p = 0,1,\dots)$ where $|\cdot|_0$ is the $L^2(\mathbb{R}^d)$-norn. We then have the Gelfand triplet

$$\mathscr{S} \subset L^2(\mathbb{R}^d) \in \mathscr{S}^*$$

(the injection being continuous with dense images). Also, $\mathscr{S}$ is a nuclear space. Let $H = L^2(\mathbb{R}^d)$ and let n be a Gaussian cylinder probability on $(H,\mathscr{C})$ with covariance form Q, where Q is a symmetric, positive definite bilinear form on H. (See example 1.4). Further, we assume that

the mapping $\phi \to Q(\phi,\phi)$ from $\mathscr{S} \to \mathbb{R}$ (4.47)

is continuous in $\mathscr{S}$-topology.

Condition (4.47) implies that there exists a constant $\delta > 0$ and an integer $p \geq 0$ such that

$$Q(\phi,\phi) \le \delta|\phi|_p^2, \tag{4.48}$$

for all $\phi \in \mathcal{S}$.

From the assumption (4.47) and the Theorem of Minlos, it follows that there exists a countably additive (Gaussian) measure μ on $(\mathcal{S}^*, \mathcal{B}(\mathcal{S}))$ such that

$$\int_{\mathcal{S}^*} \exp(i\xi[\phi])d\mu(\xi) = \exp(-\tfrac{1}{2} Q(\phi,\phi)), \quad \phi \in \mathcal{S}. \tag{4.49}$$

By the nuclearity of $\mathcal{S}$, there exists an integer $r > p$ such that if $(\phi_j) \subset \mathcal{S}$ is a CONS in $\mathcal{S}_r$, then

$$\sum_{j=1}^{\infty} |\phi_j|_p^2 < \infty \tag{4.50}$$

and hence from (4.48)

$$\sum_{j=1}^{\infty} Q(\phi_j,\phi_j) < \infty. \tag{4.51}$$

Now, in view of (4.49), $\xi \to \xi[\phi]$ is a normal random variable on $(\mathcal{S}^*, \mathcal{B}(\mathcal{S}), \mu)$ with mean zero and variance $Q(\phi,\phi)$, so that

$$\int \sum_{j=1}^{\infty} (\xi[\phi_j])^2 d\mu(\xi) = \sum_{j=1}^{\infty} Q(\phi_j,\phi_j) < \infty \tag{4.52}$$

by (4.51). Thus

$$\sum_{j=1}^{\infty} (\xi[\phi_j])^2 < \infty \quad \mu\text{--a.s.} \tag{4.53}$$

For $h \in \mathcal{S}_r$, define

$$L(h)(\xi) = \sum_{j=1}^{\infty} (h,\phi_j)_r \xi[\phi_j], \quad \text{if the series converges} \tag{4.54}$$
$$= 0 \qquad\qquad \text{otherwise.}$$

(For each h, the series in (4.54) converges μ--a.s.) It is now easy to see that $E_\mu(L(h))^2 = Q(h,h)$ for $h \in \mathcal{S}_r$.

Let us add the further assumption that Q is $|\cdot|_0$-continuous (e.g., if $Q(h,h) = |h|_0^2$). Then since $\mathcal{S}$ is dense in $H(= \mathcal{S}_0)$, for $h \in H$, we can get $h_k \in \mathcal{S}$ with $|h_k-h|_0 \to 0$. Continuity of Q in $|\cdot|_0$ norm implies $Q(h_j-h_k, h_j-h_k) \to 0$ as $j,k \to \infty$, so that $E_\mu(L(h_j)-L(h_k))^2 = Q(h_j-h_k, h_j-h_k) \to 0$ as $j,k \to \infty$. Since $\mathcal{L}^2(\mathcal{S}^*,\mathcal{B}(\mathcal{S}),\mu)$ is complete, there exists an element which we denote by L(h), such that

$$E_m(L(h_k) - L(h))^2 \to 0.$$

Then

$$E_\mu \exp(iL(h)) = \lim_{k\to\infty} E_\mu \exp(iL(h_k)) \tag{4.55}$$
$$= \lim_{k\to\infty} \exp(-\tfrac{1}{2} Q(h_k,h_k)) = \exp(-\tfrac{1}{2} Q(h,h))$$

as Q is $|\cdot|_0$ continuous and $|h-h_k|_0 \to 0$.

Thus, (L,μ) is a representation of the cylinder probability n, with the underlying representation space $(\mathcal{S}^*,\mathcal{B}(\mathcal{S}))$.

In particular, when $Q(h,h) = |h|_0^2$, the corresponding cylinder probability n is the canonical Gauss measure m on H. Thus we have obtained a representation (L,μ) of m with

$(\mathcal{S}^*, \mathcal{B}(\mathcal{S}))$ as the representation space. The measure μ obtained in this manner is called the measure of Gaussian white noise.

<u>Example 4</u>: Let $H = L^2[0,T] \otimes L^2(\mathbb{R}^d)$. Let Q be a symmetric, positive definite bilinear form on $L^2(\mathbb{R}^d)$ assumed to be continuous in the $\mathcal{S}$-topology.

Using a version of Kolmogorov's consistency theorem, we obtain on some probability space $(\Omega_1, \mathcal{A}_1, \Pi_1)$, a Gaussian family of random variables $\{Z_t(\phi): \phi \in L^2(\mathbb{R}^d),\ t \in [0,T]\}$ with the following properties: For $s,t \in [0,T]$, $\phi_1, \phi_2 \in L^2(\mathbb{R}^d)$,

$$E_{\Pi_1} Z_t(\phi_1) = 0 \tag{4.56}$$

$$E_{\Pi_1} Z_s(\phi_1) \cdot Z_t(\phi_2) = (t \wedge s) Q(\phi_1, \phi_2). \tag{4.57}$$

The relation (4.57) implies $E_{\Pi} Z_0^2(\phi_1) = 0$ so that we have

$$Z_0(\phi_1) = 0 \quad \text{a.s. } \Pi_1, \quad \text{for all } \phi_1 \in L^2(\mathbb{R}^d). \tag{4.58}$$

Also, using (4.57) it is easy to check that

$$E_{\Pi_1} (Z_t(a_1\phi_1 + a_2\phi_2) - (a_1 Z_t(\phi_1) + a_2 Z_t(\phi_2)))^2 = 0$$

for $a_1, a_2 \in \mathbb{R}$ and $\phi_1, \phi_2 \in L^2(\mathbb{R}^d)$ so that

$$Z_t(a_1\phi_1 + a_2\phi_2) = a_1 Z_t(\phi_1) + a_2 Z_t(\phi_2). \tag{4.59}$$

For each fixed $\phi \in L^2(\mathbb{R}^d)$, we can obtain a continuous version of $Z_t(\phi)$, $0 \le t \le T$ also denoted by the same symbol.

Thus, for all $\phi \in L^2(\mathbb{R}^d)$

$$Z_t(\phi) \text{ is a Wiener process with variance parameter } Q(\phi,\phi). \tag{4.60}$$

If $Q(\phi,\phi) = \|\phi\|_0^2$, ($|\cdot|_0$ being the $L^2(\mathbb{R}^d)$ norm), the family $\{Z_t(\phi)\}$ satisfying (4.59) and (4.60) is usually called a *cylindrical Brownian motion* on $L^2(\mathbb{R}^d)$.

Since Q is continuous in the $\mathcal{S}$-topology, there exists a constant $\delta > 0$ and $p \geq 1$ such that

$$Q(\phi,\phi) \leq \delta|\phi|_p^2, \quad \text{for all } \phi \in \mathcal{S}. \tag{4.61}$$

By the nuclearity of $\mathcal{S}$, there is a number $r > p$ such that if $\{\phi_j\} \subset \mathcal{S}$ is a CONS for $\mathcal{S}_r$, then

$$\sum_{j=1}^{\infty} |\phi_j|_p^2 < \infty. \tag{4.62}$$

By Doob's inequality

$$\sum_{j=1}^{\infty} E \sup_{0\leq t\leq T} [Z_t(\phi_j)]^2 \leq 4T \sum_{j=1}^{\infty} Q(\phi_j,\phi_j) < \infty, \tag{4.63}$$

by (4.61) and (4.62). Define

$$W_t(\omega) = \sum_{j=1}^{\infty} Z_t(\phi_j)(\omega)\hat{\phi}_j, \quad \omega \in \Omega_1, \tag{4.64}$$

where $\{\hat{\phi}_j\}$ is the CONS in $\mathcal{S}_{-r}$ dual to the CONS $\{\phi_j\}$ in $\mathcal{S}_r$. Using (4.63), it is easy to see that the series in (4.64) converges in $\mathcal{S}_{-r}$ for each t (a.s. Π_1) and that

$\sup_{0\leq t\leq T} \|W_t(\omega)\|_{-r}^2 < \infty$ a.s. Π_1.

Write

$$W_t^N(\omega) = \sum_{j=1}^{N} Z_t(\phi_j)(\omega)\hat{\phi}_j .$$

Then for all N,

$$W_\cdot^N(\omega) \in C_0([0,T],\mathcal{S}_{-r}) \text{ a.s.}$$

and

$$E \sup_{0\le t\le T} \|W_t(\omega)-W_t^N(\omega)\|_{-r}^2 \to 0 \quad \text{by (4.63).}$$

Hence $W_\cdot(\omega) \in C_0([0,T],\mathcal{S}_{-r})$ for a.e. ω. The map $W_\cdot$ induces a probability measure $\mu = PW_\cdot^{-1}$ on $\mathcal{X} = C_0([0,T];\mathcal{S}^*)$ which is a Gaussian measure with the properties

$$\int_{\mathcal{X}} \xi_t[\phi]d\mu(\xi) = 0, \quad \text{for all } t \in [0,T],\ \phi \in \mathcal{S} \tag{4.65}$$

and for $s,t \in [0,T]$, $\phi_1,\phi_2 \in \mathcal{S}$,

$$\int_{\mathcal{X}} \xi_t[\phi_1]\xi_s[\phi_2]d\mu(\xi) = (t\wedge s)Q(\phi_1,\phi_2). \tag{4.66}$$

The coordinate process (ξ_t) is called an $\mathcal{S}^*$-valued Wiener process. The support of μ is actually the Polish space $C_0([0,T];\mathcal{S}_{-r})$.

We shall now obtain a representation (L,μ) for the canonical Gauss measure m (on $H = L^2([0,t])\otimes L^2(\mathbb{R}^d)$) on the representation space $(\mathcal{X}.\mathcal{B}(\mathcal{X}))$.

Thus, let $Q(\phi,\phi) = |\phi|_0^2$ in the above discussion. Now for $\phi \in \mathcal{S}(\subset L^2(\mathbb{R}^d))$ and $f \in L^2([0,T])$, let

$$L(f\otimes\phi)(\xi) = \int_0^T f(s)d\xi_s[\phi] \tag{4.67}$$

where the integral appearing in (4.67) is the Wiener integral defined earlier in this section. Here, $\xi_s[\phi]$ is a Wiener process with variance parameter $Q(\phi,\phi) = |\phi|_0^2$ and thus

$$E_\mu(\exp(iL(f\otimes\phi))) = \exp(-\tfrac{1}{2}(\int_0^T f^2(s)ds)\cdot|\phi|_0^2) \tag{4.68}$$

$$= \exp(-\tfrac{1}{2}|f|^2_{L^2([0,T])}\cdot|\phi|_0^2).$$

Since $\mathcal{S}$ is dense in $L^2(\mathbb{R}^d)$, for $\phi \in L^2(\mathbb{R}^d)$, let $\phi_k \in \mathcal{S}$, $\phi_k \to \phi$ in $|\cdot|_0$. It follows from (4.68) that $L(f\otimes\phi_k)$ converges in $\mathcal{L}^2(\mathcal{X}.\mathcal{B}(\mathcal{X}),\mu)$ to a random variable, which we denote by $L(f\otimes\phi)$, and further that (4.68) continues to hold for all $\phi \in L^2(\mathbb{R}^d)$. Since

$$\|f\otimes\phi\|_H = |f|_{L^2([0,T])}\cdot|\phi|_0,$$

(4.68) can be rewritten as

$$E_\mu(\exp(iL(f\otimes\phi))) = \exp(-\tfrac{1}{2}\|f\otimes\phi\|_H^2). \tag{4.69}$$

Also, it is easy to see that L is linear and (4.69) gives

$$E_\mu(L(f\otimes\phi))^2 = \|f\otimes\phi\|_H^2$$

and hence L can be extended as a linear mapping from $H \to \mathcal{L}^2(\mathcal{X}.\mathcal{B}(\mathcal{X}),\mu)$ such that

$$E_\mu(L(h))^2 = \|h\|_H^2.$$

Hence (4.69) implies that

$$E_\mu \exp(iL(h)) = \exp(-\tfrac{1}{2}\|h\|_H^2).$$

Thus (L,μ) is the desired representation of m (on $H = L^2([0,T])\otimes L^2(\mathbb{R}^d))$ on the representation space $(\mathcal{X}.\mathcal{B}(\mathcal{X}),\mu)$, where $\mathcal{X} = C_0([0,T],\mathcal{S}^*)$.

Example 4.5: Let $\mathcal{H}$ be the RKHS of a one dimensional Wiener process and let $(\gamma_1,\mathcal{H},C_0([0,T]))$ be the abstract Wiener space corresponding to the choice of sup norm on $\mathcal{H}$. (See Example 4.2 with N = 1).

Let K be an arbitrary separable Hilbert space and let (γ_2,K,B) be an abstract Wiener space, where B is a Banach space which is the completion of K under some measurable norm and γ_2: $K \to B$ is the injection map.

Let $H = \mathcal{H}\otimes K$. Then $(\gamma,\mathcal{H}\otimes K,C_0([0,T],B))$ is an abstract Wiener space, γ being a one to one continuous map taking $\mathcal{H}\otimes K$ into $\mathcal{X} = C_0([0,T],B)$, the space of B-valued continuous functions vanishing at 0. The abstract Wiener measure μ on $(\mathcal{X}.\mathcal{B}(\mathcal{X}))$ is such that the coordinate family of random variables $\{x(t)\}$ is a B-valued Gaussian process with continuous sample paths, and for $f_1,f_2 \in B^*$

$$E_\mu f_1[x(t)] = 0 \tag{4.71}$$

and

$$E_m f_1[x(t)]\cdot f_2[x(t)] = (t\wedge s)(\gamma_2^*(f_1),\gamma_2^*(f_2))_K, \tag{4.72}$$

such a process is called a B-valued Wiener process.

Next, define L: $H \to \mathcal{L}(\mathcal{X}.\mathcal{B}(\mathcal{X}),\mu)$ as follows. First, for $\eta \in \mathcal{H}$, $f \in B^*$, let

$$L(\eta\otimes(\gamma_2^*(f)))(x) = \int_0^T \eta'(s)df[x(s)]. \tag{4.73}$$

Here $f[x(s)]$ is a one dimensional Wiener process with a variance parameter $|\gamma_2^*(f)|_K^2$ and the integral appearing above is the Wiener integral.

For $k \in K$ $(= K^*)$, there exists $\{f_j\} \subseteq B^*$ such that $\gamma^*(f_j) \to k$ in K and then we have

$$E_\mu(L(\eta\otimes\gamma_2^* f_j) - L(\eta\otimes\gamma_2^* f_i))^2 = |\eta|_{\mathcal{H}}^2 \cdot |\gamma_2^* f_j - \gamma_2^* f_i|_K^2 \tag{4.74}$$

$$\to 0 \quad \text{as } i,j \to \infty.$$

Hence there exists an element, say $L(\eta\otimes k)$, in $\mathcal{L}^2(\mathcal{X}.\mathcal{B}(\mathcal{X}),\mu)$ such that

$$E_\mu(L(\eta\otimes\gamma_2^* f_j) - L(\eta\otimes k))^2 \to 0. \tag{4.75}$$

Then we have

$$\begin{aligned} E_\mu \exp(iL(\eta\otimes k)) &= \lim_{j\to\infty} E_\mu \exp(iL(\eta\otimes\gamma_2^* f_j)) \\ &= \lim_{j\to\infty} \exp(-\tfrac{1}{2}|\eta|_{\mathcal{H}}^2 \cdot |\gamma_2^* f_j|_K^2) \\ &= \exp(-\tfrac{1}{2}|\eta\otimes k|_H^2) \end{aligned} \tag{4.76}$$

amd in particular

$$E_\mu(L(\eta\otimes k))^2 = |\eta\otimes k|_H^2. \tag{4.77}$$

Using (4.77) and the linearity of L, we can extend L to all of H such that L is a linear map and

$$E_\mu(L(h))^2 = |h|_H^2. \tag{4.78}$$

Then, we will also have

$$E_\mu \exp(iL(h)) = \exp(-\tfrac{1}{2}|h|_H^2). \tag{4.79}$$

Hence (L,μ) is a representation of the canonical Gauss measure m on H, with the representation space $(\mathfrak{X}.\mathfrak{B}(\mathfrak{X}))$ -- the sample space for a B-valued Wiener process.

5. RELATION TO THE DUNFORD-SCHWARTZ THEORY

In the previous sections, we have defined integral with respect to a cylinder probability. We will refer to it as cylinder integral for convenience in this section. A cylinder probability is, by definition, also a finitely additive measure, and a theory of integration with respect to finitely additive measures already exists and is due to several authors (see Dunford-Schwartz, [20]). We will describe this theory and then discuss its relation to the cylinder integral. To make a distinction, we will denote the Dunford-Schwartz integral of f with respect to μ by $⨍ f d\mu$.

We will show that there exist functions f on H for which $\int f dm$ is defined but $⨍ f dm$ is not defined.

We then show that a cylinder probability n on $(H,\mathcal{C})$ can be canonically extended as a finitely additive probability measure $\hat{n}$ on a field $\hat{\mathcal{C}} \supset \mathcal{C}$, such that $\int f dn$ is defined if and only if $⨍ f d\hat{n}$ is and in that case both are equal. (Here the field $\hat{\mathcal{C}}$ depends on the underlying cylinder probability n).

We begin this section by describing the general theory of integration with respect to finitely additive measures as given in Dunford and Schwartz [20]. The definition of the integral is given here for finite measures only, which enables us to give a simpler treatment than in Dunford-Schwartz who consider the general case (of a possibly infinite measure). Though the definitions look different, they are the same as in Dunford-Schwartz. Thus the material up to and including Lemma 4.3 is essentially from Dunford-Schwartz. Also, we use a slightly different notation.

Let X be a non-empty set, Σ be a field of subsets of X and let μ be a finitely additive probability measure on (X,Σ).

For any subset A of X, let $\mu^*(A)$ and $\mu_*(A)$ be defined by

$$\mu^*(A) = \inf\{\mu(B): B \supseteq A,\ B \in \Sigma\}$$

and

$$\mu_*(A) = \sup\{\mu(C): A \supseteq C;\ C \in \Sigma\}.$$

It is easy to see that $\mu_*(A) = 1 - \mu^*(A)$, $A \subseteq X$.

<u>Definition 5.1</u>: Let $\{f_k\}$ be a sequence of real valued functions on H. Say that f_k *converges in μ-probability to (a function)* f, written as $f_k \underset{\mu}{\to} f$, if for all $\epsilon > 0$,

$$\lim_{k\to\infty} \mu^*(|f_k - f| > \epsilon) = 0.$$

A function f of the form

$$f(x) = \sum_{j=1}^{k} a_j 1_{A_j}(x) \tag{5.1}$$

where $a_j \in \mathbb{R}$, $A_j \in \Sigma$, $k \geq 1$ is called a Σ-simple function.

Definition 5.2: Let $\ell^0(X,\Sigma,\mu)$ be the class of all real valued functions f on X such that there exists a sequence of Σ-simple functions $\{f_k\}$ such that $f_k \underset{\mu}{\to} f$.

The class $\ell^0(X,\Sigma,\mu)$ is the class of all 'measurable functions' on (X,Σ,μ). Clearly, $\ell^0(X,\Sigma,\mu)$ is a vector space of functions. It can be proved that if $f_1,f_2,\dots,f_k$ belong to $\ell^0(X,\Sigma,\mu)$ and $g\colon \mathbb{R}^k \to \mathbb{R}$ is continuous, then $g(f_1,\dots,f_k) \in \ell^0(X,\Sigma,\mu)$. The proof of this statement is similar to the proof of Theorem 2.1.

For a simple function $f\colon X \to \mathbb{R}$ given by (5.1) we define its integral with respect to μ, $\oint f d\mu$ by

$$\oint f d\mu := \sum_{j=1}^{k} a_j \mu(A_j). \tag{5.2}$$

We will state some inequalities whose proof is obvious (follows from (5.2)) and hence the proof is omitted.

Lemma 5.1: Let f be a simple function, $a \in \mathbb{R}^+$. Then

$$|\oint f d\mu| \leq \oint |f| d\mu, \tag{5.3}$$

$$\mu(|f| \geq a) \leq \frac{1}{a} \oint |f| d\mu, \tag{5.4}$$

and if $|f| \leq M$, then

$$\oint f d\mu \leq a + M\mu(|f| \geq a). \tag{5.5}$$

From (5.5), it follows that if $\{f_k\}$ is uniformly bounded, $f_k \underset{\mu}{\to} f$, then $\oint f_k d\mu$ is a Cauchy sequence and it is natural to define $\oint f d\mu$ as the limit of $\oint f_k d\mu$. We will need to check that this way, the integral is well-defined--i.e., does not depend on the choice of $\{f_k\}$ and that it is defined for all bounded functions in $\ell^0(X,\Sigma,\mu)$. This is the crux of the next proposition.

Proposition 5.2: Let $f \in \ell^0(X,\Sigma,\mu)$, $|f| \leq M$. Then

(a) there exists a sequence $\{f_k\}$ of Σ-simple functions such that $f_k \underset{\mu}{\to} f$, $|f_k| \leq M$,

(b) if $\{f_k\}$ are as in (a), then $\lim_{k\to\infty} \oint f_k d\mu$ exists,

(c) if $\{f_k\}$ and $\{g_k\}$ are as in (a), then

$$\lim_{k\to\infty} \oint f_k d\mu = \lim_{k\to\infty} \oint g_k d\mu.$$

Proof: (a) Since $f \in \ell^0(X,\Sigma,\mu)$, there exists a sequence $\{f_k'\}$ of Σ-simple functions such that $f_k' \underset{\mu}{\to} f$. Let f_k be defined by

$$\begin{aligned} f_k(x) &= f_k'(x) && \text{if } |f_k'(x)| \leq M \\ &= M && \text{if } f_k(x) > M \\ &= -M && \text{if } f_k(x) < -M. \end{aligned}$$

Then, since $|f(x)| \leq M$, it follows that

$$|f_k(x) - f(x)| \leq |f_k'(x) - f(x)|$$

and hence $f_k \underset{\mu}{\to} f$. By construction, $|f_k| \leq M$.

(b) If $\{f_k\}$ is as in (a) then for all $\epsilon > 0$,

$$\begin{aligned} |\oint f_k d\mu - \oint f_j d\mu| &\leq \oint |f_k - f_j| d\mu && \text{by (5.3)} \\ &\leq \frac{\epsilon}{2} + M\cdot\mu(|f_k - f_j| > \frac{\epsilon}{2}) && \text{by (5.5)} \\ &\leq \frac{\epsilon}{2} + M\mu(|f_k - f| > \frac{\epsilon}{4}) + M\mu(|f_j - f| > \frac{\epsilon}{4}). \end{aligned}$$

So choosing k_0 such that $\mu(|f_k - f| > \frac{\epsilon}{4}) < \frac{\epsilon}{4M}$ for $k \geq k_0$, we get that for $k, j \geq k_0$

$$|\oint f_k d\mu - \oint f_j d\mu| \leq \epsilon.$$

Thus $\{\oint f_k d\mu\}$ is a Cauchy sequence of real numbers and hence converges. The last part follows easily from (b). □

We are now in a position to define the integral with respect to a finitely additive probability measure.

Definition 5.3: Let $f \in \ell^0(X, \Sigma, \mu)$, $|f| \leq M$. Let $\{f_k\}$ be any sequence of Σ-simple functions such that $|f_k| \leq M$, $f_k \underset{\mu}{\rightarrow} f$. Then we define $\oint f d\mu$ by

$$\oint f d\mu := \lim_{k \to \infty} \oint f_k d\mu. \tag{5.6}$$

Definition 5.4: Let $f \in \ell^0(X, \Sigma, \mu)$ be positive. Then we define

$$\oint f d\mu := \lim_{k \to \infty} \oint (f \wedge k) d\mu. \tag{5.7}$$

Definition 5.5: Let

$$\ell^1(X, \Sigma, \mu) := \{f \in \ell^0(X, \Sigma, \mu) : \oint |f| d\mu < \infty\}$$

and for $f \in \ell^1(X, \Sigma, \mu)$, we define

$$\oint f d\mu := \oint f^+ d\mu - \oint f^- d\mu$$

where $f^+ = f \vee 0$, $f^- = -(f \wedge 0)$ as usual.

From the definition, it is easy to see that the integral has the usual properties like linearity, monotonicity (namely, $f_1 \leq f_2$ implies $\oint f_1 d\mu \leq \oint f_2 d\mu$). The following is an analogue of the Lemma 5.1 for $f \in \ell^0(X,\Sigma,\mu)$.

Lemma 5.3: Let $f \in \ell^1(X,\Sigma,\mu)$, $f \geq 0$.

(i) If $1_A \leq f$, $A \subseteq X$, then

$$\mu^*(A) \leq ⨍ f d\mu. \tag{5.8}$$

(ii) If $f \leq 1_B$, $B \subseteq X$, then

$$⨍ f d\mu \leq \mu^*(B). \tag{5.9}$$

(iii) For any $a > 0$,

$$\mu^*(|f| \geq a) \leq \frac{1}{a} ⨍ f d\mu. \tag{5.10}$$

(iv) If $|f| \leq M$, then for any $a > 0$,

$$⨍ f d\mu \leq a + M \cdot \mu^*(|f| \geq a). \tag{5.11}$$

Proof: (i) Without loss of generality, we can assume that f is bounded (otherwise work with, say $f \wedge 2$). Now, given $\epsilon > 0$, get Σ-simple function $g \geq 0$ such that $\mu^*(|f-g| > \epsilon) < \epsilon$ and $|\int f d\mu - \int g d\mu| < \epsilon$. Then (for $\epsilon < 1$)

$$A \subseteq \{|g| \geq 1-\epsilon\} \cup \{|f-g| > \epsilon\}$$

and thus

$$\begin{aligned} \mu^*(A) &\leq \mu^*(|g| \geq 1-\epsilon) + \mu^*(|f-g| > \epsilon\} \\ &\leq \frac{1}{1-\epsilon} ⨍ g d\mu + \epsilon \quad \text{(by Lemma 5.1 as g is simple)} \\ &\leq \frac{1}{1-\epsilon} (⨍ f d\mu + \epsilon) + \epsilon. \end{aligned}$$

Since $\epsilon > 0$ is arbitrary, this proves (i). For (ii), if

$C \in \Sigma$ is such that $B \subseteq C$, then $⨍ f d\mu \leq \mu(C)$. Thus (ii) holds. (iii) follows from (i) by linearity. The relation (5.11) follows from the analogous relation (5.5) for simple functions and an ϵ argument as in (i) above. □

We now introduce completion of a finitely additive probability space and show that if $f \in \ell^0(X,\Sigma,\mu)$, then for all $a \in \mathbb{R} - N_f$ (where N_f is a countable set depending on f), $\{f \leq a\}$ belongs to the μ-completion of Σ. Let

$$\overline{\Sigma} = \{A \in \Sigma\colon \mu^*(A) = \mu_*(A)\} \tag{5.12}$$

and let $\overline{\mu}\colon \overline{\Sigma} \to [0,1]$ be defined by

$$\overline{\mu}(A) = \mu^*(A), \ A \in \overline{\Sigma}. \tag{5.13}$$

Clearly, $\Sigma \subseteq \overline{\Sigma}$ and if $A \in \overline{\Sigma}$, then $A^c \in \overline{\Sigma}$ as $\mu^*(A^c) = 1-\mu_*(A)$ and $\mu_*(A^c) = 1-\mu^*(A)$. It can be checked that $\overline{\Sigma}$ is a field and that $\overline{\mu}$ is a finitely additive measure on $(X,\overline{\Sigma})$.

It is easy to see that $(\overline{\mu})^* = \mu^*$ and $(\overline{\mu})_* = \mu_*$ so that $(X,\overline{\Sigma},\overline{\mu})$ is complete. $(X,\overline{\Sigma},\overline{\mu})$ is called the completion of (X,Σ,μ). It can be checked that

$$\ell^0(X,\overline{\Sigma},\overline{\mu}) = \ell^0(X,\Sigma,\mu)$$

and

$$\ell^1(X,\overline{\Sigma},\overline{\mu}) = \ell^1(X,\Sigma,\mu).$$

<u>Theorem 5.4</u>: Let (X,Σ,μ) be a complete finitely additive probability space and let $f \in \ell^0(X,\Sigma,\mu)$.

(i) There exists a countable subset N_f of $\mathbb{R}$ such that for all $a \notin N_f$, $\{f \leq a\} \in \Sigma$.

(ii) There exists a countably additive probability measure λ on $(\mathbb{R}, \mathcal{B}(\mathbb{R}))$ such that for all $B \in \mathcal{B}(\mathbb{R})$ with $\lambda(\partial B) = 0$, $\{f \in B\} \in \Sigma$ and

$$\mu(f \in B) = \lambda(B). \tag{5.14}$$

(Here ∂B denotes the boundary B, $\partial B = \overline{B} - B^0$, $\overline{B}$ being the closure of B and B^0 being the interior of B.)

(iii) The measure λ is characterized by

$$\int g d\lambda = ⨍ g(f) d\mu$$

for all $g \in C_b(\mathbb{R})$.

<u>Proof</u>: Clearly (i) follows from (ii) by taking $N_f = \{a \in \mathbb{R}: \lambda(\{a\}) > 0\}$. We will prove (ii). First observe that for all $\epsilon > 0$, $\exists M < \infty$ such that

$$\mu^*(|f| \geq M) < \epsilon. \tag{5.16}$$

Indeed, get a simple function f_0 such that

$$\mu^*(|f-f_0| \geq 1) < \epsilon$$

and put $M = [\sup|f_0| + 1]$.

As observed earlier, $g(f) \in \ell^0(X,\Sigma,\mu)$ for $g \in C_b(\mathbb{R})$ and since $g(f)$ is bounded, $g(f) \in \ell^1(X,\Sigma,\mu)$ as well. Let $T: C_b(\mathbb{R}) \to \mathbb{R}$ be defined by

$$T(g) = ⨍ g(f) d\mu.$$

Clearly, T is a positive linear functional. We will prove

that

$$g_k \in C_b(\mathbb{R}),\ g_k \downarrow 0 \text{ implies } T(g_k) \to 0. \tag{5.17}$$

If $\{g_k\}$ are as in (5.17), then $g_k \to 0$ uniformly on $[-M,M]$ for any $M < \infty$ (a version of Dini's theorem). Now given $\epsilon > 0$, get $M < \infty$ such that (5.16) holds. Then get k_0 such that for $k \geq k_0$, $|g_k(x)| < \epsilon$ for $|x| \leq M$. Now for $k \geq k_0$,

$$\{|g_k(f)| > \epsilon\} \subseteq \{|f| > M\}$$

and hence by choice of M, (5.16), we get

$$\mu^*\{|g_k(f)| > \epsilon\} < \epsilon.$$

Thus, $g_k(f) \underset{\mu}{\to} 0$ and since $\{g_k\}$ is uniformly bounded (by the bound for g_1), the relation (5.11) gives

$$T(g_k) = ⨍ g_k(f) \to 0.$$

This proves (5.17).

By Daniell's Theorem, it follows that there exists a countably additive measure λ on $(\mathbb{R},\mathcal{B}(\mathbb{R}))$ such that $T(g) = \int g d\lambda$.

Let $B \in \mathcal{B}(\mathbb{R})$ be such that $\lambda(\partial B) = 0$. Let $g_k \in C_b(\mathbb{R})$, $0 \leq g_k \leq 1$, $1_{\overline{B}} \leq g_k$, $g_k \downarrow 1_{\overline{B}}$. Then

$$\mu^*(f \in B) \leq ⨍ g_k(f) d\mu \qquad \text{(by Lemma 4.3)} \tag{5.18}$$

$$= \int g_k d\lambda.$$

Since $0 \leq g_k \leq 1$, $g_k \downarrow 1_{\overline{B}}$, we get $\int g_k d\lambda \to \lambda(\overline{B})$ which is equal to $\lambda(B)$. Thus from (5.18) we get

$$\mu^*(f \in B) \leq \lambda(B). \tag{5.19}$$

The relation (5.19) holds for all B with $\lambda(\partial B) = 0$. If $\lambda(\partial B) = 0$, then $\lambda(\partial(B^c) = \partial B)$ and hence $\lambda(\partial B) = 0$ also implies

$$\mu^*(f \in B^c) \leq \lambda(B^c).$$

Now

$$\mu_*(f \in B) = 1 - \mu^*(f \in B^c) \geq 1 - \lambda(B^c) = \lambda(B).$$

This and (5.19) show that if $\lambda(\partial B) = 0$, then

$$\mu_*(f \in B) = \mu^*(f \in B) = \lambda(B).$$

Since Σ is assumed to be complete, this implies $(f \in B) \in \Sigma$ and that (5.14) holds. □

Let us now return to the Gauss measure m on $(H, \mathscr{C})$. We had proved (see Example 2.1) that $f(h) = |Ah|^2$ belongs to $\mathscr{L}^1(H, \mathscr{C}, m)$, where A is a self adjoint Hilbert Schmidt operator on H. We will show that this f does not belong to $\ell^0(H, \mathscr{C}, m)$.

<u>Proposition 5.5</u>: Let A be a self adjoint Hilbert Schmidt operator on H. Suppose that range A is infinite dimensional. Let $f(h) = |Ah|^2$. Then

$$m^*(f \geq a) = 1 \quad \text{for all } a. \tag{5.20}$$

As a consequence, $f \notin \ell^0(H, \mathscr{C}, m)$.

<u>Proof</u>: Let $C \in \mathscr{C}$ be such that

$$C \subseteq \{f < a\}.$$

We will prove that $C = \phi$. This will prove $m_*(f < a) = 0$ which is the same as (5.20). Suppose $C \neq \phi$. Let $P \in \mathcal{P}$ be such that $C \in \mathcal{C}_P$, let $H_0 = PH$ and let $H_1 = P^{\perp}H$. Let $h_0 \in C$. Since range A is infinite dimensional, $\exists\ h_1 \in H_1$ such that $f(h_1) \neq 0$. (For if $f \equiv 0$ on H_1, then $A \equiv 0$ on H_1, so that $\dim(\text{range } A) \leq \dim H_0 < \infty$).

Let $h_k = h_0 + kh_1$, $k \geq 2$. Then $Ph_k = Ph_0 + kPh_1 = h_0$, so that $h_k \in \mathcal{C}$ for all k since $C \in \mathcal{C}_P$. Thus $h_k \in C \subseteq \{f > a\}$ and so $f(h_k) < a$, for all $k \geq 2$. But

$$
\begin{aligned}
f(h_0 + kh_1) &= |A(h_0 + kh_1)|^2 \\
&= |Ah_0|^2 + k^2|Ah_1|^2 + 2k(Ah_0, Ah_1) \\
&= f(h_0) + k^2 f(h_1) + 2k(Ah_0, Ah_1),
\end{aligned}
$$

Since $f(h_1) > 0$, this implies $\lim_{k\to\infty} f(h_k) = \infty$ which contradicts $f(h_k) < a$! Thus $C = \phi$. This proves (5.20) as observed earlier.

The relation (5.20) also shows that $f \notin \ell^0(H, \mathcal{C}, m)$. For if $f \in \ell^0(H, \mathcal{C}, m)$, then $\lim_{a\to\infty} m^*(f \geq a) = 0$ (as seen in (5.16)) which contradicts (5.20). □

We have seen that the cylinder integral $\int f dn$ and the Dunford-Schwartz integral $⨍ f dn$ are not equivalent. We will now show that a cylinder probability n on $(H, \mathcal{C})$ can be extended as a finitely additive probability measure $\hat{n}$ on a field $\hat{\mathcal{C}} \supset \mathcal{C}$ such that

$$
\ell^0(H, \hat{\mathcal{C}}, \hat{n}) = \mathcal{L}(H, \mathcal{C}, n) \tag{5.21}
$$

$$\ell^1(H,\hat{\mathscr{C}},\hat{n}) = \mathscr{L}^1(H,\mathscr{C},n) \tag{5.22}$$

and for $f \in \ell^1(H,\hat{\mathscr{C}},\hat{n})$,

$$⨍ f d\hat{n} = \int f dn. \tag{5.23}$$

This will give the exact relationship between the two integrals and help us in understanding the cylinder integral. However, it should be noted that the field $\hat{\mathscr{C}}$ depends upon the cylinder measure n and thus in the later sections, for our purposes, it is convenient to use the cylinder integral.

Let n be a fixed cylinder probability on $(H,\mathscr{C})$. Let

$$\hat{\mathscr{C}} = \{C \subseteq H: 1_C \in \mathscr{L}(H,\mathscr{C},n)\} \tag{5.24}$$

and for $C \in \hat{\mathscr{C}}$, define

$$\hat{n}(C) = \int 1_C dn. \tag{5.25}$$

<u>Lemma 5.6</u>: Let $f \in \ell^0(H,\hat{\mathscr{C}},\hat{n})$. Then $f \in \mathscr{L}(H,\mathscr{C},n)$.

<u>Proof</u>: Given $\epsilon > 0$ get $\hat{\mathscr{C}}$-simple function g, given by

$$g = \sum_{i=1}^{k} a_i 1_{B_i}, \quad B_i \in \hat{\mathscr{C}} \tag{5.26}$$

such that

$$(\hat{n})^*(|f-g| > \tfrac{\epsilon}{3}) < \tfrac{\epsilon}{6}. \tag{5.27}$$

Get $B_0 \in \hat{\mathscr{C}}$ such that

$$\{|f-g| > \tfrac{\epsilon}{3}\} \subset B_0, \quad \hat{n}(B_0) < \tfrac{\epsilon}{6}.$$

From the definition of $\hat{\mathscr{C}}$, $1_{B_i} \in \mathscr{L}(H,\mathscr{C},n)$, $0 \le i \le k$. Let $P_0 \in \mathscr{P}$ be such that for $P_0 \le P_1$, $P_0 \le P_2$, $P_1, P_2 \in \mathscr{P}$, $1 \le i \le k$, we have

$$\int |1_{B_i} \circ P_1 - 1_{B_i} \circ P_2| dn < \frac{\epsilon}{3k} \tag{5.28}$$

and

$$n(P_1^{-1} B_0) = \int 1_{B_0} \circ P_1 dn \le \hat{n}(B_0) + \frac{\epsilon}{6} < \frac{\epsilon}{3}. \tag{5.29}$$

(Recall that $\int 1_{B_0} \circ P dn \to \int 1_{B_0} dn$ by definition of the integral). Now for $P_0 \le P_1$, $P_0 \le P_2$, we have

$$\begin{aligned} n(|f \circ P_1 - f \circ P_2| > \epsilon) &\le n(|f \circ P_1 - g \circ P_1| > \tfrac{\epsilon}{3}) + n(|f \circ P_2 - g \circ P_2| > \tfrac{\epsilon}{3}) \\ &\quad + n(|g \circ P_1 - g \circ P_2| > \tfrac{\epsilon}{3}) \\ &\le n(P_1^{-1} B_0) + n(P_2^{-1} B_0) + \sum_{i=1}^{k} \int |1_{B_i} \circ P_1 - 1_{B_i} \circ P_2| dn \\ &\le \frac{\epsilon}{3} + \frac{\epsilon}{3} + \frac{\epsilon}{3} = \epsilon. \end{aligned}$$

Thus $f \in \mathscr{L}(H,\mathscr{C},n)$. □

To prove the converse of this lemma, we need the following result.

Let $\Lambda = \mathscr{P} \times \mathscr{P}$ and let $(P', P'') \le (P_1', P_1'')$ if $P' \le P_1'$ and $P'' \le P_1''$. Then $(\Lambda, \le)$ is a directed set.

<u>Proposition 5.7</u>: Let $f \in \mathscr{L}(H,\mathscr{C},n)$. The net $\{n \circ (f \circ P', f \circ P'')^{-1} : (P', P'') \in \mathscr{P} \times \mathscr{P}\}$ of countably additive probability measures on

$(\mathbb{R}^2, \mathcal{B}(\mathbb{R}^2))$ converges weakly to the measure λ_1 on $(\mathbb{R}^2, \mathcal{B}(\mathbb{R}^2))$ given by

$$\lambda_1(B) = \int_{\mathbb{R}} 1_B(x,x) d\lambda(x), \quad B \in \mathcal{B}(\mathbb{R}^2) \tag{5.30}$$

where $\lambda = n \circ f^{-1}$ (see Proposition 2.3).

Proof: Let $\mu_{P',P''} = n \circ [f \circ P', f \circ P'']^{-1}$ and

$$\phi_{P',P''}(t_1, t_2) = \int_{\mathbb{R}^2} e^{it_1x_1 + it_2x_2} d\mu_{P',P''}(x_1, x_2).$$

Using the inequality

$$|e^{it_1x_1 + it_2x_2} - e^{it_1y_1 + it_2y_2}|$$

$$\leq |e^{it_1x_1} - e^{it_1y_1}| + |e^{it_2x_2} - e^{it_2y_2}|$$

for $t_1, t_2, x_1, x_2, y_1, y_2 \in \mathbb{R}$, it follows that

$$|\phi_{P',P''}(t_1, t_2) - \phi_{P_1',P_1''}(t_1, t_2)|$$

$$\leq \int |e^{it_1 f \circ P_1'(h)} - e^{it_1 f \circ P_1'(h)}| dn(h)$$

$$+ \int |e^{it_2 f \circ P''(h)} - e^{it_2 f \circ P_1''(h)}| dn(h)$$

and hence as in the proof of Proposition 2.3, it follows that the net of characteristic functions $\{\phi_{P',P''}(t_1, t_2): (P', P'') \in \mathcal{P} \times \mathcal{P}\}$ converges uniformly on compact subsets of $\mathbb{R}^2$ to a characteristic function, say ϕ_1, of a probability measure μ_1 on $(\mathbb{R}^2, \mathcal{B}(\mathbb{R}^2))$. Thus, $\mu_{P',P''} \to \mu_1$ weakly.

Also

$$\phi_1(t_1,t_2) = \lim_{P\in\mathcal{P}} \phi_{P,P}(t_1,t_2)$$

and

$$\phi_{P,P}(t_1,t_2) = \int e^{it_1 f\circ P(h)+it_2 f\circ P(h)} dn(h) = \phi_P(t_1+t_2)$$

where ϕ_P is as in Proposition 2.3. Since $\phi_P \to \phi$, the characteristic function of $\lambda = n\circ f^{-1}$, we get

$$\phi_1(t_1,t_2) = \phi(t_1 + t_2).$$

This implies that ϕ_1 is also the characteristic function of λ_1 given by (5.30) and thus $\lambda_1 = \mu_1$. □

Corollary 5.8: Let $B \in \mathcal{B}(\mathbb{R}^2)$ be such that $\lambda_1(\partial B) = 0$. Then

$$\lim_{(P',P'')\in\mathcal{P}\times\mathcal{P}} n(h: (f\circ P'(h), f\circ P''(h) \in B)) = \lambda_1(B). \quad (5.31)$$

Proof: The weak convergence of $\mu_{P',P''}$ to λ_1 implies that for all B such that $\lambda_1(\partial B) = 0$, $\mu_{P',P''}(B) \to \lambda_1(B)$. This proves (5.31). □

Theorem 5.9: Let n be a cylinder probability on $(H,\mathcal{C})$. Let $(\hat{\mathcal{C}},\hat{n})$ be defined by (5.24), (5.25). Then

(i) $\ell^0(H,\hat{\mathcal{C}},\hat{n}) = \mathcal{L}(H,\mathcal{C},n)$

(ii) $\ell^1(H,\hat{\mathcal{C}},\hat{n}) = \mathcal{L}^1(H,\mathcal{C},n)$

(iii) for $f \in \ell^1(H,\hat{\mathcal{C}},\hat{n})$, $⨏ f d\hat{n} = \int f dn$.

Proof: (i) We have proved that $\ell^0(H,\hat{\mathcal{C}},\hat{n}) \subseteq \mathcal{L}(H,\mathcal{C},n)$ (see Lemma 5.6). To prove the reverse inclusion, let $f \in \mathcal{L}(H,\mathcal{C},n)$ and let $\lambda = n\circ f^{-1}$.

Let $A \in \mathcal{B}(\mathbb{R})$ be such that $\lambda(\partial B) = 0$. Then we claim that

$$\{h: f(h) \in A\} \in \hat{\mathscr{C}} \quad \text{and} \quad \hat{n}(f \in A) = \lambda(A). \tag{5.32}$$

We will prove this statement first. Let B = A×A. Then since

$$\partial B = ((\partial A) \times \mathbb{R}) \cup (\mathbb{R} \times (\partial A)),$$

$\lambda_1(\partial B) \leq 2\lambda(\partial A) = 0$ so that $\lambda_1(\partial B) = 0$ (where λ_1 is given by (5.30)). Now by Corollary 5.8, we have

$$\lim_{(P',P'')\in\mathscr{P}\times\mathscr{P}} n(h: (f\circ P'(h), f\circ P''(h)) \in A\times A) = \lambda_1(A\times A) = \lambda(A). \tag{5.33}$$

Also since $\partial(A^c) = \partial(A)$, we also have ((5.31) for $B = A^c\times A^c$),

$$\lim_{(P',P'')\in\mathscr{P}\times\mathscr{P}} n(h: (f\circ P'(h), f\circ P''(h)) \in A^c\times A^c) = \lambda_1(A^c) \tag{5.34}$$

The relations (5.33), (5.34) and the fact that n(H) = 1, $\lambda(\mathbb{R}) = 1$ together yield

$$\lim_{(P',P'')\in\mathscr{P}\times\mathscr{P}} n(h: (f\circ P'(h), f\circ P''(h)) \in (A\times A^c) \cup (A^c\times A)) = 0. \tag{5.35}$$

Let $D = f^{-1}A$. Then (5.35) is the same as

$$\lim_{(P',P'')\in\mathscr{P}\times\mathscr{P}} n(|1_D\circ P' - 1_D\circ P''| > 0) = 0. \tag{5.36}$$

This implies $1_D \in \mathscr{L}(H,\mathscr{C},n)$. Also (5.33) implies

$$\lim_{\mathscr{P}\in\mathscr{P}} n(h: f\circ P(h) \in A) = \lambda(A), \quad \text{i.e.,} \quad \lim_{\mathscr{P}\in\mathscr{P}} n(P^{-1}D) = \lambda(A).$$

Thus

$$\hat{n}(D) = \lambda(A).$$

This proves (5.32).

Now for each $k \geq 1$, get $\{a_0^k, a_1^k, \ldots, a_{j_k}^k\}$ such that

$$\lambda(\{a_i^k\}) = 0, \quad 0 \leq i \leq j_k, \; k \geq 1 \tag{5.37}$$

$$\lambda((-\infty, a_0^k) \cup [a_{j_k}^k, \infty)) < \frac{1}{k}, \quad k \geq 1 \tag{5.38}$$

$$a_{i+1}^k - a_i^k < \frac{1}{k}, \quad 0 \leq i \leq j_k, \; k \geq 1. \tag{5.39}$$

Let $\quad g_k(h) = a_i^k \quad$ if $\quad a_i^k \leq f(h) < a_{i+1}^k.$

$\qquad\qquad = 0 \quad$ if $\quad f(h) < a_0^k$ or $a_{j_k}^k \leq f(h).$

Then g_k is a $\hat{\mathscr{C}}$-simple function, since each of the sets $(a_i^k \leq f < a_{i+1}^k) \in \hat{\mathscr{C}}$ by the choice of $\{a_j^k\}$ and (5.32). Also in view of (5.39),

$$\{h: |f(h) - g_k(h)| > \tfrac{1}{k}\} \subseteq \{f < a_0^k\} \cup \{f \geq a_{j_k}^k\}$$

so that

$$\begin{aligned} \hat{n}^*(|f-g_k| > \tfrac{1}{k}) &\leq \hat{n}(f < a_0^k) + \hat{n}(f \geq a_{j_k}^k) \\ &= \lambda((-\infty, a_0^k) \cup [a_{j_k}^k, \infty)) \qquad \text{(by (5.32))} \\ &< \frac{1}{k}. \qquad \text{(by (5.38))}. \end{aligned}$$

Thus $g_k \underset{\hat{n}}{\to} f$ and $f \in \mathscr{L}^0(H, \hat{\mathscr{C}}, \hat{n})$. This proves (i).

For (ii), observe that $\hat{n}(f \in A) = \lambda(A)$ for A such that

$\lambda(\partial B) = 0$ implies that (see Theorem 5.4)

$$⨍ g(f)d\hat{n} = \int g d\lambda, \quad g \in C_B(\mathbb{R}) \tag{5.40}$$

and by Proposition 2.3 (since $\lambda = n \circ f^{-1}$),

$$\int g(f)dn = \int g n\lambda, \quad g \in C_b(\mathbb{R}) \tag{5.41}$$

so that

$$\int g(f)dn = ⨍ g(f)dn, \quad g \in C_b(\mathbb{R}). \tag{5.42}$$

Thus for all $k \geq 1$, taking $g(x) = (x \vee 0) \wedge k$ (for k fixed) in (5.42), we have

$$\int f^+ \wedge k dn = ⨍ f^+ \wedge k d\hat{n}$$

and similarly

$$\int f^- \wedge k dn = ⨍ f^- \wedge k d\hat{n}.$$

Hence by the definition of integrals of positive functions, we have

$$\int f^+ dn = ⨍ f^+ dn$$

and

$$\int f^- dn = ⨍ f^- dn.$$

Clearly these two relations imply (ii), (iii). □

Remark 5.1: We had seen in Proposition 5.5 that $f(h) = |Ah|^2$ does not belong to $\ell^0(H, \mathscr{C}, m)$. Now, in view of Theorem (5.9) and Example 2.1, we get

$$f \in \ell^1(H, \hat{\mathscr{C}}, \hat{m}).$$ □

Example 5.1: We will see later in Chapter VI that functions of the form

$$f(h) = \int \exp((h,h_0) - \tfrac{1}{2}|h_0|^2)d\mu(h_0), \quad h \in H \tag{5.43}$$

where μ is a countably additive probability measure on $(H,\mathscr{B}(H))$ arise as densities of mixtures of translates of the Gauss measure m (with respect to m). We will also see that these functions arise in the expression for conditional expectation for the filtering problem. There, it will be shown that $f \in \mathscr{L}^1(H,\mathscr{C},m)$, which in light of Theorem 5.9 would imply that $f \in \ell^1(H,\mathscr{C},m)$.

We will now show that in general, f given by (5.43) does not belong to $\ell^0(H,\mathscr{C},m)$. Thus we couldn't have carried out the analysis of the filtering model (and other problems that we deal with) without either extending the measure m to $\hat{m}$ or equivalently, dealing with the cylinder integral. As explained earlier, we have opted for the latter course.

Suppose μ in (5.43) is such that $\mu(PH) < 1$ for all $P \in \mathscr{P}$ (i.e., μ is not concentrated on any finite dimensional subspace). Then, we will show that for any $a > 0$,

$$C \in \mathscr{C},\ C \subseteq \{f < a\} \text{ implies } C = \phi. \tag{5.44}$$

This as in Proposition (5.5) will prove that $f \notin \ell^0(H,\mathscr{C},m)$.

Suppose $C \in \mathscr{C}$, $C \subseteq \{f < a\}$. Let $P \in \mathscr{P}$ be such that $C \in \mathscr{C}_P$. Since $\mu(PH) < 1$, there exists $h_1 \in P^\perp H$ such that

$$\mu(h: (h_1,h) \neq 0) > 0. \tag{5.55}$$

For if $\mu(h: (h_1,h) = 0) = 1$ for all $h_1 \in P^{\perp}H$, then $\mu(h: Ph = h) = 1$ so that $\mu(PH) = 1$.

Without loss of generality, suppose (otherwise work with $-h_1$)

$$\mu(h: (h_1,h) > 0) > 0. \tag{5.56}$$

Suppose $C \neq \phi$. Get $h_0 \in C$. Let $h_k = h_0 + kh_1$, $k \geq 2$. $h_1 \in P^{\perp}H$, $h_0 \in C \in \mathscr{C}_P$ implies that $h_k \in C$, and hence

$$h_k \in \{f < a\}$$

i.e., $f(h_k) \leq a$ for all $k \geq 2$. But

$$f(h_k) = \int \exp((h_0,h) + k(h_1,h) - \tfrac{1}{2}|h|^2)d\mu(h) \tag{5.57}$$

and in view of (5.56) the integrand in (5.57) goes to ∞ on a set of positive μ-measure. Hence $f(h_k) \to \infty$. This contradicts $f(h_k) < a$ for all k. Hence $C = \phi$. This proves $f \in \ell^0(H,\mathscr{C},m)$.

□

CHAPTER IV

CONDITIONAL EXPECTATION: DEFINITION AND BASIC PROPERTIES

The notion of conditional expectation is fundamental in statistical estimation problems. Hence for the development of the white noise theory of nonlinear filtering and smoothing, it is essential to have a satisfactory definition of conditional expectation (together with the related concept of absolute continuity) in the finitely additive theory. The definitions available in the current literature on the subject are not suitable from our point of view.

In this chapter, no attempt is made to prove general theorems of existence since our interest is not in the theory of finitely additive measures *per se*. Our aim is to offer definitions which allow us to identify conditional expectations as "random variables" on the finitely additive probability space itself, and not on some larger representation space of the former. It is this circumstance that makes the definitions and the resulting theory operational. A price to pay is that, in contrast to the situation in the countably additive theory, the conditional expectation of an integrable "random variable" need not always exist.

1. ABSOLUTE CONTINUITY OF CYLINDER PROBABILITIES

The most commonly used notion of absolute continuity in the context of finitely additive probabilities is the following: Let μ_1, μ_2 be finitely additive measures on (X, Σ), where Σ is a field of subsets of X. If for any $\epsilon > 0$, there exists $\delta > 0$ such that for $A \in \Sigma$, $\mu_2(A) < \delta$ implies $\mu_1(A) < \epsilon$, then μ_1 is said to be absolutely continuous with respect to μ_2. This does not imply the existence of a Randon-Nikodym derivative. We do have the following result, known as Bochner's theorem. If μ_1 is absolutely continuous with respect to μ_2 (in the sense described above), then for every $\epsilon > 0$ there exists a Σ-simple function f_ϵ such that $|\mu_1(A) - \int_A f_\epsilon d\mu_2| < \epsilon$ for all $A \in \Sigma$. The function f_ϵ is known as the ϵ-Radon-Nikodym derivative.

The notion of ϵ-derivative is not suitable for statistical purposes. To define conditional expectation or to define likelihood ratios, we need the notion of an exact or 0-Radon-Nikodym derivative.

Balakrishnan has used the $\epsilon - \delta$ definition in the context of cylinder probabilities [5]. In his setup too, absolute continuity does not imply existence of a Radon-Nikodym derivative.

Gross's notion (see [28]) of Radon-Nikodym derivatives for cylinder probabilities (or weak distributions) is also unsuitable for statistical purposes. For his Radon-Nikodym derivative, though it always exists, is a measurable function on the representation space and not on the Hilbert space on which the given cylinder probabilities are defined.

Our definition of absolute continuity simultaneously introduces the Radon-Nikodym derivative of the kind useful in white noise calculus and its applications.

Definition 1.1: Let n_1, n_2 be cylinder probabilities on $(H, \mathscr{C})$. n_1 is said to be *absolutely continuous* with respect to n_2 (written as $n_1 \ll n_2$) if there exists a non-negative function $f \in \mathscr{L}^1(H, \mathscr{C}, n_2)$ such that for all $C \in \mathscr{C}$

$$n_1(C) = \int_C f dn_2.$$

Further, f is defined to be the Radon-Nikodym (R-N) derivative of n_1 with respect to n_2 and is written as $f = dn_1/dn_2$.

Remark 1.1: Uniqueness (up to mod[n]) of the R-N derivative. Let f and f' be elements of $\mathcal{L}(H,\mathcal{C},n_2)$ such that

$$\int_C f dn_2 = n_1(C) = \int_C f' dn_2, \qquad C \in \mathcal{C}.$$

Then

$$\int_C (f-f')dn_2 = 0, \qquad C \in \mathcal{C}$$

and hence by Theorem III.3.9, it follows that $f \equiv f'$mod[n]. Thus, the R-N derivative dn_1/dn_2 is defined uniquely up to mod[n].

Example 1.1: Let m be the canonical Gauss measure on $(H,\mathcal{C})$ and let $h_0 \in H$. Let m_1 be defined by

$$m_1(C) = m(C-h_0), \qquad C \in \mathcal{C}$$

where $C-h_0 = \{h-h_0: h \in C\}$. Then $m_1 << m$ and

$$\frac{dm_1}{dm}(h) = \exp((h,h_0) - \tfrac{1}{2}\{h_0|^2), \qquad h \in H. \tag{1.1}$$

To verify (1.1), we need to prove that the function given on the RHS of (1.1) is integrable, which is easy to see as it is a cylinder function. We must also show that for all $C \in \mathcal{C}$,

$$m_1(C) = \int_C \exp((h,h_0 - \tfrac{1}{2}|h_0|^2)dm(h)$$

This can be verified using the properties of Gaussian

measures on finite dimensional spaces. We will return to this example in Theorem VI.2.3.

The following result lists some useful properties of Radon-Nikodym derivatives in our setup.

<u>Theorem 1.1</u>: Let n, n_1, n_2 be cylinder probabilities on $(H,\mathscr{C})$ such that $n_1 \ll n$, $n_2 \ll n$.

(i) For $a_1 \geq 0$, $a_2 \geq 0$, $a_1 + a_2 = 1$, $(a_1n_1 + a_2n_2) \ll n$ and

$$\frac{d(a_1n_1+a_2n_2)}{dn} = a_1 \frac{dn_1}{dn} + a_2 \frac{dn_2}{dn} . \tag{1.2}$$

(ii) Let (L,Π_0) be a representation of n (Π_0 a probability on $(\Omega_0,\mathscr{A}_0)$) and let R_n be the corresponding n-lifting. Let Π_1 be the probability measure on $(\Omega_0,\mathscr{A}_0)$ defined by $d\Pi_1 = R_n(dn_1/dn)d\Pi_0$. Then (L,Π_1) is a representation of n_1. If R_{n_1} is the corresponding n_1-lifting, then for all cylinder functions f

$$R_n(f) = R_{n_1}(f). \tag{1.3}$$

(iii) Let $g \in \mathscr{L}(H,\mathscr{C},n)$. Then $g \in \mathscr{L}(H,\mathscr{C},n_1)$ and

$$R_{n_1}(g) = R_n(g). \tag{1.4}$$

(iv) Let $g \in \mathscr{L}^*(H,\mathscr{C},n)$. Then $g \in \mathscr{L}^*(H,\mathscr{C},n_1)$.

(v) Let $g \in \mathscr{L}(H,\mathscr{C},n)$. Then

$$g \in \mathscr{L}^1(H,\mathscr{C},n_1) \text{ if and only if } g \cdot \frac{dn_1}{dn} \in \mathscr{L}^1(H,\mathscr{C},n)$$

and in that case

$$\int g\,dn_1 = \int g \cdot \frac{dn_1}{dn} \cdot dn. \tag{1.5}$$

(vi) If $\frac{dn_1}{dn} > 0$ and $R_n\left[\frac{dn_1}{dn}\right] > 0$ a.s. Π_0, then $n \ll n_1$ and

$$\frac{dn}{dn_1} = \left[\frac{dn_1}{dn}\right]^{-1}. \tag{1.6}$$

Furthermore, $n_2 \ll n_1$ with

$$\frac{dn_2}{dn_1} = \frac{dn_2}{dn} \cdot \left[\frac{dn_1}{dn}\right]^{-1}.$$

Proof: (i) follows easily from the definition and linearity of the integral, i.e.,

$$\int f d(a_1 n_1 + a_2 n_2) = a_1 \int f dn_1 + a_2 \int f dn_2.$$

(ii) Since $\Pi_1 \ll \Pi_0$, L can also be regarded as a mapping into $\mathcal{L}(\Omega_0, \mathcal{A}_0, \Pi_1)$. If $C \in \mathcal{C}$ is given by

$$C = \{h: ((h,h_1),\dots,(h,h_k)) \in B\}, \tag{1.7}$$

for $h_i \in H$, $B \in \mathcal{B}(\mathbb{R}^k)$, then

$$\begin{aligned} \Pi_1((L(h_1),\dots,L(h_k)) \in B) &= \int 1_{\{(L(h_1),\dots,L(h_k)) \in B\}} \\ &\quad \times R_n\left[\frac{dn_1}{dn}\right] d\Pi_0 \\ &= \int R_n(1_C) R_n\left[\frac{dn_1}{dn}\right] d\Pi_0 \\ &= \int_C \frac{dn_1}{dn}\, dn = n_1(C). \end{aligned} \tag{1.8}$$

Hence (L,Π_1) is a representation of n_1. Now (1.3) follows from the definition of lifting for cylinder functions.

(iii) As observed above, for $P \in \mathcal{P}$, $R_{n_1}(g \circ P) = R_n(g \circ P)$. Since $\Pi_1 \ll \Pi_0$, convergence (of a sequence or a net) in Π_0-probability implies its convergence in Π_1-probability. Hence $g \in \mathcal{L}(H,\mathcal{C},n_1)$ implies $g \in \mathcal{L}(H,\mathcal{C},n)$ and that (1.4) holds. Assertion (iv) follows by arguments similar to those given above.

(v) $g \in \mathcal{L}^1(H,\mathcal{C},n_1) \iff R_{n_1}(g) \in \mathcal{L}^1(\Omega_0,\mathcal{A}_0,\Pi_1)$

$$\iff R_{n_1}(g)R_{n_1}\left[\frac{dn_1}{dn}\right] \in \mathcal{L}^1(\Omega_0,\mathcal{A}_0,\Pi_0)$$

$$\iff R_n\left[g\,\frac{dn_1}{dn}\right] \in \mathcal{L}^1(\Omega_0,\mathcal{A}_0,\Pi_0)$$

$$\iff g\,\frac{dn_1}{dn} \in \mathcal{L}^1(H,\mathcal{C},n).$$

Also,

$$\int g\,dn_1 = \int R_{n_1}(g)\,d\Pi_1 = \int R_{n_1}(g)R_n\left[\frac{dn_1}{dn}\right]d\Pi_1$$

$$= \int R_n\left[g\,\frac{dn_1}{dn}\right]d\Pi_1 = \int g\,\frac{dn_1}{dn}\,dn.$$

(vi) Let $f = dn_1/dn$. Since $R_n(f) > 0$ a.s. Π_0, by Theorem III.3.7, it follows that $1/f \in \mathcal{L}(H,\mathcal{C},n)$. Using (1.5) with $g = (1/f)1_C$, it follows that

$$\int (1/f)1_C\,dn_1 = \int (1/f)f1_C\,dn = \int 1_C\,dn = n(C).$$

Hence $n \ll n_1$ and $dn/dn_1 = 1/f = [dn_1/dn]^{-1}$. The remaining assertion can be verified similarly. □

As in the countably additive case, given $f \in \mathcal{L}^1(H,\mathcal{C},n)$, such that $f \geq 0$ and $\int f dn = 1$, there exists a (unique) cylindrical probability n_1 such that $dn_1/dn = f$.

Theorem 1.2: Let $f \in \mathcal{L}^1(H,\mathcal{C},n)$, $f \geq 0$, $\int f dn = 1$. Let n_1: $\mathcal{C} \to [0,1]$ be defined by

$$n_1(C) = \int_C f dn. \tag{1.9}$$

Then n_1 is a cylindrical probability on $(H,\mathcal{C})$ and $\frac{dn_1}{dn} = f$.

Proof: Let (L,Π_0) be a representation of n, R_n the corresponding n-lifting. Let Π_1 be given by $d\Pi_1 = R_n(f)d\Pi_0$. Then it can be checked that

$$n_1(h\colon ((h,h_1),\ldots,(h,h_i)) \in B) \tag{1.10}$$

$$= \Pi_1((L(h_1),\ldots,L(h_i)) \in B).$$

This implies that n_1 is a cylinder probability, i.e., n_1 is countably additive on $\mathcal{C}_P$ for all $P \in \mathcal{P}$. □

2. CYLINDRICAL MAPPINGS

In this section, we introduce a class of mappings ϕ: $(H,\mathcal{C},n) \to (H_1,\mathcal{C}_1)$, where H_1 is another Hilbert space and $\mathcal{C}_1 = \mathcal{C}(H_1)$, such that we can associate an induced measure $n\circ\phi^{-1}$ which is a cylinder probability on $(H_1,\mathcal{C}_1)$. These mappings will be called cylindrical mappings. In the next section, we will define the conditional expectation $E_n(f|\phi)$ of a function $f \in \mathcal{L}^1(H,\mathcal{C},n)$ with respect to a cylindrical mapping ϕ.

First suppose that ϕ: $(H,\mathcal{C},n) \to (H_1,\mathcal{C}_1)$ satisfies: for all $P_1 \in \mathcal{P}_1 = \mathcal{P}(H_1)$, there exists $P \in \mathcal{P}$ such that

$$\phi^{-1}(\mathscr{C}_{1,P_1}) \subseteq \mathscr{C}_P, \tag{2.1}$$

where $\mathscr{C}_{1,P_1}$ is the class of all cylinder sets in $\mathscr{C}_1$ with base P_1. If we define

$$n\circ\phi^{-1}(C_1) = n(\phi^{-1}C_1), \qquad C_1 \in \mathscr{C}_1, \tag{2.2}$$

then (2.1) implies that for all $P_1 \in \mathscr{P}_1$, the restriction of $n\circ\phi^{-1}$ to $\mathscr{C}_{1,P_1}$ is countably additive as $n|_{\mathscr{C}_P}$ is countably additive. Hence $n\circ\phi^{-1}$ is a cylinder probability.

In our earlier papers [39, 40], (2.1) was taken as the definition of a cylindrical mapping. However, this is too restrictive a definition as it does not include the case when $\phi(h) = f(h)h_1$ where $f \in \mathscr{L}(H,\mathscr{C},n)$ and f is not a cylinder function and $h_1 \in H_1$ is a fixed element. So we define a cylindrical mapping as follows.

<u>Definition 2.1</u>: A mapping ϕ: $(H,\mathscr{C},n) \to (H_1,\mathscr{C}_1)$ is called a *cylindrical mapping* if for all $h_1 \in H_1$,

$$(h_1,\phi)_1 \in \mathscr{L}(H,\mathscr{C},n)$$

where $(\cdot,\cdot)_1$ is the inner product in H_1.

If ϕ satisfies (2.1), then $(h_1,\phi)_1$ is a cylinder function on H, for if P_1 is the orthogonal projection onto span$\{h_1\}$ and P is obtained as in (2.1), then $(h_1,\phi)_1$ is $\mathscr{C}_P$ measurable. Thus if ϕ satisfies (2.1), ϕ is a cylindrical mapping.

Let ϕ be a cylindrical mapping from $(H,\mathscr{C},n)$ into $(H_1,\mathscr{C}_1)$. To define the induced measure $n\circ\phi^{-1}$, let Ψ: $H_1 \to \mathbb{C}$ be defined by

$$\Psi(h_1) := \int_H \exp(i(h_1,\phi(h))_1)dn(h). \tag{2.3}$$

(Integral of a complex valued function can be defined as usual by decomposing it into real and imaginary parts). Let R_n be the n-lifting corresponding to a representation (L,Π_0) of n. Then by Theorem III.3.5

$$\Psi(h_1) = \int R_n(\exp(i(h_1,\phi(\cdot))_1))d\Pi_0$$

$$= \int \exp(iR_n((h_1,\phi(\cdot))_1))d\Pi_0$$

by Theorem III.3.7 as $x \to \exp(ix)$ is a continuous function. Thus

$$\Psi(h_1) = \int \exp(iL_1(h_1))d\Pi_0 \tag{2.4}$$

where

$$L_1(h_1) = R_n((h_1,\phi(\cdot))_1). \tag{2.5}$$

In view of Theorem III.3.7, L_1 satisfies (III.3.1) and thus by Theorem III.3.1, there exists a unique cylinder probability n_1 on $(H_1,\mathscr{C}_1)$ such that its cylinder characteristic functional is Ψ. Also, (L_1,Π_0) is a representation of n_1.

Definition 2.2: The clinder probability n_1 is called the *measure induced by the cylindrical mapping* ϕ *under* n and will be denoted by $n_1 = n\circ\phi^{-1}$. The representation (L_1,Π_0) will be called the *representation of* n_1 *induced by* ϕ (corresponding to the representation (L,Π_0) of n).

Lemma 2.2: Let R_{n_1} be the n_1-lifting corresponding to (L_1,Π_0). Let f be a continuous cylinder function from $H_1 \to \mathbb{R}$. Then

$$R_{n_1}(f) = R_n(f\circ\phi). \tag{2.6}$$

If further f is bounded, then

$$\int_{H_1} f dn_1 = \int_H (f\circ\phi) dn. \tag{2.7}$$

Proof: Let f be given by

$$f(h_1) = f_1((h_1',h_1)_1,(h_2',h_1)_1,\ldots,(h_k',h_1)_1), \qquad h_1 \in H_1$$

for $h_j' \in H_1$, $k \geq 1$, $f_1: \mathbb{R}^k \to \mathbb{R}$ continuous. Then

$$R_{n_1}(f) = f_1(L_1(h_1'),\ldots,L_1(h_k')) \tag{2.8}$$

$$= f_1(R_n((h_1',\phi)_1),\ldots,R_n((h_k',\phi_1))) = R_n(f\circ\phi)$$

by Theorem III.3.7 as f is continuous. Now (2.7) follows from (2.6) and Theorem III.3.5. □

Remark 2.1: The relation (2.7) justifies the use of the term "measure induced by the mapping ϕ" under n. It is easy to see that if ϕ satisfies (2.1), then (2.6) holds for all cylinder functions, since we do not need to invoke Theorem III.3.7 in (2.8). As a consequence, (2.7) also holds for all bounded cylinder functions.

It is natural to ask if (2.7) holds for all $f \in \mathscr{L}^1(H_1,\mathscr{C}_1,n_1)$. It seems that the answer to this question is negative, though we do not have a counterexample. All the same there seems to be no way of proving that (2.6) holds for all $f \in \mathscr{L}(H_1,\mathscr{C}_1,n_1)$. However, we do need this property when we define conditional expectation. To get around this difficulty, we shall impose the restriction that the function in question satisfy (2.6). Let us introduce the following class for this purpose.

Let $\mathcal{U}(H,\mathcal{C},n;\phi) := \{g \in \mathcal{L}(H_1,\mathcal{C}_1,n_1)$: $g\circ\phi \in \mathcal{L}(H,\mathcal{C},n)$ and $R_{n_1}(g) = R_n(g\circ\phi)\}$.

When the space $(H,\mathcal{C},n)$ is clear from the context, we will write $\mathcal{U}(\phi)$ for $\mathcal{U}(H,\mathcal{C},n;\phi)$. So Lemma 2.1 says that continuous cylinder functions belong to $\mathcal{U}(\phi)$. Furthermore if ϕ satisfies (2.1), then all cylinder functions belong to $\mathcal{U}(\phi)$.

Since the mapping R_n is multiplicative, if $g_1,g_2 \in \mathcal{U}(\phi)$ then so does $g_1\cdot g_2$. Indeed $R_{n_1}(g_1\cdot g_2) = R_{n_1}(g_1)R_{n_1}(g_2)$ and $R_n((g_1\cdot g_2)\circ\phi) = R_n((g_1\circ\phi)(g_2\circ\phi)) = R_n(g_1\circ\phi)R_n(g_2\circ\phi)$ and thus $g_1\cdot g_2 \in \mathcal{U}(\phi)$. Similarly, if g_1, $g_2 \in \mathcal{U}(\phi)$, then $g_1 + g_2 \in \mathcal{U}(\phi)$.

Remark 2.2: Before we proceed, we would like to convince ourselves (and the reader) that the class $\mathcal{U}(\phi)$ does not depend on the choice of the representation of n. Suppose $(L,\Pi_0),(L',\Pi_0')$ are two representations of n and R_n,R_n' are the corresponding n-liftings. Let

$$L_1(h_1) = R_{n_1}((h_1,\phi(\cdot))_1)$$

and

$$L_1'(h_1) = R_{n_1}'((h_1,\phi(\cdot))_1), \quad h_1 \in H_1$$

be the representations of n_1 induced by ϕ (under (L,Π_0) and (L',Π_0') respectively) and let R_{n_1} and R_{n_1}' be the corresponding n_1-liftings.

For $\{\xi_\alpha: \alpha \in \Lambda\} \subseteq \mathcal{L}(\Omega_1,\mathcal{A}_1,\Pi_1)$ and $\{\eta_\alpha: \alpha \in \Lambda\} \subseteq \mathcal{L}(\Omega_2,\mathcal{A}_2,\Pi_2)$, where $(\Omega_i,\mathcal{A}_i,\Pi_i)$, i = 1,2 are countably additive probability spaces and Λ is an arbitrary index set, we will use the notation

$$\{\xi_\alpha;\ \alpha \in \Delta\} \overset{d}{=} \{\eta_\alpha;\ \alpha \in \Delta\}$$

to denote the equality of joint distributions of $\{\xi_\alpha:\ \alpha \in \Delta\}$ and $\{\eta_\alpha:\ \alpha \in \Delta\}$, i.e.,

$$\Pi_1((\xi_{\alpha_1},\xi_{\alpha_2},\ldots,\xi_{\alpha_j}\} \in B) = \Pi_2((\eta_{\alpha_1},\eta_{\alpha_2},\ldots,\eta_{\alpha_j}) \in B)$$

for all $\alpha_1,\alpha_2,\ldots,\alpha_j \in \Delta$, $B \in \mathcal{B}(\mathbb{R}^j)$ and $j \geq 1$.

Since (L,Π_0) and (L',Π_0') are both representations of n, we have

$$\{L(h):\ h \in H\} \overset{d}{=} (L'(h):\ h \in H\}.$$

The definition of lifting for cylinder functions now implies

$$\{R_n(f):\ f:\ H \to \mathbb{R} \text{ cylinder}\} \overset{d}{=} \{R_n'(f):\ f:\ H \to \mathbb{R} \text{ cylinder}\}$$

For $f \in \mathcal{L}(H,\mathcal{C},n)$, $R_n(f)(R_n'(f))$ is the limit in probability of $\{R_n(f\circ P):\ P \in \mathcal{P}\}$ ($\{R_n'(f\circ P):\ P \in \mathcal{P}\}$ respectively) and thus

$$\{R_n(f):\ f \in \mathcal{L}(H,\mathcal{C},n)\} \overset{d}{=} \{R_n'(f):\ f \in \mathcal{L}(H,\mathcal{C},n)\}.$$

The definitions of L_1,L_1' now imply

$$\{R_n(f):\ f \in \mathcal{L}(H,\mathcal{C},n);\ L_1(h_1):\ h_1 \in H_1\}$$

$$\overset{d}{=} \{R_n'(f):\ f \in \mathcal{L}(H,\mathcal{C},n;\ L_1'(h_1):\ h_1 \in H_1\}.$$

Proceeding as above, this gives

$$\{R_n(f):\ f \in \mathcal{L}(H,\mathcal{C},n);\ R_{n_1}(g):\ g \in \mathcal{L}(H_1,\mathcal{C}_1,n_1)\}$$

$$\stackrel{d}{=} \{R'_n(f): f \in \mathcal{L}(H,\mathcal{C},n);\ R'_{n_1}(g): g \in \mathcal{L}(H_1,\mathcal{C}_1,n_1)\}$$

In particular, for all $g \in \mathcal{L}(H_1,\mathcal{C}_1,n_1)$ such that $g\circ\phi \in \mathcal{L}(H,\mathcal{C},n)$, we have

$$\{R_n(g\circ\phi), R_{n_1}(g)\} \stackrel{d}{=} \{R'_n(g\circ\phi), R'_{n_1}(g)\}$$

and hence $R_n(g\circ\phi) = R_{n_1}(g)$ a.e. Π_0 if and only if $R'_n(g\circ\phi) = R'_{n_1}(g)$ a.s. Π'_0. Thus the class $\mathcal{U}(\phi)$ does not depend upon the choice of the representation of n used in its definition.

<u>Examples of Cylindrical Mappings</u>

Let Q be an orthogonal projection on H with Range Q = H_1. H_1 itself is a Hilbert space with inner product

$$(h_1,h_1')_1 = (h_1,h_1'), \quad h_1,h_1' \in H_1.$$

Let $\mathcal{C}_1 = \mathcal{C}(H_1)$.

<u>Lemma 2.2</u>: Q satisfies (2.1). Hence for any cylinder probability n on $(H,\mathcal{C})$, Q is a cylindrical mapping from $(H,\mathcal{C},n) \to (H_1,\mathcal{C}_1)$.

<u>Proof</u>: Let $P_1 \in \mathcal{P}_1 = \mathcal{P}(H_1)$ and let $P = P_1Q$. We will first show that $P \in \mathcal{P}$. Let $h \in H$ and $h_1 = Ph$. Then $P^2h = Ph_1 = P_1Qh_1$. Now $h_1 \in$ Range $P_1 \subseteq$ Range Q. Thus $Qh_1 = h_1$, $P_1h_1 = h_1$. Hence $P^2h = h_1 = Ph$. So $P^2 = P$. Also for $h,k \in H$

$$\begin{aligned}(Ph,k) &= (P_1Qh,k)\\ &= (P_1Qh,Qk) \quad \text{as } P_1Qh \in \text{Range } Q,\\ &= (P_1Qh,P_1Qk) \quad \text{as } P_1Qh \in \text{Range } P_1\end{aligned}$$

$$= (Ph, Pk)$$

$$= (h, Pk) \qquad \text{(retracing the steps)}.$$

Now let $C_1 \in \mathscr{C}_{1,P_1}$. Then $C_1 = P_1^{-1}B$ for some $B \in \mathscr{B}(\text{Range } P_1)$. Now

$$C = Q^{-1}C_1 = Q^{-1}P_1^{-1}B = (P_1Q)^{-1}B = P^{-1}B \in \mathscr{C}_P.$$

Thus

$$Q^{-1}\mathscr{C}_{P_1} \subset \mathscr{C}_P.$$

where $P = P_1Q$. Hence Q satisfies (2.1). □

The next lemma gives a sufficient condition for a function to belong to $\mathscr{U}(Q)$. The proof sheds some light on why $\mathscr{U}(Q) = \mathscr{U}(H, \mathscr{C}, n; Q)$ need not be equal to $\mathscr{L}(H_1, \mathscr{C}_1, n \circ Q^{-1})$.

<u>Proposition 2.3</u>: Let n be a cylinder probability on $(H, \mathscr{C})$ and Q be an orthogonal projection with Range H_1. Let $n_1 = n \circ Q^{-1}$ be the induced cylinder probability on $(H_1, \mathscr{C}_1)$ where $\mathscr{C}_1 = \mathscr{C}(H_1)$.

Let $g_1 : H_1 \to \mathbb{R}$ be any function and let $g(h) = g_1(Qh)$. If $g \in \mathscr{L}^*(H, \mathscr{C}, n)$, then $g_1 \in \mathscr{U}(Q) = \mathscr{U}(H, \mathscr{C}, n; Q)$ and $g \in \mathscr{L}^*(H_1, \mathscr{C}_1, n_1)$.

<u>Proof</u>: Let (L, Π_0) be any representation of n. The induced representation (L_1, Π_0) of n_1 is given by

$$L_1(h_1) = R_n((h_1, Q\cdot)) = R_n((Qh_1, \cdot)) \tag{2.9}$$

$$= R_n((h_1, \cdot)) = L(h_1).$$

Here R_n is the n-lifting corresponding to (L,Π_0). Let R_{n_1} be the n_1-lifting corresponding to (L_1,Π_0).

Let $P'_j \in \mathcal{P}_1 = \mathcal{P}(H_1)$ be such that $P'_j \overset{s}{\to} I_{H_1}$, where I_{H_1} is the identity in H_1. Let H_2 be the orthogonal complement of H_1 and let $Q^{\perp}$ be the orthogonal projection onto H_2.

Let $P''_j \in \mathcal{P}_2 = \mathcal{P}(H_2)$ be any sequence such that $P''_j \to I_{H_2}$ and let

$$P_j = P'_jQ + P''_jQ^{\perp}.$$

Proceeding as in the proof of Lemma 2.2 and noting that $(P'_jQ)(P''_jQ^{\perp}) = 0$, it can be shown that $P_j \in \mathcal{P}$. Also,

$$QP_j = QP'_jQ + QP''_jQ^{\perp} = P'_jQ. \tag{2.10}$$

Thus

$$g(P_jh) = g_1(QP_jh) = g_1(P'_jQh) = g_1\circ P'_j(Qh). \tag{2.11}$$

Since $g_1\circ P'_j$ is a cylinder function and Q satisfies (2.1) by Lemma 2.1 and the remark following it, we have

$$R_{n_1}(g_1\circ P'_j) = R_n(g_1\circ P'_j\circ Q) = R_n(g\circ P_j). \tag{2.12}$$

Since $P_j \overset{s}{\to} I_H$ and $g \in \mathcal{L}^*(H,\mathcal{C},n)$, it follows that $R_n(g\circ P_j) \to R_n(g)$ in Π_0-probability. In view of (2.12), this gives

$$R_{n_1}(g_1\circ P'_j) \to R_n(g) \qquad \text{in } \Pi_0\text{-probability}. \tag{2.13}$$

Since (2.13) holds for any $P'_j \in \mathcal{P}_1$, $P'_j \overset{s}{\to} I_{H_1}$, it follows that $g \in \mathcal{L}^*(H_1, \mathcal{C}_1, n_1)$ and that

$$R_{n_1}(g) = R_n(g_1) = R_n(g \circ Q).$$

Hence $g \in \mathcal{U}(Q)$. □

Suppose we try to prove that $\mathcal{L}(H_1, \mathcal{C}_1, n_1) = \mathcal{U}(Q)$. Given $g_1 \in \mathcal{L}(H_1, \mathcal{C}_1, n_1)$, all we can conclude is that $\exists P'_j \in \mathcal{P}_1$ such that for all $\overline{P}'_j \in \mathcal{P}$, $P'_j \leq \overline{P}'_j$,

$$R_n(g_1 \circ \overline{P}'_j) \to R_{n_1}(g_1).$$

This will tell us that $P_j \in \mathcal{P}$, such that for all $P_{1,j} \in \mathcal{P}$, $P_j \leq P_{1,j}$ *and* $QP_{1,j} = P_{1,j}Q$, for $g = g_1 \circ Q$, we have

$$R_n(g \circ P_{1,j}) \to R_{n_1}(g_1).$$

This can be proved following the steps in the proof given above. We cannot carry through this argument for $P_{1,j} \in \mathcal{P}$ for which $QP_{1,j} \neq P_{1,j}Q$!

We now give an example of a cylindrical mapping ϕ from H into itself for which $\mathcal{L}(H, \mathcal{C}, n \circ \phi^{-1}) = \mathcal{U}(\phi)$.

Let U be a unitary operator from $H \to H$.

<u>Proposition 2.4</u>: U satisfies (2.1). Thus for any cylinder probability n on $(H, \mathcal{C})$, $U: (H, \mathcal{C}, n) \to (H, \mathcal{C})$ is a cylindrical mapping. Further, if $n_1 = n \circ U^{-1}$, then

$$f \in \mathcal{L}(H, \mathcal{C}, n_1) \text{ if and only if } f \circ U \in \mathcal{L}(H, \mathcal{C}, n) \tag{2.14}$$

so that

$$\mathcal{U}(U) = \mathcal{L}(H, \mathcal{C}, n_1).$$

Also (2.14) remains valid if $\mathcal{L}$ is replaced by $\mathcal{L}^*$, $\mathcal{L}^1$, $\mathcal{L}^{1*}$ respectively.

Proof: Let $P \in \mathcal{P}$ and let $\overline{P}$ be defined by

$$\overline{P} = U^*PU. \tag{2.15}$$

Then $\quad \overline{P}^* = (U^*PU)^* = U^*P^*(U^*)^* = U^*PU = \overline{P}$

and $\quad (\overline{P})^2 = U^*PU \cdot U^*PU = U^*PU.$

Thus $\quad \overline{P} \in \mathcal{P}.$

We will prove that for all $P \in \mathcal{P}$,

$$U^{-1}\mathcal{C}_P \subseteq \mathcal{C}_{\overline{P}} \quad \text{(in fact, equality holds)} \tag{2.16}$$

where $\overline{P}$ is given by (2.15). Let $C \in \mathcal{C}_P$ be given by

$$C = \{h: ((h,h_1),\ldots,(h,h_k)) \in B\}$$

for $h_i \in PH$, $B \in \mathcal{B}(\mathbb{R}^k)$. Then

$$U^{-1}C = \{h: Uh \in C\} = \{h: ((Uh,h_1),\ldots,(Uh,h_k)) \in B\}$$
$$= \{h: ((h,U^*h_1),\ldots,(h,U^*h_k) \in B\}.$$

Now $\overline{P}U^*h_j = U^*PU(U^*h_j) = U^*Ph_j = U^*h_j$. So $U^*h_j \in \overline{P}H$ and hence $U^{-1}C \in \mathcal{C}_{\overline{P}}$. Let R_n be the n-lifting corresponding to a representation (L,Π_0) of n and let (L_1,Π_0) be the representation induced by U with R_{n_1} the corresponding lifting. Since U satisfies (2.1), by Lemma 2.1 and the remark following it, we have

$$R_{n_1}(g) = R_n(g\circ U) \tag{2.17}$$

for all cylinder functions $g: H \to \mathbb{R}$. Suppose $f \in \mathcal{L}(H,\mathcal{C},n_1)$. Then

$$\begin{aligned} R_{n_1}(f\circ P) &= R_n(f\circ P\circ U) && \text{by (2.17)} \\ &= R_n(f\circ U\circ \overline{P}) && \text{for } \overline{P} \text{ as in (2.15).} \end{aligned} \tag{2.18}$$

Since $f \in \mathcal{L}(H,\mathcal{C},n_1)$, there exists $P_i \subseteq \mathcal{P}$, $P_i \uparrow I$ such that for $P_i' \in \mathcal{P}$, $P_i \leq P_i'$, we have (see Theorem III.3.6)

$$R_{n_1}(f\circ P_i') \to R_{n_1}(f) \quad \text{in } \Pi_0\text{-probability.} \tag{2.19}$$

If $\overline{P}_i = U^*P_iU$, then $\overline{P}_i \uparrow I$ and if $\overline{P}_i' \in P$, $\overline{P}_i \leq \overline{P}_i'$, then $P_i' = U\overline{P}_i'U^*$ satisfy $P_i \leq P_i'$ so that (2.19) holds. Hence

$$\begin{aligned} R_n(f\circ U\circ \overline{P}_i') &= R_{n_1}(f\circ P_i') && \text{by (2.18)} \\ &\to R_{n_1}(f) && \text{in } \Pi_0\text{-probability.} \end{aligned} \tag{2.20}$$

By Theorem III.3.6, this implies $f\circ U \in \mathcal{L}(H,\mathcal{C},n)$ and

$$R_n(f\circ U) = R_{n_1}(f). \tag{2.21}$$

Hence

$$\mathcal{L}(H,\mathcal{C},n_1) = \mathcal{U}(Q).$$

We have proved one half of (2.14), namely

$$f \in \mathcal{L}(H,\mathcal{C},n_1) \Rightarrow f\circ U \in \mathcal{L}(H,\mathcal{C},n). \tag{2.22}$$

For the other part, observe that

$$U^*: (H,\mathscr{C},n_1) \to (H,\mathscr{C})$$

if also a cylindrical mapping (as it is a unitary operator) and $n_1 \circ U^{*-1} = n$. Thus, (2.22) (for U^*) gives

$$g \in \mathscr{L}(H,\mathscr{C},n) \Rightarrow g \circ U^* \in \mathscr{L}(H_1,\mathscr{C}_1,n_1). \tag{2.23}$$

Using (2.23) for $g = f \circ U$

$$f \circ U \in \mathscr{L}(H,\mathscr{C},n) \Rightarrow f = f \circ U \circ U^* \in \mathscr{L}(H,\mathscr{C},n_1). \tag{2.24}$$

Now (2.22) and (2.24) imply (2.14).

The other assertions about $\mathscr{L}^*$, $\mathscr{L}^1$, $\mathscr{L}^{1*}$ can be proved similarly. □

Let m be the canonical Gauss measure on H and U, a unitary operator. Let $m_1 = m \circ U^{-1}$. Then

$$\int e^{i(h,h_1)} dm_1 = \int e^{i(h,Uh_2)} dm(h_2) = \int e^{i(U^*h,h_2)} dm(h_2)$$

$$= e^{-|U^*h|^2} = e^{-|h|^2}.$$

Thus $m_1 = m$. In other words, $m \circ U^{-1} = m$ for all unitary operators $U: H \to H$, or m is invariant under 'rotations'. We will now prove that $\mathscr{L}(H,\mathscr{C},m)$ is also invariant under rotations.

<u>Proposition 2.5</u>: Let U be a unitary mapping $H \to H$. Then

$$f \in \mathscr{L}(H,\mathscr{C},m) \Rightarrow f \circ U \in \mathscr{L}(H,\mathscr{C},m). \tag{2.25}$$

Also (2.25) holds if $\mathscr{L}$ is replaced by $\mathscr{L}^*$, $\mathscr{L}^1$, $\mathscr{L}^{1*}$.

Proof: As already observed, $m \circ U^{-1} = m$. Now all the assertions follow from Proposition 2.4. □

Remark 2.3: Let H_1, H_2 be two Hilbert spaces and let $\mathscr{C}_i = \mathscr{C}(H_i)$, $\mathscr{P}_i = \mathscr{P}(H_i)$, $i = 1,2$. Let I be an isometry between H_1 and H_2, $I: H_1 \to H_2$. Let I^{-1} be the inverse of I. Propositions 2.4 and 2.5 are true if U is replaced by I and H is replaced by H_1 or H_2 suitably. We state these results in this context.

(i) $I: (H_1, \mathscr{C}_1, n_1) \to (H_2, \mathscr{C}_2)$ is a cylindrical mapping.

(ii) For a cylinder probability n_1 on $(H_1, \mathscr{C}_1)$, if $n_2 = n_1 \circ (I)^{-1}$, then

$$f \in \mathscr{L}(H_1, \mathscr{C}_1, n_1) \text{ if and only if } f \circ I \in \mathscr{L}(H_2, \mathscr{C}_2, n_2) \tag{2.26}$$

and

$$\mathscr{U}(I) = \mathscr{L}(H_2, \mathscr{C}_2, n_2).$$

(iii) The relation (2.26) is valid if $\mathscr{L}$ is replaced by $\mathscr{L}^*$, $\mathscr{L}^1$, $\mathscr{L}^{1*}$.

(iv) Let m_i be the canonical Gauss measure on H_i, $i = 1,2$. Then $m_2 = m_1 \circ (I)^{-1}$.

These results can be proved exactly as in the case $H_1 = H_2$ and $I = U$. The role of U^* is played by I^{-1}. The rest of the proof is essentially the same as before.

We now look at the composition of two cylindrical mappings. Is the composition a cylindrical mapping? Also, how are the $\mathscr{U}$ classes related? In general, the composition of two cylindrical mappings need not be a cylindrical mapping, but is so under the conditions imposed in the next result.

<u>Theorem 2.6</u>: Let ϕ_1: $(H,\mathcal{C},n) \to (H_1,\mathcal{C}_1)$ be a cylindrical mapping with $n_1 = n \circ \phi_1^{-1}$. Let ϕ_2: $(H_1,\mathcal{C}_1,n_1) \to (H_2,\mathcal{C}_2)$ be a cylindrical mapping and $n_2 = n_1 \circ \phi_2^{-1}$. Suppose that for all $h_2 \in H_2$ (the inner product in H_2 is $(\cdot,\cdot)_2$),

$$(h_2,\phi_2(\cdot))_2 \in \mathcal{U}(\phi_1) = \mathcal{U}(H,\mathcal{C},n;\phi_1). \tag{2.27}$$

Then $\phi = \phi_2 \circ \phi_1$ is a cylindrical mapping and $n \circ \phi^{-1} = n_2$. Further, if $f \in \mathcal{U}(\phi_2)$ and $f \circ \phi_2 \in \mathcal{U}(\phi_1)$, then

$$f \in \mathcal{U}(\phi).$$

Also, (2.27) holds if both ϕ_1,ϕ_2 satisfy (2.1).

<u>Proof</u>: Let (L,Π_0) be a representation of n, let (L_1,Π_0) be the representation of n_1 induced by ϕ_1 and let (L_2,Π_0) be the representation of n_2 induced by ϕ_2. Since by assumption $g(h_1) = (h_2,\phi_2(h_1))_2 \in \mathcal{U}(\phi_1)$, we have

$$g \circ \phi_1 \in \mathcal{L}(H,\mathcal{C},n)$$

and

$$R_n(g \circ \phi_1) = R_{n_1}(g).$$

But $g \circ \phi_1(h) = (h_2,\phi_2 \circ \phi_1(h))_2 = (h_2,\phi(h))_2$. Thus ϕ is a cylindrical mapping and

$$R_n((h_2,\phi(\cdot))_2) = R_n(g \circ \phi_1) = R_{n_1}(g) = L_2(h_2).$$

so that (L_2,Π_0) is also the representation induced by ϕ. This implies $n \circ \phi^{-1} = n_2 = n_1 \circ \phi_1^{-1}$. Now if $f \in \mathcal{U}(\phi_2)$ and

$f \circ \phi_2 \in \mathscr{U}(\phi_1)$, then

$$R_{n_2}(f) = R_{n_1}(f \circ \phi_2) = R_n(f \circ \phi_2 \circ \phi_1) = R_n(f \circ \phi).$$

Thus, $f \in \mathscr{U}(\phi)$.

If ϕ_1, ϕ_2 satisfy (2.1), then $(h_2, \phi_2(\cdot))$ is a cylinder function on H_1 and hence belongs to $\mathscr{U}(\phi_1)$. Hence (2.27) holds. □

3. CONDITIONAL EXPECTATION

We shall motivate the definition by an informal discussion of the options open to us.

Let $\phi: (H, \mathscr{C}, n) \to (H_1, \mathscr{C}_1)$ be a cylindrical mapping and let $g \in \mathscr{L}^1(H, \mathscr{C}, n)$. We want to define the conditional expectation of g given ϕ. Let $n_1 = n \circ \phi^{-1}$.

To begin with, let us assume that ϕ satisfies (2.1). As in the countably additive probability theory, if there exists $g_1 \in \mathscr{L}^1(H_1, \mathscr{C}_1, n_1)$ such that for all $C_1 \in \mathscr{C}_1$,

$$\int_H g(h) 1_{C_1}(\phi(h)) dn(h) = \int_{H_1} g_1(h_1) 1_{C_1}(h_1) dn_1(h_1), \tag{3.1}$$

then we may define $g_1 \circ \phi$ to be the conditional expectation of g given ϕ. Alternatively, if there exists a measurable function $g_1: H_1 \to \mathbb{R}$ such that $g_1 \circ \phi \in \mathscr{L}^1(H, \mathscr{C}, n)$ and for all $C_1 \in \mathscr{C}_1$,

$$\int_H g(h) 1_{C_1}(\phi(h)) dn(h) = \int_H g_1 \circ \phi(h) 1_{C_1}(\phi(h)) dn(h), \tag{3.2}$$

then we can define $g_1 \circ \phi$ to be the conditional expectation of g given ϕ. In the countably additive probability theory, the conditions (3.1) and (3.2) are equivalent because of the change of variable formula. We have seen in Section 2 that in general

$$\int g_1(\phi)\cdot 1_{C_1}(\phi)dn = \int g_1\cdot 1_{C_1}dn_1 \tag{3.3}$$

may not be true for all $g_1 \in \mathcal{L}^1(H_1,\mathcal{C}_1,n_1)$ and hence the two conditions (3.1) and (3.2) may not be equivalent. Note that if $g_1 \in \mathcal{U}(\phi)$ and $g_1 \in \mathcal{L}^1(H_1,\mathcal{C}_1,n_1)$ then (3.3) holds. In our earlier papers, we had adopted (3.1) as the defining relation. However, with this definition, we cannot assert, for example, that $g_1(\phi)$ minimizes $E_n(g-f(\phi))^2$, $f \in \mathcal{L}^1(H_1,\mathcal{C}_1,n_1)$, if $E_n(g^2) < \infty$. The latter is a desirable property and its proof depends on both (3.1) and (3.2). Also, in any case given g,ϕ, we cannot assert the existence of g_1 satisfying (3.1). Thus we can as well add the requirement that $g_1 \in \mathcal{U}(\phi)$ in the definition so that both (3.1) and (3.2) are equivalent as observed earlier.

In the preceding discussion, we had assumed that ϕ satisfies (2.1) so that $1_{C_1}(\phi) \in \mathcal{L}^1(H,\mathcal{C},n)$ for all $C_1 \in \mathcal{C}_1$ and $1_{C_1} \in \mathcal{U}(\phi)$. However, for a general cylindrical mapping as defined in Section 2, $1_{C_1}(\phi)$ may not belong to $\mathcal{L}^1(H,\mathcal{C},n)$ for a cylinder set C_1 in H_1. In this case, the role of $1_{C_1}(\phi)$ in (3.1) will be played by $f(\phi)$, where f is a bounded, continuous cylinder function on H_1. That this determines g_1 uniquely (up to mod[n]) is the content of the next lemma.

<u>Lemma 3.1</u>: Let n be a cylinder probability on $(H,\mathcal{C})$ and let

R_n be an n-lifting corresponding to a representation (L,Π_0) of n.

(i) Let $C \in \mathscr{C}$. Then there exists a sequence of continuous cylinder functions $\{f_j\}$ on H, $0 \leq f_j \leq 1$ such that

$$R_n(f_j) \to R_n(1_C) \quad \text{a.s. } \Pi_0. \tag{3.4}$$

(ii) Let $g_1, g_2 \in \mathscr{L}^1(H,\mathscr{C},n)$ be such that for all bounded continuous cylinder functions f on H,

$$\int g_1 f dn = \int g_2 f dn. \tag{3.5}$$

Then for all $C \in \mathscr{C}$, we have

$$\int_C g_1 dn = \int_C g_2 dn \tag{3.6}$$

and as a consequence, $g_1 = g_2 \bmod[n]$.

<u>Proof</u>: (i) Let $C \in \mathscr{C}_P$ be of the form

$$C = \{h: ((h,h_1),\ldots,(h,h_k)) \in B\}$$

where $h_i \in H$, $B \in \mathscr{B}(\mathbb{R}^k)$ and $k \geq 1$. Then by definition,

$$R_n(1_C) = 1_D$$

where

$$D = \{\omega: (L(h_1)(\omega),\ldots,L(h_k)(\omega)) \in B\}.$$

Let $\mu = \Pi_0 \circ (L(h_1),\ldots,L(h_k))^{-1} \in \mathscr{M}_0(\mathbb{R}^k)$. Now, by Luzin's Theorem we can get a sequence of continuous functions F_j on $\mathbb{R}$ such that $F_j(x) \to 1_B(x)$ a.e. μ. Without loss of generality, we can assume that $0 \leq F_j(x) \leq 1$. Now let $f_j(h) =$

$F_j((h,h_1),\ldots,(h,h_k))$. Then

$$R_n(f_j) = F_j(L(h_1),\ldots,L(h_k)).$$

Now, the choice of μ and $\{F_j\}$ implies the required assertion (3.4).

For (ii), given $C \in \mathscr{C}$, let $\{f_j\}$ be as in the part (i). Then, by (3.5) for f_j, we have (using Theorem III.3.5)

$$\int R_n(g_1)R_n(f_j)d\Pi_0 = \int R_n(g_2)R_n(f_j)d\Pi_0.$$

Here, $R_n(g_1)$, $R_n(g_2) \in \mathscr{L}^1(\Omega_0,\mathscr{A}_0,\Pi_0)$ and $0 \le R_n(f_j) \le 1$ as $0 \le f_j \le 1$. Hence by the dominated convergence theorem, we have

$$\int R_n(g_1)R_n(1_C)d\Pi_0 = \int R_n(g_2)R_n(1_C)d\Pi_0$$

which is the same as (3.6), again invoking Theorem III.3.5. As proved in Theorem III.3.9, (3.6) implies $g_1 = g_2$ mod[n].

□

We are now in a position to make the following definition.

<u>Definition 3.1</u>: Let ϕ: $(H,\mathscr{C},n) \to (H_1,\mathscr{C}_1)$ be a cylindrical mapping and let $n_1 = n\circ\phi^{-1}$. Let $g \in \mathscr{L}^1(H,\mathscr{C},n)$.

If there exists $g_1 \in \mathscr{U}(H,\mathscr{C},n;\phi)$ with $g_1 \in \mathscr{L}^1(H_1,\mathscr{C}_1,n_1)$ such that for all bounded continuous cylinder functions f on H,

$$\int_H g(h)f(\phi(h))dn(h) = \int_{H_1} g_1(h_1)f(h_1)dn_1(h_1), \tag{3.7}$$

then $g_1\circ\phi$ is defined to be the conditional expectation of g given ϕ and is expressed as

$$E_n(g|\phi) = g_1 \circ \phi. \tag{3.8}$$

<u>Remark 3.1</u>: It should be noted that we are not asserting that the conditional expectation $E_n(g|\phi)$ exists for all $g \in \mathscr{L}^1(H,\mathscr{C},n)$. We do not have an example of a function $f \in \mathscr{L}^1(H,\mathscr{C},n)$ for which $E_n(g|\phi)$ does not exist, but we strongly believe that, in general, it is not possible to prove that conditional expectation as defined above always exists.

<u>Remark 3.2</u> (Uniqueness): Let $g \in \mathscr{L}^1(H,\mathscr{C},n)$ be such that $E_n(g|\phi)$ exists. Then it is unique up to mod[n]. For if $g_1, g_2 \in \mathscr{U}(\phi)$ satisfy (3.7), then by Lemma 3.1, $g_1 = g_2$ mod$[n_1]$. By Theorem III.3.9, we have $R_{n_1}(g_1) = R_{n_1}(g_2)$. Since both g_1, g_2 belong to $\mathscr{U}(\phi)$, we conclude that $R_n(g_1 \circ \phi) = R_n(g_2 \circ \phi)$ and as a consequence $g_1 \circ \phi = g_2 \circ \phi$ mod[n] (again we need to invoke Theorem III.3.9).

<u>Remark 3.3</u>: In view of the requirement $g_1 \in \mathscr{U}(\phi)$, condition (3.7) is equivalent to

$$\int_H g(h) f(\phi(h)) dn(h) = \int_H g_1(\phi(h)) f(\phi(h)) dn(h) \tag{3.9}$$

which may therefore be taken as the defining relation.

<u>Remark 3.4</u>: Let ϕ satisfy (2.1). Then the relation (3.7) implies the relation (3.1). To see this, let R_n, R_{n_1} be the n,n_1-liftings as in Lemma 2.2. Given $C_1 \in \mathscr{C}_1$, let $\{f_j\}$ be a sequence of bounded continuous cylinder functions such that $R_{n_1}(f_j) \to R_{n_1}(1_{C_1})$ a.s. Π_0 (the existence of such a sequence is proved in Lemma 3.1). By Lemma 2.2, we have for all j

$$R_n(f_j \circ \phi) = R_{n_1}(f_j).$$

Since ϕ satisfies (2.1), by Remark 2.1, we also have

$$R_n(1_{C_1}\circ\phi) = R_{n_1}(1_{C_1}) \tag{3.10}$$

and hence by the choice of $\{f_j\}$, we have

$$R_n(f_j\circ\phi) \to R_n(1_{C_1}\circ\phi). \tag{3.11}$$

Now (3.11) and the dominated convergence theorem give

$$\int R_n(g)R_n(f_j\circ\phi)d\Pi_0 \to \int R_n(g)R_n(1_{C_1}\circ\phi)d\Pi_0$$

which in view of Theorem III.3.5 is the same as

$$\int gf_j\circ(\phi)dn \to \int g1_{C_1}(\phi)dn. \tag{3.12}$$

We also have, as seen in Lemma 3.1,

$$\int g_1f_jdn_1 \to \int g_1 1_{C_1}dn_1. \tag{3.13}$$

Now the fact that (3.7) holds for $f = f_j$ and (3.12), (3.13) imply that (3.1) holds. It is easy to see that the converse is true, i.e., (3.1) implies (3.7). □

Before we go on to establish some important properties of the conditional expectation operation, we will connect our notion of conditional expectation with the corresponding notion in the countably additive probability theory.

Let R_n be the n-lifting corresponding to a representation (L,Π_0) of n with the underlying representation space $(\Omega_0,\mathcal{A}_0)$. Let ϕ: $(H,\mathcal{C},n) \to (H_1,\mathcal{C}_1)$ be a cylindrical mapping, $n_1 = n\circ\phi^{-1}$ the induced measure and (L_1,Π_0) the representation of n_1 induced by ϕ (corresponding to (L,Π_0)). Let

$$\mathcal{F} = \{R_n(f\circ\phi) : \text{ f continuous cylinder function on } H_1\}$$

and

$$\mathcal{D}_\phi := \sigma(\mathcal{F}),$$

i.e., $\mathcal{D}_\phi$ is the smallest σ-field on Ω_0 with respect to which the family $\mathcal{F}$ is measurable. Let $\overline{\mathcal{D}}_\phi$ be the σ-field generated by $\mathcal{D}_\phi$ and Π_0-null sets.

Theorem 3.2:

(i) If $g_1 \in \mathcal{L}(H_1, \mathcal{C}_1, n_1)$, then $R_{n_1}(g_1)$ is $\overline{\mathcal{D}}_\phi$-measurable.

(ii) If $g_1 \in \mathcal{U}(H, \mathcal{C}, n; \phi)$, then $R_n(g_1\circ\phi)$ is $\overline{\mathcal{D}}_\phi$-measurable.

(iii) Let $g \in \mathcal{L}^1(H, \mathcal{C}, n)$ be such that $E_n(g|\phi)$ exists. Then we have

$$R_n[E_n(g|\phi)] = E_{\Pi_0}[R_n(g)|\overline{\mathcal{D}}_\phi], \quad \text{a.s. } \Pi_0. \tag{3.14}$$

(iv) Let $g \in \mathcal{L}^1(H, \mathcal{C}, n)$. If there exists a $g_1 \in \mathcal{U}(\phi)$ such that

$$R_n[g_1\circ\phi] = E_{\Pi_0}[R_n(g)|\overline{\mathcal{D}}_\phi], \quad \text{a.s. } \Pi_0, \tag{3.15}$$

then $E_n(g|\phi)$ exists and is equal to $g_1\circ\phi$.

Proof: Let f be a continuous cylinder function on H_1. Then by Lemma 2.2, $R_{n_1}(f) = R_n(f\circ\phi)$ and hence $R_{n_1}(f)$ is $\mathcal{D}_\phi$-measurable. Now, by Lemma 3.1, given $C_1 \in \mathcal{C}_1$, we can get bounded continuous cylinder functions f_j such that $R_{n_1}(f_j) \to R_{n_1}(1_{C_1})$ a.s. Π_0 and hence $R_{n_1}(1_{C_1})$ is $\overline{\mathcal{D}}_\phi$-measurable. Thus,

$L_1(h_1)$ is $\overline{\mathscr{D}}_\phi$-measurable for all $h_1 \in H_1$. This implies that $R_{n_1}(f)$ is $\overline{\mathscr{D}}_\phi$-measurable for all cylinder functions f on H_1. Now, given $g_1 \in \mathscr{L}(H_1,\mathscr{C}_1,n_1)$, $g_1 \circ P$ is a cylinder function for $P \in \mathscr{P}_1 = \mathscr{P}(H_1)$ and hence $R_{n_1}(g \circ P)$ is $\overline{\mathscr{D}}_\phi$-measurable. By Theorem III.3.6, we can get a sequence $\{P_k\} \subseteq \mathscr{P}_1$ such that $R_{n_1}(g_1 \circ P_k)$ converges to $R_{n_1}(g_1)$ a.s. Π_0 which shows that $R_{n_1}(g_1)$ is itself $\overline{\mathscr{D}}_\phi$-measurable. This proves (i).

For (ii), if $g_1 \in \mathscr{U}(H,\mathscr{C},n;\phi)$, then $R_n(g_1 \circ \phi) = R_{n_1}(g_1)$ and hence $R_n(g_1 \circ \phi)$ is $\mathscr{D}_\phi$-measurable.

Let $g \in \mathscr{L}^1(H,\mathscr{C},n)$ be such that $E_n(g|\phi)$ exists. Let $E_n(g|\phi) = g_1 \circ \phi$. By part (ii), $R_n(E_n(g|\phi))$ is $\mathscr{D}_\phi$-measurable. By the definition of conditional expectation (and Theorem III.3.5), we have for all bounded continuous cylinder functions f,

$$\int R_n(g)R_n(f \circ \phi)d\Pi_0 = \int R_{n_1}(g_1)R_{n_1}(f)d\Pi_0$$

which gives

$$\int R_n(g)R_{n_1}(f)d\Pi_0 = \int R_{n_1}(g_1)R_{n_1}(f)d\Pi_0. \tag{3.16}$$

Given $C_1 \in \mathscr{C}_1$, by Lemma 3.1, we can get bounded continuous cylinder functions $\{f_j\}$ such that $R_{n_1}(f_j)$ converges to $R_{n_1}(1_{C_1})$ a.s. Π_0. The relation (3.16) and the dominated convergence theorem now give

$$\int R_n(g)R_{n_1}(1_{C_1})d\Pi_0 = \int R_{n_1}(g_1)R_{n_1}(1_{C_1})d\Pi_0. \tag{3.17}$$

Let

$$\mathcal{D}_\phi^0 = \{D \in \mathcal{A}_0 \colon 1_D = R_{n_1}(1_{C_1}) \text{ a.s. } \Pi_0 \text{ for some } C_1 \in \mathcal{C}_1\}.$$

We have seen that $\mathcal{D}_\phi^0 \subseteq \overline{\mathcal{D}}$ in the proof of part (i). Linearity and the multiplicative property of R_{n_1} (Theorem III.3.7) imply that $\mathcal{D}_\phi^0$ is a field. As in the proof of part (i), it follows that $R_{n_1}(g_1)$ for $g_1 \in \mathcal{L}^1(H_1,\mathcal{C}_1,n_1)$ is $\sigma(\mathcal{D}_\phi^0)$ measurable and hence for continuous cylinder functions f, $R_n(f\circ\phi) = R_{n_1}(f)$ is $\sigma(\mathcal{D}_\phi^0)$ measurable. Thus $\overline{\mathcal{D}}_\phi = \sigma(\mathcal{D}_\phi^0)$, as $\mathcal{D}_\phi^0$ contains Π_0-null sets. The relation (3.17) now yields

$$\int_D R_n(g)d\Pi_0 = \int_D R_{n_1}(g_1)d\Pi_0 \tag{3.18}$$

for all $D \in \mathcal{D}_\phi^0$. Since $\mathcal{D}_\phi^0$ is a field, we conclude that (3.18) holds for all $D \in \overline{\mathcal{D}}_\phi$. Thus,

$$E_{\Pi_0}(R_n(g)|\overline{\mathcal{D}}_\phi) = R_{n_1}(g_1) = R_{n_1}(E_n(g|\phi)).$$

For (iv), observe that (3.15) implies that (3.16) holds for all bounded continuous cylinder functions f, since by part (ii) above, $R_{n_1}(f) = R_n(f\circ\phi)$ is $\mathcal{D}_\phi$ measurable. Now (3.16) is the same as (3.7) and by assumption, $g_1 \in \mathcal{U}(\phi)$. This proves that $g_1\circ\phi = E_n(g|\phi)$. □

The following result gives some important properties of conditional expectations.

Theorem 3.3: Let ϕ be a cylindrical mapping from $(H,\mathcal{C},n)$ into $(H_1,\mathcal{C}_1)$ and let $n_1 = n\circ\phi^{-1}$. Let $g,g' \in \mathcal{L}^1(H,\mathcal{C},n)$ be such that $E_n(g|\phi)$ and $E_n(g'|\phi)$ exist.

(i) Let $a, a' \in \mathbb{R}$. Then $E_n(ag+a'g'|\phi)$ exists and

$$E_n(ag+a'g'|\phi) = aE_n(g|\phi) + a'E_n(g'|\phi). \tag{3.19}$$

(ii) Let $f \in \mathcal{U}(\phi)$ be such that $f(\phi)g \in \mathcal{L}^1(H,\mathcal{C},n)$. Then

$$E_n(g|\phi)f(\phi) \in \mathcal{L}^1(H,\mathcal{C},n),$$

$$\int g\cdot f(\phi)dn = \int E_n(g|\phi)\cdot f(\phi)dn, \tag{3.20}$$

$E_n(g\cdot f(\phi)|\phi)$ exists and

$$E_n(g\cdot f(\phi)|\phi) = E_n(g|\phi)\cdot f(\phi). \tag{3.21}$$

(iii) If $g^2 \in \mathcal{L}^1(H,\mathcal{C},n)$, then $(E_n(g|\phi))^2 \in \mathcal{L}^1(H,\mathcal{C},n)$ and

$$\int[g-E_n(g|\phi)]^2dn = \min_{f_0\in\mathcal{U}(\phi)} \int[g-f_0(\phi)]^2dn. \tag{3.22}$$

<u>Proof</u>: (i) Let $g_1 = E_n(g|\phi)$ and $g_1' = E_n(g'|\phi)$ and let $g_2 = ag_1 + a'g_1'$. By linearity of the class $\mathcal{U}(\phi)$, it follows that $g_2 \in \mathcal{U}(\phi)$. Now for any bounded continuous cylinder function f on H_1,

$$\begin{aligned}\int_H(ag+a'g')\cdot f(\phi)dn &= a\int g\cdot f(\phi)dn + a'\int g'\cdot f(\phi)dn \\ &= a\int g_1 f dn_1 + a'\int g_1' f dn_1 \\ &= \int(ag_1+a'g_1')\cdot f dn_1 \\ &= \int g_2\cdot f dn_1.\end{aligned}$$

Hence $E_n(ag+a'g'|\phi)$ exists and equals g_2. This proves (i). For (ii), first note that the assumption $f(\phi)\cdot g \in \mathcal{L}^1(H,\mathcal{C},n)$ implies $R_n(f(\phi).g) = R_n(f(\phi))R_n(g) \in \mathcal{L}^1(\Omega_0,\mathcal{A}_0,\Pi_0)$. By Theorem 3.2, $R_n(f(\phi))$ is $\mathcal{D}_\phi$-measurable. Hence

$$E_{\Pi_0}(R_n(f(\phi)\cdot g)|\mathscr{D}_\phi) = R_n(f(\phi))E_{\Pi_0}(g|\mathscr{D}_\phi) \tag{3.23}$$
$$= R_n(f(\phi))R_n(g_1(\phi))$$
$$= R_n((f\cdot g_1)(\phi))$$

where $g_1 = E_n(g|\phi)$. In particular, $R_n((f\cdot g_1)(\phi)) \in \mathscr{L}^1(\Omega_0,\mathscr{A}_0,\Pi_0)$ so that $(f\cdot g_1)(\phi) \in \mathscr{L}^1(H,\mathscr{C},n)$. Now, $g_1 \in \mathscr{U}(\phi)$ as $g_1 = E_n(g|\phi)$ and $f \in \mathscr{U}(\phi)$, thus $f\cdot g_1 \in \mathscr{U}(\phi)$ as the class $\mathscr{U}(\phi)$ is multiplicative. Now, (3.23), the observation that $f\cdot g_1 \in \mathscr{U}(\phi)$ and part (iii) in Theorem 3.2 yield the required conclusion that $E_n(f(\phi)\cdot g|\phi)$ exists and that (3.20) holds.

For (iii), let $Z = R_n(g)$ and $Z_1 = R_n(E_n(g|\phi))$. Then $g^2 \in \mathscr{L}^1(H,\mathscr{C},n)$ gives $R_n(g^2) = Z^2 \in \mathscr{L}^1(H,\mathscr{C},n)$. Now by properties of conditional expectations on countably additive probability spaces, recalling that $Z_1 = E_{\Pi_0}(Z|\overline{\mathscr{D}}_\phi)$ in view of Theorem 3.2, we have

$$Z_1^2 \leq E_{\Pi_0}(Z^2|\overline{\mathscr{D}}_\phi) \tag{3.24}$$

and for all $X \in \mathscr{L}^2(\Omega_0,\overline{\mathscr{D}}_\phi,\Pi_0)$

$$E_{\Pi_0}[Z-Z_1]^2 \leq E_{\Pi_0}(Z-X)^2. \tag{3.25}$$

The relation (3.24) implies that $Z_1^2 \in \mathscr{L}^1(\Omega_0,\mathscr{A}_0,\Pi_0)$ which gives $[E_n(g|\phi)]^2 \in \mathscr{L}^1(H,\mathscr{C},n)$. For $f \in \mathscr{U}(\phi)$, $R_n(f(\phi))$ is $\mathscr{D}_\phi$-measurable as observed in Theorem 3.2 and hence by (3.25),

$$E_{\Pi_0}[Z-Z_1]^2 \leq E_{\Pi_0}[Z-R_n(f(\phi))]^2 \tag{3.26}$$

which gives

$$E_n(g-E_n(g|\phi)]^2 \leq E_n[g-f(\phi)]^2. \tag{3.27}$$

Since (3.27) holds for all $f \in \mathcal{U}(\phi)$, we conclude that (3.22) is true. This completes the proof. □

CHAPTER V

QUASI-CYLINDER PROBABILITIES

In this chapter, we construct a "product" $(E,\mathcal{E})$ of a measurable space $(\Omega,\mathcal{A})$ and $(H,\mathcal{C})$ and introduce the notion of a quasi-cylinder probability (QCP) on $(E,\mathcal{E})$, which is a generalization of cylinder probability. After defining the integral, representation and lifting for a QCP we introduce quasi-cylindrical mappings (QCM) and the concept of conditional expectation with respect to a QCM. The definitions are similar to those discussed in Chapters III and IV and will be given without further elaboration. The proofs of the results are similar to the proofs in the special case treated earlier and in several instances will be either omitted or given only in outline.

1. QUASI-CYLINDER PROBABILITIES AND INTEGRATION

Let $(\Omega,\mathcal{A})$ be a measurable space. Let $H,\mathcal{P},\mathcal{C}_P,\mathcal{C}$ be as in Section III.1. Let

$$E := \Omega\times H \tag{1.1}$$

and for $P \in \mathcal{P}$, let

$$\mathcal{E}_P := \mathcal{A}\otimes\mathcal{C}_P, \tag{1.2}$$

and

$$\mathcal{E} := \bigcup_{P\in\mathcal{P}} \mathcal{E}_P. \tag{1.3}$$

G

Here, $\mathcal{A}\otimes\mathcal{C}_P$ denotes the product σ-field on $\Omega\times H = E$. Thus $\mathcal{E}_P$ is a σ-field. $\mathcal{E}$, being the union of $\mathcal{E}_P$'s, is a field but not a σ-field (recall that we are only considering infinite dimensional Hilbert spaces H). Symbolically, we express the relations (1.1), (1.2), and (1.3) by

$$(E,\mathcal{E}) := (\Omega,\mathcal{A}) \odot (H,\mathcal{C}).$$

Definition 1.1: A *quasi-cylinder probability* (QCP) β on $(E,\mathcal{E})$ is a finitely additive non-negative set function on $\mathcal{E}$ with $\beta(E) = 1$ such that the restriction β_P of β to $\mathcal{E}_P$ is countable additive for every $P \in \mathcal{P}$.

Example 1.1: Let Π be a countably additive probability measure on $(\Omega,\mathcal{A})$ and n be a cylinder probability on $(H,\mathcal{C})$. For each $P \in \mathcal{P}$, let β_P be the countably additive probability measure on $\mathcal{E}_P \otimes \mathcal{C}_P$ given by

$$\beta_P = \Pi \otimes n_P \tag{1.4}$$

where n_P is the restriction of n to $\mathcal{C}_P$. (Recall that n_P is countably additive on $\mathcal{C}_P$.) Here $\Pi \otimes n_P$ denotes the product of the countably additive measure Π and n_P. We will first show that the family $\{\beta_P\colon P \in \mathcal{P}\}$ is consistent. Suppose $F \in \mathcal{E}_{P_i}$, $i = 1,2$. Let $P \in \mathcal{P}$ be such that $P_1 \le P$ and $P_2 \le P$. Now $n_{P_1} = n_P$ on $\mathcal{C}_{P_1}$ and $n_{P_2} = n_P$ on $\mathcal{C}_{P_2}$. Thus $\beta_{P_1} = \Pi \otimes n_{P_1}$ is equal to $\beta_P = \Pi \otimes n_P$ on $\mathcal{E}_{P_1}$. Similarly $\beta_{P_2} = \beta_P$ on $\mathcal{E}_{P_2}$. Thus $\beta_{P_1}(F) = \beta_P(F) = \beta_{P_2}(F)$. Define a set function β on $\mathcal{E}$ by

$$\beta(F) := \beta_P(F),\ F \in \mathcal{E}_P,\ P \in \mathcal{P}. \tag{1.5}$$

Since the family $\{\beta_P\colon P \in \mathcal{P}\}$ is consistent, the relation

(1.5) unambiguously defines the set function β. Clearly, β is a quasi-cylinder probability as its restriction to $\mathcal{E}_P$ is β_P which is a countably additive probability measure. The QCP β so defined is called the product of Π and n and is denoted by $\beta = \Pi \odot n$. Also we write

$$(E,\mathcal{E},\beta) = (\Omega,\mathcal{A},\Pi) \odot (H,\mathcal{C},n)$$

if (1.1)-(1.5) hold.

<u>Integration With Respect to a QCP</u>

<u>Definition 1.2</u>: Let $(\Omega',\mathcal{A}')$ be a measurable space. A function $f: E \to \Omega'$ is called a *cylinder function* if f can be expressed as

$$f(\omega,h) = f_1(\omega,(h,h_1),\ldots,(h,h_k)) \tag{1.6}$$

for some $k \geq 1$, $h_i \in H$, $1 \leq i \leq k$ and a $\mathcal{A} \otimes \mathcal{B}(\mathbb{R}^k)/\mathcal{A}'$ measurable function $f_1: \Omega\times\mathbb{R}^k \to \Omega'$.

The following result provides an alternative description of cylinder functions and is proved using the arguments of Proposition III.2.1.

<u>Proposition 1.1</u>: Suppose $\mathcal{A}'$ separates points in Ω', i.e., $\omega_1',\omega_2' \in \Omega'$; $\omega_1' \neq \omega_2'$ implies the existence of a set $A' \in \mathcal{A}'$ such that $\omega_1' \in A'$ and $\omega_2' \notin A'$. Then $f: (E,\mathcal{E}) \to (\Omega',\mathcal{A}')$ is a cylinder function if and only if f is $\mathcal{E}_P/\mathcal{A}'$-measurable for some $P \in \mathcal{P}$.

For a function $f: E \to \Omega'$ and $P \in \mathcal{P}$, let $f_P: E \to \Omega'$ be defined by

$$f_P(\omega,h) = f(\omega,Ph). \tag{1.7}$$

It is easy to see that if f is $\mathcal{A} \otimes \mathcal{B}(H)/\mathcal{A}'$ measurable, then f_P is a cylinder function for all $P \in \mathcal{P}$.

Definition 1.3: We will say that a real valued, $\mathcal{A} \otimes \mathcal{B}(H)$ measurable function f on E belongs to the class $\mathcal{L}(E,\mathcal{E},\beta)$ if for all $\epsilon > 0$, $\delta > 0$, there exists a $P_0 \in \mathcal{P}$ such that for $P_1, P_2 \in \mathcal{P}$, $P_0 \leq P_i$, $i = 1,2$, we have

$$\beta((\omega,h): |f_{P_1}(\omega,h) - f_{P_2}(\omega,h)| > \delta) < \epsilon. \tag{1.8}$$

A function $f: E \to \mathbb{R}$ will be referred to as an *accessible random variable* or simply a *random variable* if it belongs to $\mathcal{L}(E,\mathcal{E},\beta)$. The following result is the analogue of Theorem III.2.2 and its proof is exactly the same as that of Theorem III.2.2.

Theorem 1.2: Let $f_1, f_2, \ldots, f_k \in \mathcal{L}(E,\mathcal{E},\beta)$ and let $g: \mathbb{R}^k \to \mathbb{R}$ be a continuous function. Then $g(f_1, f_2, \ldots, f_k) \in \mathcal{L}(E,\mathcal{E},\beta)$.

As a consequence, we have that the class $\mathcal{L}(E,\mathcal{E},\beta)$ is closed under addition, multiplication and pointwise maximum of two functions.

Remark 1.1: Note that the set H can be identified with $E = \Omega \times H$ where Ω is a singleton, $\Omega = \{0\}$, say. Here the identification is $h \to (0,h)$. If A is a subset of H, then A can be identified with $\{0\} \times A$, a function f on H can be identified with the function $(0,h) \to f(h)$ on $\Omega \times H$. The singleton set Ω admits only one σ-field $\mathcal{A}$, namely the trivial σ-field $\mathcal{A} = \{\phi, \{0\}\}$ and only one probability measure Π given by $\Pi\{0\} = 1$. Then, $(H,\mathcal{C},n)$ can be identified with $(E,\mathcal{E},\beta) = (\Omega,\mathcal{A},\Pi) \odot (H,\mathcal{C},n)$. Under this identification, the class $\mathcal{L}(E,\mathcal{E},\beta)$ is equal to $\mathcal{L}(H,\mathcal{C},n)$ defined in Section III.2.

Similarly the definitions of integral with respect to β, the induced measure $\beta \circ f^{-1}$ which will be given below reduce to the analogous notions given in Chapter III if we take Ω to be a singleton.

Let $f \in \mathcal{L}(E, \mathcal{E}, \beta)$ be bounded. Then proceeding as in Section III.2., it can be checked that the net of real numbers $\{\int f_P d\beta \colon P \in \mathcal{P}\}$ is Cauchy. Note that f_P is a cylinder function and hence $\int f_P d\beta$ is well defined as a Lebesgue integral. In view of this, we make the following definition.

Definition 1.4:

(i) Let $f \in \mathcal{L}(E, \mathcal{E}, \beta)$ be bounded. Then define

$$\int f d\beta := \lim_{P \in \mathcal{P}} \int f_P d\beta. \tag{1.9}$$

(ii) Let $f \in \mathcal{L}(E, \mathcal{E}, \beta)$ be non-negative. Then define

$$\int f d\beta := \lim_{k \to \infty} \int (f \wedge k) d\beta. \tag{1.10}$$

(iii) Let $\mathcal{L}^1(E, \mathcal{E}, \beta)$ be the class of f in $\mathcal{L}(E, \mathcal{E}, \beta)$ such that $\int f^+ d\beta < \infty$ and $\int f^- d\beta < \infty$. Then for $f \in \mathcal{L}^1(E, \mathcal{E}, \beta)$, define

$$\int f d\beta = \int f^+ d\beta - \int f^- d\beta. \tag{1.11}$$

(Here, f^+ and f^- denote, respectively, the positive and negative part of f.)

The Induced Measure $\beta \circ f^{-1}$

Let $f \in \mathcal{L}(E, \mathcal{E}, \beta)$. Proceeding exactly as in the proof of Proposition III.2.3, it can be shown that the net of countably additive probability measures $\{\beta\circ f_P^{-1}, \quad P \in \mathcal{P}\}$ converges in the sense of weak convergence of probability measures to a countably additive probability measure λ on $(\mathbb{R},\mathcal{B}(\mathbb{R}))$. λ is defined to be the distribution of f under β and is denoted by $\lambda = \beta\circ f^{-1}$. Further, it can also be seen that

$$f \in \mathcal{L}^1(E,\mathcal{E},\beta) \text{ if and only if } \int_{\mathbb{R}}|x|d\beta\circ f^{-1}(x) < \infty \tag{1.12}$$

and for $f \in \mathcal{L}^1(E,\mathcal{E},\beta)$

$$\int_E f d\beta = \int_{\mathbb{R}} x d\beta\circ f^{-1}(x). \tag{1.13}$$

Equality Modulo β

Definition 1.4: Let $f,g \in \mathcal{L}(E,\mathcal{E},\beta)$. We say that f is equal to g modulo β, written as $f \equiv g \bmod[\beta]$ if for all $\epsilon > 0$, there exists a $P_0 \in \mathcal{P}$, such that for all $P \in \mathcal{P}$, $P_0 \in \mathcal{P}$, we have

$$\beta((\omega,h): |f_P(\omega,h) - g_P(\omega,h)| > \epsilon) < \epsilon. \tag{1.14}$$

It is easily seen that equality modulo β is an equivalence relation on $\mathcal{L}(E,\mathcal{E},\beta)$. Proceeding as in the proof of Theorem III.2.4, it can be proved that for $f,g \in \mathcal{L}(E,\mathcal{E},\beta)$,

$$f \equiv g \bmod[\beta] \text{ if and only if } f - g \equiv 0 \bmod[\beta] \tag{1.15}$$

and

$$f \equiv 0 \bmod[\beta] \text{ if and only if } \beta\circ f^{-1} = \delta_{\{0\}}, \tag{1.16}$$

where $\delta_{\{0\}}$ is the unit mass at 0.

Relationship to the Dunford-Schwartz Integral

The connection between the integral given in Definition 1.4 and the Dunford-Schwartz integral introduced in Section III.5 can be easily deduced. Using the definitions and notation of Section III.5, we have the following results.

Let $\hat{\mathcal{E}} := \{F \subseteq E: 1_F \in \mathcal{L}(E,\mathcal{E},\beta)\}$ and for $F \in \hat{\mathcal{E}}$, let $\hat{\beta}(F) := \int 1_F d\beta$. Then $\hat{\beta}$ is a finitely additive probability measure on the field $\hat{\mathcal{E}}$ and $\hat{\beta}$ is equal to β on $\mathcal{E}$. Further,

$$\ell^0(E,\hat{\mathcal{E}},\hat{\beta}) = \mathcal{L}(E,\mathcal{E},\beta), \tag{1.17}$$

$$\ell^1(E,\hat{\mathcal{E}},\hat{\beta}) = \mathcal{L}^1(E,\mathcal{E},\beta) \tag{1.18}$$

and for $f \in \mathcal{L}^1(E,\mathcal{E},\beta)$

$$⨏ f d\beta = \int f d\beta. \tag{1.19}$$

2. REPRESENTATION OF A QCP AND THE LIFTING MAP

Throughout this section, β will denote a QCP on $(E,\mathcal{E}) = (\Omega,\mathcal{A})\odot(H,\mathcal{C})$.

Definition 2.1: A *representation* of β is a triplet $(\rho,L,\tilde{\Pi})$, where $\tilde{\Pi}$ is a countably additive probability measure on a measurable space $(\tilde{\Omega},\tilde{\mathcal{A}})$, ρ is a measurable mapping from $(\tilde{\Omega},\tilde{\mathcal{A}})$ into $(\Omega,\mathcal{A})$ and L is a mapping from H into $\mathcal{L}(\tilde{\Omega},\tilde{\mathcal{A}},\tilde{\Pi})$ such that for all $A \in \mathcal{A}$ and $C \in \mathcal{C}$ of the form

$$C = \{h: ((h,h_1),\dots,(h,h_j)) \in B\}, \tag{2.1}$$

for $j \geq 1$, $h_i \in H$, $1 \leq i \leq j$ and $B \in \mathcal{B}(\mathbb{R}^j)$, we have

$$\beta(A\times C) = \tilde{\Pi}(\tilde{\omega}:\ \rho(\tilde{\omega})\in A;\ (L(h_1)(\tilde{\omega}),\ldots,L(h_j(\tilde{\omega}))) \in B\}. \quad (2.2)$$

The measurable space $(\tilde{\Omega},\tilde{\mathscr{A}})$ will be called the underlying representation space.

Remark 2.1: The condition (2.2) implies that L is linear, i.e., L satisfies the equation III.3.1. If $L: H \to \mathscr{L}(\tilde{\Omega},\tilde{\mathscr{A}},\tilde{\Pi})$ is given to be linear and (2.1) and (2.2) hold for every orthonormal set $\{h_1,h_2,\ldots,h_j\}$, then (2.1) and (2.2) also hold for all $h_1,h_2,\ldots,h_j \in H$. To see this, given $B \in \mathscr{B}(\mathbb{R}^j)$ and $h_1,h_2,\ldots,h_j \in H$, let $\{e_1,e_2,\ldots,e_k\}$ be an orthonormal basis of the linear span of $\{h_1,\ldots,h_j\}$ and let $F: \mathbb{R}^k \to \mathbb{R}^j$ be defined by

$$F(x_1,x_2,\ldots,x_k) = \Big(\sum_{r=1}^{k}(h_i,e_r)x_r\Big)_{1\le i\le j}.$$

Then we have

$$F((h,e_1),\ldots,(h,e_k)) = \Big(\sum_{r=1}^{k}(h_i,e_r)(h,e_r)\Big)_{1\le i\le j}$$

$$= ((h,h_1),\ldots,(h,e_j))$$

as $\{e_1,e_2,\ldots,e_k\}$ is an orthonormal basis of $\text{span}\{h_1,\ldots,h_j\}$. Also, using linearity of L, it follows that

$$\sum_{r=1}^{k}(h_i,e_r)L(e_r) = L(h_i) \quad \text{a.s. } \tilde{\Pi}$$

and hence

$$F(L(e_1),L(e_2),\ldots,L(e_r)) = (L(h_1),\ldots,L(h_j)) \quad \text{a.s. } \tilde{\Pi}.$$

Taking $B_1 = F^{-1}B$, we can write C given by (2.1) as

$$C = \{h: ((h,h_1),\dots,(h,h_j)) \in B\}$$
$$= \{h: F((h,e_1),\dots,(h,e_k)) \in B\}$$
$$= \{h: (h,e_1),\dots,(h,e_k)) \in B_1\}.$$

Since we have assumed that (2.1), (2.2) hold for all $k \geq 1$, $e_1, e_2, \dots, e_k \in H$ orthonormal, $B_1 \in \mathscr{B}(\mathbb{R}^k)$, it follows that

$$\beta(A\times C) = \tilde{\Pi}(\tilde{\omega}: \rho(\tilde{\omega}) \in A;\ (L(e_1),\dots,L(e_k)) \in B_1\}$$
$$= \tilde{\Pi}(\tilde{\omega}: \rho(\tilde{\omega}) \in A;\ F(L(e_1),\dots,L(e_k)) \in B\}$$
$$= \tilde{\Pi}(\tilde{\omega}: \rho(\tilde{\omega}) \in A;\ (L(h_1),\dots,L(h_j)) \in B\}.$$

Thus (2.1), (2.2) for orthonormal $\{h_1,\dots,h_k) \subseteq H$ and linearity of L imply that $(\rho, L, \tilde{\Pi})$ is a representation of β.

Existence of a Representation

If the QCP β is a product $\Pi\Theta n$ as in Example 1.1, then a representation of β can be easily constructed as follows. Let (L_0, Π_0) be a representation of n with the underlying representation space $(\Omega_0, \mathscr{A}_0)$ and let

$$(\tilde{\Omega}, \tilde{\mathscr{A}}, \tilde{\Pi}) = (\Omega, \mathscr{A}, \Pi) \otimes (\Omega_0, \mathscr{A}_0, \Pi_0). \tag{2.3}$$

For $\tilde{\omega} = (\omega, \omega_0) \in \tilde{\Omega}$ and $h \in H$, let

$$\rho(\tilde{\omega}) := \omega \tag{2.4}$$

and

$$L(h)(\tilde{\omega}) := L_0(h)(\omega_0). \tag{2.5}$$

Then it is easy to see that $(\rho, L, \tilde{\Pi})$ is a representation of β on the representation space $(\tilde{\Omega}, \tilde{\mathcal{A}})$.

To prove a general existence result, we need the following result, which is a slight generalization of Kolmogorov's consistency theorem.

<u>Theorem 2.1</u>: Let $(\Omega, \mathcal{A})$ be a measurable space and let $(\tilde{\Omega}, \mathcal{A}_1) = (\Omega, \mathcal{A}) \otimes (\mathbb{R}^\infty, \mathcal{B}(\mathbb{R}^\infty))$. For $\omega \in \Omega$ and $\underline{x} = (x_1, x_2, \ldots) \in \mathbb{R}^\infty$, let

$$\theta(\omega, \underline{x}) = \omega$$

and

$$X_j(\omega, \underline{x}) = x_j, \quad j \geq 1.$$

Let $\mathcal{F}_0 = \sigma(\theta)$ and for $k \geq 1$, let $\mathcal{F}_k = \sigma(\theta, X_j, 1 \leq j \leq k)$. For each $k \geq 0$, let μ_k be a countably additive probability measure on $\mathcal{F}_k$ such that for all $k \geq 0$,

$$\mu_{k+1}(F) = \mu_k(F), \quad F \in \mathcal{F}_k. \tag{2.6}$$

Then there exists a countably additive probability measure μ on $\sigma(\bigcup_k \mathcal{F}_k) = \mathcal{A}_1$ such that $\mu = \mu_k$ on $\mathcal{F}_k$.

<u>Proof</u>: For $k \geq 1$, consider the probability space $(\Omega_1, \mathcal{F}_k, \mu_k)$ and let $\nu_k(\omega_1, B)$ be a regular conditional probability distribution of $(X_1, X_2, \ldots, X_k)$ given $\mathcal{F}_0$. This means

for all $B \in \mathcal{B}(\mathbb{R}^k)$, $\omega_1 \to \nu_k(\omega_1, B)$ is an $\mathcal{F}_0$-measurable function on Ω_1; (2.7) for all $\omega_1 \in \Omega_1$, $B \to \nu_k(\omega_1,B)$ is a countably additive probability measure on $(\mathbb{R}^k,\mathcal{B}(\mathbb{R}^k))$; (2.8)

and for all $B \in \mathcal{B}(\mathbb{R}^k)$,
$$E_{\mu_k}[1_{\{(X_1,X_2,\ldots,X_k)\in B\}}|\mathcal{F}_0](\omega_1) = \nu_k(\omega_1,B) \quad \text{a.s. } \mu_k. \tag{2.9}$$

Note that the restriction of μ_k to $\mathcal{F}_0$ is μ_0 and both expressions in (2.9) are $\mathcal{F}_0$-measurable random variables. Thus, (2.9) holds a.s. μ_0. The relation (2.9) implies that for $B \in \mathcal{B}(\mathbb{R}^k)$,

$$\nu_k(\omega_1,B) = \nu_{k+1}(\omega_1, B\times\mathbb{R}) \quad \text{a.s. } \mu_0 \tag{2.10}$$

as $1_{\{(X_1,X_2,\ldots,X_k) \in B\}} = 1_{\{(X_1,X_2,\ldots,X_{k+1}) \in B\times\mathbb{R}\}}$. Let $\mathcal{D}_k$ be any countable field on $\mathbb{R}^k$ such that $\sigma(\mathcal{D}_k) = \mathcal{B}(\mathbb{R}^k)$. It can be taken to be the field generated by the rectangles $\{\underline{x} = (x_1,x_2,\ldots,x_k)\colon a_i < x_i \le b_i\}$, a_i,b_i rationals, $1 \le i \le k$. Let

$$N_k = \{\omega_1\colon \nu_k(\omega_1,B) \ne \nu_{k+1}(\omega_1, B\times\mathbb{R}) \quad \text{for some } B \in \mathcal{D}_k\}.$$

Since $\mathcal{D}_k$ is countable, (2.10) implies $\mu_0(N_k) = 0$. Fix $\omega_1 \notin N_k$. Then

$$\mathcal{D}_k^0 := \{B \in \mathcal{B}(\mathbb{R}^k)\colon \nu_k(\omega_1,B) = \nu_{k+1}(\omega_1, B\times\mathbb{R})\}$$

is clearly a σ-field and contains $\mathcal{D}_k$ which shows that $\mathcal{D}_k^0 = \mathcal{B}(\mathbb{R}^k)$. Thus for $\omega_1 \notin N_k$,

$$\nu_k(\omega_1,B) = \nu_{k+1}(\omega_1, B\times\mathbb{R}) \quad \text{for all } B \in \mathcal{B}(\mathbb{R}^k). \tag{2.11}$$

Let $N = \bigcup_k N_k$. Then $\mu_0(N) = 0$ and for $\omega_1 \notin N$, (2.11) holds for all $k \ge 1$. By Kolmogorov's consistency theorem for $\omega_1 \notin N$, we get a countably additive probability measure $\nu(\omega_1,\cdot)$ on $\mathcal{B}(\mathbb{R}^\infty)$ such that for all $k \geq 1$ and $B \in \mathcal{B}(\mathbb{R}^k)$

$$\nu(\omega_1, B\times\mathbb{R}^\infty) = \nu_k(\omega_1,B) \tag{2.12}$$

where $B\times\mathbb{R}^\infty = \{\underline{x} \in \mathbb{R}^\infty: (x_1,x_2,\ldots,x_k) \in B\}$ for $B \subseteq \mathbb{R}^k$. For $\omega_1 \in N$, take $\nu(\omega_1,\cdot)$ to be equal to λ, where λ is some fixed probability measure on $\mathbb{R}^\infty$. Let

$$\mathcal{G} := \{F \in \mathcal{B}(\mathbb{R}^\infty): \omega_1 \to \nu(\omega_1,F) \text{ is } \mathcal{F}_0\text{-measurable}\}.$$

It is easy to see that $\mathcal{G}$ is a σ-field. Also, in view of (2.7), (2.12) and the fact that $N \in \mathcal{F}_0$, it follows that $\mathcal{G}$ contains $\mathcal{G}_0$, where

$$\mathcal{G}_0 = \{B_k\times\mathbb{R}^\infty: B_k \in \mathcal{B}(\mathbb{R}^k),\ k \geq 1\}.$$

Now $\mathcal{G}_0$ is a field and $\sigma(\mathcal{G}_0) = \mathcal{B}(\mathbb{R}^\infty)$ and hence $\mathcal{G} = \mathcal{B}(\mathbb{R}^\infty)$. Finally, define μ on $\mathcal{A}_1$ by

$$\mu(D) = \int\nu((\omega,\underline{x}),D_\omega)d\mu_0(\omega,\underline{x}) \tag{2.13}$$

for $D \in \mathcal{A}_1$. The integral appearing in (2.13) is well defined as a Lebesgue integral since $\mathcal{G} = \mathcal{B}(\mathbb{R}^\infty)$ so that $\omega_1 \to \nu(\omega_1,F)$ is $\mathcal{F}_0$-measurable for all $F \in \mathcal{B}(\mathbb{R}^\infty)$. In (2.13), D_ω denotes the ω-section of D, i.e., $D_\omega = \{\underline{x} \in \mathbb{R}^\infty: (\omega,\underline{x}) \in D\}$. Using the fact that for all $\omega_1 \in \Omega_1$, $\nu(\omega_1,\cdot)$ is a countably additive probability measure, it can be verified that μ defined by (2.13) is itself a countably additive probability measure on $\mathcal{A}_1$. To complete the proof, we will show that $\mu = \mu_k$ on $\mathcal{F}_k$. Fix $k \geq 1$ and let $D \in \mathcal{F}_k$ be given by

$$D = \{(\omega,\underline{x}):\ \omega \in A,\ (x_1,x_2,\ldots,x_k) \in B\} \tag{2.14}$$

for some $A \in \mathcal{A}$ and $B \in \mathcal{B}(\mathbb{R}^k)$. Then $D_\omega = B\times\mathbb{R}^\infty$ if $\omega \in A$ and $D_\omega = \phi$ if $\omega \in A^c$. Thus

$$\begin{aligned}\mu(D) &= \int \nu((\omega,\underline{x}),D_\omega)d\mu_0(\omega,\underline{x})\\ &= \int \nu((\omega,\underline{x}),\ B\times\mathbb{R}^\infty)1_A(\omega)d\mu_0(\omega,\underline{x})\\ &= \int \nu_k((\omega,\underline{x}),B)1_A(\omega)d\mu_0(\omega,\underline{x}) \quad \text{(by (2.12) and } \mu(N)=0)\\ &= \int \nu_k((\omega,\underline{x}),B)1_A(\theta(\omega,\underline{x}))d\mu_k(\omega,\underline{x}) \quad (\text{since } \mu_0=\mu_k \text{ on } \mathcal{F}_0)\\ &= E_{\mu_k}[1_A(\theta)E_{\mu_k}(1_{\{X_1,X_2,\ldots,X_k)\in B\}}|\mathcal{F}_0)] \quad \text{(by (2.9))}\\ &= E_{\mu_k}[1_A(\theta)\cdot 1_{\{(X_1,X_2,\ldots,X_k)\in B\}}]\\ &= \mu_k(D).\end{aligned}$$

Thus

$$\mu_k(D) = \mu(D) \tag{2.15}$$

for all sets of the form (2.14). These sets constitute a field $\mathcal{F}_k^0$ and $\mathcal{F}_k = \sigma(\mathcal{F}_k^0)$. Thus (2.15) holds for all $D \in \mathcal{F}_k$, i.e., $\mu = \mu_k$ on $\mathcal{F}_k$. The proof is complete. □

We are now in a position to prove the general existence result.

<u>Theorem 2.2</u>: Let β be any quasi-cylindrical probability on $(E,\mathcal{E})$. Then a representation $(\rho,L,\tilde{\Pi})$ of β exists.

Proof: Let $\tilde{\Omega} = \Omega \times \mathbb{R}^H$ and $\tilde{\mathcal{A}} = \mathcal{A} \otimes (\bigotimes_{h \in H} \mathcal{B}(\mathbb{R}))$. For $\omega \in \Omega$ and $r \in \mathbb{R}^H$, let $\rho(\omega, r_\cdot) = \omega$ and for $h_0 \in H$, let $L(h_0)(\omega, r_\cdot) = r_{h_0}$. Let $\tilde{\Pi}$ be the set function defined on finite dimensional cylinder sets in $\tilde{\mathcal{A}}$ by

$$\tilde{\Pi}((\rho, L(h_1), L(h_2), \ldots, L(h_k)) \in D) \tag{2.16}$$

$$= \beta((\omega, h) : (\omega, (h, h_1), \ldots, (h, h_k)) \in D)$$

for $D \in \mathcal{A} \otimes \mathcal{B}(\mathbb{R}^k)$.

We will show that for any fixed sequence $\{h_i\} \subseteq H$, there exists a countably additive probability measure Π' on $\sigma(\rho, L(h_i); 1 \leq i < \infty\}$ agreeing with $\tilde{\Pi}$ on cylinder sets in $\tilde{\mathcal{A}}$. Since

$$\tilde{\mathcal{A}} = \bigcup_{\{h_i\} \subseteq H} \sigma(\rho, L(h_i); 1 \leq i < \infty),$$

it will imply that $\tilde{\Pi}$ has an extension to $\tilde{\mathcal{A}}$ as a countably additive probability measure. This step is similar to the proof of Kolmogorov's consistency theorem for an uncountable index set and will complete the proof of the theorem as (2.2) implies (2.16) and thus $(\rho, L, \tilde{\Pi})$ is a representation of β.

So fix a sequence $\{h_i\} \subseteq H$. For $k \geq 1$, let $\hat{\mu}_k(D)$ for $D \in \mathcal{A} \otimes \mathcal{B}(\mathbb{R}^k)$ be equal to the right hand side in (2.16). Then $\hat{\mu}_k$ is a countably additive probability measure. Recall the notation in Theorem 2.1 and for $k \geq 1$, let μ_k be defined by

$$\mu_k((\theta, X_1, X_2, \ldots, X_k) \in D) = \hat{\mu}_k(D). \tag{2.17}$$

Also, let μ_0 be the restriction to $\mathcal{F}_0$ of μ_1. It is easy to see that the sequence $\{\mu_k\}$ satisfies (2.6). Let μ be the measure obtained in Theorem 2.1 which is an extension of μ_k. Let Π' on $\sigma(\rho, L(h_i); 1 \leq i < \infty)$ be defined by

$$\Pi'((\rho, L(h_1), L(h_2), \ldots) \in F) = \mu(F) \tag{2.18}$$

for $F \in \mathcal{A}_1 = \mathcal{A} \otimes \mathcal{B}(\mathbb{R}^\infty)$. Then Π' is a countably additive probability measure on $\sigma(\rho, L(h_i); 1 \leq i < \infty)$ and it agrees with $\tilde{\Pi}$ on their common domain of definition.

This completes the proof. □

Remark 2.2: In [40], we had asserted the existence of a representation in some special cases. The proofs of Theorem 2.1 and 2.2 given above are due to B.V. Rao.

The Lifting Map

We will now define the β-lifting for elements in $\mathcal{L}(E, \mathcal{E}, \beta)$. Since the definition of β-lifting and its properties are very similar to those of n-lifting introduced in Section III.3, we will content ourselves with merely giving the definitions and stating the results. All these results can be proved following the arguments given in the proofs of their analogues in Section III.3—with obvious modifications like writing f_P for $f \circ P$, E for H, $\mathcal{E}$ for $\mathcal{C}$, β for n and R_β for R_n.

Let $(\rho, L, \tilde{\Pi})$ be a representation of β with the underlying representation space $(\tilde{\Omega}, \tilde{\mathcal{A}})$, held fixed for the rest of the section.

Definition 2.2: Let f be a real valued cylinder function given by (1.6) (with $\Omega' = \mathbb{R}$, $\mathcal{A}' = \mathcal{B}(\mathbb{R})$). Then the β-lifting of f is defined by

$$R_\beta(f)(\tilde{\omega}) = f_1(\rho(\tilde{\omega}), L(h_1)(\tilde{\omega}), \ldots, L(h_k)(\tilde{\omega})) \tag{2.19}$$

Lemma 2.3: Let $f: E \to \mathbb{R}$ be a cylinder function. Then

$$\beta \circ f^{-1} = \tilde{\Pi} \circ [R_\beta(f)]^{-1}. \tag{2.20}$$

Theorem 2.4: For every $f \in \mathcal{L}(E, \mathcal{E}, \beta)$, there exists a random variable $X_f \in \mathcal{L}(\tilde{\Omega}, \tilde{\mathcal{A}}, \tilde{\Pi})$ such that the net $\{R_\beta(f_P): P \in \mathcal{P}\}$ converges to X_f in $\tilde{\Pi}$-probability, i.e., for every $\epsilon > 0$, there exists a $P_1 \in \mathcal{P}$ such that $P \in \mathcal{P}$, $P_1 \leq P$ implies

$$\Pi(|R_\beta(f_P) - X_f| > \epsilon) < \epsilon. \tag{2.21}$$

Definition 2.3: For $f \in \mathcal{L}(E, \mathcal{E}, \beta)$, define the *$\beta$-lifting of* $f, R_\beta(f)$ by

$$R_\beta(f) := \lim_{P \in \mathcal{P}} \text{ in } \tilde{\Pi}\text{-probability } R_\beta(f_P). \tag{2.22}$$

The β-lifting R_β is a mapping from $\mathcal{L}(E, \mathcal{E}, \beta)$ into $\mathcal{L}(\tilde{\Omega}, \tilde{\mathcal{A}}, \tilde{\Pi})$. The following lemma shows that the integral defined in Section 1 is the same as the one defined in our earlier paper [40].

Lemma 2.5: Let $f \in \mathcal{L}(E, \mathcal{E}, \beta)$. Then

$$\beta \circ f^{-1} = \tilde{\Pi} \circ [R_\beta(f)]^{-1} \tag{2.23}$$

Further, $f \in \mathcal{L}^1(E, \mathcal{E}, \beta)$ if and only if $\int |R_\beta(f)| d\tilde{\Pi} < \infty$ and for

$f \in \mathcal{L}^1(E,\mathcal{E},\beta)$

$$\int f d\beta = \int R_\beta(f) d\tilde{\Pi}. \tag{2.24}$$

The next result gives a characterization of $\mathcal{L}(E,\mathcal{E},\beta)$ and R_β in terms of sequential convergence.

Theorem 2.6: Let $f: E \to \mathbb{R}$ be an $\mathcal{A} \otimes \mathcal{B}(H)$ measurable function. Then the following are equivalent.

(a) $f \in \mathcal{L}(E,\mathcal{E},\beta)$.

(b) There exists $\{P_k\} \subseteq \mathcal{P}$, $P_k \uparrow I$ such that for all $\{P_k'\} \subseteq \mathcal{P}$, $P_k \leq P_k'$, $R_\beta(f_{P_k'})$ converges a.s. $\tilde{\Pi}$.

(c) There exists $\{P_k\} \subseteq \mathcal{P}$, $P_k \uparrow I$ such that for all $\{P_k'\} \subseteq \mathcal{P}$, $P_k \leq P_k'$, $R_\beta(f_{P_k'})$ converges in $\tilde{\Pi}$-probability.

Moreover, if $f \in \mathcal{L}(E,\mathcal{E},\beta)$, then for $\{P_k'\}$ as in (b), [as in (c)] $R(f_{P_k'})$ converges to $R_\beta(f)$ a.s. $\tilde{\Pi}$ (in $\tilde{\Pi}$-probability).

Theorem 2.7: For $k \geq 1$, let $f_1, f_2, \ldots, f_k \in \mathcal{L}(E,\mathcal{E},\beta)$, and let $g: \mathbb{R}^k \to \mathbb{R}$ be a continuous function. Then we have the following.

(i) For $a_1, a_2 \in \mathbb{R}$, $a_1 f_2 + a_2 f_2 \in \mathcal{L}(E,\mathcal{E},\beta)$ and

$$R_\beta(a_1 f_1 + a_2 f_2) = a_1 R_\beta(f_1) + a_2 R_\beta(f_2) \quad \text{a.s. } \tilde{\Pi}. \tag{2.25}$$

(ii) $f_1 \cdot f_2 \in \mathcal{L}(E,\mathcal{E},\beta)$ and

$$R_\beta(f_1 \cdot f_2) = R_\beta(f_1) \cdot R_\beta(f_2) \quad \text{a.s. } \tilde{\Pi}. \tag{2.26}$$

(iii) $g(f_1, f_2, \ldots, f_k) \in \mathcal{L}(E,\mathcal{E},\beta)$ and

$$R_\beta(g(f_1,f_2,\ldots,f_k)) = g(R_\beta(f_1),R_\beta(f_2),\ldots,R_\beta(f_k)) \text{ a.s. } \tilde{\Pi}. \tag{2.27}$$

(iv) If $f_1 > 0$ and $\tilde{\Pi}(R_\beta(f_1) > 0) = 1$, then

$$\frac{1}{f_1} \in \mathcal{L}(E,\mathcal{E},\beta) \text{ and } R_\beta(\frac{1}{f_1}) = \frac{1}{R(f_1)} \quad \text{a.s. } \tilde{\Pi}. \tag{2.28}$$

(v) If $f_1 \leq f_2$, then

$$R_\beta(f_1) \leq R_\beta(f_2) \quad \text{a.s. } \tilde{\Pi}. \tag{2.29}$$

<u>Definition 2.4</u>: Let $\mathcal{L}^*(E,\mathcal{E},\beta)$ consist of $\mathcal{A} \otimes \mathcal{B}(H)$ measurable functions $f: E \to \mathbb{R}$ such that for all $\{P_i\} \subseteq \mathcal{P}$, $P_i \overset{s}{\to} I$, $R_\beta(f_{P_i})$ converges in $\tilde{\Pi}$-probability.

<u>Definition 2.5</u>: Let $\mathcal{L}^{1*}(E,\mathcal{E},\beta)$ consist of $\mathcal{A} \otimes \mathcal{B}(H)$ measurable functions $f: E \to \mathbb{R}$ such that for all $\{P_i\} \subseteq \mathcal{P}$, $P_i \overset{s}{\to} I$, $R_\beta(f_{P_i})$ converges in $\mathcal{L}^1(\tilde{\Omega},\tilde{\mathcal{A}},\tilde{\Pi})$.

<u>Theorem 2.8</u>:

(i) $\mathcal{L}^*(E,\mathcal{E},\beta) \subseteq \mathcal{L}(E,\mathcal{E},\beta)$ and for $f \in \mathcal{L}^*(E,\mathcal{E},\beta)$,

$$R_\beta(f_{P_i}) \to R_\beta(f) \text{ in } \tilde{\Pi}\text{-probability} \tag{2.30}$$

for all $\{P_i\} \subseteq P$, with $P_i \overset{s}{\to} I$.

(ii) $\mathcal{L}^{1*}(E,\mathcal{E},\beta) \subseteq \mathcal{L}^1(E,\mathcal{E},\beta)$ and for $f \in \mathcal{L}^{1*}(E,\mathcal{E},\beta)$,

$$R_\beta(f_{P_i}) \to R_\beta(f) \quad \text{in } \mathcal{L}^1(\tilde{\Omega},\tilde{\mathcal{A}},\tilde{\Pi}). \tag{2.31}$$

Theorem 2.9:

(i) Let $f \in \mathscr{L}(E,\mathscr{E},\beta)$. Then

$$f \equiv 0 \bmod[\beta] \text{ if and only if } R_\beta(f) = 0 \text{ a.s. } \tilde{\Pi}. \tag{2.32}$$

(ii) Let $f \in \mathscr{L}^1(E,\mathscr{E},\beta)$. Then

$$\int_F f d\beta = 0 \text{ for all } F \in \mathscr{E} \text{ implies } f \equiv 0 \bmod[\beta]. \tag{2.33}$$

3. POLISH SPACE VALUED MAPPINGS ON $(E,\mathscr{E},\beta)$

Let S be a Polish space, i.e., a complete separable metric space and let d be a metric under which S is complete. Let $\mathscr{B}(S)$ be the Borel σ-field on S.

Throughout this section, β will be a fixed quasi-cylinder probability on $(E,\mathscr{E}) = (\Omega,\mathscr{A}) \odot (H,\mathscr{C})$ and $(\rho,L,\tilde{\Pi})$ will denote a representation of β with representation space $(\tilde{\Omega},\tilde{\mathscr{A}})$. Recall the notation used in Section 1: For a function $f: E \to \Omega'$ and $P \in \mathscr{P}$, f_P denotes the function $f_P(\omega,h) = f(\omega,Ph)$, $(\omega,h) \in E$.

We now define the class of S-valued random variables on $(E,\mathscr{E},\beta)$. The definition is an obvious generalization of Definition 1.3, with the distance function d on S replacing the distance function on $\mathbb{R}$.

Definition 3.1: Define $\mathscr{L}(E,\mathscr{E},\beta;S)$ to be the class of $\mathscr{A}\otimes\mathscr{B}(H)/\mathscr{B}(S)$ measurable functions $f: E \to S$ such that for all $\epsilon > 0$, $\delta > 0$, there exists a $P_0 \in \mathscr{P}$ such that for $P_1,P_2 \in \mathscr{P}$, $P_0 \le P_i$, $i = 1,2$, we have

$$\beta((\omega,h): d(f_{P_1}(\omega,h),f_{P_2}(\omega,h)) > \delta) < \epsilon. \tag{3.1}$$

We will refer to elements of $\mathcal{L}(E,\mathcal{E},\beta;S)$ as S-valued *accessible random variables or random variables*.

Remark 3.1: Note that if $f: E \to S$ is $\mathcal{A}\otimes\mathcal{B}(H)/\mathcal{B}(S)$ measurable, then for $P \in \mathcal{P}$, f_P is $\mathcal{E}_P/\mathcal{B}(S)$ measurable. For $P_1, P_2 \in \mathcal{P}$, if $P_3 \in \mathcal{P}$ is such that $P_i \le P_3$, $i = 1,2$, then f_{P_i}, $i = 1,2$, are $\mathcal{E}_{P_3}/\mathcal{B}(S)$ measurable and hence by separability of S, $d(f_{P_1}(\omega,h), f_{P_2}(\omega,h))$ is $\mathcal{E}_{P_3}$-measurable and thus the left hand side in (3.1) is well defined.

Remark 3.2: Recall Remark 1.1, where we had seen that if Ω is a singleton, then $(E,\mathcal{E},\beta) = (\Omega,\mathcal{A},\Pi)\odot(H,\mathcal{C},n)$ can be identified with $(H,\mathcal{C},n)$. Thus we define $\mathcal{L}(H,\mathcal{C},n;S)$ to be equal to $\mathcal{L}(E,\mathcal{E},\beta;S)$ where Ω is a singleton. The same applies to later definitions and results as well--by taking Ω to be a singleton, we will have the corresponding definition or result for elements in $\mathcal{L}(H,\mathcal{C},n;S)$ as well.

Remark 3.3: Note that if $S = \mathbb{R}$, Definition 3.1 reduces to (1.3) so that $\mathcal{L}(E,\mathcal{E},\beta;\mathbb{R}) = \mathcal{L}(E,\mathcal{E},\beta)$ as defined earlier.

Definition 3.2: Let $f,g \in \mathcal{L}(E,\mathcal{E},\beta;S)$. We say that f is equal to g modulo β, written as $f \equiv g \bmod[\beta]$ if for all $\epsilon > 0$, there exists a $P_0 \in \mathcal{P}$ such that for $P \in \mathcal{P}$, $P_0 \le P$, we have

$$\beta((\omega,h)\colon d(f_P(\omega,h),g_P(\omega,h)) > \epsilon) < \epsilon. \tag{3.2}$$

From the above definition, it follows that

$$f \equiv g \bmod[\beta] \text{ if and only if } d(f,g) \equiv 0 \bmod[\beta]. \tag{3.3}$$

The Lifting Map

Recall the definition (1.2) of a $(\Omega',\mathcal{A}')$ valued cylinder function, where $(\Omega',\mathcal{A}')$ is a measurable space.

Definition 3.3: Let $f: E \to \Omega'$ be a cylinder function given by (1.6). Then the β-lifting $R_\beta(f)$ of f is defined by

$$R_\beta(f) := f_1(\rho, L(h_1), \ldots, L(h_k)). \tag{3.4}$$

Thus $R_\beta(f)$ is an $\tilde{\mathcal{A}}/\mathcal{A}'$ measurable mapping from $\tilde{\Omega}$ into Ω'. The following result is proved exactly as its special case Lemma III.3.3 was proved. The proof is omitted.

Lemma 3.1: Let $f: E \to \Omega'$ be a cylinder function. Then

$$\beta \circ f^{-1} = \tilde{\Pi} \circ [R_\beta(f)]^{-1}. \tag{3.5}$$

Here $\beta \circ f^{-1}$ and $\tilde{\Pi} \circ [R_\beta(f)]^{-1}$ are countably additive measures on $(\Omega', \mathcal{A}')$.

Our next result is an analogue of Theorem 2.4:

Theorem 3.2: For every $f \in \mathcal{L}(E, \mathcal{E}, \beta; S)$, there exists a random variable $X_f \in \mathcal{L}(\tilde{\Omega}, \tilde{\mathcal{A}}, \tilde{\Pi}; S)$ such that the net $\{R_\beta(f_P): P \in \mathcal{P}\}$ converges in $\tilde{\Pi}$-probability to X_f, i.e., for all $\epsilon > 0$, there exists a $\tilde{P} \in \mathcal{P}$ with the property that $\tilde{P} \leq P$ implies

$$\tilde{\Pi}(d(f_P, X_f) > \epsilon) < \epsilon. \tag{3.6}$$

Proof: The proof is similar to that of Theorem III.3.4 and hence we will only give an outline.

Using (3.1), get a sequence $\{P_k\} \subseteq \mathcal{P}$ such that P_1', $P_2' \in \mathcal{P}$, $P_k \leq P_1'$, $P_k \leq P_2'$ implies

$$\beta((\omega,h): d(f_{P_1'}(\omega,h), f_{P_2'}(\omega,h)) > \frac{1}{2^k}) < \frac{1}{2^k}. \tag{3.7}$$

In view of (3.5), this can be written as

$$\tilde{\Pi}(d(R_\beta(f_{P_1'}), R_\beta(f_{P_2'})) > \frac{1}{2^k}) < \frac{1}{2^k} . \tag{3.8}$$

Without loss of generality, we can assume that $P_k \leq P_{k+1}$ for all k. Then (3.8) implies that for all k

$$\tilde{\Pi}(d(R_\beta(f_{P_k}), R_\beta(f_{P_{k+1}})) > \frac{1}{2^k}) < \frac{1}{2^k} . \tag{3.9}$$

From the Borel-Cantelli lemma, it now follows that

$$\tilde{\Pi}(d(R_\beta(f_{P_k}), R_\beta(f_{P_{k+1}})) > \frac{1}{2^k} \text{ infinitely often}) = 0. \tag{3.10}$$

Let N be the set appearing in the left hand side of (3.10) so that $\tilde{\Pi}(N) = 0$. For $\tilde{\omega} \notin N$, we can get $k_0(\tilde{\omega}) = k_0$ such that for $k \geq k_0$, we have

$$d(R_\beta(f_{P_k})(\tilde{\omega}), R_\beta(f_{P_{k+1}})(\tilde{\omega})) \leq \frac{1}{2^k}$$

and hence by the triangle equality, for $k \geq k_0$ and $j > 1$,

$$\begin{aligned} &d(R_\beta(f_{P_k})(\tilde{\omega}), R_\beta(f_{P_{k+j}})(\tilde{\omega})) \\ &\leq \sum_{i=0}^{j-1} d(R_\beta(f_{P_{k+i}})(\tilde{\omega}), R_\beta(f_{P_{k+i+1}})(\tilde{\omega}) \qquad (3.11) \\ &\leq \sum_{i=0}^{j-1} \frac{1}{2^{k+i}} \leq \frac{1}{2^{k-1}}. \end{aligned}$$

The relation (3.11) shows that for $\tilde{\omega} \notin N$, the sequence $\{R_\beta(f_{P_k})(\tilde{\omega})\}$ of elements in S is Cauchy in the d-metric.

Since S is complete under the metric d, this sequence converges. Thus, define

$$X_f(\tilde{\omega}) = \lim_{k\to\infty} R_\beta(f_{P_k})(\tilde{\omega}) \quad \text{if } \tilde{\omega} \notin N$$

$$= s \quad \text{if } \omega \in N$$

where s is some fixed element of S.

Now as in the proof of Theorem III.3.4, it can be proved that (3.6) holds for this choice of X_f. □

We can now define $R_\beta(f)$ for f in the class $\mathcal{L}(E,\mathcal{E},\beta;S)$.

Definition 3.4: For $f \in \mathcal{L}(E,\mathcal{E},\beta;S)$, define the *β-lifting of* f, $R_\beta(f)$ by

$$R_\beta(f) = \lim_{P\in\mathcal{P}} \text{ in } \tilde{\Pi}\text{-probability } R_\beta(f_P). \tag{3.12}$$

Remark 3.4: Definition (3.12) is easily seen to be consistent with the earlier Definition 2.3 of $R_\beta(f)$ given for the case $S = \mathbb{R}$.

For each Polish space S, R_β is a mapping from $\mathcal{L}(E,\mathcal{E},\beta;S)$ into $\mathcal{L}(\tilde{\Omega},\tilde{\mathcal{A}},\tilde{\Pi};S)$, but to avoid complex notation, we use the same notation R_β for all these mappings.

The following result can be proved by suitably modifying the arguments given in the proof of Theorem III.3.6, namely, by writing d(x,y) for $|x-y|$, f_P for $f\circ P$ and $(E,\mathcal{E},\beta)$ for $(H,\mathcal{C},n)$. We will state the result for future reference.

Theorem 3.3: Let $f: E \to S$ be a $\mathcal{A}\otimes\mathcal{B}(H)/\mathcal{B}(S)$ measurable function. Then the following are equivalent.

(a) $f \in \mathcal{L}(E,\mathcal{E};\beta;S)$.

(b) There exists $\{P_k\} \subseteq \mathcal{P}$, $P_k \uparrow I$ such that for all $\{P'_k\} \subseteq \mathcal{P}$, $P_k \leq P'_k$,

$$R_\beta(f_{P'_k}) \text{ converges a.s. } \tilde{\Pi}. \tag{3.13}$$

Further, if $f \in \mathcal{L}(E,\mathcal{E},\beta;S)$ and $\{P'_k\}$ are as in (b) above, then $R_\beta(f_{P'_k})$ converges a.s. $\tilde{\Pi}$ to $R_\beta(f)$.

We will now deduce some properties of the lifting map.

<u>Theorem 3.4</u>:

(i) Let S_1,S_2 be Polish spaces, and let g: $S_1 \to S_2$ be a continuous function. Suppose $f \in \mathcal{L}(E,\mathcal{E},\beta;S)$. Then $g\circ f \in \mathcal{L}(E,\mathcal{E},\beta;S_2)$ and

$$R_\beta(g\circ f) = g(R_\beta(f)) \quad \text{a.s. } \tilde{\Pi}. \tag{3.14}$$

(ii) Let S_1,S_2,f be as in (i). Suppose $U \in \mathcal{B}(S_1)$ is such that (Range f) $\subseteq U$ and $R_\beta(f) \in U$ a.s. $\tilde{\Pi}$. Let g: $U \to S$ be a continuous function. Then $g\circ f \in \mathcal{L}(E,\mathcal{E},\beta;S_2)$, and (3.14) is true as well.

(iii) Let $S_1,S_2,\ldots,S_i$ be Polish spaces and let $S = S_1 \times S_2 \times \ldots \times S_i$. Let $f_j \in \mathcal{L}(E,\mathcal{E},\beta;S_j)$, $1 \leq j \leq i$ and let f: $E \to S$ be defined by

$$f(\omega,h) = (f_1(\omega,h),f_2(\omega,h),\ldots,f_i(\omega,h)).$$

Then $f \in \mathcal{L}(E,\mathcal{E},\beta;S)$ and

$$R_\beta(f) = (R_\beta(f_1),R_\beta(f_2),\ldots,R_\beta(f_i)) \quad \text{a.s. } \tilde{\Pi}. \tag{3.15}$$

(iv) Let $\{S_j\}$ be a sequence of Polish spaces and let $S = X_{j=1}^{\infty} S_j$. Then S is a Polish space. Let $f_j \in \mathcal{L}(E,\mathcal{E},\beta;S_j)$

for $j \geq 1$ and let $f: E \to S$ be defined by

$$f(\omega,h) = (f_1(\omega,h), f_2(\omega,h), \ldots)$$

Then $f \in \mathcal{L}(E,\mathcal{E},\beta;S)$ and

$$R_\beta(f) = (R_\beta(f_1), R_\beta(f_2), \ldots) \quad \text{a.s. } \tilde{\Pi}. \tag{3.16}$$

Proof: (i) First note that (3.14) holds if f is a cylinder function. This follows from Definition 3.3. Now, let $\{P_k\} \subseteq \mathcal{P}$ be such that (3.13) holds for all $\{P_k'\} \subseteq \mathcal{P}$, $P_k \leq P_k'$. This can be done in view of Theorem 3.3 since $f \in \mathcal{L}(E,\mathcal{E},\beta;S)$. Then, for any $\{P_k'\} \subseteq \mathcal{P}$, $P_k \leq P_k'$,

$$R_\beta((g\circ f)_{P_k'}) = R_\beta(g\circ f_{P_k'}) = g(R_\beta(f_{P_k'})) \tag{3.17}$$

as $f_{P_k'}$ is a cylinder function. Now, (3.13), (3.17) and continuity of g together imply that

$$R_\beta((g\circ f)_{P_k'}) \to g(R_\beta(f)) \quad \text{a.s. } \Pi_0. \tag{3.18}$$

Again from Theorem 3.3, we have (i).

To prove (ii), note that since (Range f) $\subseteq U$, $R_\beta(f_P) \in U$ a.s. $\tilde{\Pi}$ for any $P \in \mathcal{P}$. Also we have assumed that $R_\beta(f) \in U$ a.s. $\tilde{\Pi}$. If $\{P_k\}$, $\{P_k'\}$ are as in (i) above, we have

$$R_\beta(f_{P_k'}) \to R_\beta(f) \quad \text{a.s. } \tilde{\Pi}$$

and hence

$$g(R_\beta(f_{P_k'})) \to g(R_\beta(f)) \quad \text{a.s. } \tilde{\Pi}$$

as g is continuous on U, $R_\beta(f_{P'_k}) \in U$ a.s. $\tilde{\Pi}$ and $R_\beta(f) \in U$ a.s. $\tilde{\Pi}$. The remaining steps are the same as in (i) above.

For (iii), first note that if d_j is a complete metric on S_j, then

$$d((s_1,s_2,\ldots,s_k),\ (s'_1,s'_2,\ldots,s'_i)) := \sum_{j=1}^{i} d_j(s_j,s'_j) \tag{3.19}$$

for $(s_1,s_2,\ldots,s_k) \in S$, $(s'_1,s'_2,\ldots,s'_k) \in S$ is a complete metric on S and under this metric S is a Polish space.

Invoking Theorem 3.3, for each j, $1 \le j \le i$, let $\{P_{k,j}\} \subseteq \mathcal{P}$ be such that for $\{P'_k\} \subseteq \mathcal{P}$, $P_{k',j} \le P'_k$, we have

$$R_\beta(f_{j,P'_k}) \text{ converges to } R_\beta(f_j) \quad \text{a.s. } \tilde{\Pi} \text{ (in } S_j). \tag{3.20}$$

Choose a sequence $\{P_k\} \subseteq \mathcal{P}$, $P_k \uparrow I$ such that $P_{k,j} \le P_k$ for all $k \ge 1$ for $1 \le j \le i$. Then (3.20) holds for all $\{P'_k\}$, $P_k \le P'_k$, $1 \le j \le i$, so that

$$\begin{aligned} R_\beta(f_{P'_k}) &= (R_\beta(f_{1,P'_k}),\ldots,R_\beta(f_{i,P'_k})) \qquad (3.21)\\ &\to (R_\beta(f_1),\ldots,R_\beta(f_i)) \quad \text{a.s. } \tilde{\Pi} \quad \text{(in S)}. \end{aligned}$$

The first equality in (3.21) follows from Definition 3.3 as $f_{P'_k}$ is a cylinder function and the convergence follows from (3.20) and the form of the metric (3.19). This completes the proof of (iii), again using the equivalence of (a) and (b) in Theorem 3.3.

For (iv), if d_j is a complete metric on S_j, then d defined by

$$d(\{s_j\},\{s_j'\}) = \sum_{j=1}^{\infty} [d_j(s_j,s_j') \wedge 1] \cdot \frac{1}{2^j} \tag{3.22}$$

is a metric on S, under which S is a complete separable metric space. Given $f_j \in \mathcal{L}(E,\mathcal{E},\beta;S)$, for each $j = 1,2,\ldots$ choose $\{P_{k,j}\} \subseteq \mathcal{P}$ such that (3.20) holds for all $\{P_k'\} \subseteq \mathcal{P}$, $P_{k,j} \leq P_k'$. Let $P_1 = P_{1,1}$ and for $k \geq 2$ inductively choose $P_k \in \mathcal{P}$ such that $P_{k-1} \leq P_k$ and $P_{k,j} \leq P_k$, $1 \leq j \leq k$. Hence it follows that if $\{P_k'\} \subseteq \mathcal{P}$ is such that $P_k \leq P_k'$ for all k, then (3.20) holds for all $j \geq 1$. The rest of the proof is similar to that of part (iii) given above. □

The first part of the following result is an analogue of Proposition III.2.3. It could be proved without using the lifting map, but the proof is quite complicated. So we have chosen to give a simpler proof using the lifting map.

<u>Theorem 3.5</u>: Let $f \in \mathcal{L}(E,\mathcal{E},\beta;S)$.

(i) The net of countably additive measures $\{\beta\circ f_P^{-1}: P \in \mathcal{P}\}$ on $(S,\mathcal{B}(S))$ converges weakly to a countably additive probability measure λ. The measure λ is called the measure induced by f under β and is denoted by $\lambda = \beta\circ f^{-1}$. Further, for all $g \in C_b(S)$,

$$\int_E g(f)d\beta = \int_S g(x)d(\beta\circ f^{-1})(x). \tag{3.23}$$

(ii) We also have

$$\beta\circ f^{-1} = \tilde{\Pi}\circ[R_\beta(f)]^{-1}. \tag{3.24}$$

<u>Proof</u>: As we have observed earlier (see (3.5)),

$$\beta\circ f_P^{-1} = \tilde{\Pi}\circ[R_\beta(f_P)]^{-1} \tag{3.25}$$

for all $P \in \mathcal{P}$, since f_P is a cylinder function. Now, the net $\{R_\beta(f_P): P \in \mathcal{P}\}$ converges to $R_\beta(f)$ in $\tilde{\Pi}$-probability and hence

$$\tilde{\Pi}\circ[R_\beta(f_P)]^{-1} \to \tilde{\Pi}\circ[R_\beta(f)]^{-1} \tag{3.26}$$

in the sense of weak convergence. The relations (3.25), (3.26) imply

$$\beta\circ f_P^{-1} \to \tilde{\Pi}\circ[R_\beta(f)]^{-1}. \tag{3.27}$$

This proves all assertions, except (3.23). Let $g \in C_b(S)$. Then using Theorem 3.4,

$$\int g(f)d\beta = \int R_\beta(g(f))d\tilde{\Pi} = \int g(R_\beta(f))d\tilde{\Pi}$$

$$= \int g d\tilde{\Pi}\circ[R_\beta(f)]^{-1} = \int g d\beta\circ f^{-1}. \qquad \square$$

Let us now give an example of an H-valued random variable defined on $(H,\mathcal{C},m)$. Following the example, we will put in the perspective of this chapter the notion of an abstract Wiener space by recasting Theorem III.4.1. (See Theorem 3.6 below.)

<u>Example 3.1</u>: Let m be the canonical Gauss measure on $(H,\mathcal{C})$. Let $A: H \to H$ be a self-adjoint Hilbert-Schmidt operator on H. We will show that

$$A \in \mathcal{L}(H,\mathcal{C},m;H). \tag{3.28}$$

Indeed, it was shown in Example III.4.1 that if $\{\alpha_j\}$ are the eigenvalues of A and $\{e_j\}$ are the corresponding eigenvectors, then denoting by P_k the orthogonal projection on $\text{span}\{e_1,e_2,\dots,e_k\}$, we have, for any $P' \in \mathcal{P}$, $P_k \perp P'$, and for any $\delta > 0$

$$m(h: |AP'h| > \delta) \leq \frac{1}{\delta^2} \sum_{j=k+1}^{\infty} \alpha_j^2. \tag{3.29}$$

Now given $\epsilon > 0$, choose k such that $\frac{1}{\delta^2} \Sigma_{j=k+1}^{\infty} \alpha_j^2 < \epsilon$ which can be done since $\Sigma_{j=1}^{\infty} \alpha_j^2 < \infty$. Then for any $P_1', P_2' \in \mathcal{P}$, $P_k \leq P_1'$, $P_k \leq P_2'$, we have, writing $P' = P_1' - P_2'$,

$$\begin{aligned} m(h: |AP_1'h - AP_2'h| > \delta) &= m(h: |AP'h| > \delta) & (3.30) \\ &\leq \frac{1}{\delta^2} \sum_{j=k+1}^{\infty} \alpha_j^2 \quad \text{(by (3.29) as } P' \perp P_k) \\ &< \epsilon. \end{aligned}$$

Note that for the mapping A, $A_P(h) = APh$, where A_P is given by (1.7). Thus (3.30) shows that $A \in \mathcal{L}(H, \mathcal{C}, m; H)$.

As before, let m be the canonical Gauss measure on $(H, \mathcal{C})$. Let $\|\cdot\|_1$ be a measurable norm on H (see Definition III.4.1), let B be the completion of H under $\|\cdot\|_1$ and let γ be the injection from H into B.

We will give a proof of Theorem III.4.1 using the notion of a B-valued random variable on $(H, \mathcal{C}, m)$. Let (L, Π_0) be a representative of m on a representative spaqce $(\Omega_0, \mathcal{A}_0)$ and let R_m be the corresponding m-lifting.

<u>Theorem 3.6</u>:

(i) $\gamma \in \mathcal{L}(H, \mathcal{C}, m; B)$.

(ii) Let $\lambda := \Pi_0 \circ [R_m(\gamma)]^{-1}$, so that λ is a countably additive probability measure on $(B, \mathcal{B}(B))$. Then λ is an extension of the finitely additive probability $m \circ \gamma^{-1}$ defined on cylinder sets in B.

Proof: Given $\epsilon > 0$, $\delta > 0$, let $\epsilon' = \epsilon\wedge\delta$. Let P_C be such that for $P_0' \in \mathcal{P}$, $P_0' \perp P_0$, we have

$$m(h: \|P_0'h\|_1 > \epsilon') < \epsilon'.$$

Such a choice is possible because $\|\cdot\|_1$ is a measurable norm. Now if $P', P'' \in \mathcal{P}$, $P_0 \in P'$, $P_0 \leq P''$, then

$$\begin{aligned} m(h: \|\gamma_{P'}(h)-\gamma_{P''}(h)\|_1 > \epsilon') &= m(h: \|P'h-P''h\|_1 > \epsilon') \\ &= m(h: \|(P'-P'')h\|_1 > \epsilon') \\ &= m(H: \|P_0'h\|_1 > \epsilon') \quad (\text{where } P_0' = P'-P'') \\ &< \epsilon' \qquad\qquad \text{as } P_0' \perp P_0. \end{aligned}$$

Since $\epsilon' = \epsilon\wedge\delta$, we have

$$\begin{aligned} m(h: \|\gamma_{P'}(h)-\gamma_{P''}(h)\|_1 > \delta) &\leq m(h: \|\gamma_{P'}(h(-\gamma_{P''}(h)\|_1 > \epsilon') \\ &< \epsilon' \leq \epsilon. \end{aligned}$$

Thus, $\gamma \in \mathcal{L}(H,\mathcal{C},m;B)$.

For (ii), note that $R_m(\gamma)$ by definition is an element of $\mathcal{L}(\Omega_0,\mathcal{A}_0,\Pi_0;B)$ and thus λ is a countably additive measure on $(B,\mathcal{B}(B))$. For $f \in B^*$,

$$R_m(f\circ\gamma) = f(R_m(\gamma)) \quad \text{a.s. } \Pi_0 \tag{3.31}$$

as f is continuous. On the other hand,

$$f\circ\gamma(h) = f[\gamma(h)] = (\gamma^*(f),h)$$

and hence

$$R_m(f\circ\gamma) = L(\gamma^*(f)). \tag{3.32}$$

Therefore,

$$f(R_m(\gamma)) = L(\gamma^*(f)) \quad \text{a.s. } \Pi_0. \tag{3.33}$$

Noting that $\lambda = \Pi_0 \circ [R_m(\gamma)]^{-1}$ we have for $f \in B^*$

$$\begin{aligned} \int \exp\{if[x]\} d\lambda(x) &= \int \exp\{if[R_m(\gamma)]\} d\Pi_0 \\ &= \int \exp\{iL(\gamma^*(f))\} d\Pi_0 \\ &= \exp(-\tfrac{1}{2}|\gamma^*(f)|^2) \end{aligned} \tag{3.34}$$

since L is a representation of m. Now (3.34) and (IV.4.9) imply that λ agrees with $m \circ \gamma^{-1}$ on cylinder sets. This proves the required assertion. □

Remark 3.4: The measure λ obtained in Theorem 3.6 above is the abstract Wiener measure on $(B, \mathcal{B}(B))$, which was denoted by μ in Section III.4. The proof of Theorem III.4.1 given there is based on the same idea as the one given above but we had not defined the notion of a B-valued random variable then and hence had to work harder. In the notation established earlier in Theorem 3.5, $\lambda = m \circ \gamma^{-1}$. We avoided using this notation in the statement of Theorem 3.6 because $m \circ \gamma^{-1}$ is also being used to denote the induced measure on cylinder sets.

Our next result is a consequence of Theorem 3.6. Let S be a Polish space.

Theorem 3.7: Let $g: B \to S$ be a continuous function and let f be its restriction to H (i.e., $f(h) = g(\gamma(h))$). Then

(i) $f \in \mathcal{L}(H, \mathcal{C}, m; S)$,

(ii) $R_m(f) = g(R_m(\gamma))$, and

(iii) $m \circ f^{-1} = \mu \circ g^{-1}$.

Proof: Since g is continuous on B and $\gamma \in \mathcal{L}(H,\mathcal{C},m;B)$, (i) and (ii) follow from part (i) in Theorem 3.4. Assertion (iii) follows from the relations

$$\begin{aligned} m \circ f^{-1} &= \Pi_0 \circ [R_m(f)]^{-1} \qquad \text{by Theorem 3.5} \\ &= \Pi_0 \circ [g(R_m(\gamma))]^{-1} \\ &= [\Pi_0 \circ [R_m(\gamma)]^{-1}] \circ g^{-1} \\ &= \mu \circ g^{-1}. \end{aligned}$$

□

4. ABSOLUTE CONTINUITY FOR QCP'S, QUASI-CYLINDRICAL MAPPINGS AND CONDITIONAL EXPECTATION

As in the case of cylindrical probabilities, we have the following definition of absolute continuity for QCP's. Throughout this section, $(E,\mathcal{E})$ will be as in Section 1, given by (1.1), (1.2), (1.3).

Definition 4.1: Let β_1, β_2 be quasi-cylinder probabilities on $(E,\mathcal{E})$. β_1 is said to be absolutely continuous with respect to β_2 (written as $\beta_1 \ll \beta_2$) if there exists a function $f \in \mathcal{L}^1(E,\mathcal{E},\beta_2)$, $f \geq 0$, such that for all $f \in \mathcal{E}$

$$\beta_2(F) = \int_F f d\beta_2. \tag{4.1}$$

The function f is called the Radon-Nikodym derivative of β_1 with respect to β_2 and is denoted by $d\beta_1/d\beta_2$.

Remark 4.1: If $\beta_1 \ll \beta_2$, the relation (4.2) determines $d\beta_1/d\beta_2$ uniquely up to equality modulo β_2 (see Theorem 2.9).

The next result is an analogue of Theorem IV.1.1 and can be proved using the arguments in the proof of the latter.

Theorem 4.1: Let β_1, β_2 be QCP's on $(E, \mathcal{E})$ with $\beta_1 \ll \beta_2$.

(i) If $g \in \mathcal{L}(E, \mathcal{E}, \beta_2)$, then $g \in \mathcal{L}(E, \mathcal{E}, \beta_1)$.

(ii) Let $g \in \mathcal{L}(E, \mathcal{E}, \beta_2)$. Then

$$g \in \mathcal{L}^1(E, \mathcal{E}, \beta_1) \text{ if and only if } g \cdot \frac{d\beta_1}{d\beta_2} \in \mathcal{L}(E, \mathcal{E}, \beta_2)$$

and in that case

$$\int_E g d\beta_1 = \int_E g \cdot \frac{d\beta_1}{d\beta_2} \cdot d\beta_2. \tag{4.2}$$

(iii) If $d\beta_1/d\beta_2 > 0$ and $R_{\beta_2}(d\beta_1/d\beta_2) > 0$ a.s. $\tilde{\Pi}$ (where R_{β_2} is a β_2-lifting on a representation space $(\tilde{\Omega}, \tilde{\mathcal{A}}, \tilde{\Pi})$), then $\beta_2 \ll \beta_1$ and

$$\frac{d\beta_2}{d\beta_1} = \left[\frac{d\beta_1}{d\beta_2}\right]^{-1}. \tag{4.3}$$

Quasi-Cylindrical Mapping

Let β be a QCP on $(E, \mathcal{E}) = (\Omega, \mathcal{A}) \odot (H, \mathcal{C})$. Let $(\Omega_1, \mathcal{A}_1)$ be a measurable space and H_1 a real separable Hilbert space with inner product $(\cdot,\cdot)_1$, $\mathcal{C}_1 = \mathcal{C}(H_1)$, $\mathcal{P}_1 = \mathcal{P}(H_1)$. Let $(E_1, \mathcal{E}_1) := (\Omega_1, \mathcal{A}_1) \odot (H_1, \mathcal{C}_1)$. Let $\phi_1: E \to \Omega_1$ and $\phi_1: E \to H_1$ be mappings and let $\phi = (\phi_1, \phi_2): E \to E_1 = \Omega_1 \times H_1$.

Definition 4.2: ϕ is said to be a *quasi-cylindrical mapping* (QCM) from $(E, \mathcal{E}, \beta)$ into $(E_1, \mathcal{E}_1)$ if

$$\phi_1 \text{ is a cylinder function} \tag{4.4}$$

and

$$(h_1, \phi_2)_1 \in \mathscr{L}(E, \mathscr{E}, \beta), \text{ for all } h_1 \in H_1. \tag{4.5}$$

Here and in the sequel $(h_1, \phi_2)_1$ denotes the function $(\omega, h) \to (h_1, \phi_2(\omega, h))_1$ on E.

Remark 4.2: As in the case when $E = H$ and $E_1 = H_1$, it is easy to see that if ϕ_2 satisfies the condition

$$\text{for all } P_1 \in \mathscr{P}_1, \text{ there exists a } P \in \mathscr{P} \text{ such that } \phi_2^{-1}(\mathscr{C}_{1,P_1}) \subseteq \mathscr{E}_P, \tag{4.6}$$

then ϕ_2 satisfies (4.5). Thus if $\phi = (\phi_1, \phi_2)$ satisfies (4.4) and (4.6), then ϕ is a QCM. Indeed, if ϕ satisfies (4.4) and (4.6), then

$$\text{for all } P_1 \in \mathscr{P}_1, \text{ there exists a } P' \in \mathscr{P} \text{ such that } \phi_2^{-1}(\mathscr{E}_{1,P_1}) \subseteq \mathscr{E}_{P'}. \tag{4.7}$$

To see this, note that since ϕ_1 is a cylinder function,

$$\phi_1^{-1}(\mathscr{A}) \subseteq \mathscr{E}_{P''}$$

for some $P'' \in \mathscr{P}$. Given $P_1 \in \mathscr{P}_1$, choose $P \in \mathscr{P}$ such that (4.6) holds and take $P' \in \mathscr{P}$ such that $P \leq P'$ and $P'' \leq P'$. Then

$$\phi_2^{-1}(\mathscr{C}_{1,P_1}) \subseteq \mathscr{E}_{P'}$$

and

$$\phi_1^{-1}(\mathscr{A}) \subseteq \mathscr{E}_{P'}$$

which implies

$$\phi^{-1}(\mathcal{E}_{1,P_1}) = (\phi_1,\phi_2)^{-1}(\mathcal{A}\otimes\mathcal{C}_{1,P_1}) \subseteq \mathcal{E}_{P'} \ .$$

For the rest of the section, let us fix a QCM ϕ: $(E,\mathcal{E},\beta) \to (E_1,\mathcal{E}_1)$ and a representation $(\rho,L,\tilde{\Pi})$ of β on a representative space $(\tilde{\Omega},\tilde{\mathcal{A}},\tilde{\Pi})$ and let R_β be the corresponding β-lifting. Let $\mathcal{F}(E_1,\mathcal{E}_1)$ be the class of functions of the form

$$f(\omega_1,h_1) = f^1(\omega_1)\cdot f^2(h_1) \tag{4.9}$$

where $f^1$is an $\mathcal{A}_1$-measurable real valued function and f^2 is a continuous cylinder function on H_1.

Let ρ_1: $\tilde{\Omega} \to \Omega'$ and L_1: $H_1 \to \mathcal{L}(\tilde{\Omega},\tilde{\mathcal{A}},\tilde{\Pi})$ be defined by

$$\rho_1(\tilde{\omega}) := R_\beta(\phi_1)(\tilde{\omega}), \quad \tilde{\omega} \in \tilde{\Omega} \tag{4.10}$$

and

$$L_1(h_1) := R_\beta((h_1,\phi_2)_1), \quad h_1 \in H_1. \tag{4.11}$$

Note that ρ_1 and L_1 are well defined as ϕ_1 is a cylinder function and $(h_1,\phi_2)_1 \in \mathcal{L}(E,\mathcal{E},\beta)$.

We have the following result which shows the existence of the induced measure $\beta\circ\phi^{-1}$ and gives some of its properties.

<u>Theorem 4.2</u>: There exists a unique QCP β_1 on $(E_1,\mathcal{E}_1)$ such that for all bounded functions f belonging to $\mathcal{F}(E_1,\mathcal{E}_1)$ we have

$$\int_{E_1} f d\beta_1 = \int_E f\circ\phi d\beta. \tag{4.12}$$

Further, $(\rho_1, L_1, \tilde{\Pi})$ is a representation of β_1 and if R_{β_1} is the corresponding β_1-lifting, then for all $f \in \mathscr{F}(E_1, \mathscr{E}_1)$

$$R_{\beta_1}(f) = R_\beta(f \circ \phi). \tag{4.13}$$

<u>Proof</u>: For $P_1 \in \mathscr{P}_1$, let $F_{P_1} : \tilde{\Omega} \to P_1H_1$ be defined by

$$F_{P_1}(\tilde{\omega}) := \sum_{j=1}^{k} L_1(e_j)(\tilde{\omega})e_j \tag{4.14}$$

where $e_1, e_2, \ldots, e_k$ is an orthonormal basis of P_1H_1 and let

$$\lambda_{P_1}(B) := \tilde{\Pi}(\tilde{\omega}\colon (\rho_1, (\tilde{\omega}), F_{P_1}(\tilde{\omega})) \in B),\ B \in \mathscr{A}\otimes\mathscr{B}(P_1H_1). \tag{4.15}$$

Let $P_1, P_2 \in \mathscr{P}_1$ be such that $P_1 \leq P_2$. Denoting by P_1 again, the restriction of P_1 to P_2H_1, using the linearity of L_1, it can be checked that

$$P_1[F_{P_2}(\tilde{\omega})] = F_{P_1}(\tilde{\omega}). \tag{4.16}$$

For $P_1 \in \mathscr{P}_1$, let

$$\beta_{1,P_1}((\omega_1, h_1)\colon (\omega_1, P_1h_1) \in B) := \lambda_{P_1}(B),\ B \in \mathscr{A}\otimes\mathscr{B}(P_1H_1). \tag{4.17}$$

Then (4.17) defines a countably additive probability measure on $\mathscr{E}_{1,P_1}$ and the relation (4.16) implies that for $P_1 \leq P_2$, the restriction of β_{1,P_2} to $\mathscr{E}_{1,P_1}$ is equal to β_{1,P_1}. Thus $\{\beta_{1,P_1} : P_1 \in \mathscr{P}_1\}$ is a consistent family and

$$\beta_1(D) = \beta_{1,P_1}(D) \quad \text{if } D \in \mathscr{E}_{1,P_1} \tag{4.18}$$

unambiguously defines a set function on $\mathcal{E}_1$, which is a quasi-cylinder probability. From the relations (4.14), (4.15), (4.17) and (4.18), it follows that for $k \geq 1$, $e_1, e_2, \ldots, e_k$ orthonormal, $A \in \mathcal{A}$ and $B_1 \in \mathcal{B}(\mathbb{R}^k)$, we have

$$\beta_1((\omega_1, h_1)\colon \omega_1 \in A,\ ((h_1, e_1)_1, \ldots, (h_1, e_k)_1) \in B_1)$$
$$= \tilde{\Pi}(\tilde{\omega}\colon \rho_1(\tilde{\omega}) \in A,\ (L_1(e_1), \ldots, L_1(e_k))(\tilde{\omega}) \in B_1). \tag{4.19}$$

By linearity of L_1, it follows that (4.19) holds even if $e_1, e_2, \ldots, e_k$ are arbitrary elements in H_1 and thus $(\rho_1, L_1, \tilde{\Pi})$ is a representation of β_1 (see Remark 2.1).

Let $f \in \mathcal{F}(E_1, \mathcal{E}_1)$ be given by (4.9). Then

$$f \circ \phi = f^1(\phi_1) \cdot f^2(\phi_2). \tag{4.20}$$

Since ϕ_1 is a cylinder function, so if $f^1(\phi_1)$ and it follows from the definition of lifting for cylinder functions that

$$R_\beta(f^1(\phi_1)) = f^1(R_\beta(\phi_1)) = f^1(\rho_1). \tag{4.21}$$

Recall that f^2 is a continuous cylinder function, and thus for some $P_1 \in \mathcal{P}_1$,

$$f^2(h_1) = f^2(P_1 h_1).$$

Now $P_1\phi_2 = \Sigma_{j=1}^k (e_j, \phi_2) e_j$, if $e_1, e_2, \ldots, e_k$ is an orthonormal basis of $P_1 H_1$, and hence

$$R_\beta(P_1\phi_2) = \sum_{j=1}^k R_\beta((e_j, \phi_2)) e_j \qquad \text{a.s. } \tilde{\Pi}$$

$$= \sum_{j=1}^{k} L_1(e_j)e_j \qquad \text{a.s. } \tilde{\Pi}$$

$$= F_{P_1}.$$

Since f^2 is continuous, we have

$$R_\beta(f^2 \circ \phi_2) = R_\beta(f^2(P_1\phi_2)) \tag{4.22}$$

$$= f^2(R_\beta(P_1\phi_2)) \qquad \text{a.s. } \tilde{\Pi}$$

$$= f^2(F_{P_1}) \qquad \text{a.s. } \tilde{\Pi}.$$

Thus

$$R_\beta(f\circ\phi) = R_\beta((f^1\circ\phi_1)\cdot(f^2\circ\phi_2)) \tag{4.23}$$

$$= R_\beta(f^1\circ\phi_1)\cdot R_\beta(f^2\circ\phi_2) \qquad \text{a.s. } \tilde{\Pi}$$

$$= f^1(\rho_1)\cdot f^2(F_{P_1}) \qquad \text{a.s. } \tilde{\Pi}.$$

Since f is a cylinder function on $(E_1,\mathcal{E}_1)$, it follows that

$$R_{\beta_1}(f) = f^1(\rho_1)\cdot f^2(F_{P_1}). \tag{4.24}$$

This proves (4.13). The relation (4.12) follows from (4.13) and the definition of the integral. It remains to prove that the relation (4.12) uniquely determines β_1.

Suppose β_1 and β_1' are QCP's on $(E_1,\mathcal{E}_1)$ such that for all $f \in \mathcal{F}(E_1,\mathcal{E}_1)$, f bounded

$$\int f d\beta_1 = \int f d\beta_1'. \tag{4.25}$$

Fix $P_1 \in \mathcal{P}_1$. Then the restrictions β_{1,P_1}, β'_{1,P_1} of β_1, β'_1 to $\mathcal{E}_{1,P_1}$ are countably additive. Let $A \in \mathcal{A}_1$ and $C \in \mathcal{C}_{1,P_1}$. We will show that

$$\beta_{1,P_1}(A\times C) = \beta'_{1,P_1}(A\times C). \tag{4.26}$$

This would give the equality of β_{1,P_1} and β'_{1,P_1} (on $\mathcal{E}_{1,P_1}$) and as P_1 is arbitrary, would show that $\beta_1 = \beta'_1$.

Let C be given by

$$C = \{h_1 \in H_1: ((h_1,e_1)_1,\dots,(h_1,e_k)_1) \in B\} \tag{4.27}$$

for $k \geq 1$, $e_1, e_2, \dots, e_k \in H_1$, $B \in \mathcal{B}(\mathbb{R}^k)$. Let

$$\mu(B) = \tfrac{1}{2}\beta_{1,P_1}(\Omega_1\times C) + \tfrac{1}{2}\beta'_{1,P_1}(\Omega_1\times C) \tag{4.28}$$

where $B \in \mathcal{B}(\mathbb{R}^k)$ and C are related to (4.27). Then μ is a countably additive probability measure on $\mathbb{R}^k$. Now by using Luzin's theorem, there exists a sequence of continuous functions g_k on $\mathbb{R}^k$, $0 \leq g_k \leq 1$ such that

$$g_k \to 1_B \quad \text{a.s. } \mu. \tag{4.29}$$

Let $f_k \in \mathcal{F}(E_1, \mathcal{E}_1)$ be defined by

$$f_k(\omega_1,h_1) = 1_A(\omega_1)\cdot g_k((h_1,e_1)_1,\dots,(h_1,e_k)_1). \tag{4.30}$$

Then in view of (4.28) and (4.29),

$$f_k(\omega_1,h_1) \to 1_{A\times C}(\omega_1,h_1) \tag{4.31}$$

a.s. β_{1,P_1} and a.s. β'_{1,P_1}. Now (4.25) holds for f_k since

$f_k \in \mathcal{F}(E_1,\mathcal{E}_1)$. Also f_k is $\mathcal{E}_{1,P_1}$ measurable. So we have

$$\int f_k d\beta_{1,P_1} = \int f_k d\beta'_{1,P_1}. \tag{4.32}$$

Now (4.31), (4.32) and the dominated convergence theorem give us

$$\beta_{1,P_1}(A\times C) = \beta'_{1,P_1}(A\times C).$$

As noted earlier, this completes the proof. □

Remark 4.3: Let $\phi = (\phi_1,\phi_2)$ satisfy (4.4) and (4.6). Then, as noted in Remark 4.2, (4.7) holds. In this case, the measure β_1 appearing in (4.12) can be directly defined by

$$\beta_1(F) = \beta((\omega,h)\colon \phi(\omega,h) \in F) \text{ for } F \in \mathcal{E}_1 \tag{4.33}$$

and then it follows that (4.12) holds for all bounded real valued cylinder functions. Also, it can be verified that (4.13) holds for all cylinder functions.

Definition 4.3: The QCP β_1 on $(E_1,\mathcal{E}_1)$ (given by (4.12)) is defined to be the *measure induced by the QCM ϕ under β* and will be denoted by $\beta_1 = \beta\circ\phi^{-1}$. The representation $(\rho_1,L_1,\tilde{\Pi})$ defined by (4.10) and (4.11) will be called the *representation of β_1 induced by ϕ* (corresponding to the representation $(\rho,L,\tilde{\Pi})$ of β).

Theorem 4.3: Let β be a quasi-cylinder probability on $(E,\mathcal{E})$ and let $(\rho,L,\tilde{\Pi})$ be a representation of β on a representation space $(\tilde{\Omega},\tilde{\mathcal{A}})$. Let R_β denote the corresponding β-lifting.

(a) Let $A \in \mathscr{A}$, $C \in \mathscr{C}$. Then there exists a sequence $\{f_k\} \subseteq \mathscr{F}(E,\mathscr{E})$, $0 \le f_k \le 1$, such that

$$R_\beta(f_k) \to R_\beta(1_{A\times C}) \quad \text{a.s. } \tilde{\Pi}.$$

(b) Let $g_1, g_2 \in \mathscr{L}^1(E,\mathscr{E},\beta)$ be such that for all bounded f belonging to $\mathscr{F}(E,\mathscr{E})$

$$\int g_1 f d\beta = \int g_2 f d\beta. \tag{4.34}$$

Then $g_1 \equiv g_2 \mathrm{mod}\ [\beta]$.

Proof: (a) Let $C \in \mathscr{C}$ be of the form

$$C = \{h: ((h,h_1),\ldots,(h,h_i)) \in B\}, \tag{4.35}$$

$i \ge 1$, $h_1,\ldots,h_i \in H$ and $B \in \mathscr{B}(\mathbb{R}^i)$. Then by the definition of β-lifting for cylinder functions, we have

$$R_\beta(1_{A\times C}) = 1_D \tag{4.36}$$

where

$$D = \{\tilde{\omega}: \rho(\tilde{\omega}) \in A,\ (L(h_1)(\tilde{\omega}),\ldots,L(h_i)(\tilde{\omega})) \in B\}. \tag{4.37}$$

Let $\mu = \tilde{\Pi}\circ[L(h_1),\ldots,L(h_i)]^{-1}$. Let $\{g_k\}$ be a sequence of continuous function on $\mathbb{R}^i$, such that $0 \le g_k \le 1$ and

$$g_k(x) \to 1_B(x) \quad \text{a.s. } \mu.$$

(This can be done once again using Luzin's Theorem). Let $f_k \in \mathscr{F}(E,\mathscr{E})$ be defined by

$$f_k(\omega,h) = 1_A(\omega)\cdot g_k((h,h_1),\ldots,(h,h_k)).$$

Then

$$R_\beta(f_k)(\tilde{\omega}) = 1_A(\rho(\tilde{\omega}))\cdot g_k(L(h_1)(\tilde{\omega}),\ldots,L(h_k)(\tilde{\omega}))$$

$$\to 1_D(\tilde{\omega}) \quad \text{a.s. } \tilde{\Pi}$$

by the choice of the sequence $\{g_k\}$. This proves (a).

(b) The relation (4.34) can be expressed as

$$\int R_\beta(g_1)\cdot R_\beta(f)d\tilde{\Pi} = \int R_\beta(g_2)\cdot R_\beta(f)d\tilde{\Pi}.$$

Using part (a) and the dominated convergence theorem, it follows that for $D \in \tilde{\mathscr{A}}$ given by (4.37),

$$\int_D R_\beta(g_1)d\tilde{\Pi} = \int_D R_\beta(g_2)d\tilde{\Pi}. \tag{4.38}$$

Let $\mathscr{D}_1$ be the collection of sets D of the form (4.37) for $i \geq 1$, $h_1,\ldots,h_i \in H$ and $B \in \mathscr{B}(\mathbb{R}^i)$ and let $\mathscr{D}_2$ be the field generated by $\mathscr{D}_1$. Then $\mathscr{D}_2$ is given by

$$\mathscr{D}_2 = \{ \bigcup_{j=1}^{k} D_j : D_j \in \mathscr{D}_1;\ D_1,D_2,\ldots,D_k \text{ disjoint, } k \geq 1\}$$

Thus (4.38) holds for all $D \in \mathscr{D}_2$. Standard arguments in measure theory now give that (4.38) holds for all $D \in \sigma(\mathscr{D}_2) := \mathscr{D}_3$. It is easy to see that the mappings $\{\rho, L(h_i);\ h \in H\}$ are $\mathscr{D}_3$ measurable and hence $\{R_\beta(g):\ g \in \mathscr{L}(E,\mathscr{E},\beta)\}$ is a collection of $\overline{\mathscr{D}}_3$ measurable functions. This gives $R_\beta(g_1) = R_\beta(g_2)$ a.s. $\tilde{\Pi}$ and then Theorem 2.9 yields the required assertion, $g_1 \equiv g_2 \bmod [\beta]$.

<u>Remark 4.4</u>: Let $\mathscr{D}_3$ be as in the proof given above. Then it is easy to see that $\mathscr{D}_3$ is the σ-field generated by the

family $\{\rho, L(h)\colon h \in H\}$. Part (a) above also shows that $\overline{\mathcal{D}}_3$ is the smallest σ-field with respect to which the family $\{R_\beta(f)\colon f \in \mathcal{F}(E,\mathcal{E})\}$ is measurable.

Let S be a Polish space. Let $\beta_1, \rho_1, L_1, \tilde{\Pi}, R_{\beta_1}$ be as in Theorem 4.2, also fixed for the rest of the section. Let

$$\mathcal{U}(E,\mathcal{E},\beta;\phi,S) := \{f \in \mathcal{L}(E_1,\mathcal{E}_1,\beta_1;S)\colon f\circ\phi \in \mathcal{L}(E,\mathcal{E},\beta;S)$$

$$\text{and } R_{\beta_1}(f) = R_\beta(f\circ\phi)\}.$$

As in Section IV.2, we will write $\mathcal{U}(\phi)$ for $\mathcal{U}(E,\mathcal{E},\beta;\phi,S)$ when there is no possibility of confusion. Also, we will write $\mathcal{U}(E,\mathcal{E},\beta;\phi)$ for $\mathcal{U}(E,\mathcal{E},\beta;\phi,\mathbb{R})$.

It is easy to see that if g_1 and $g_2 \in \mathcal{U}(\phi)$ then g_1+g_2, $g_1\cdot g_2$ also belong to $\mathcal{U}(\phi)$. In view of Theorem 4.2, we have

$$\mathcal{F}(E_1,\mathcal{E}_1) \subseteq \mathcal{U}(E,\mathcal{E},\beta;\phi).$$

If ϕ satisfies (4.4), (4.6), then by Remark 4.3, all S-valued cylinder functions on $(E_1,\mathcal{E}_1)$ also belong to $\mathcal{U}(E,\mathcal{E},\beta;\phi,S)$.

Proceeding as in Remark 2.2, it can be verified that the class $\mathcal{U}(E,\mathcal{E},\beta;\phi,S)$ does not depend upon the choice of the representation of β.

By taking Ω or Ω_1 to be a singleton, the above definitions cover the cases of mappings from $(H,\mathcal{C},n)$ into $(E_1,\mathcal{E}_1)$ and from $(E,\mathcal{E},\beta)$ into $(H_1,\mathcal{C}_1)$ (see Remark 1.1).

The following result corresponds to Theorem IV.2.6 and deals with the composition of two QCM's. Let $(E_2,\mathcal{E}_2) = (\Omega_2,\mathcal{A}_2)\odot(H_2,\mathcal{C}_2)$ where $(\Omega_2,\mathcal{A}_2)$ is a real separable Hilbert space with inner product $(\cdot,\cdot)_2$, $\mathcal{C}_2 = \mathcal{C}(H_2)$, $\mathcal{P}_2 = \mathcal{P}(H_2)$.

Let $\phi' = (\phi_1',\phi_2')$ be a QCM from $(E_1,\mathcal{E}_1,\beta_1)$ into $(E_2,\mathcal{E}_2)$

and let $\beta_2 = \beta_1 \circ [\phi']^{-1}$. Recall that ϕ is a QCM from $(E,\mathcal{E},\beta)$ into $(E_1,\mathcal{E}_1)$ and $\beta_1 = \beta \circ \phi^{-1}$.

Theorem 4.4: Suppose that

$$\phi_1' \circ \phi \text{ is a cylinder function on } (E,\mathcal{E}) \tag{4.39}$$

and

$$(h_2, \phi_2') \in \mathcal{U}(E,\mathcal{E},\beta;\phi) \text{ for each } h_2 \in H_2. \tag{4.40}$$

Then $\phi' \circ \phi$ is a QCM from $(E,\mathcal{E},\beta)$ into $(E_2,\mathcal{E}_2)$ with $\beta_2 = \beta \circ [\phi' \circ \phi]^{-1}$. Further, if $f\colon E_2 \to S$ is such that

$$f \in \mathcal{U}(E_1,\mathcal{E}_1,\beta_1;\phi',S) \tag{4.41}$$

and

$$f \circ \phi' \in \mathcal{U}(E,\mathcal{E},\beta;\phi,S) \tag{4.42}$$

then

$$f \in \mathcal{U}(E,\mathcal{E},\beta;\phi' \circ \phi,S). \tag{4.43}$$

If both ϕ,ϕ' satisfy (4.4) and (4.6), then (4.39) and (4.40) are satisfied.

Proof: Condition (4.40) implies that

$$(h_2, \phi_2' \circ \phi)_2 \in \mathcal{L}(E,\mathcal{E},\beta) \text{ for all } h_2 \in H_2. \tag{4.44}$$

Thus $\phi' \circ \phi = (\phi_1' \circ \phi,\ \phi_2' \circ \phi)$ satisfies (4.39), (4.44) and hence is a QCM from $(E,\mathcal{E},\beta)$ into $(E_2,\mathcal{E}_2)$. Then as in the proof of Threorem IV.2.6, it can be shown that $\beta_2 = \beta \circ [\phi' \circ \phi]^{-1}$ and that (4.43) holds.

If both ϕ,ϕ' satisfy (4.4) and (4.6), then ϕ is itself a cylinder mapping as noted in Remark 4.2 and hence it follows that $\phi_1'\circ\phi$ is a cylinder mapping. Also, (h_2,ϕ_2') is a cylinder mapping because ϕ_2' satisfies (4.6) and hence belongs to $\mathcal{U}(\phi)$ as observed in Remark 4.3. This completes the proof. □

Remark 4.5: The definition of a QCM can be generalized in several ways. First, if Ω_1 is a Polish space and $\mathcal{A}_1 = \mathcal{B}(\Omega_1)$, then the condition (4.4) can be replaced by

$$\phi_1 \in \mathcal{L}(E,\mathcal{E},\beta;\Omega_1).$$

In general, (4.4) can be replaced by the following. Suppose that for some Polish space S, ϕ_1 can be written as $\phi_1 = \Phi\circ g$, where $g \in \mathcal{L}(E,\mathcal{E},\beta;S)$ and Φ is a measurable mapping from S into Ω_1. In this case, the induced measure can be defined by (4.12), but the class $\mathcal{F}(E_1,\mathcal{E}_1)$ has to be modified suitably -- it would depend on S. We do not have an occasion to use this and hence we will not go into details.

Conditional Expectation

The following definition is patterned after Definition IV.3.1.

Definition 4.4: Let ϕ be a QCM from $(E,\mathcal{E},\beta)$ into $(E_1,\mathcal{E}_1)$ and $\beta_1 = \beta\circ\phi^{-1}$. Suppose $g \in \mathcal{L}^1(E,\mathcal{E},\beta)$. If there exists $g_1 \in \mathcal{U}(E,\mathcal{E},\beta;\phi)$, $g_1 \in \mathcal{L}^1(E_1,\mathcal{E}_1,\beta_1)$ such that for all $f \in \mathcal{F}(E_1,\mathcal{E}_1)$,

$$\int_E g(\omega,h)f(\phi(\omega,h))d\beta(\omega,h) = \int_{E_1} g_1(\omega_1,h_1)f(\omega_1,h_1)d\beta_1(\omega_1,h_1) \tag{4.45}$$

then $g_1\circ\phi$ is defined to be the *conditional expectation* of g given ϕ, and is written as

$E_\beta(g|\phi) = g_1 \circ \phi$.

Remark 4.6: The existence of $E_\beta(g|\phi)$ for every $g \in \mathscr{L}^1(E,\mathscr{E},\beta)$ is not asserted. It may not exist for some g, ϕ and β.

Remark 4.7: (Uniqueness). Let $g \in \mathscr{L}^1(E,\mathscr{E},\beta)$ be such that $E_\beta(g|\phi)$ exists. Then it is determined uniquely up to equality modulo β. For if $g_1, g_2 \in \mathscr{U}(\phi)$ satisfy (4.45), then by Theorem 4.3, $g_1 = g_2 \bmod [\beta_1]$. Using $g_1, g_2 \in \mathscr{U}(\phi)$, it follows that $g_1 \circ \phi = g_2 \circ \phi \bmod[\beta]$.

Remark 4.8: In view of the requirement $g_1 \in \mathscr{U}(\phi)$, the condition (4.45) is equivalent to

$$\int_E g \cdot f(\phi) d\beta = \int_E g_1(\phi) \cdot f(\phi) d\beta. \tag{4.46}$$

Let $\mathscr{D}_\phi$ be the smallest σ-field on $\tilde{\Omega}$ with respect to which the family $\{R_\beta(f \circ \phi): f \in \mathscr{F}(E_1,\mathscr{E}_1)\}$ is measurable. Since $\mathscr{F}(E_1,\mathscr{E}_1) \subseteq \mathscr{U}(\phi)$, $\mathscr{D}_\phi$ can also be described as the smallest σ-field on $\tilde{\Omega}$ with respect to which the family $\{R_{\beta_1}(f): f \in \mathscr{F}(E_1,\mathscr{E}_1)\}$ is measurable. Let $\overline{\mathscr{D}}_\phi$ be the σ-field generated by $\mathscr{D}_\phi$ and Π_0-null sets.

The following result corresponds to Theorem IV.3.2 in the set of the present chapter.

Theorem 4.5:

(i) $\overline{\mathscr{D}}_\phi$ is the smallest σ-field on $\tilde{\Omega}$ with respect to which the family $\{\rho_1, L_1(h_1), h_1 \in H_1\}$ and $\tilde{\Pi}$-null sets in $\tilde{\mathscr{A}}$ are measurable. Further, for all $g_1 \in \mathscr{L}(E_1,\mathscr{E}_1,\beta_1)$, $R_{\beta_1}(g_1)$ is $\overline{\mathscr{D}}_\phi$ measurable.

(ii) If $g \in \mathcal{U}(E,\mathcal{E},\beta;\phi)$, then $R_\beta(g\circ\phi)$ is $\overline{\mathcal{D}}_\phi$ measurable.

(iii) Let $g \in \mathcal{L}^1(E,\mathcal{E},\beta)$ be such that $E_\beta(g|\phi)$ exists. Then

$$R_\beta[E_\beta(g|\phi)] = E_{\Pi_0}[R_\beta(g)|\overline{\mathcal{D}}_\phi], \quad \text{a.s. } \tilde{\Pi}. \tag{4.47}$$

(iv) Let $g \in \mathcal{L}^1(E,\mathcal{E},\beta)$. If there exists a $g_1 \in \mathcal{U}(\phi)$ such that

$$R_\beta(g_1\circ\phi) = E_{\Pi_0}[R_\beta(g)|\overline{\mathcal{D}}_\phi], \quad \text{a.s. } \tilde{\Pi}. \tag{4.48}$$

then $E_\beta(g|\phi)$ exists and is equal to $g_1\circ\phi$.

<u>Proof</u>: Part (i) is a restatement of Remark 4.4. The rest of the proof can be completed by following the proof of Theorem IV.3.2.

The following properties can be deduced from the definition of conditional expectation and Theorem 4.5 exactly as in the proof of Theorem IV.3.3.

<u>Theorem 4.6</u>: Let ϕ be a QCM from $(E,\mathcal{E},\beta)$ into $(E_1,\mathcal{E}_1)$ and let $\beta_1 = \beta\circ\phi^{-1}$. Let $g,g' \in \mathcal{L}^1(E,\mathcal{E},\beta)$ be such that $E_\beta(g|\phi)$ and $E_\beta(g'|\phi)$ exist.

(i) Let $a,a' \in \mathbb{R}$. Then $E_\beta(ag+a'g'|\phi)$ exists and

$$E_\beta(ag+a'g'|\phi) = aE_\beta(g|\phi) + a'E_\beta(g'|\phi). \tag{4.49}$$

(ii) Let $f \in \mathcal{U}(\phi)$ be such that $f(\phi)\cdot g \in \mathcal{L}^1(E,\mathcal{E},\beta)$. Then $E_\beta(g\cdot f(\phi)|\phi)$ exists,

$$E_\beta(g|\phi)\cdot f(\phi) \in \mathcal{L}^1(E,\mathcal{E},\beta),$$

$$\int g\cdot f(\phi)d\beta = \int E_\beta(g|\phi)\cdot f(\phi)d\beta \tag{4.50}$$

and

$$E_\beta(g\cdot f(\phi)|\phi) = f(\phi)\cdot E_\beta(g|\phi). \tag{4.51}$$

(iii) If $g^2 \in \mathcal{L}^1(E,\mathcal{E},\beta)$, then $[E_\beta(g|\phi)]^2 \in \mathcal{L}^1(E,\mathcal{E},\beta)$ and

$$\int[g-E_\beta(g|\phi)]^2 d\beta = \min_{f_0\in\mathcal{U}(\phi)} \int[g-f_0(\phi)]^2 d\beta.$$

The most important example of a QCM to be considered later is of the type ϕ: $E \to H_1$, where H_1 is a Hilbert space. This case is covered by earlier definitions, by taking Ω_1 to be a singleton, in which case $E_1 = \Omega_1\times H_1$ is identified with H_1.

Let us state for future use the definition of a QCM and conditional expectation for such a QCM.

Let $(E,\mathcal{E},\beta)$ be a QCP, H_1 be a separable Hilbert space $\mathcal{C}_1 = \mathcal{C}(H_1)$. A mapping ϕ: $E \to H_1$ is a QCM from $(E,\mathcal{E},\beta)$ into $(H_1,\mathcal{C}_1)$ if for all $h_1 \in H_1$,

$$(\phi,h_1) \in \mathcal{L}(E,\mathcal{E},\beta). \tag{4.52}$$

The class $\mathcal{F}(E_1,\mathcal{E}_1)$ where $E_1 = \Omega_1\times H_1$, Ω_1 a singleton, reduces to the class of all bounded continuous cylinder functions on H_1. Thus, for $g \in \mathcal{L}^1(E,\mathcal{E},\beta)$,

$$E_\beta(g|\phi) = g_1\circ\phi \tag{4.53}$$

if and only if $g_1 \in \mathcal{U}(E,\mathcal{E},\beta;\phi)$ and

$$\int g\cdot f\circ\phi d\beta = \int(g_1\circ\phi)(f\circ\phi)d\beta \tag{4.54}$$

for all bounded continuous cylinder functions f on H_1.

Suppose that ϕ has the property that for all $P_1 \in \mathcal{P}(H_1)$, $\exists\ P \in \mathcal{P}(H)$, such that

$$\phi^{-1}(\mathcal{C}_1, P_1) \subseteq \mathcal{E}_P. \tag{4.55}$$

Then it is easy to check that ϕ satisfies (4.52) and hence that ϕ is a QCM. In this case,

$$1_{C_1}(\phi) \in \mathcal{L}^1(E, \mathcal{E}, \beta)$$

for all $C_1 \in \mathcal{C}_1$. The definition of conditional expectation can be recast in this case as follows. For $g \in \mathcal{L}^1(E, \mathcal{E}, \beta)$, $E_\beta(g|\phi)$ exists and equals $g_1 \circ \phi$ if and only if there exists a $g_1 \in \mathcal{U}(E, \mathcal{E}, \beta; \phi)$ such that

$$\int g 1_{C_1}(\phi) d\beta = \int (g_1 \circ \phi) \cdot 1_{C_1}(\phi) d\beta \tag{4.56}$$

for all $C_1 \in \mathcal{C}_1$.

To verify this statement, we need to check that the validity of (4.53) for all bounded continuous cylinder functions f on H_1 is equivalent to the validity of (4.56) for all $C_1 \in \mathcal{C}_1$. That the first part implies the second can be proved using arguments given in Remark IV.3.4. For the other part, if (4.56) holds for all $C_1 \in \mathcal{C}_1$, it can be checked that

$$E_{\tilde{\Pi}}(R_\beta(g) | \mathcal{D}_\phi) = R_\beta(g_1 \circ \phi) \tag{4.57}$$

where $\mathcal{D}_\phi$ is as in Theorem 4.4. Then (iv) of the said Theorem implies that $E_\beta(g|\phi) = g_1(\phi)$.

5. INDEPENDENCE

We will now define the notion of independence for random variables on $(E, \mathcal{E}, \beta)$ and for a random variable and a QCM. Our aim is to prove several results which should convince the reader that the notion of independence introduced here and that of conditional expectation introduced in the previous section are the right analogues of the corresponding notions in the countably additive theory.

Throughout this section, we will use the notation introduced in the previous section. The letter S with or without a subscript will denote a Polish space.

<u>Definition 5.1</u>: Let $f_i \in (E, \mathcal{E}, \beta; S_i)$, $1 \le i \le k$. Then $f_1, f_2, \ldots, f_k$ is said to be an *independent collection* (of random variables) if for all $g_i \in C_b(S_i)$, $1 \le i \le k$,

$$\int \prod_{i=1}^{k} g_i(f_i) d\beta = \prod_{i=1}^{k} \int g_i(f_i) d\beta. \tag{5.1}$$

f_1 is said to be independent of f_2 if $\{f_1, f_2\}$ is an independent collection, i.e., if (5.1) holds with $k = 2$.

The following result gives some equivalent conditions for $\{f_1, f_2, \ldots, f_k\}$ to be an independent collection.

<u>Theorem 5.1</u>: Let $f_i \in (E, \mathcal{E}, \beta; S_i)$, $1 \le i \le k$. Then the following are equivalent.

(i) $\{f_1, f_2, \ldots, f_k\}$ is an independent collection on $(E, \mathcal{E}, \beta)$.

(ii) $\{R_\beta(f_1), R_\beta(f_2), \ldots, R_\beta(f_k)\}$ is an independent collection on $(\tilde{\Omega}, \tilde{\mathcal{A}}, \tilde{\Pi})$.

(iii) Let $f\colon E \to S = S_1 \times S_2 \times \ldots \times S_k$ be defined by $f(\omega, h) = (f_1(\omega, h), \ldots, (f_k(\omega, h))$, $(\omega, h) \in E$. Then

$$\beta\circ f^{-1} = \bigotimes_{i=1}^{k} \beta\circ f_i^{-1}. \tag{5.2}$$

Proof: Let $g_i \in C_b(S_i)$, $1 \leq i \leq k$. Then by Theorem 3.4,

$$R_\beta(g_i(f_i)) = g_i(R_\beta(f_i))$$

and hence (5.1) is equivalent to

$$\int \prod_{i=1}^{k} g_i(R_\beta(f_i))d\tilde{\Pi} = \prod_{i=1}^{k} \int g_i(R_\beta(f_i))d\tilde{\Pi}. \tag{5.3}$$

Now (5.3) holds for all $g_i \in C_b(S_i)$ if and only if $\{R_\beta(f_1), \dots, R_\beta(f_k)\}$ is an independent collection on $(\tilde{\Omega}, \tilde{\mathcal{A}}, \tilde{\Pi})$. This proves the equivalence of (i) and (ii).

We will now prove that (ii) and (iii) are equivalent. For this note that by the definition of the induced measures $\beta\circ f^{-1}$, $\beta\circ f_i^{-1}$, we have

$$\beta\circ f^{-1} = \tilde{\Pi}\circ[R_\beta(f)]^{-1} \tag{5.4}$$

and

$$\beta\circ f_i^{-1} = \tilde{\Pi}\circ[R_\beta(f_i)]^{-1}. \tag{5.5}$$

Thus (5.2) is equivalent to

$$\tilde{\Pi}\circ[R_\beta(f)]^{-1} = \bigotimes_{i=1}^{k} \tilde{\Pi}\circ[R_\beta(f_i)]^{-1}. \tag{5.6}$$

Clearly (5.6) is equivalent to (ii). This completes the proof. □

Though we have not defined the characteristic function of a real or vector valued random variables on $(E, \mathcal{E}, \beta)$, it can be defined in an obvious manner. The following result is an analogue of a standard result in countably additive probability theory.

Theorem 5.2: Let $f_j \in \mathscr{L}(E,\mathscr{E},\beta)$, $1 \le j \le k$. Then $\{f_1, f_2, \ldots, f_k\}$ is an independent collection if and only if for all $t_1, t_2, \ldots, t_k \in \mathbb{R}$,

$$\int \exp i\{\sum_{j=1}^{k} t_j f_j\} d\beta = \prod_{j=1}^{k} \int \exp\{i t_j f_j\} d\beta. \tag{5.7}$$

Proof: Note that $x \to \exp\{ix\}$ is a bounded continuous function and hence the integrals appearing in (5.7) are well defined by Theorem 3.4. Also from Theorem 3.4,

$$R_\beta(\exp\{i \sum_{j=1}^{k} t_j f_j\}) = \exp\{i \sum_{j=1}^{k} t_j R_\beta(f_j)\}$$

and

$$R_\beta(\exp\{i t_j f_j\}) = \exp\{i t_j R_\beta(f_j)\}.$$

Thus (5.7) is equivalent to

$$\int \exp\{i \sum_{j=1}^{k} t_j R_\beta(f_j)\} d\tilde{\Pi} = \sum_{j=1}^{k} \int \exp\{i t_j R_\beta(f_j)\} d\tilde{\Pi}. \tag{5.8}$$

By the result quoted above, (5.8) holds for all $t_1, t_2, \ldots, t_k \in \mathbb{R}$ if and only if $\{R_\beta(f_1), \ldots, R_\beta(f_k)\}$ is an independent collection. This completes the proof. □

We now turn to the definition of independence of a random variable and a QCM.

Let $\phi: (E,\mathscr{E},\beta) \to (E_1,\epsilon_1)$ be a QCM. Note that for $G \in \mathscr{F}(E_1,\mathscr{E}_1)$, $G(\phi) \in \mathscr{L}(E,\mathscr{E},\beta)$.

Definition 5.2: Let $f \in \mathscr{L}(E,\mathscr{E},\beta;S)$. Then f is said to be *independent of* ϕ if for all $G \in \mathcal{F}(E_1, \mathcal{E}_1)$, the random variables f and $G(\phi)$ are independent of each other.

<u>Theorem 5.3</u>: Let $f \in \mathcal{L}(E, \mathcal{E}, \beta; S)$. Then f is independent of ϕ if and only if for all $g \in C_b(S)$, $E_\beta(g(f)|\phi)$ exists and

$$E_\beta(g(f)|\phi) = \int g(f)d\beta. \tag{5.9}$$

<u>Proof</u>: Suppose f and ϕ are independent. Then for all $G \in \mathcal{F}(E_1, \mathcal{E}_1)$, f and $G(\phi)$ are independent. For $G \in \mathcal{F}(E_1, \mathcal{E}_1)$ bounded, say by M, the definition of independence of f and $G(\phi)$ gives

$$\int g(f)\cdot g'(G(\phi))d\beta = \int g(f)d\beta \cdot \int g'(G(\phi))d\beta.$$

for all $g \in C_b(S)$, $g' \in C_b(\mathbb{R})$. Taking $g' \in C_b(\mathbb{R})$ such that $g'(x) = x$ for $|x| \leq M$, we get

$$\begin{aligned} \int g(f)\cdot G(\phi)d\beta &= \int g(f)d\beta \cdot \int G(\phi)d\beta \\ &= \int g(f)d\beta \cdot \int G d\beta_1 \end{aligned} \tag{5.10}$$

where $\beta_1 = \beta \circ \phi^{-1}$. The relation (5.10) shows that if we define $g_1 := \int g(f)d\beta$, then g_1 satisfies the requirements for being the conditional expectation of g(f) given ϕ (being a constant, it belongs to $\mathcal{U}(\phi)$) and thus (5.9) holds.

Conversely, suppose (5.9) holds. Then from Theorem 4.5, we have

$$E_{\tilde{\Pi}}(R_\beta(g(f))|\overline{\mathcal{D}}_\phi) = E_{\tilde{\Pi}}(R_\beta(g(f))). \tag{5.11}$$

Let $g' \in C_b(\mathbb{R})$ and $G \in \mathcal{F}(E_1, \mathcal{E}_1)$. By the definition of the σ-field $\mathcal{D}_\phi$, $R_\beta(G(\phi))$ is $\mathcal{D}_\phi$ measurable and hence $R_\beta(g'(G(\phi))) = g'(R_\beta(G(\phi)))$ is $\mathcal{D}_\phi$ measurable. Thus from (5.11) we conclude

$$E_{\tilde{\Pi}}[R_\beta(g(f))R_\beta(g'(G(\phi)))] = E_{\tilde{\Pi}}[R_\beta(g(f))]E_{\tilde{\Pi}}[R_\beta(g'(G(\phi)))] \tag{5.12}$$

which is the same as

$$\int g(f)\cdot g'(G(\phi))d\beta = \int g(f)d\beta\cdot\int g'(G(\phi))d\beta. \tag{5.13}$$

Since $g \in C_b(S)$ and $g' \in C_b(\mathbb{R})$ are arbitrary in (5.13), it follows that f and $G(\phi)$ are independent. □

CHAPTER VI

THE ABSTRACT STATISTICAL MODEL AND BAYES' FORMULA

This chapter introduces the abstract form of the model suitable for filtering, prediction and smoothing problems in the presence of (finitely additive) Gaussian white noise. The basic result that is central to our theory is the finitely additive version of the Bayes formula presented in Section 3. The first two sections contain preparatory material on measurable representations and canonical Gauss measures on Hilbert spaces.

1. MEASURABLE REPRESENTATION

Let β be a QCP on $(E,\mathcal{E}) = (\Omega,\mathcal{A})\odot(H,\mathcal{C})$ and let $(\rho,L,\tilde{\Pi})$ be a representation of β on a representation space $(\tilde{\Omega},\tilde{\mathcal{A}})$. Then L is a mapping from H into $L(\tilde{\Omega},\tilde{\mathcal{A}},\tilde{\Pi})$. Recall that elements of $L(\tilde{\Omega},\tilde{\mathcal{A}},\tilde{\Pi})$ are equivalence classes of measurable functions where f_1 and f_2 are identified if $f_1 = f_2$ a.s. $\tilde{\Pi}$. Thus, for each $h\in H$, L(h) is an equivalence class of measurable functions. Following the usual convention, we also use L(h) to denote an arbitrary element of this equivalence class. In some cases, a specific choice can be made such that the mapping $h \to L(h)$ has some additional properties. One such situation is the following.

<u>Definition 1.1</u>: The representation $(\rho,L,\tilde{\Pi})$ of β is said to

admit a *measurable version* if for every $h \in H$, we can choose an element $L_1(h)$ belonging to $L(h)$ such that the mapping

$$(h,\tilde{\omega}) \to L_1(h)(\tilde{\omega}) \text{ from } H \times \tilde{\Omega} \text{ into } \mathbb{R} \tag{1.1}$$

is $\mathcal{B}(H)\otimes\tilde{\mathcal{A}}$ measurable.

The triple $(\rho, L_1, \tilde{\Pi})$ will be called a *measurable representation* of β.

Theorem 1.1: Let m be the canonical Gauss measure on $(H,\mathcal{C})$ and let (L_0', Π_0) be an arbitrary representation of m on a representation space $(\Omega_0, \mathcal{A}_0)$. Then (L_0', Π_0) admits a measurable version (L_0, Π_0).

Proof: Let $\{e_j\}$ be a CONS in H. Then for $k \geq 1$, and $a_1, a_2, \ldots, a_k \in \mathbb{R}$,

$$E_{\Pi_0} \exp\{i \sum_{j=1}^{k} a_j L_0'(e_j)\} = E_{\Pi_0} \exp\{iL'(\sum_{j=1}^{k} a_j e_j)\} \tag{1.2}$$

$$= \int \exp\{i(h, \sum_{j=1}^{k} a_j e_j)\} dm(h)$$

$$= \exp(-\tfrac{1}{2}|\sum_{j=1}^{k} a_j e_j|^2)$$

$$= \exp(-\tfrac{1}{2}\sum_{j=1}^{k} a_j^2),$$

and hence $\{L_0'(e_j): j \geq 1\}$ is a sequence of independent normal random variables with mean zero and variance one. Thus for any sequence $\{a_j\}$ of real numbers such that $\sum_{j=1}^{\infty} a_j^2 < \infty$, we have

$$\sum_{j=1}^{k} a_j L_0'(e_j) \quad \text{converges a.s. } \Pi_0 \text{ as } k \to \infty \tag{1.3}$$

and as a consequence, for all $h \in H$,

$$Z_k(h)(\omega_0) := \sum_{j=1}^{k} (h, e_j) L_0'(e_j)(\omega_0) \text{ converges a.s. } \Pi_0. \tag{1.4}$$

For $(h, \omega_0) \in H \times \tilde{\Omega}$, define $L_0(h)(\omega_o)$ by

$$L_0(h)(\omega_0) = \lim_{k\to\infty} Z_k(h)(\omega_0) \quad \text{if the limit exists} \tag{1.5}$$

$$= 0 \qquad \text{otherwise.}$$

Now, for each $k \geq 1$, Z_k is $\mathcal{B}(H)\otimes\mathcal{A}_0$ measurable and hence L_0 defined by (1.5) is $\mathcal{B}(H)\otimes\mathcal{A}_0$ measurable. Also, in view of (1.4), we have for all $h \in H$

$$Z_k(h)(\omega_0) \to L_0(h)(\omega_0) \qquad \text{a.s. } \Pi_0 \tag{1.6}$$

On the other hand, by linearity of L_0',

$$E_{\Pi_0}(L_0'(h) - Z_k(h))^2 = E_{\Pi_0}(L_0'(h-P_k h))^2 = |h-P_k h|^2 \tag{1.7}$$

where P_k is the orthogonal projection onto span $\{e_1, e_2, \dots, e_k\}$. Hence $Z_k(h)$ converges to $L_0'(h)$ in Π_0-probability. This and (1.5) imply that $L_0(h) = L_0'(h)$ a.s. Π_0, which means that $L_0(h)$ belongs to the equivalence class $L_0'(h)$. Thus (L_0, Π_0) is the desired measurable version of (L_0', Π_0). □

2. MORE ON CANONICAL GAUSS MEASURE

The material presented in this section is in preparation for the Bayes formula given in the next section.

For the remaining part of this chapter, we fix a measurable representation (L_0,Π_0) of the canonical Gauss measure m, Π_0 being a probability measure on $(\Omega_0,\mathcal{A}_0)$ and let the corresponding m-lifting be R_m.

We begin with a result from measure theory, proved here for the sake of completeness.

<u>Proposition 2.1</u>: Let $(X_i,\mathcal{D}_i,\lambda_i)$ $(i=1,2)$ be countably additive probability spaces and let

$$(X,\mathcal{D},\lambda) = (X_1,\mathcal{D}_1,\lambda_1)\otimes(X_2,\mathcal{D}_2,\lambda_2).$$

Let Z_k, Z be positive Borel measurable integrable functions on $(X,\mathcal{D},\lambda)$ such that for all $x_1 \in X_1$,

$$Z_k(x_1,\cdot) \to Z(x_1,\cdot) \quad \text{in } \lambda_2\text{-probability} \tag{2.1}$$

and

$$\int_{X_1}[\int_{X_2} Z_k(x_1,x_2)d\lambda_2(x_2)]d\lambda_1(x_1) \tag{2.2}$$

$$\to \int_{X_2}[\int_{X_1} Z(x_1,x_2)d\lambda_2(x_2)]d\lambda_1(x_1).$$

Then

$$\int_{X_1} |Z_k(x_1,x_2)-Z(x_1,x_2)|d\lambda_1(x_1) \to 0 \tag{2.3}$$

$$\text{in } \mathcal{L}^1(X_2,\mathcal{D}_2,\lambda_2)$$

and as a consequence

$$\int Z_k(x_1,x_2)d\lambda_1(x_1) \to \int Z(x_1,x_2)d\lambda_1(x_1) \quad \text{in } \mathcal{L}^1(X_2,\mathcal{D}_2,\lambda_2).$$

<u>Proof:</u> For any $\epsilon > 0$, by Fubini's theorem

$$\lambda((x_1,x_2):|Z_k(x_1,x_2)-Z(x_1,x_2)| > \epsilon) \tag{2.4}$$

$$= \int \lambda_2(x_2 : |Z_k(x_1,x_2) - Z(x_1,x_2)| > \epsilon) d\lambda_1(x_1) \to 0 \quad \text{as } k \to \infty$$

by (2.1) and the dominated convergence theorem (recall that λ_1, λ_2 are probability measures. We are using the version of dominated convergence with convergence in probability.) Thus

$$Z_k \to Z \quad \text{in } \lambda\text{-probability.} \tag{2.5}$$

By the Fubini theorem and assumption (2.2) we have

$$\int Z_k d\lambda \to \int Z d\lambda. \tag{2.6}$$

Now

$$\int |Z_k - Z| d\lambda = \int Z_k d\lambda + \int Z d\lambda - 2\int (Z_k \wedge Z) d\lambda. \tag{2.7}$$

Since Z_k, Z are positive, $Z_k \wedge Z$ is dominated by Z which is λ-integrable. Also (2.5) implies that $Z_k \wedge Z$ converges to Z in λ-probability. Hence by the dominated convergence theorem,

$$\int (Z_k \wedge Z) d\lambda \to \int Z d\lambda. \tag{2.8}$$

Now (2.6), (2.7), (2.8) imply

$$\int |Z_k - Z| d\lambda \to 0 \quad \text{as} \quad k \to \infty. \tag{2.9}$$

This and Fubini's theorem imply (2.3). □

Let $\mathcal{M}_0(H)$ be the class of countably additive probability measures on $(H, \mathcal{B}(H))$ and let $\mathcal{I}(H)$ be the class of real valued functions f on H of the form

$$f(h) = \int_H \exp((k,h) - \tfrac{1}{2}|k|^2) d\nu(k), \qquad h \in H \tag{2.10}$$

for some $\nu \in \mathcal{M}_0(H)$. Observe that since $(k,h) - \frac{1}{2}|k|^2 \leq \frac{1}{2}|h|^2$, the integral appearing in (2.10) is finite for all $h \in H$.

Theorem 2.2:

(i) $\mathcal{S}(H) \subseteq \mathcal{L}^{1*}(H,\mathcal{C},m)$.

(ii) For $f \in \mathcal{S}(H)$ given by (2.10),

$$R_m(f) = \int_H \exp(L_0(k) - \tfrac{1}{2}|k|^2)d\nu(k). \tag{2.11}$$

and

$$\int f dm = 1.$$

Proof: Fix $\{P_i\} \subseteq \mathcal{P}$, $P_i \overset{s}{\to} I$. Let $\{e_1,e_2,\ldots,e_j\}$ be an orthonormal basis of Range P_i (i fixed for the moment). Then

$$(k,P_ih) = \sum_{r=1}^{j} (k,e_r)(h,e_r)$$

and hence

$$f\circ P_i(h) = \int \exp((k,P_ih) - \tfrac{1}{2}|k|^2)d\nu(k)$$

$$= \int \exp(\sum_{r=1}^{j} (k,e_r)(h,e_r) - \tfrac{1}{2}|k|^2)d\nu(k).$$

Hence by the definition of lifting for cylinder functions,

$$R_m(f\circ P_i) = \int \exp(\sum_{r=1}^{j} (k,e_r)L_0(e_r) - \tfrac{1}{2}|k|^2)d\nu(k) \tag{2.12}$$

$$= \int \exp(L_0(P_ik) - \tfrac{1}{2}|k|^2)d\nu(k)$$

by linearity of L_0 and since $P_ik = \sum_{r=1}^{j} (k,e_r)e_r$.

Let $Z_i,Z\colon H \times \Omega_0 \to \mathbb{R}^+$ be defined by

$$Z_i(k,\omega_0) = \exp(L_0(P_ik)(\omega_0) - \tfrac{1}{2}|k|^2) \tag{2.13}$$

and

$$Z(k,\omega_0) = \exp(L_0(k)(\omega_0)-\tfrac{1}{2}|k|^2), \quad (k,\omega_0) \in H\times\Omega_0. \tag{2.14}$$

Since L_0 is a measurable representation, Z_i, Z are $\mathcal{B}(H)\times\mathcal{A}_0$ measurable. For every $k \in H$,

$$E_{\Pi_0}(L_0(P_ik) - L_0(k))^2 = |k-P_ik|^2 \to 0 \quad \text{as } k \to \infty$$

and thus $L_0(P_ik)(\cdot)$ converges to $L_0(k)(\cdot)$ in Π_0-probability. Hence, for all $k \in H$,

$$Z_i(k,\omega_o) \to Z(k,\omega_0) \quad \text{in } \Pi_0\text{-probability}. \tag{2.15}$$

Since $L_0(P_ik)$ has a normal distribution with mean zero and variance $|P_ik|^2$, we have

$$\int\int Z_i(k,\omega_0)d\Pi_0(\omega)d\nu(k) \tag{2.16}$$

$$= \int\exp(\tfrac{1}{2}|P_ik|^2 - \tfrac{1}{2}|k|^2)d\nu(k) \to 1 \text{ as } i \to \infty$$

by the dominated convergence theorem as $|P_ik|^2 \leq |k|^2$ and $|P_ik|^2 \to |k|^2$ as $i \to \infty$. Also, for all $k \in H$, $\int Z(k,\omega_0)d\Pi_0(\omega_0) = 1$ as $L_0(k)$ has a normal distribution with mean zero and variance one, so that

$$\int\int Z(k,\omega_0)d\Pi_0 d\nu(k) = \int 1 d\nu(k) = 1. \tag{2.17}$$

Hence by Proposition 2.1,

$$R_m(f\circ P_i) = \int Z_i(k,\cdot)d\nu(k) \tag{2.18}$$

$$\to \int Z(k,\cdot)d\nu(k)$$

in $\mathscr{L}^1(\Omega_0,\mathscr{A}_0,\Pi_0)$. Since (2.18) holds for all $\{P_i\} \subseteq \mathscr{P}$, $\mathscr{P}_i \overset{s}{\to} I$, we conclude that $f \in \mathscr{L}^{1*}(H,\mathscr{C},m)$ and that (2.11) holds. Since $\int Z d\Pi_0 = 1$ as observed, we have $\int f dm = 1$.

Absolute Continuity of Translates of the Gauss Measure. On a finite dimensional space, the canonical Gauss measure, also known as the standard normal distribution, is countably additive and has the property that its translation by any vector is mutually absolutely continuous with respect to itself.

We will now prove that the canonical Gauss measure has this property in the infinite dimensional case as well.

For $C \subseteq H$ and $k \in H$, let

$$C - k = \{h : h+k \in C\}.$$

Thus

$$h \in [C-k] \text{ if and only if } h + k \in C.$$

Theorem 2.3: Let $k \in H$ be fixed. Let $m' \colon \mathscr{C} \to [0,1]$ be defined by

$$m'(C) = m(C-k). \tag{2.19}$$

Then m' is a cylinder probability, $m' \ll m$ and $\frac{dm'}{dm} = f$ where

$$f(h) = \exp((k,h) - \tfrac{1}{2}|k|^2). \tag{2.20}$$

Proof: This is an easy consequence of properties of the normal distribution on finite dimensional spaces. We will give a proof for the sake of completeness.

Observe that f is a cylinder function and is integrable; indeed, $\int f dm = 1$ (see Theorem III.1.3). Let

$$m_1(C) = \int_C f(h)dm(h), \quad C \in \mathscr{C}. \tag{2.21}$$

That m_1 is a cylinder probability on $(H,\mathscr{C})$ follows from Theorem IV.1.2. To complete the proof, it suffices to prove that $m' = m_1$, since $m_1 \ll m$ and $\frac{dm_1}{dm} = f$ by the definition of m_1.

Let $C \in \mathscr{C}$ be fixed and let $P \in \mathscr{P}$ be such that $C \in \mathscr{C}_P$ and $k \in$ Range P. If $\{e_1,\ldots,e_j\}$ is an orthonormal basis of Range P, C can be expressed as (see Proposition III.1.1)

$$C = \{h: ((h,e_1),\ldots,(h,e_j)) \in B\} \tag{2.22}$$

for some $B \in \mathscr{B}(\mathbb{R}^j)$. Let $t_i = (k,e_i)$, $\underline{t} = (t_1,\ldots,t_j) \in \mathbb{R}^j$ and let $B_1 \in \mathscr{B}(\mathbb{R}^j)$ be given by $B_1 = \{\underline{x}: \underline{x}+\underline{t} \in B\}$. Then

$$\begin{aligned} C-k &= \{h: h+k \in C\} \\ &= \{h: ((h+k,e_1),\ldots,(h+k,e_j)) \in B\} \\ &= \{h: ((h,e_1)+t_1, \ldots, (h,e_j)+t_j) \in B\} \\ &= \{h: ((h,e_1),\ldots,(h,e_j)) \in B_1\}. \end{aligned} \tag{2.23}$$

Since $k \in PH$, $(k,h) = \sum_{i=1}^{j} (h,e_i)(k,e_i) = \sum_{i=1}^{j} t_i(h,e_i)$ and $|k|^2 = \sum_{i=1}^{j} t_i^2$ so that

$$\begin{aligned} m_1(C) &= \int 1_C(h)\exp((k,h) - \tfrac{1}{2}|k|^2)dm(h) \\ &= \int 1_B((h,e_1),\ldots,(h,e_j))\exp(\sum_{i=1}^{j} t_i(h,e_i) - \tfrac{1}{2}\sum_{i=1}^{j} t_i^2)dm(h) \end{aligned} \tag{2.24}$$

$$= \int 1_B(\underline{x})\exp(\sum_{i=1}^{j} t_i x_i - \tfrac{1}{2}\sum_{i=1}^{j} t_i^2)\exp(-\tfrac{1}{2}\sum_{i=1}^{j} x_i^2)d\lambda_j(\underline{x})$$

by Theorem III.1.3, where λ_j is Lebesgue measure on $\mathbb{R}^j$. From (2.24),

$$m_1(C) = \int_{\mathbb{R}^j} 1_B(\underline{x})\exp(-\tfrac{1}{2}\sum_{i=1}^{j}(x_i-t_i)^2)d\lambda_j(\underline{x}) \tag{2.25}$$

$$= \int_{\mathbb{R}^j} 1_{B_1}(\underline{x}-\underline{t})\exp(-\tfrac{1}{2}\sum_{i=1}^{j}(x_i-t_i)^2)d\lambda_j(\underline{x})$$

$$= \int_{\mathbb{R}^j} 1_{B_1}(\underline{y})\exp(-\tfrac{1}{2}\sum_{i=1}^{j} y_i^2)d\lambda_j(\underline{y})$$

$$= \int 1_{B_1}((h,e_1),\ldots,(h,e_j))dm(h)$$

$$= m(C-k) = m'(C).$$

We have used Theorem III.1.3 and the relation (2.23) in the steps given above. Thus $m_1=m'$. This completes the proof.

□

<u>Remark 2.1</u>: For any $C \in \mathscr{C}$, the mapping $k \to m(C - k)$ is Borel measurable. To see this note that by Theorem 2.3

$$m(C-k) = \int 1_C(h)\exp((h,k) - \tfrac{1}{2}|k|^2)dm(h)$$

$$= \int R_m(1_C)\exp(L_0(k) - \tfrac{1}{2}|k|^2)d\Pi_0.$$

The measurability of $k \to m(C - k)$ now follows from Fubini's theorem and the fact that L_0 is a measurable representation. Indeed, it can be shown that the said mapping is continuous, but we only need measurability.

Theorem 2.4: Let $\mu \in \mathcal{M}_0(H)$. Let m^μ: $\mathscr{C} \to [0,1]$ be defined by

$$m^\mu(C) = \int_H m(C-k)d\mu(k). \tag{2.26}$$

Then m^μ is a cylinder probability, $m^\mu \ll m$, and $dm^\mu/dm = g$ where

$$g(h) = \int_H \exp((k,h) - \tfrac{1}{2}|k|^2)d\mu(k). \tag{2.27}$$

Proof. Let m_1 be the cylinder probability on $(H,\mathscr{C})$ given by (see Theorem IV.1.2)

$$m_1(C) = \int_C g dm.$$

Recall that g given by (2.27) belongs to $\mathscr{L}^{1*}(H,\mathscr{C},m)$, $g \geq 0$ and $\int g dm = 1$. Hence, using Fubini's theorem,

$$\begin{aligned} m_1(C) &= \int R_m(1_C)\cdot R_m(g)d\Pi_0 \qquad (2.28) \\ &= \int_{\Omega_0} R_m(1_C)\cdot\int_H \exp(L_0(k) - \tfrac{1}{2}|k|^2)d\mu(k)d\Pi_0 \\ &= \int_H[\int_{\Omega_0} R_m(1_C)\exp(L_0(k) - \tfrac{1}{2}|k|^2)d\Pi]d\mu(k) \\ &= \int_H[\int_H 1_C\exp((h,k) - \tfrac{1}{2}|k|^2)dm(h)]d\mu(k) \\ &= \int_H m(C - k)d\mu(k) = m^\mu(C). \end{aligned}$$

This completes the proof. □

Remark 2.2: The fact that m' defined by (2.19) and m^μ defined by (2.26) are cylinder probabilities can be proved directly without appealing to Radon-Nikodym derivatives. To see this first observe that if $C \in \mathscr{C}_P$, $P \in \mathscr{P}$, then

$$[C-k] = [C - Pk] \in \mathscr{C}_P. \tag{2.29}$$

If m_P, m'_P, m^μ_P denote the restrictions of m, m', m^μ to $\mathscr{C}_P$ respectively, then we get

$$m'_P = m_P(C - Pk) \tag{2.30}$$

$$m^\mu_P = \int m_P(C - Pk)d\mu(k). \tag{2.31}$$

[The measurability of $k \to m_P(C - Pk)$ can be deduced from the expression for it in terms of a finite dimensional Lebesgue integral given by Theorem III.1.3.]

The relations (2.30)-(2.31) imply that m'_P, m^μ_P are countably additive on $\mathscr{C}_P$ as m_P is, so that m', m^μ are cylinder probabilities.

We will also need the following result.

<u>Theorem 2.5</u>: Let m^μ be given by (2.26).

(i) $\mathscr{I}(H) \subseteq \mathscr{L}^*(H,\mathscr{C},m^\mu)$.

(ii) $m << m^\mu$.

(iii) m^μ admits a measurable representation.

(iv) If (L_1,Π_1) is any measruable representation of m^μ and if R_{m^μ} is the corresponding m^μ-lifting, then for $f \in \mathscr{I}(H)$ given by (2.10) we have

$$R_{m^\mu}(f) = \int_H \exp(L_1(k) - \tfrac{1}{2}|k|^2)d\upsilon(k) \tag{2.32}$$

so that $R_{m^\mu}(f) > 0$ a.s.

<u>Proof</u>: Since $m^\mu \ll m$ (i) follows from Theorem IV.1.1 and Theorem 2.2. For (ii), observe that $\frac{dm^\mu}{dm} = g$, given by (2.27) belongs to $\mathcal{I}(H)$. Now $g > 0$ and by the relation (2.11), $R_m(g) > 0$ a.s. Hence by Theorem IV.1.1, it follows that $m \ll m^\mu$ and $\frac{dm}{dm^\mu} = \frac{1}{g}$. To prove (iii), recall that (L_0, Π_0) is a measurable representation of m. Let Π_0' on $(\Omega_0, \mathcal{A}_0)$ be defined by $d\Pi_0' = R_m(g)\cdot d\Pi_0$. Then by Theorem IV.1.1, (L_0, Π_0') is a representation of m^μ which is measurable by choice of (L_0, Π_0). For (iv), let $(\Omega_1, \mathcal{A}_1)$ be the representation space of (L_1, Π_1). Let Π_1' on $(\Omega_1, \mathcal{A}_1)$ be given by $d\Pi_1' = R_{m^\mu}(\frac{dm}{dm^\mu})\cdot d\Pi_1$. Then by Theorem IV.1.1, (L_1, Π_1') is a (measurable) representation of m and if R_m is the corresponding m-lifting, then for all $f \in L(H, \mathcal{C}, m^\mu; \mathbb{R})$,

$$R_m(f) = R_{m^\mu}(f). \tag{2.33}$$

By Theorem 2.2,

$$R_m(f) = \int_H \exp(L_1(k) - \tfrac{1}{2}|k|^2)d\nu(k). \tag{2.34}$$

Now (2.33) and (2.34) imply (2.32). □

Let Q be an orthonormal projection on H with range $Q = H_1$. Recall that by Lemma IV.2.2, Q is a cylindrical mapping satisfying (IV.2.1).

<u>Theorem 2.6</u>:

(i) Let $m_1 = m \circ Q^{-1}$. Then m_1 is the canonical Gauss measure on $(H,\mathcal{C})$.

(ii) Let m^{μ} be given by (2.26). Then

$$m^{\mu}\circ Q^{-1} = m_1^{\mu_1}; \ \mu_1 = \mu\circ Q^{-1} \tag{2.35}$$

where, for $C_1 \in \mathcal{C}_1$,

$$m_1^{\mu_1}(\mathcal{C}_1) = \int_{H_1} m_1(C_1-k)d\mu_1(k). \tag{2.36}$$

Hence, $m_1^{\mu_1} << m_1$ and

$$\frac{dm_1^{\mu_1}}{dm_1}(h) = \int_{H_1} \exp((h,k) - \tfrac{1}{2}|k|^2)d\mu_1(k). \tag{2.37}$$

<u>Proof</u>: For $h \in H_1$, we have

$$\int_{H_1} e^{i(k,h)}dm_1(k) = \int_H e^{i(Qk,h)}dm(k). \tag{2.38}$$

as $m_1 = m\circ Q^{-1}$, $e^{i(k,h)}$ is a bounded continuous cylinder function and hence belongs to $\mathcal{U}(Q)$.

Since $h \in H_1$, $(Qk,h) = (k,h)$ for all $k \in H$ and

$$\int_{H_1} e^{i(k,h)}dm_1(k) = \int_H e^{i(k,h)}dm(k) = e^{-\frac{1}{2}|h|^2}. \tag{2.39}$$

Thus m_1 is the canonical Gauss measure on H_1. For the proof of (ii), observe that for $C_1 \in \mathcal{C}_1 = \mathcal{C}(H_1)$ and $h_0 \in H$, we have

$$Q^{-1}C_1 - h_0 = \{h: h+h_0 \in Q^{-1}C_1\} = \{h: Qh+Qh_0 \in C_1\} \tag{2.40}$$

$$= \{h: Qh \in C_1 - Qh_0\} = Q^{-1}(C_1 - Qh_0).$$

Thus, for $C_1 \in \mathscr{C}_1$,

$$m^{\mu}\circ Q^{-1}(C_1) = m^{\mu}(Q^{-1}C_1) = \int_H m(Q^{-1}C_1-k)d\mu(k)$$

$$= \int_H m(Q^{-1}(C_1-Qk))d\mu(k) = \int_H m_1(C_1-Qk)d\mu(k)$$

$$= \int_{H_1} m_1(C_1-k')d\mu_1(k') \quad (\text{where } \mu_1 = \mu\circ Q^{-1}),$$

$$= m_1^{\mu_1}(C_1). \qquad \square$$

3. THE ABSTRACT STATISTICAL MODEL AND THE BAYES FORMULA

Let $(\Omega,\mathscr{A},\Pi)$ be a countably additive probability space and let ξ: $(\Omega,\mathscr{A},\Pi) \to (H,\mathscr{B}(H))$ be a measurable mapping. Let

$$(E,\mathscr{E},\alpha) = (\Omega,\mathscr{A},\Pi) \odot (H,\mathscr{C},m).$$

Recall that (L_0,Π_0) is a fixed measurable representation of m. Let $(\tilde{\Omega},\tilde{\mathscr{A}},\tilde{\Pi})$, ρ, L be given by (V.2.3)-(V.2.5). Then $(\rho,L,\tilde{\Pi})$ is a representation of α. It is easy to see that $(\rho,L,\tilde{\Pi})$ is a measurable representation as L_0 and hence L satisfies (1.1). We will work with this fixed representation of α. Let R_α be the corresponding α-lifting.

Definition 3.1: Let e be the identity mapping on H. Considered as a mapping from $(H,\mathscr{C},m)$ into $(H,\mathscr{C})$, e will be called *Gaussian white noise*.

We shall adopt the convention of regarding a function defined on either H or Ω as defined on $E = \Omega\times H$, so that

$\xi(\omega,h) = \xi(\omega)$ and $e(\omega,h) = e(h) = h$. Let $y\colon E \to h$ be defined by

$$y = \xi + e \tag{3.1}$$

so that for $(\omega,h) \in E$, $y(\omega,h) = \xi(\omega) + h$. We shall call (3.1) the *abstract statistical model*.

Remark 3.1: The model (3.1) is suitable for situations where the observation is a functional ξ of the signal process, corrupted by a noise -- assumed to be white noise and independent of the signal process. It is easy to check that in the model (3.1), the H-valued random variable ξ on $(E,\mathcal{E},\alpha)$ and the H-valued QCM e on $(E,\mathcal{E},\alpha)$ are independent, in the sense of Section V.5. This is so because we have chosen α to be a product measure. Thus, in situations where noise is assumed to be white and independent of the signal process, (3.1) is the right choice for the observation model.

In the next chapter, by making a specific choice of the Hilbert space H and the mapping ξ, we obtain the white noise model for nonlinear filtering. We will prove below that y is a QCM and that for all integrable functions g on Ω, $E_\alpha(g|y)$ exists. We will obtain a formula for the conditional expectation, which is an analogue of the Kallianpur-Striebel Bayes formula in the conventional approach to nonlinear filtering theory.

Theorem 3.1:

(i) y is a QCM from $(E,\mathcal{E},\alpha)$ into $(H,\mathcal{C})$.

(ii) Let $n = \alpha \circ y^{-1}$. Then $n = m^{\mu}$ where $\mu = \Pi \circ \xi^{-1} \in \mathcal{M}_0(H)$ and m is given by (2.26).

Proof: To prove (i), we will show that for all $P \in \mathcal{P}$,

$$y^{-1}(\mathcal{C}_P) \subseteq \mathcal{E}_P = \mathcal{A}\otimes\mathcal{C}_P. \tag{3.2}$$

This would imply that y is a QCM in view of Remark V.4.2. Fix $P \in \mathcal{P}$ and let $C \in \mathcal{C}_P$ be of the form

$$C = \{h: ((h,h_1),\ldots,(h,h_j)) \in B\}$$

for some $j \geq 1$, $h_1,\ldots,h_j \in PH$ and $B \in \mathcal{B}(\mathbb{R}^j)$. Then

$$\{(\omega,h): y(\omega,h) \in C\} = \{(\omega,h): \xi(\omega)+h \in C\} \tag{3.3}$$
$$= \{(\omega,h): ((\xi(\omega)+h,h_1),\ldots,(\xi(\omega)+h,h_j)) \in B\}.$$

Now note that for $1 \leq i \leq j$, $\omega \to (\xi(\omega),h_i)$ is $\mathcal{A}$ measurable and $h \to (h,h_i)$ is $\mathcal{C}_P$ measurable (as $h_i \in PH$) so that

$$(\omega,h) \to ((\xi(\omega) + h,h_1),\ \ldots,\ (\xi(\omega) + h,h_j))$$

is a $\mathcal{A}\otimes\mathcal{C}_P/\mathcal{B}(\mathbb{R}^j)$ measurable mapping from E into $\mathbb{R}^j$. In view of (3.3), this gives

$$\{\omega,h): y(\omega,h) \in C\} \in \mathcal{A}\otimes\mathcal{C}_P = \mathcal{E}_P. \tag{3.4}$$

Since $C \in \mathcal{C}$ is arbitrary in (3.4), we conclude that (3.2) holds.

(ii) For $P \in \mathcal{P}$, if α_P, m_P denote the restrictions of α, m to $\mathcal{E}_P, \mathcal{C}_P$ respectively, then recall that $\alpha_P = \Pi\otimes m_P$. Thus for all $C \in \mathcal{C}_P$ we have, using Fubini's theorem for the third step

$$n(C) = \alpha(y \in C) = \alpha_P(y \in C) = \int m_P(h: y(\omega,h) \in C)d\Pi(\omega)$$
$$= \int m_P(h: \xi(\omega)+h \in C)d\Pi(\omega)$$

$$= \int m(C - \xi(\omega))d\Pi(\omega)$$

$$= \int m(C - k)d\mu(k)$$

$$= m^{\mu}(C)$$

where $\mu = \Pi\circ\xi^{-1}$.

Theorem 3.2: Let $f \in \mathcal{I}(H)$. Then

(i) $f(y) \in \mathcal{L}^{*}(E,\mathcal{E},\alpha)$, and
(ii) $f \in \mathcal{U}(E,\mathcal{E},\alpha;y)$.

Proof: As before, let $(\rho,L,\tilde{\Pi})$ be the representation of α given by (V.2.3)-(V.2.5), (L_0,Π_0) being a representation of m. Here $(\tilde{\Omega},\tilde{A},\tilde{\Pi}) = (\Omega,\mathcal{A},\Pi)\otimes(\Omega_0,\mathcal{A}_0,\Pi_0)$. We will denote a generic point $\tilde{\omega} \in \tilde{\Omega}$ by $\tilde{\omega} = (\omega,\omega_0)$. Let $(L_1,\tilde{\Pi})$ be the representation of $n = \alpha\circ y^{-1}$ induced by y. Then L_1 is given by

$$L_1(k) = R_{\alpha}((y,k)) = R_{\alpha}((\xi(\omega),k)+(h,k))$$

$$= (\xi(\rho),k)+L(k), \quad (k \in H),$$

i.e.,

$$L_1(k)(\tilde{\omega}) = (\xi(\omega),k) + L_0(k)(\omega_0), \quad \tilde{\omega} = (\omega,\omega_0). \tag{3.5}$$

Let R_n be the corresponding n-lifting. If $f \in \mathcal{I}(H)$ is given by

$$f(h) = \int \exp((h,k) - \tfrac{1}{2}|k|^2)d\nu(k), \tag{3.6}$$

then, by Theorem 2.5, $R_n(f) = \int \exp(L_1(k)-\frac{1}{2}|k|^2)d\nu(k)$, so that

$$R_n(f)(\tilde{\omega}) = \int \exp((\xi(\omega),k)+L_0(k)(\omega_0)-\tfrac{1}{2}|k|^2)d\nu(k), \tag{3.7}$$

where $\tilde{\omega} = (\omega,\omega_0) \in \tilde{\Omega}$.

Let $\{P_i\} \subseteq \mathcal{P}$, $P_i \overset{s}{\to} I$ be arbitrary. Let $g = f \circ y$ so that $g(\omega,h) = f(y(\omega,h)) = f(\xi(\omega)+h)$. Then

$$g_{P_i}(\omega,h) = g(\omega,P_i h) = f(\xi(\omega) + P_i h)$$

$$= \int \exp((\xi(\omega),k) + (P_i h,k) - \tfrac{1}{2}|k|^2)d\nu(k).$$

Proceeding as in the proof of Theorem 2.2, it follows that

$$R_\alpha(g_{P_i})(\tilde{\omega}) = \int \exp((\xi(\omega),k)+L_0(P_i k)(\omega_0)-\tfrac{1}{2}|k|^2)d\nu(k). \quad (3.8)$$

We will prove that

$$R_\alpha(g_{P_i}) \to R_n(f) \qquad \text{in } \tilde{\Pi}\text{-probability}. \quad (3.9)$$

This will imply that $g \in \mathcal{L}^*(E,\mathcal{E},\alpha)$ since $\{P_i\} \subseteq \mathcal{P}$ is an arbitrary sequence converging strongly to the identity. Furthermore, (3.9) also implies

$$R_\alpha(g) = R_\alpha(f \circ y) = R_n(f) \quad \text{a.s. } \tilde{\Pi}, \quad (3.10)$$

so that $f \in \mathcal{U}(E,\mathcal{E},\alpha;y)$. Thus the proof will be complete once we prove (3.9). Let Z_i, Z be defined by

$$Z_i(k,\tilde{\omega}) := \exp((\xi(\omega),k) + L_0(P_i k)(\omega_0) - \tfrac{1}{2}|k|^2) \quad (3.11)$$

$$Z(k,\tilde{\omega}) := \exp((\xi(\omega),k) + L_0(k)(\omega_0) - \tfrac{1}{2}|k|^2), \quad (3.12)$$

$k \in H$, $\tilde{\omega} = (\omega,\omega_0) \in \tilde{\Omega}$. Then Z_i, Z are measurable functions from $(H\times\tilde{\Omega}, \mathscr{B}(H)\otimes\tilde{\mathscr{A}}) \to (\mathbb{R},\mathscr{B}(\mathbb{R}))$ as L_0 is a measurable representation. We have

$$R_\alpha(g_{P_i})(\tilde{\omega}) = \int Z_i(k,\tilde{\omega})d\upsilon(k) \tag{3.13}$$

by (3.8) and

$$R_n(f)(\tilde{\omega}) = \int Z(k,\tilde{\omega})d\upsilon(k) \tag{3.14}$$

by (3.5) and Theorem 2.5. It is easy to see that $L_0(P_i k)$ converges to $L_0(k)$ in Π_0-probability. Hence from (3.11), (3.12) and the fact that $\tilde{\Pi} = \Pi\otimes\Pi_0$, it follows that

$$\text{for all } k\in H,\ Z_i(k,\cdot) \to Z(k,\cdot) \text{ in } \tilde{\Pi}\text{-probability.} \tag{3.15}$$

Thus (3.9) would follow from Proposition 2.1 if we can prove that

$$\int_H\int_{\tilde{\Omega}} Z_i d\tilde{\Pi}d\upsilon \to \int_H\int_{\tilde{\Omega}} Z d\tilde{\Pi}d\upsilon < \infty. \tag{3.16}$$

However,

$$\int_H\int_{\tilde{\Omega}} Z d\tilde{\Pi}d\upsilon = \int_H\int_\Omega \exp((\xi(\omega),k))d\Pi(\omega)d\upsilon(k) \tag{3.17}$$

which may not be finite since we do not have any conditions on ξ or on υ (nor would we like to impose such conditions).

We can circumvent this difficulty if we can show the existence of a countably additive probability measure Π' on $(\tilde{\Omega},\tilde{A})$ such that $\Pi' \ll \tilde{\Pi}$, $\tilde{\Pi} \ll \Pi'$ and (3.16) holds for Π' instead of $\tilde{\Pi}$. For such a measure Π', (3.15) is true with Π' in place of $\tilde{\Pi}$ as $\Pi' \ll \tilde{\Pi}$. Proposition 2.1 would then give the convergence of $R_\alpha(g_{P_i})$ to $R_n(f)$ in Π' probability and finally (3.9) as $\tilde{\Pi} \ll \Pi'$. To obtain such a measure, first observe that for all $k \in H$,

$$\int_{\Omega_0} \exp(L_0(k)(\omega_0) - \tfrac{1}{2}|k|^2) d\Pi_0(\omega_0) = 1 \tag{3.18}$$

since $L_0(k)$ is normally distributed with zero mean and variance $|k|^2$ under Π_0. Thus for all $A \in \mathcal{A}$,

$$\int_\Omega \int_{\Omega_0} 1_A(\omega)\exp(L_0(k-\xi(\omega))(\omega_0)-\tfrac{1}{2}|k-\xi(\omega)|^2) d\Pi_0(\omega_0) d\Pi(\omega) = 1. \tag{3.19}$$

This implies

$$E_{\tilde{\Pi}}(\exp(L_0(k-\xi) - \tfrac{1}{2}|k-\xi|^2)|\mathcal{D}) = 1 \tag{3.20}$$

where $\mathcal{D} := \mathcal{A}\otimes\{\phi,\Omega_0\}$. Let Π' be the countably additive measure on $(\tilde{\Omega},\tilde{\mathcal{A}})$ given by

$$\frac{d\Pi'}{d\tilde{\Pi}}(\omega,\omega_0)$$
$$:= a\cdot\exp(-\tfrac{1}{2}|\xi(\omega)|^2)\cdot\exp(-L_0(\xi(\omega))(\omega_0) - \tfrac{1}{2}|\xi(\omega)|^2),$$

where a is the normalizing constant given by

$$\int a\cdot\exp(-\tfrac{1}{2}|\xi(\omega)|^2) d\Pi(\omega) = 1. \tag{3.21}$$

In view of (3.20) and (3.21), it follows that Π' is a probability measure on $(\tilde{\Omega},\tilde{\mathcal{A}})$. Clearly $\Pi' \ll \tilde{\Pi}$ and $\tilde{\Pi} \ll \Pi'$. Using the relations (3.20), (3.21) and writing $\tilde{\omega} = (\omega,\omega_0)$ for a generic point in $\tilde{\Omega}$, we have

$$\int_H\int_{\tilde{\Omega}} Z(k,\tilde{\omega})d\Pi'(\tilde{\omega})d\nu(k) = \int_H\int_{\tilde{\Omega}} Z(k,\tilde{\omega})\frac{d\Pi'}{d\tilde{\Pi}}(\tilde{\omega})d\tilde{\Pi}(\tilde{\omega})d\nu(k) \tag{3.22}$$

$$= \int_H\int_{\tilde{\Omega}} a\cdot\exp(-\tfrac{1}{2}|\xi(\omega)|^2)\exp(L_0(k-\xi(\omega))(\omega_0)$$

$$- \tfrac{1}{2}|k-\xi(\omega)|^2)d\tilde{\Pi}(\tilde{\omega})d\nu(k)$$

$$= \int_H\int_{\tilde{\Omega}} a\cdot\exp(-\tfrac{1}{2}|\xi(\omega)|^2)d\tilde{\Pi}(\tilde{\omega})d\nu(k) = 1.$$

Similarly

$$\int_H\int_{\tilde{\Omega}} Z_i(k,\tilde{\omega})d\Pi'(\tilde{\omega})d\nu(k) \tag{3.23}$$

$$= \int_H\int_{\tilde{\Omega}} a\cdot\exp(-\tfrac{1}{2}|\xi(\omega)|^2)\exp((\xi(\omega),P_i^{\perp}k) - \tfrac{1}{2}|P_i^{\perp}k|^2)$$

$$\cdot\exp(L_0(P_ik-\xi(\omega))(\omega_0) - \tfrac{1}{2}|P_ik-\xi(\omega)|^2)d\tilde{\Pi}(\tilde{\omega})d\nu(k)$$

$$= \int_H\int_{\tilde{\Omega}} a\cdot\exp(-\tfrac{1}{2}|\xi(\omega)|)^2\exp((\xi(\omega),P_i^{\perp}k)-\tfrac{1}{2}|P_i^{\perp}k|^2)d\tilde{\Pi}(\tilde{\omega})d\nu(k)$$

$$= \int_H\int_{\tilde{\Omega}} a\cdot\exp\{-\tfrac{1}{2}\ |\xi(\omega) - P_i^{\perp}k|^2\}d\tilde{\Pi}(\tilde{\omega})d\nu(k).$$

Since $P_i \overset{s}{\to} I$, $|P_i^{\perp}k|^2 \to 0$ for all $k \in H$. Thus the integrand above goes to a $\exp(-\frac{1}{2}|\xi(\omega)|^2)$ pointwise. Also the integrand is bounded by a. Hence, by the dominated convergence theorem,

$$\int_H\int_{\tilde{\Omega}} Z_i(k,\tilde{\omega})d\tilde{\Pi}'(\tilde{\omega})d\nu(k) \to \int_H\int_{\tilde{\Omega}} a\ \exp(-\tfrac{1}{2}|\xi(\omega)|^2)d\tilde{\Pi}(\tilde{\omega})d\nu(k) \tag{3.24}$$

$$= 1.$$

It also follows from $\Pi' \ll \tilde{\Pi}$ and (3.15) that

$$\text{for all } k\in H,\ Z_i(k,\cdot) \to Z(k,\cdot) \text{ in } \Pi'\text{-probability.} \tag{3.25}$$

Now (3.14), (3.15), (3.22), (3.24) and (3.25) and Proposition 2.1 imply that

$$R_\alpha(g_{P_i}) \to R_n(f) \quad \text{in } \Pi'\text{-probability.} \tag{3.26}$$

The desired conclusion (3.9) is now a consequence of (3.26) and the fact that $\tilde{\Pi} \ll \Pi'$. □

Let Q be an orthogonal projection on H with range H_1.

<u>Theorem 3.3</u>:

(i) $Qy: (E,\mathcal{E},\alpha) \to (H_1,\mathcal{C}_1)$ is a QCM.

(ii) Let $n_1 = \alpha\circ(Qy)^{-1}$. Then $n_1 = n\circ Q^{-1}$, where $n = \alpha\circ y^{-1}$.

Hence, $n_1 = m_1^{\mu_1}$, where $\mu_1 = \Pi\circ(Q\xi)^{-1}$, and $m_1^{\mu_1}$ is given by (2.36).

(iii) Let $f_1 \in \mathcal{J}(H_1)$. Then

$$f_1 \in \mathcal{U}(Qy) = \mathcal{U}(E,\mathcal{E},\alpha;Qy) \tag{3.27}$$

and

$$f_1(Qy) \in \mathcal{L}^*(E,\mathcal{E},\alpha). \tag{3.28}$$

<u>Proof</u>: We have seen in Theorem 3.1 that $y: (E,\mathcal{E},\alpha) \to (H,\mathcal{C})$ is a QCM satisfying (V.4.6) and in Lemma IV.2.2 it was proved that Q is a QCM also satisfying (IV.2.1). Thus, by Theorem V.4.4, Qy is a QCM and

$$\alpha\circ(Qy)^{-1} = (\alpha\circ y^{-1})\circ Q^{-1} = n\circ Q^{-1}$$

so that $n_1 = n \circ Q^{-1}$. Now $n = m^{\mu}$, for $\mu = \Pi \circ \xi^{-1}$ as seen in Theorem 3.1. The assertion (ii) follows from Theorem 2.6. To prove (iii), first observe that $f_1 \in \mathcal{I}(H_1)$ implies $f \in \mathcal{I}(H)$ if $f(h) = f_1(Qh)$.

By Theorem 2.2, $f \in \mathcal{L}^*(H, \mathcal{C}, m^{\mu}) = \mathcal{L}^*(H, \mathcal{C}, n)$. Thus by Proposition IV.2.3,

$$f_1 \in \mathcal{U}(Q) = \mathcal{U}(H, \mathcal{C}, n; Q). \tag{3.29}$$

Also, since $f_1 = f \circ Q \in \mathcal{I}(H)$, by Theorem 3.2,

$$f(y) \in \mathcal{L}^*(E, \mathcal{E}, \alpha) \tag{3.30}$$

and

$$f = f_1 \circ Q \in \mathcal{U}(y) = \mathcal{U}(E, \mathcal{E}, \alpha; y). \tag{3.31}$$

Now (3.29), (3.31), the fact that $\alpha \circ y^{-1} = n$ and Theorem V.4.4 together imply that

$$f_1 \in \mathcal{U}(Qy). \tag{3.32}$$

The assertion (3.28) is the same as (3.30). This completes the proof. □

We are now in a position to prove our main result--the Bayes formula.

<u>Theorem 3.4 (Bayes formula)</u>:
Let g be an integrable function on $(\Omega, \mathcal{A}, \Pi)$. Then the conditional expectation of g given Qy exists and

$$E_{\alpha}(g|Qy) = \frac{\sigma_Q(g, Qy)}{\sigma_Q(1, Qy)} \tag{3.33}$$

where for $h \in H$, and $f \in \mathscr{L}^1(\Omega,\mathscr{A},\Pi)$,

$$\sigma_Q(f,h) := \int_\Omega f(\omega)\exp((h,Q\xi(\omega)) - \tfrac{1}{2}|Q\xi(\omega)|^2)d\Pi(\omega).$$

Proof: By linearity of the maps $g \to E_\alpha(g|Qy)$ and $f \to \sigma_Q(f,h)$, it suffices to prove (3.33) for $g \geq 0$, $\int g d\Pi = 1$. So, fix $g \geq 0$ and $\int g d\Pi = 1$.

It is easy to see that $\sigma_Q(g,\cdot) \in \mathscr{I}(H_1)$ and $\sigma_Q(1,\cdot) \in \mathscr{I}(H_1)$ so that by Theorem 3.3, $\sigma_Q(g,\cdot) \in \mathscr{U}(Qy)$ and $\sigma_Q(1,\cdot) \in \mathscr{U}(Qy)$. Further, by Theorem 2.5

$$R_{n_1}(\sigma_Q(1,\cdot)) > 0. \tag{3.34}$$

Hence if

$$v(h) = \frac{\sigma_Q(g,h)}{\sigma_Q(1,h)}, \quad h \in H_1$$

then

$$v \in \mathscr{U}(Qy). \tag{3.35}$$

Now, if $n_1 = \alpha\circ(Qy)^{-1}$ then by Theorem 2.6 and 3.3

$$\frac{dn_1}{dm_1}(h) = \int_{H_1} \exp((h,k) - \tfrac{1}{2}|k|^2)d\mu_1(k) \tag{3.36}$$

with $\mu_1 = \Pi\circ(Q\xi)^{-1}$. Hence

$$\frac{dn_1}{dm_1}(h) = \sigma_Q(1,h). \tag{3.37}$$

Let Π' be the probability measure on $(\Omega,\mathscr{A})$ given by $d\Pi' = g d\Pi$ and let $\alpha' = \Pi'\Theta m$. Then for $C_1 \in \mathscr{C}_1$

$$n_1'(C_1) := \int g 1_{C_1}(Qy)d\alpha = \int 1_{C_1}(Qy)d\alpha' \tag{3.38}$$

$$= \alpha' \circ (Qy)^{-1}(C_1).$$

Hence as in (3.36), it follows that

$$\frac{dn_1'}{dm_1}(h) = \int_{H_1} \exp((h,k) - \tfrac{1}{2}|k|^2)d\mu_1'(k) \quad \text{with } \mu_1' = \Pi' \circ (Q\xi)^{-1} \tag{3.39}$$

$$= \int_{\Omega} \exp((h, Q\xi(\omega)) - \tfrac{1}{2}|Q\xi(\omega)|^2)d\Pi'(\omega)$$

$$= \int_{\Omega} g(\omega)\exp((h, Q\xi(\omega) - \tfrac{1}{2}|Q\xi(\omega)|^2)d\Pi(\omega)$$

$$= \sigma_Q(g,h).$$

From (3.37), (3.39), (3.34) and Theorem IV.1.1, it follows that $n_1' << n_1$ and

$$\frac{dn_1'}{dm_1} = \left(\frac{dn_1'}{dm_1}\right)\cdot\left(\frac{dn_1}{dm_1}\right)^{-1} = \frac{\sigma_Q(g,\cdot)}{\sigma_Q(1,\cdot)} . \tag{3.40}$$

Thus

$$n_1'(C_1) = \int_{H_1} \frac{\sigma_Q(g,h)}{\sigma_Q(1,h)}dn_1(h). \tag{3.41}$$

Now (3.38), (3.41) and (3.35) imply that

$$E_{\alpha}(g|Qy) = v(Qy). \qquad \square$$

CHAPTER VII

DIFFERENTIAL EQUATIONS OF FILTERING: FINITE DIMENSIONAL NOISE AND SIGNAL

The theory of quasicylindrical probability measures and white noise calculus developed in the preceding chapters will now be applied to nonlinear estimation problems. In this chapter we consider the important and widely studied problem of obtaining the nonlinear filter for the additive Gaussian white noise model where the signal is a d-dimensional Markov process. From the Bayes formula for this model we derive the f.a. version of the Zakai equation for the unnormalized conditional expectation. Specializing to the case where the signal process is $\mathbb{R}^d$-valued diffusion, we obtain a parabolic PDE whose coefficients involve the observation process. The main result of this chapter is concerned with showing that the Cauchy problem posed by the PDE has a unqiue solution which is the unnormalized conditional density for the filtering problem. The treatment includes the case of unbounded coefficients. A particular application is to the well known cubic sensor problem.

1. NONLINEAR FILTERING MODEL AND THE BAYES FORMULA

We begin with a comment on H-valued white noise where

$$H = L^2([0,T]: \mathbb{R}^N), \text{ i.e.;}$$

$$H = \{\phi=(\phi^1,\ldots,\phi^N): [0,T] \to \mathbb{R}^N;\ ||\phi||^2 = \sum_{j=1}^{N} \int_0^T (\phi_s^i)^2 ds < \infty\}. \tag{1.1}$$

Then H is a real separable Hilbert space with the inner product given by

$$(\phi,\hat{\phi}) = \sum_{j=1}^{N} \int_0^T \phi_s^i \hat{\phi}_s^i ds \tag{1.2}$$

and norm $||\phi|| = (\phi,\phi)^{\frac{1}{2}}$, where $\phi = (\phi^1,\ldots,\phi^N)$, $\hat{\phi} = (\hat{\phi}^1,\ldots,\hat{\phi}^N)$ $\in$ H.

Let m be the canonical Gauss measure on $(H,\mathscr{C})$ and let e be the identity mapping on H, i.e., $e(\phi) = \phi$. Then e as a mapping from $(H,\mathscr{C},m)$ into $(H,\mathscr{C})$ is defined to be the H-valued white noise.

<u>Remark 1.1</u>: Recall that elements of H are equivalence classes of functions, with $\phi = \hat{\phi}$ if $\phi_s = \hat{\phi}_s$ a.s. Thus, we cannot define $e_t(\phi) = \phi_t$, $\phi \in H$, since $\phi \equiv \hat{\phi}$ does not imply $\phi_t = \hat{\phi}_t$ for all t. However, we can define $e_t: H \to \mathbb{R}^N$ such that $e(\phi) = (e_t(\phi))_{i\le t\le T}$. To see this, let

$$e_t^j(\phi) = \limsup_{k\to\infty} k \int_t^{(t+\frac{1}{k})\wedge T} \phi_s^j ds \tag{1.3}$$

for $\phi \in H$, $0 \le t \le T$, $1 \le j \le N$. Then by Lebesgue's Theorem, $e_t^j(\phi) = \phi_t^j$ for a.e. t and also $\phi = \hat{\phi}$ implies $e_t^j(\phi) = e_t^j(\phi)$. Thus for all t,

$$e_t(\phi) = (e_t^1(\phi),\ldots,e_t^N(\phi)),\ \phi \in H \tag{1.4}$$

defines a mapping from $H \to \mathbb{R}^N$ with the property that

$$e(\phi) = \phi \equiv (e_t(\phi))_{0 \leq t \leq T}. \tag{1.5}$$

Thus, $(e_t(\phi))_{o \leq t \leq T}$ is well defined.

In our treatment of the filtering problem, we will not use $e_t(\phi)$ explicitly.

Let the signal process $(X_t)_{o \leq t \leq T}$ be an $\mathbb{R}^d$-valued process on some countably additive probability space $(\Omega, \mathcal{A}, \Pi)$. We assume that the paths of (X_t) are measurable. Let h: $[0,T] \times \mathbb{R}^d \to \mathbb{R}^N$ be a Borel measurable function. Further assume that

$$\int_0^T |h_t(X_t(\omega))|^2 dt < \infty \quad \text{a.s. } \Pi. \tag{1.6}$$

Here, the function h is written as $h_t(x)$: $t \in [0,T]$, $x \in \mathbb{R}^N$ and $|\cdot|$ denotes the Euclidean norm on $\mathbb{R}^N$; $|x|^2 = \Sigma_{j=1}^N (x^j)^2$, $x = (x^1, \ldots, x^N) \in \mathbb{R}^N$. Without loss of generality (by redefining X_t on a Π-null set) we will assume that (1.6) holds for all $\omega \in \Omega$.

Let $\xi_s(\omega) = h_s(X_s(\omega))$, $0 \leq s \leq T$. Then $\xi_s(\omega) \in H$. Thus $\omega \to \xi(\omega) = (\xi_s(\omega))$ defines a mapping from Ω into H. Further, for $\phi \in H$,

$$(\phi, \xi\ (\omega)) = \sum_{j=1}^{N} \int_0^T \phi_t^j h_t^j (X_t(\omega)) dt. \tag{1.7}$$

Since h is Borel measurable and the paths of X_t are measurable, $(t,\omega) \to h_t(X_t(\omega))$ is $\mathcal{B}([0,T]) \otimes \mathcal{A}$ measurable. Thus by Fubini's Theorem, $\omega \to (\phi, \xi(\omega))$ is $\mathcal{A}$-measurable. The Borel σ-field $\mathcal{B}(H)$ on H is the smallest σ-field on H such that the mappings $\hat{\phi} \to (\phi,\hat{\phi})$ are measurable for all $\phi \in H$. Thus, $\mathcal{A}$-measurability of $\omega \to (\phi,\xi(\omega))$ implies that $\omega \to \xi(\omega)$ is $\mathcal{A}/\mathcal{B}(H)$-measurable.

For our filtering model, we want to realize the signal process (X_t) and the noise (e_t) on a probability space such that (X_t) and (e_t) are independent. Thus, we take

$$(E,\mathcal{E},\alpha) = (\Omega,\mathcal{A},\Pi)\odot(H,\mathcal{C},m). \tag{1.8}$$

Regarded as r.v.'s on $(E,\mathcal{E},\alpha)$, $X = (X_t)$ and $e = (e_t)$ are independent. Here, the abstract statistical model

$$y = \xi + e \tag{1.9}$$

takes the form

$$y_t = h_t(X_t) + e_t, \quad 0 \leq t \leq T \tag{1.10}$$

which is our model for nonlinear filtering.

For $0 \leq t \leq T$, let

$$H_t = \{\phi \in H: \textstyle\int_t^T |\phi_s|^2 ds = 0\} \tag{1.11}$$

and let Q_t be the orthogonal projection on H with Range $Q_t = H_t$. Then

$$(Q_t\phi) = \phi\cdot 1_{[0,t]}, \quad \phi \in H. \tag{1.12}$$

Thus $Q_t y$ represents the observations $\{y_s: 0 \leq s \leq t\}$ over the time interval $[0,t]$.

Applying the Bayes formula (Theorem VI.3.4) for $Q = Q_t$, we get the following result.

<u>Theorem 1.1</u>: Let $g \in \mathcal{L}^1(\Omega,\mathcal{A},\Pi)$, $0 \leq t \leq T$. Then $E_\alpha(g|Q_t y)$ exists and is given by

$$E_\alpha(g|Q_t y) = \frac{\tilde{\sigma}_t(g,y)}{\tilde{\sigma}_t(1,y)} \tag{1.13}$$

where for $\phi \in H$, and $f \in \mathscr{L}^1(\Omega,\mathscr{A},\Pi)$

$$\tilde{\sigma}_t(f,\phi) = \int_\Omega f(\omega)\exp((\phi,Q_t\xi(\omega)) - \tfrac{1}{2}\|Q_t\xi\|^2)d\Pi(\omega) \tag{1.14}$$

$$= \int_\Omega f(\omega)\exp(\sum_{j=1}^N \int_0^t \phi_s^j h_s^j(X_s(\omega))ds$$

$$- \tfrac{1}{2}\int_0^t |h_s(X_s(\omega))|^2 ds)d\Pi(\omega).$$

Taking $g = 1_{(X_t \in A)}$, $A \in \mathscr{B}(\mathbb{R}^d)$, the formula (1.13) determines the conditional distribution of X_t given $Q_t y$, i.e., given $\{y_s: s \leq t\}$.

For a Borel measurable function $f: \mathbb{R}^d \to \mathbb{R}$ such that $f(X_t)$ is integrable, we will write

$$\sigma_t(f,\phi) = \tilde{\sigma}_t(f(X_t),\phi). \tag{1.15}$$

In view of (1.13), $\sigma_t(f,y)$ is called unnormalized conditional expectation of $f(X_t)$ given $Q_t y$.

2. WHITE NOISE ANALOGUES OF THE ZAKAI AND FKK EQUATIONS

In this section we assume that the signal process (X_t) is an $\mathbb{R}^d$-valued Markov process. Further, we assume that the paths of (X_t) are $(\mathscr{F}_t^X)$-progressively measurable and that the Markov process (X_t) admits a transition probability function.

Let L be the weak generator of (X_t) and let $\mathscr{D}$ be its domain. (See Section II.2.) Recall that $\mathscr{D}$ is a subset of

$J([0,\infty)\times\mathbb{R}^d,\ \mathcal{B}([0,\infty))\otimes\mathcal{B}(\mathbb{R}^d))$ and for $f \in \mathcal{D}$, LF also belongs to $J([0,\infty)\times\mathbb{R}^d,\ \mathcal{B}([0,\infty))\otimes\mathcal{B}(\mathbb{R}^d))$ and thus is a bounded function.

Let $\mathcal{D}_0$ be the class of functions f from $\mathbb{R}^d$ into $\mathbb{R}$ such that f_1 defined by

$$f_1(s,x) = f(x),\ (s,x) \in [0,\infty) \times \mathbb{R}^d \tag{2.1}$$

belong to $\mathcal{D}$. For $f \in \mathcal{D}_0$, $0 \leq t < \infty$, define L_t by

$$(L_t f)(x) = (Lf_1)(t,x) \tag{2.2}$$

where f_1 is given by (2.1). Thus for $f \in \mathcal{D}_0$, $f(x)$, $(L_t f)(x)$ are uniformly bounded by a constant.

Recall that by Theorem II.2.2,

$$f_1(t,X_t) - \int_0^t (Lf_1)(s,X_s)ds \tag{2.3}$$

is a martingale, which means that

$$f(X_t) - \int_0^t (L_s f)(X_s)ds \tag{2.4}$$

is a martingale. We now derive an evolution equation for the unnormalized conditional expectation $\sigma_t(f,y)$.

<u>Theorem 2.1</u> (Zakai equation): Suppose that h satisfies

$$E\int_0^T |h_s(X_s)|^2 ds < \infty. \tag{2.5}$$

For all $f \in \mathcal{D}_0$, for all $y \in H$, we have

$$\frac{d}{dt}\sigma_t(f,y) = \sigma_t(L_t f,y) + \sum_{j=1}^N \sigma_t(h_t^j f,y)y_t^j \tag{2.6}$$

$$- \tfrac{1}{2}\sigma_t(|h_t|^2 f, y) \text{ for a.e. } t$$

Proof: For $\phi \in H$, $\omega \in \Omega$ and $0 < t < \infty$ define

$$q_t(\phi,\omega) := \exp\left[\sum_{j=1}^{N} \int_0^t \phi_s^j h_s^j(X_s(\omega))ds - \tfrac{1}{2}\int_0^t |h_s(X_s(\omega))|^2 ds\right]. \tag{2.7}$$

Observe that for all $\omega \in \Omega$, $q_t(\phi,\omega) \le \exp(\int_0^t |\phi_s|^2 ds)$. Then $\tilde{\sigma}_t(g,\phi)$ and $\sigma_t(f,\phi)$ can be written as

$$\tilde{\sigma}_t(g,\phi) = \int g(\omega) q_t(\phi,\omega) d\Pi(\omega) \tag{2.8}$$

and

$$\sigma_t(f,\phi) = \int f(X_t) q_t(\phi,\omega) d\Pi(\omega). \tag{2.9}$$

In view of our assumption that the paths of (X_t) are progressively measurable $q_t(\phi,\cdot)$ is $\mathscr{F}_t^X$ measurable. Hence for $g \in \mathscr{L}^1(\Omega,\mathscr{A},\Pi)$

$$\begin{aligned} \tilde{\sigma}_t(g,\phi) &= \int g(\omega) q_t(\phi,\omega) d\Pi(\omega) \\ &= \int E_\Pi(g q_t(\phi,\cdot)|\mathscr{F}_t^X) d\Pi = \int E_\Pi(g|\mathscr{F}_t^X)\cdot q_t(\phi,\cdot) d\Pi \\ &= \tilde{\sigma}_t(E_\Pi(g|\mathscr{F}_t^X),\phi). \end{aligned} \tag{2.10}$$

Fix $f \in \mathscr{D}_0$ and let $0 \le t \le T$. Let $g_t\colon \Omega \to \mathbb{R}$ be defined by

$$g_t(\omega) := f(X_T(\omega)) - \int_t^T (L_s f)(X_s(\omega))ds. \tag{2.11}$$

From (2.4), it follows that

$$E_\Pi(f(X_T(\omega)) - \int_0^T (L_s f)(X_s(\omega))ds\,|\mathscr{F}_t^X)$$

$$= f(X_t(\omega)) - \int_0^t (L_s f)(X_s(\omega))ds \quad \text{a.s.}$$

and hence

$$E_\Pi(g_t|\mathcal{F}_t^X) = f(X_t) \quad \text{a.s.} \tag{2.12}$$

By (2.10) for $g = g_t$, we get using (2.12)

$$\tilde{\sigma}_t(g_t,\phi) = \tilde{\sigma}_t(f(X_t),\phi) = \sigma_t(f,\phi). \tag{2.13}$$

From their definitions, it follows that for all ω,ϕ, $g_t(\omega)$ and $q_t(\phi,\omega)$ are absolutely continuous and for a.e. t

$$\frac{d}{dt} g_t(\omega) = (L_t f)(X_t(\omega)) \tag{2.14}$$

and

$$\frac{d}{dt} q_t(\phi,\omega) = \left[\sum_{j=1}^N h_t^j(X_t(\omega))\phi_t^j - \tfrac{1}{2}|h_t(X_t(\omega))|^2\right] q_t(\phi,\omega). \tag{2.15}$$

Hence $g_t(\omega)q_t(\phi,\omega)$ is also an absolutely continuous function of t and we have

$$\begin{aligned} g_t(\omega)q_t(\phi,\omega) &= g_0(\omega)q_0(\phi,\omega) + \int_0^t \frac{d}{ds}(g_s(\omega)\cdot q_s(\phi,\omega))ds \\ &= g_0(\omega)q_0(\phi,\omega) + \int_0^t (\frac{d}{ds} g_s(\omega))q_s(\phi,\omega)ds \\ &\quad + \int_0^t g_s(\omega)\frac{d}{ds} q_s(\phi,\omega)ds. \end{aligned} \tag{2.16}$$

Thus

$$\sigma_t(f,\phi) = \sigma_t(g_t,\phi) = E_\Pi g_t(\omega)q_t(\phi,\omega) = I + II + III, \text{ say,} \tag{2.17}$$

where

$$I = E_\Pi g_0(\omega)q_0(\phi,\omega) = E_\Pi g_0(\omega), \tag{2.18}$$

$$II = E_{\Pi} \int_0^t (\frac{d}{ds} g_s(\omega)) q_s(\phi,\omega)ds \tag{2.19}$$

and

$$III = E_{\Pi} \int_0^t g_s(\omega) \frac{d}{ds} q_s(\phi,\omega)ds. \tag{2.20}$$

Now

(2.21)

$$II = E_{\Pi} \int_0^t (L_s f)(X_s(\omega)) q_s(\phi,\omega)ds \quad \text{(by (2.19) and (2.16))}$$

$$= \int_0^t E_{\Pi}(L_s f(X_s(\omega))) q_s(\phi,\omega)ds \quad \text{(by Fubini's theorem as, the integrand is bounded)}$$

$$= \int_0^t \sigma(L_s f,\phi)ds$$

and

(2.22)

$$III = E_{\Pi} \int_0^t g_s(\omega) \left[\sum_{j=1}^N h_s^j(X_s(\omega))\phi_t^j - \tfrac{1}{2}|h_s(X_s)|^2 \right] q_s(\phi,\omega)ds$$

$$= \int_0^t \left[\sum_{j=1}^N \tilde{\sigma}_s(g_s h_s^j(X_s),\phi)\phi_s^j - \tfrac{1}{2}\tilde{\sigma}_s(g_s|h_s(X_s)|^2,\phi) \right] ds$$

by Fubini's Theorem as the integrand is bounded by a constant times $|h_s(X_s)|^2$, which is assumed to be integrable. Also,

(2.23)

$$\tilde{\sigma}_s(g_s h_s^j(X_s),\phi) = \tilde{\sigma}_s(E_{\Pi}(g_s h_s^j(X_s)|\mathcal{F}_s^X),\phi) \quad \text{(by (2.10))}$$

$$= \tilde{\sigma}_s(h_s^j(X_s)E_{\Pi}(g_s|\mathcal{F}_s^X),\phi)$$

$$= \tilde{\sigma}_s(h_s^j(X_s)f(X_s),\phi)$$

$$= \sigma_s(h_s^j f,\phi).$$

Similarly

$$\tilde{\sigma}_s(g_s|h_s(X_s)|^2,\phi) = \sigma_s(|h_s|^2 f,\phi). \tag{2.24}$$

Finally, from (2.17), (2.18), (2.21), and (2.23), we get

$$\sigma_t(f,\phi) = E_\Pi(g_0) + \int_0^t \sigma_s(L_s f,\phi)ds \tag{2.25}$$

$$+ \sum_{j=1}^N \int_0^t \sigma_s(h_s^j f,\phi)\phi_s^j ds - \tfrac{1}{2}\int_0^t \sigma_s(|h_s|^2 f,\phi)ds.$$

This proves (2.6), by substituting y for ϕ. □

Remark 2.1: Several authors (e.g., [13]) while dealing with problems on robust nonlinear filtering use the "Stratonovich form" of the Zakai equation of the Itô theory (see Appendix), and then make the heuristic change from dY_t to $y_t dt$. The equation thus obtained is precisely the equation (2.6).

Remark 2.2: Let (X_t) be a diffusion process with diffusion and drift coefficients a,b (see Section II.3). Let L_t be defined by (II.3.4). Then as noted in Section II.3, for $f \in C_0^2(\mathbb{R}^d)$,

$$f(X_t) - \int_0^t (L_u f)(X_u)du$$

is a martingale. Thus (2.12) above holds. Also, $(L_u f)(x)$ is uniformly bounded in (u,x) as f has compact support and a,b satisfy (II.3.2). Thus it follows that the Zakai equation (2.6) is valid for all $f \in C_0^2(\mathbb{R}^d)$, where L_t is defined by (II.3.4).

For $f: \mathbb{R}^d \to \mathbb{R}$ such that $E_\Pi|f(X_t)| < \infty$, let

$$\pi_t(f,\phi) := \frac{\sigma_t(f,\phi)}{\sigma_t(1,\phi)}, \quad \phi \in H \tag{2.26}$$

so that

$$E_\alpha(f(X_t)|Q_t y) = \pi_t(f,y).$$

Using the fact that σ_t satisfies (2.6), we obtain an equation for π_t. This is analogous to the Fujisaki-Kallianpur-Kunita equation.

Theorem 2.2: Let h satisfy (2.5). For all $f \in \mathscr{D}$, $y \in H$, we have

$$\frac{d}{dt}\pi_t(f,y) = \pi_t(L_t f,y) + \sum_{j=1}^N \pi_t(fh_t^j,y)y_t^j - \tfrac{1}{2}\pi_t(f|h_t|^2,y)$$
$$- \pi_t(f,y)\{\sum_{j=1}^N \pi_t(h_t^j,y)y_t^j - \tfrac{1}{2}\pi_t(|h_t|^2,y)\} \text{ a.e. } t. \tag{2.27}$$

Proof: This is a direct consequence of (2.6). First note that the function 1 (i.e., $f(x) = 1$) belongs to $\mathscr{D}$ and $L1 = 0$. Hence

$$\frac{d}{dt}\sigma_t(1,y) = \sum_{j=1}^N \sigma_t(h_t^j,y)y_t^j - \tfrac{1}{2}\sigma_t(|h_t|^2,y) \quad \text{a.e. } t. \tag{2.28}$$

Therefore

$$\frac{d}{dt}\sigma_t(f,y) = \frac{\{\frac{d}{dt}\sigma_t(f,y)\}\sigma_t(1,y) - \sigma_t(f,y)\{\frac{d}{dt}\sigma_t(1,y)\}}{[\sigma_t(1,y)]^2}$$
$$= \frac{[\frac{d}{dt}\sigma_t(f,y)]}{[\sigma_t(1,y)]} - \pi_t(f,y)\frac{[\frac{d}{dt}\sigma_t(1,y)]}{[\sigma_t(1,y)]}. \tag{2.29}$$

The assertion (2.27) follows from (2.29) using (2.6) and (2.28) for $\frac{d}{dt}\sigma_t(f,y)$ and $\frac{d}{dt}\sigma_t(1,y)$ respectively and the definition (2.26) of $\pi_t(f,y)$. □

Remark 2.3: Clearly, $\pi_t(f,y)$ is an absolutely continuous function of t since $\sigma_t(f,y)$ is and hence we can integrate the right hand side of (2.27) from 0 to t to obtain $\pi_t(f,y)$.

3. ZAKAI EQUATION FOR THE UNNORMALIZED DENSITY: UNIQUENESS OF SOLUTION (h BOUNDED)

In this section, we consider the case when the signal process (X_t) in the filtering model is an $\mathbb{R}^d$-valued diffusion process. We will show that under suitable conditions on the process (X_t), the conditional distribution of X_t given $\{y_s: 0 \leq s \leq t\}$ admits a density, which can be characterized (up to a normalizing constant depending on y) as the unique solution to a partial differential equation (PDE). This PDE is a perturbation of Kolmogorov's forward equation for the (unconditional) density of (X_t).

Before proceeding with the proof of these results, let us recall some definitions and results on parabolic partial differential equations. The definitions and the results stated without proof are from Friedman [17].

Let $a^0_{ij}(t,x)$, $b^0_i(t,x)$, $c^0(t,x)$ be measurable functions on $[0,T] \times \mathbb{R}^d$, $1 \leq i, j \leq d$. Further, suppose that for some $\lambda_0 > 0$

$$\sum_{i,j=1}^{d} a^0_{ij}(t,x)z^i z^j \geq \lambda_0 \sum_{i=1}^{d} (z^i)^2 \tag{3.1}$$

for all real $z^1, z^2, \ldots, z^d$, for all $(t,x) \in [0,T]\times\mathbb{R}^d$ and that

$$a^0_{ij}(t,x) = a^0_{ji}(t,x) \tag{3.2}$$

for all i, j, $(t,x) \in [0,T]\times\mathbb{R}^d$. For $0 \le t \le T$, let M_t be the linear partial differential operator given by

$$M_t f(x) := \tfrac{1}{2} \sum_{i,j=1}^{d} a^0_{ij}(t,x)\frac{\partial^2 f}{\partial x^i \partial x^j}(x) \tag{3.3}$$

$$+ \sum_{i=1}^{d} b^0_i(t,x)\frac{\partial f}{\partial x^i}(x) + c^0(t,x)f(x)$$

for $f \in C^2(\mathbb{R}^d)$. Let $0 < t_0 \le T$ and let g_0 be a given function on $\mathbb{R}^d$. Consider the problem of finding a function v such that

$$\frac{\partial v}{\partial t} + M_t v = 0 \tag{3.4}$$

for $0 \le t < t_0$ and

$$v(t_0, z) = g_0(z). \tag{3.5}$$

The equations (3.4), (3.5) will be together referred to as Cauchy problem (I).

Definition 3.1: A function $v\colon [0,t_0] \times \mathbb{R}^d \to \mathbb{R}$ is said to be a *classical solution* to the Cauchy problem (I) if

$$v \in C^{1,2}([0,t_0)\times\mathbb{R}^d) \cap C([0,t_0]\times\mathbb{R}^d)$$

and the relations (3.4), (3.5) hold pointwise for (t,z), $0 \le t < t_0$, $z \in \mathbb{R}^d$. It is assumed here that the functions a^0, b^0, c^0 and g_0 are continuous.

Definition 3.2: A *fundamental solution* of (3.4) in $[0,T]\times\mathbb{R}^d$

is a function $G(t,z,s,x)$ defined for all $(s,x) \in [0,T]\times\mathbb{R}^d$, $(t,z) \in [0,T]\times\mathbb{R}^d$, $s < t$, which satisfies the following conditions:

For fixed $(t_0,z) \in (0,T]\times\mathbb{R}^d$, it satisfies the equation (3.4) pointwise as a function of $(s,x) \in [0,t_0)\times\mathbb{R}^d$; (3.6)

For every continuous function f on $\mathbb{R}^d$ with compact support, (3.7)

$$\lim_{s\uparrow t_0} \int G(t_0,z,s,x)f(z)dz = f(x).$$

Suppose that $a^0_{ij}(t,\cdot) \in C^2(\mathbb{R}^d)$ and $b^0_i(t,\cdot) \in C^1(\mathbb{R}^d)$. Then the *formal adjoint* M^*_t of M_t is the linear partial differential operator given by

$$M^*_t f(x) := \tfrac{1}{2} \sum_{i,j=1}^{d} \frac{\partial^2}{\partial x^i \partial x^j}(a^0_{ij}(t,\cdot)f(\cdot))(x) - \sum_{i=1}^{d} \frac{\partial}{\partial x^i}(b^0_i(t,\cdot)f(\cdot))(x) + c^0(t,x)f(x) \tag{3.8}$$

for $f \in C^2(\mathbb{R}^d)$. It is easy to see that

$$M^*_t f(x) := \tfrac{1}{2} \sum_{i,j=1}^{d} a^0_{ij}(t,x)\frac{\partial^2 f}{\partial x^i \partial x^j}(x) + b^{0*}_i(t,x)\frac{\partial f}{\partial x^i}(x) + c^{0*}(t,x)f(x) \tag{3.9}$$

for $f \in C^2(\mathbb{R}^d)$, where

$$b_i^{0*}(t,x) = -b_i^0(t,x) + \sum_{j=1}^{d} \frac{\partial a_{ij}^0(t,x)}{\partial x^j} \tag{3.10}$$

$$c^{0*}(t,x) = c^0(t,x) - \sum_{i=1}^{d} \frac{\partial b_i^0(t,x)}{\partial x^i} + \frac{1}{2} \sum_{i,j=1}^{d} \frac{\partial^2 a_{ij}^0(t,x)}{\partial x^i \partial x^j} . \tag{3.11}$$

Suppose that for each t, $c^0(t,\cdot)$ is integrable on every compact subset of $\mathbb{R}^d$. From the integration by parts formula and the defining relation (3.8), it follows that for $f,g \in C^2(\mathbb{R}^d)$ with either f or g having compact support,

$$\int g(x)[M_t^* f(x)]dx = \int [M_t g(x)]f(x)dx. \tag{3.12}$$

The above identity justifies the name formal adjoint for M_t^*. It should be noted that M_t^* is defined only when a_{ij}^0 is twice continuously differentiable and b_i^0 is continuously differentiable.

Now let $0 \leq s_0 < T$ and $p_0(x): \mathbb{R}^d \to \mathbb{R}$ be given. The problem of finding a function u such that

$$\frac{\partial u}{\partial t} = M_t^* u \tag{3.13}$$

for $s_0 < t \leq T$ and

$$u(s_0,x) = p_0(x) \tag{3.14}$$

will be referred to as the Cauchy problem (II).

Note that the term "Cauchy problem" has been used above in a somewhat more general sense that is commonly used in the literature.

Definition 3.3: A function $u: [s_0,T] \times \mathbb{R}^d \to \mathbb{R}$ is said to be a *classical solution solution* to the Cauchy problem (II) if

$$u \in C^{1,2}((s_0,T]\times\mathbb{R}^d) \cap C([s_0,T]\times\mathbb{R}^d)$$

and the relations (3.13), (3.14) hold pointwise, $s_0 < t \leq T$, $z \in \mathbb{R}^d$. It is assumed that c^0 and p_0 are continuous.

Definition 3.4: A *fundamental solution* of (3.13) in $[0,T]\times\mathbb{R}^d$ is a function $G^*(s,x,t,z)$ defined for all $(s,x) \in [0,T]\times\mathbb{R}^d$, $(t,z) \in [0,T]\times\mathbb{R}^d$, $s < t$, which satisfies the following conditions

For fixed $(s_0,x) \in (0,T)\times\mathbb{R}^d$, it satisfies the equation (3.13) pointwise as a function of $(t,z) \in (s_0,T]\times\mathbb{R}^d$; (3.15)

For every continuous function f on $\mathbb{R}^d$ with compact support, (3.16)

$$\lim_{s\downarrow t_0} \int G^*(s_0,x,t,z)f(x)dx = f(z).$$

We will now state a result from Friedman [17] on the existence of fundamental solutions for the equations (3.4), (3.13) and of classical solutions for the Cauchy problems (I), (II).

Recall that a function $f: [0,T]\times\mathbb{R}^d \to \mathbb{R}$ is said to be Hölder continuous (with exponent α) if for some constants $A < \infty$, $\alpha > 0$,

$$|f(t_1,x_1) - f(t_2,x_2)| \leq A(|x_1-x_2|^\alpha + |t_1-t_2|^\alpha) \quad (3.17)$$

for all $t_1, t_2 \in [0,T]$, $x_1, x_2 \in \mathbb{R}^d$.

A function $f: [0,T]\times\mathbb{R}^d \to \mathbb{R}$ is said to be locally Hölder continuous if for every compact set K in $\mathbb{R}^d$, there exist constants $A < \infty$, $\alpha > 0$ (depending on K) such that (3.17) holds for all $t_1, t_2 \in [0,T]$, $x_1, x_2 \in K$.

<u>Theorem 3.1</u>: Let M_t be given by (3.3). Suppose that a^0 satisfies (3.1) and (3.2). Further, assume that

$$a^0_{ij},\ \frac{\partial}{\partial x^i} a^0_{ij},\ \frac{\partial^2}{\partial x^i \partial x^j} a^0_{ij},\ b^0_i,\ \frac{\partial}{\partial x^i} b^0_i \tag{3.18}$$

are bounded Hölder continuous functions on $[0,T]\times\mathbb{R}^d$ and

c^0 is also a bounded Hölder continuous function on $[0,T]\times\mathbb{R}^d$. (3.19)

Then (i) equations (3.4) and (3.13) admit fundamental solutions $G(t,z,s,x)$ and $G^*(s,x,t,z)$ $(s < t;\ s,z \in \mathbb{R}^d)$ satisfying the following conditions:

(a) $G(t,z,s,x) = G^*(s,x,t,z);\ s < t,\ x,z \in \mathbb{R}^d$;

(b) there exist positive constants K_1, K_2 such that for all $s < t;\ x,z \in \mathbb{R}^d$, we have

$$|G(t,z,s,x)| \leq K_1 (t-s)^{-d/2} \exp(-K_2 \frac{|z-x|^2}{(t-s)}) \tag{3.20}$$

$$|\frac{\partial G}{\partial x^i}(t,z,s,x)| \leq K_1 (t-s)^{-(d+1)/2} \exp(-K_2 \frac{|z-x|^2}{(t-s)}) \tag{3.21}$$

$$|\frac{\partial G}{\partial z^i}(t,z,s,x)| \leq K_1 (t-s)^{-(d+1)/2} \exp(-K_2 \frac{|z-x|^2}{(t-s)}). \tag{3.22}$$

(ii) Let $0 < t_0 \leq T$ and let $g_0: \mathbb{R}^d \to \mathbb{R}$ be a continuous function such that $|g_0(x)| \leq \exp(K|x|^{2-\epsilon})$ for some $K < \infty$, $\epsilon > 0$. Then the function $v: [0,t_0]\times\mathbb{R}^d \to \mathbb{R}$ defined by $v(t_0,x) = g_0(x)$ and

$$v(s,x) = \int G(t_0,z,s,x)g_0(z)dz, \quad s < t_0 \tag{3.23}$$

is a classical solution to the Cauchy problem (I). Moreover, v given above is the unique classical solution to the Cauchy problem (I) in the class of $C^{1,2}([0,t_0)\times\mathbb{R}^d) \cap C([0,t_0]\times\mathbb{R}^d)$ functions f satisfying

$$\int_0^{t_0}\int_{\mathbb{R}^d} |f(t,x)|\exp(-K_3|x|^2)dxdt < \infty \tag{3.24}$$

for some finite constant K_3.

(iii) Let $0 \leq s_0 < T$ and $p_0: \mathbb{R}^d \to \mathbb{R}$ be a continuous function such that $|p_0(x)| \leq \exp(K|x|^{2-\epsilon})$ for some $K < \infty$, $\epsilon > 0$. Then the function $u: [s_0,T]\times\mathbb{R}^d \to \mathbb{R}$ defined by $u(s_0,x) = p_0(x)$ and

$$u(t,z) = \int G^*(s_0,x,t,z)p_0(x)dx, \quad t > s_0 \tag{3.25}$$

is the unique classical solution to the Cauchy problem (II) in the class of $C^{1,2}((s_0,T]\times\mathbb{R}^d) \cap C([s_0,T]\times\mathbb{R}^d)$ functions f satisfying (3.24). Further, u satisfies

$$|u(s,x)| \leq \exp(K_3'|x|^2) \tag{3.26}$$

for some finite constant K_3'.

Generalized Solution to the Cauchy Problem (II)

We have defined a classical solution to the Cauchy problem (II) only when a^0, b^0, c^0 are sufficiently smooth and further, under the requirement that the solution itself be smooth.

We will now consider a generalized solution to this problem. To motivate the definition, suppose that the coefficients a^0, b^{0*}, c^{0*} appearing in the expression (3.9) for the differential operator M_t^* are continuous and that the Cauchy problem (II) admits a classical solution u. Then for $\phi \in C_0^\infty(\mathbb{R}^d)$ we have

$$\int_{\mathbb{R}^d} u(t_0,x)\phi(x)dx - \int_{\mathbb{R}^d} p_0(x)\phi(x)dx = \int_{\mathbb{R}^d} (u(t_0,x)-p_0(x))\phi(x)dx$$

$$= \int_{\mathbb{R}^d} [\int_{s_0}^{t_0} \frac{\partial}{\partial t} u(t,x)dt]\phi(x)dx$$

$$= \int_{\mathbb{R}^d} [\int_{s_0}^{t_0} M_t^* u(t,x)dt]\phi(x)dx$$

$$= \int_{s_0}^{t_0} \int_{\mathbb{R}^d} [M_t^* u(t,x)]\phi(x)dxdt$$

$$= \int_{s_0}^{t_0} \int_{\mathbb{R}^d} u(t,x)[M_t\phi(x)]dxdt$$

by (3.12). Also we have used Fubini's theorem which is justified as ϕ has compact support and the functions u, a^0, b^{0*}, c^{0*} are continuous so that $M_t^* u(t,x)\cdot\phi(x)$ is a bounded function with compact support. Thus, if u is a classical solution to the Cauchy problem (II), then for all $\phi \in C_0^\infty(\mathbb{R}^d)$, $s_0 < t_0 \leq T$,

$$\int_{\mathbb{R}^d} u(t_0,x)\phi(x)dx - \int_{\mathbb{R}^d} p_0(x)\phi(x)dx \tag{3.27}$$

$$= \int_{s_0}^{t_0} \int_{\mathbb{R}^d} u(t,x)\cdot[M_t\phi(x)]dxdt.$$

Note that the expression (3.27) does not involve M_t^*, so that (3.27) makes sense even when a^0, b^0 are not differentiable. Also, the function u need not be differentiable either for (3.27) to hold.

This leads us to the following definition of a generalized solution.

<u>Definition 3.5</u>: A measurable function $u: [s_0,T]\times\mathbb{R}^d \to \mathbb{R}$ is said to be a *generalized solution* to the Cauchy problem (II) if for all $\phi \in C_0^\infty(\mathbb{R}^d)$

$$\int_{\mathbb{R}^d} u(t,x)\phi(x)dx = \int_{\mathbb{R}^d} p_0(x)\phi(x)dx + \int_{s_0}^{t} \int_{\mathbb{R}^d} u(s,x)M_s\phi(x)dxds \tag{3.28}$$

for all $t \in [s_0,T]$. In (3.28) above, it is assumed that the integrals appearing on both the sides exist and are finite.

A generalized solution may exist even when a^0,b^0 are not smooth enough and M_t^* cannot be defined as a differential operator. We have already observed that a classical solution to the Cauchy problem (II) is also a generalized solution. We will now prove that under the conditions of the previous theorem, the Cauchy problem (II) admits a unique generalized solution and as a consequence the generalized solution is smooth and is a classical solution as well. This is an important step in identifying the unique classical solution of the Zakai equation as the unnormalized conditional density later in this section. We will not attempt to prove the uniqueness of the generalized solution under very general conditions on a^0, b^0, c^0 but will do so with an eye on the later applications to Theorem 3.8, 3.9.

<u>Lemma 3.2</u>: Suppose that u is a generalized solution to the Cauchy problem (II). Suppose further that

$$a^0(t,x),\ b^0(t,x) \text{ are continuous functions} \tag{3.29}$$

and that for compact subsets $K \subseteq \mathbb{R}^d$,

$$\sup_{s_0 \leq t \leq T} \int_K |u(t,x)|dx < \infty \tag{3.30}$$

$$\int_{s_0}^{T} \int_K |c^0(t,x)|\,|u(t,x)|dxdt < \infty. \tag{3.31}$$

(i) Then the equation (3.28) is valid for all $\phi \in C_0^2(\mathbb{R}^d)$.

(ii) Let $\phi \in C_0(\mathbb{R}^d)$. Then $f: [s_0,T] \to \mathbb{R}$ defined by

$$f(t) = \int u(t,x)\phi(x)dx \tag{3.32}$$

is a continuous function.

(iii) Let $\psi \in C_0([s_0,t_0]\times\mathbb{R}^d)$, $s_0 < t_0 \leq T$. Then

$$g(t) = \int u(t,x)\psi(t,x)dx \tag{3.33}$$

is a continuous function from $[s_0,t_0]$ into $\mathbb{R}$.

Proof: Given $\phi \in C_0^2(\mathbb{R}^d)$, we can get $\{\phi_j\} \subseteq C_0^\infty(\mathbb{R}^d)$ with $\{\text{support } \phi_j\} \subseteq K$ (a fixed compact set in $\mathbb{R}^d$) such that $\phi_j \to \phi$, $\frac{\partial}{\partial x^i}\phi_j \to \frac{\partial}{\partial x^i}\phi$ and $\frac{\partial}{\partial x^i \partial x^\ell}\phi_j \to \frac{\partial}{\partial x^i \partial x^\ell}\phi$ uniformly as $j \to \infty$. The equation (3.28) holds for each ϕ_j and the conditions (3.29), (3.30) and (3.31) permit us to take the limit as $j \to \infty$ inside the integrals in (3.28). This proves (i). For (ii), let $\{\phi_j\} \subseteq C_0^\infty(\mathbb{R}^d)$ be such that $\phi_j \to \phi$ uniformly and $\{\text{support } \phi_j\} \subseteq K$, a fixed compact set. Let f_j be defined by (3.32) for $\phi = \phi_j$. Then we have

$$\sup_{s_0 \leq t \leq T} |f_j(t)-f(t)| \leq \sup_{s_0 \leq t \leq T} \int |u(t,x)|\,|\phi_j(x)-\phi(x)|dx \tag{3.34}$$

$$\leq \sup_x |\phi_j(x)-\phi(x)| \sup_{s_0 \leq t \leq T} \int_K |u(t,x)|dx$$

and hence f_j converges uniformly to f. Since ϕ_j satisfies (3.28), f_j is a continuous function of t and hence so is f. This proves (ii). Observe that if $\{\phi_j\} \subseteq C_0(\mathbb{R}^d)$ converge uniformly to ϕ with $\{\text{support } \phi_j\} \subseteq K$, then for $t_j \to t$, we have

$$\left|\int u(t_j,x)\phi_j(x)dx - \int u(t,x)\phi(x)dx\right| \tag{3.35}$$

$$\leq \int |u(t_j,x)|\,|\phi_j(x)-\phi(x)|dx$$

$$+ \left|\int u(t_j,x)\phi(x)dx - \int u(t,x)\phi(x)dx\right|$$

$$\to 0 \quad \text{as } j \to \infty$$

because the first term on the right hand side is bounded by

$$(\sup_t \int_K |u(t,x)|dx) \sup_x |\phi_j(x)-\phi(x)|$$

and hence goes to zero as $j \to \infty$ and the second term goes to zero by part (ii) above. For $\psi \in C_0([s_0, t_0]\times\mathbb{R}^d)$, if $t_j \to t$, then taking $\phi_j(x) = \psi(t_j, x)$ and $\phi(x) = \psi(t, x)$ in (3.35), we can conclude that $g(t_j) \to g(t)$ (g is given by (3.33)) and hence that g is a continuous function. □

Lemma 3.3: Suppose that the conditions of the previous Lemma are satisfied. Let $s_0 < t \le T$ and $g \in C_0^{1,2}([s_0, t]\times\mathbb{R}^d)$. Then we have

$$\int u(t,x)g(t,x)dx - \int u(s_0,x)g(s_0,x)dx \tag{3.36}$$

$$= \int_{s_0}^{t} \int_{\mathbb{R}^d} [(\frac{\partial}{\partial s} + M_s)g(s,x)]u(s,x)dxds.$$

Proof: Using (3.28) for $\phi(x) = g(t,x)$ (by part (i) of Lemma 3.2) we conclude that

$$\int u(t,x)g(t,x)dx - \int u(s_0,x)g(t,x)dx \tag{3.37}$$

$$= \int_{s_0}^{t} \int_{\mathbb{R}^d} u(s,x)M_s g(t,s)dxds$$

$$= \int_{s_0}^{t} \int_{\mathbb{R}^d} u(s,x)M_s g(s,x)dxds$$

$$+ \int_{s_0}^{t} \int_{\mathbb{R}^d} u(s,x)[M_s g(t,x) - M_s g(s,x)]dxds$$

where the last relation follows from the previous one by writing $g(t,x) = g(s,x) + [g(t,x) - g(s,x)]$. Now

$$M_s g(t,x) - M_s g(s,x) = \int_s^t \frac{\partial}{\partial \tau}[M_s g(\tau,x)]d\tau = \int_s^t M_s[\frac{\partial}{\partial \tau} g(\tau,x)]d\tau$$

because $g \in C_0^{1,2}([s_0,t]\times\mathbb{R}^d)$ and as a consequence, we can interchange $\frac{\partial}{\partial\tau}$ and $\frac{\partial}{\partial x^i}$, $\frac{\partial^2}{\partial x^i\partial x^j}$. Thus

$$\int_{s_0}^{t}\int_{\mathbb{R}^d}u(s,x)[M_s g(t,x)-M_s g(s,x)]dxds \tag{3.38}$$

$$=\int_{s_0}^{t}\int_{\mathbb{R}^d}\int_{s}^{t}u(s,x)M_s[\frac{\partial}{\partial\tau}g(\tau,x)]d\tau dxds$$

$$=\int_{s_0}^{t}\int_{s_0}^{\tau}\int_{\mathbb{R}^d}u(s,x)M_s[\frac{\partial}{\partial\tau}g(\tau,x)]dxdsd\tau$$

by Fubini's theorem, which is justified in view of the assumptions (3.29)-(3.31). The relation (3.28) for $\phi(x) = \frac{\partial}{\partial\tau}g(\tau,x)$ gives

$$\int_{s_0}^{\tau}\int_{\mathbb{R}^d}u(s,x)M_s[\frac{\partial}{\partial\tau}g(\tau,x)]dxds \tag{3.39}$$

$$=\int_{\mathbb{R}^d}u(\tau,x)\frac{\partial}{\partial\tau}g(\tau,x)dx - \int_{\mathbb{R}^d}u(s_0,x)\frac{\partial}{\partial\tau}g(\tau,x)dx.$$

Finally, by Fubini's theorem, we have

$$\int_{s_0}^{t}\int_{\mathbb{R}^d}u(s_0,x)\frac{\partial}{\partial\tau}g(\tau,x)dxd\tau \tag{3.40}$$

$$=\int_{\mathbb{R}^d}u(s_0,x)\int_{s_0}^{t}\frac{\partial}{\partial\tau}g(\tau,x)d\tau dx$$

$$=\int_{\mathbb{R}^d}u(s_0,x)g(t,x)dx - \int_{\mathbb{R}^d}u(s_0,x)g(s_0,x)dx.$$

The equations (3.37)-(3.40) prove the validity of (3.36).

□

Theorem 3.4: Suppose that the conditions of Theorem 3.1 are satisfied and that p_0 is as in part (iii) of the said theorem.

(i) Let u_1, u_2 be generalized solution to the Cauchy problem (II) satisfying (3.24) and (3.30). Then for all t,

$$u_1(t,x) = u_2(t,x) \quad \text{a.e. } x.$$

(ii) Let u_1 be as in (i) and u be the classical solution to the Cauchy problem (II) given by (3.25). Then for all t

$$u_1(t,x) = u(t,x) \quad \text{a.e. } x.$$

(Here and in the proof below, a.e. refers to the Lebesgue measure).

Proof: (i) Let $f(t,x) = u_1(t,x) - u_2(t,x)$ for $s_0 \leq t \leq T$, $x \in \mathbb{R}^d$. Then f is a generalized solution to the Cauchy problem (II) with $p_0 = 0$ and f satisfies (3.24) and (3.30). Since c^0 is assumed to be continuous, the conditions of Lemma 3.2, 3.3 are satisfied. Note that $f(s_0,x) = 0$ a.e. x.

Fix $s_0 < t_0 \leq T$ and $g_0 \in C_0(\mathbb{R}^d)$. Let v be given by (3.23). Thus, v is the classical solution to the Cauchy problem (I) and $v \in C^{1,2}([s_0,t_0)\times\mathbb{R}^d) \cap C([s_0,t_0]\times\mathbb{R}^d)$. If we could use (3.36) for this v, we could complete the proof easily, but v may not belong to the class $C_0^{1,2}([s_0,t_0]\times\mathbb{R}^d)$ and hence (3.36) may not be valid for v. Instead, we proceed as follows. Let R be such that

$$\{\text{support } g_0\} \subseteq \{x \in \mathbb{R}^d\colon\ |x| \le R\}$$

and choose a function $\phi \in C_0^\infty(\mathbb{R}^d)$ such that $0 \le \phi \le 1$, $\phi = 1$ on $\{x \in \mathbb{R}^d\colon\ |x| \le R\}$, $\phi = 0$ on $\{x \in \mathbb{R}^d\colon\ |x| > R+1\}$ and

$$\sum_{i,j=1}^{d} \left|\frac{\partial^2}{\partial x^i \partial x^j} \phi(x)\right| + \sum_{i=1}^{d} \left|\frac{\partial}{\partial x^i} \phi(x)\right| \le K_4$$

for some constant K_4 (not depending on R) (see [17, p. 30]) and define $v_1\colon [s_0, t_0]\times\mathbb{R}^d \to \mathbb{R}$ by

$$v_1(s,x) = v(s,x)\phi(x).$$

Then

$$\begin{aligned}
&(\tfrac{\partial}{\partial s}+M_s)v_1(s,x) \\
&= \phi(x)[\tfrac{\partial}{\partial s}+M_s]v(s,x) + \tfrac{1}{2}\sum_{i,j=1}^{d} a_{ij}^0(s,x)\frac{\partial^2}{\partial x^i \partial x^j}\phi(x)v(s,x) \\
&+ \sum_{i=1}^{d} b_i^0(s,x)\frac{\partial}{\partial x^i}\phi(x)v(s,x) + \sum_{i,j=1}^{d} a_{ij}^0(s,x)\frac{\partial}{\partial x^i}\phi(x)\frac{\phi}{\partial x^j}v(s,x)
\end{aligned}$$

and hence using the bounds of ϕ and the fact that $(\frac{\partial}{\partial s}+M_s)v(s,x) = 0$, we get

$$\begin{aligned}
|(\tfrac{\partial}{\partial s}+M_s)v_1(s,x)| \le K_5(&|v(s,x)| \\
&+ \sum_{i=1}^{d} |\frac{\phi}{\partial x^i}v(s,x)|)1_{\{R\le|x|\le R+1\}}
\end{aligned} \tag{3.41}$$

for some finite constant K_5. Using the bound (3.20) on G, the inequality $|x-z|^2 \ge \frac{1}{2}|x|^2 - |z|^2$ and the fact that g_0 has compact support, we obtain the inequality

$$|v(s,x)| \le K_6(t_0-s)^{-d/2}\exp\left(\frac{-K_2|x|^2}{2(t_0-s)}\right), \tag{3.42}$$

K_6 being a constant. The bound (3.21) on $\frac{\partial G}{\partial x^i}$ implies that

$$\frac{\partial}{\partial x^i} v(s,x) = \int \frac{\partial}{\partial x^i} G(t_0,z,s,x)g_0(z)dz$$

and hence using (3.21), we get as in (3.42),

$$\left|\frac{\partial}{\partial x^i} v(s,x)\right| \le K_7(t_0-s)^{-\frac{d+1}{2}}\exp\left(\frac{-K_2|x|^2}{2(t_0-s)}\right). \tag{3.43}$$

Combining (3.41), (3.42) and (3.43), we get

$$|(\frac{\partial}{\partial s}+M_s)v_1(s,x)| \le K_8(t_0-s)^{-\frac{d+1}{2}}\exp\left[\frac{-K_2|x|^2}{2(t_0-s)}\right]1_{\{R\le|x|\le R+1\}} \tag{3.44}$$

for a suitable constant K_8. Let K_9 be a constant such that for $\theta > 0$,

$$\theta^{\frac{d+1}{2}} \le K_9 \exp\left[\frac{K_2}{4}\cdot\theta\right].$$

Then for $s_0 \le s < t_0$, $|x| \ge 1$, we have

$$(t_0-s)^{-\frac{d+1}{2}} \le K_9 \exp\left[\frac{K_2}{4(t_0-s)}\right] \le K_9 \exp\left[\frac{K_2|x|^2}{4(t_0-s)}\right]$$

so that (3.44) implies

$$|(\frac{\partial}{\partial s}+M_s)v_1(s,x)| \le K_{10} \exp\left[\frac{-K_2|x|^2}{4(t_0-s)}\right]\cdot 1_{\{R\le|x|\le R+1\}} \tag{3.45}$$

where $K_{10} = K_9\cdot K_8$. By its construction,

$v_1 \in C_0^{1,2}([s_0,t_0)\times\mathbb{R}^d) \cap C_0([s_0,t_0]\times\mathbb{R}^d)$ and hence

$$F(t) = \int_{\mathbb{R}^d} f(t,x)v_1(t,x)dx$$

is a continuous function from $[s_0,t_0]$ into $\mathbb{R}$. Let $s_0 < t < t_0$. Then $v_1 \in C_0^{1,2}([s_0,t]\times\mathbb{R}^d)$ and hence by Lemma 3.3, we have

$$\int_{\mathbb{R}^d} f(t,x)v_1(t,x)dx = \int_{s_0}^{t} \int_{\mathbb{R}^d}[(\frac{\partial}{\partial s}+M_s)v_1(s,x)]f(s,x)dx$$

since $f(s_0,x) = 0$ a.e. x. Using the estimate (3.45) on v_1, this yields

$$\left|\int_{\mathbb{R}^d} f(t,x)v_1(t,x)dx\right| \tag{3.46}$$

$$\leq K_{10} \int_{s_0}^{t} \int_{\mathbb{R}^d} \exp\left[\frac{-K_2|x|^2}{4(t_0-s)}\right] |f(s,x)| 1_{\{R\leq|x|R+1\}}dxds.$$

Let K_3 be a constant such that (3.24) holds for f and define $\delta = K_2/4K_3$. If t_0 is such that $t_0-s_0 \leq \delta$ then for $s_0 \leq s \leq t < t_0$,

$$\frac{K_2}{4(t_0-s)} \geq \frac{K_2}{4(t_0-s_0)} \geq \frac{K_2}{4\delta} = K_3$$

and hence in view of (3.24), we have

$$\int_{s_0}^{t} \int_{\mathbb{R}^d} \exp\left[\frac{-K_2|x|^2}{4(t_0-s)}\right] |f(s,x)|dxds < \infty.$$

Therefore the right hand side of (3.46) goes to zero as $R \to \infty$. Since the left hand side of (3.46) does not depend upon R as long as $\{\text{support } g_0\} \subseteq \{x: |x| \leq R\}$, this gives

$$|\int f(t,x)v_1(t,x)dx| = 0, \quad s_0 < t < t_0. \tag{3.47}$$

The continuity of the function F defined above yields $F(t_0) = 0$, which gives, recalling that $v_1(t_0,x) = g_0(x)$,

$$\int f(t_0,x)g_0(x)dx = 0. \tag{3.48}$$

Since $g_0 \in C_0(\mathbb{R}^d)$ is arbitrary, this implies

$$f(t_0,x) = 0 \quad \text{a.e. } x \tag{3.49}$$

for all $t_0 \in [s_0,s_0+\delta]$. Repeating this argument successively we can conclude that (3.49) holds for all $t_0 \in [s_0+\delta,s_0+2\delta]$, $t_0 \in [s_0+2\delta,s_0+3\delta]$, and so on. Note that the number δ depends only on the function f and coefficients a^0,b^0,c^0 (through constants K_2,K_3) and not on s_0. Thus in finitely many stages, we can conclude that (3.49) holds for all $t_0 \in [s_0,T]$. This proves (i).

We have already noted that a classical solution is also a generalized solution. As noted in Theorem 3.1, the classical solution u satisfies (3.24) and (3.26) and hence (3.30) as well. Now, (ii) follows from (i). □

The proof given above is a modification of the proof of uniqueness of the classical solution given in Friedman [17, p. 30]. The essential idea is that if the "adjoint equation" (3.4) admits a fundamental solution, then the given equation has a unqiue (generalized) solution. We will now show that we can relax the condition (3.19) on c^0 and still prove uniqueness of the generalized solution. This will be useful in the context of the Zakai equation, as this will enable us to characterize the unnormalized conditional density as the unique generalized solution of the Zakai equation *for all observation paths* $y \in H$.

First we prove a result on approximation and then go on to the uniqueness part referred to in the earlier paragraph.

Lemma 3.5: Let c be a bounded measurable function from $[0,T]\times\mathbb{R}^d$ into $\mathbb{R}$ having compact support. Then there exists $\{c_i\} \subseteq C_0^\infty([0,T]\times\mathbb{R}^d)$ such that $\{c_i\}$ is uniformly bounded by the bound for c and

$$\lim_{i\to\infty} \int_0^T\int_{\mathbb{R}^d} |c(t,x)-c_i(t,x)|dxdt = 0.$$

Proof: Let ρ: $\mathbb{R}^{d+1} \to \mathbb{R}$ be given by $\rho(z) = K\exp(-(|z|^2-1)^{-1})$ for $|z| < 1$ and equal to zero for $|z| \geq 1$ and K is chosen such that $\int\rho(z)dz = 1$. For $i \geq 1$, define

$$c_i(z) = (i)^{d+1}\int_{\mathbb{R}^{d+1}} c(z)\rho(i(z_1-z))dz_1$$

where $z = (t,x) \in \mathbb{R}^{d+1}$. This sequence satisfies the required approximation property (see [18, p. 94]).

Theorem 3.6: Let M_t be given by (3.3). Suppose that a^0, b^0 satisfy (3.1), (3.2), (3.18) and that p_0 is an integrable function. Suppose that c^0 is a measurable function satisfying

$$c^0(t,x) \leq \theta(t), \quad x \in \mathbb{R}^d,\ 0 \leq t \leq T \tag{3.50}$$

where θ satisfies $\theta(t) \geq 0$ and $\int_0^T\theta(t)dt < \infty$.

Then if u_1, u_2 are generalized solutions to the Cauchy problem (II) satisfying (3.30) and

$$\int_{s_0}^{T}\int_{\mathbb{R}^d}[1 + |c^0(t,x)|]|u_i(t,x)|dxdt < \infty, \quad i = 1,2. \tag{3.51}$$

Then for all $t \in [s_0,T]$, we have

$$u_1(t,x) = u_2(t,x) \quad \text{a.e. } x.$$

Proof: Let $f(t,x) = u_1(t,x) - u_2(t,x)$ for all $(t,x) \in [s_0,T]\times\mathbb{R}^d$. Then f is a generalized solution to the Cauchy problem (II) with $p_0 = 0$ and further, f satisfies (3.30) and (3.51). Thus the conditions of Lemma 3.2 are satisfied by f. For each integer $k \geq 1$, let us define $c^{k,0}\colon [s_0,T]\times\mathbb{R}^d \to \mathbb{R}$ by

$$c^{k,0}(t,x) = c^0(t,x)1_{\{|x|\leq k\}}\cdot 1_{\{|c^0(t,x)|\leq k\}}.$$

Then $c^{k,0}$ is a measurable function with compact support and

$$|c^{k,0}(t,x)| \leq k.$$

Further

$$c^{k,0}(t,x) \leq \theta(t), \quad (t,x) \in [s_0,T]\times\mathbb{R}^d. \tag{3.52}$$

Let $\{c^{k,i}\}$ be the sequence of approximating functions for $c^{k,0}$ given by Lemma 3.5, i.e., $\{c^{k,i}\} \subseteq C_0^\infty([s_0,T]\times\mathbb{R}^d)$,

$$|c^{k,i}(t,x)| \leq k \tag{3.53}$$

and

$$\lim_{i\to\infty}\int_{s_0}^{T}\int_{\mathbb{R}^d}|c^{k,i}(t,x) - c^{k,0}(t,x)|dxdt = 0. \tag{3.54}$$

For $k \geq 1$, $i \geq 1$, let $M^{k,i}$ be the differential operator given by (3.3) for $(a^0, b^0, c^{k,i})$ (i.e., $c^{k,i}$ instead of c^0 in (3.3)). Then $M^{k,i}$ satisfies the conditions of Theorem 3.1 for $k \geq 1$, $i \geq 1$.

Given $g_0 \in C_0(\mathbb{R}^d)$, $s_0 < t_0 \leq T$, let $v^{k,i}$ be the solution to the Cauchy problem (I) for $M^{k,i}$ (instead of M). Then

$$(\frac{\partial}{\partial s}+M_s)v^{k,i} = (\frac{\partial}{\partial s}+M_s^{k,i})v^{k,i} + (c^0-c^{k,i})v^{k,i} \tag{3.55}$$

$$= (c^0-c^{k,i})v^{k,i}.$$

Let R, ϕ be as in the proof of Theorem 3.4 and let

$$v_1^{k,i}(s,x) = v^{k,i}(s,x)\phi(x).$$

Then, using (3.55), we can conclude

$$|(\frac{\partial}{\partial s}+M_s)v_1^{k,i}| \leq |c^0-c^{k,i}|\cdot|v^{k,i}| + K_{10}\cdot 1_{\{R\leq|x|\leq R+1\}}. \tag{3.56}$$

This inequality can be obtained following the steps leading to (3.45). We need to use (3.55) instead of $(\frac{\partial}{\partial s} + M_s v) = 0$. Also in (3.56), the constant K_{10} may depend on $c^{k,i}$.

As in the proof of Theorem 3.4, $v_1^{k,i} \in C_0^{1,2}([s_0,t_0)\times\mathbb{R}^d)$ and hence we can use (3.36) with $g = v_1^{k,i}$ for any $t < t_0$. Remembering that $f(s_0,x) = 0$ a.e. x, we get for any $t < t_0$,

$$|\int_{\mathbb{R}^d} f(t,x)v_1^{k,i}(t,x)dx| \leq \int_{s_0}^{t}\int_{\mathbb{R}^d}|f(s,x)|\cdot|(\frac{\partial}{\partial s}+M_s)v_1^{k,i}(s,x)|dxds$$

$$\leq \int_{s_0}^{t} \int_{\mathbb{R}^d} |f|\cdot|c^0-c^{k,i}|\cdot|v^{k,i}|dxds$$

$$+ K_{10} \int_{s_0}^{t} \int_{\mathbb{R}^d} |f(s,x)| 1_{\{R\leq|x|\leq R+1\}} dxds$$

by (3.56). Taking limit as $R \to \infty$, the second term goes to zero as f satisfies (3.51) and we conclude

$$|\int f(t,x)v_1^{k,i}(t,x)dx| \leq \int_{s_0}^{t} \int_{\mathbb{R}^d} |f|\cdot|c^0-c^{k,i}|\cdot|v^{k,i}|dxds \tag{3.57}$$

for $s_0 < t < t_0$. Since $v_1^{k,i} \in C_0([s_0,t_0]\times\mathbb{R}^d)$, the left hand side in the inequality above is a continuous function of t (by Lemma 3.2) and hence taking limit as $t \to t_0$, we get

$$|\int f(t_0,x)g_0(x)dx| \leq \int_{s_0}^{t_0} \int_{\mathbb{R}^d} |f|\cdot|c^0-c^{k,i}|\cdot|v^{k,i}|dxds. \tag{3.58}$$

We want to conclude from this that the left hand side in (3.58) is zero. Since g_0 has compact support, we can conclude (as in (3.42)) that $v^{k,i}$ is bounded. Hence by the Feynman-Kac's formula Theorem II.4.1, we have for $k\geq 1$, $i\geq 1$

$$v^{k,i}(s,x) = E_{Q_{s,x}}[g_0(Z_{t_0})\exp(\int_s^{t_0} c^{k,i}(u,Z_u)du)]. \tag{3.59}$$

Here (Z_t) is the coordinate process on $\Omega_d = C([0,\infty),\mathbb{R}^d)$ and $\{Q_{s,x}\}$ is the solution to the martingale problem for (a^0,b^0) starting at (s,x). Thus

$$|v^{k,i}(s,x)| \leq K \exp(kT),$$

K being an upper bound of $|g_0|$. Since for each $u > s$, the measure $Q_{s,x} \circ Z_u^{-1}$ is absolutely continuous with respect to Lebesgue measure (see Remark 3.2), condition (3.54) implies

$$\int_s^t c^{k,i}(u,Z_u)du \to \int_s^t c^{k,0}(u,Z_u)du$$

in $Q_{s,x}$ probability as $i \to \infty$. Hence

$$v^{k,i}(s,x) \to v^{k,0}(s,x) \quad \text{for each } (s,x) \in [s_0,T]\times\mathbb{R}^d$$

where $v^{k,0}$ is defined by (3.59) for $i = 0$. Hence taking limit as $i \to \infty$ in (3.58) and remembering that f satisfies (3.51), we conclude

$$|\int f(t_0,x)g_0(x)dx| \leq \int_{s_0}^{t_0} \int_{\mathbb{R}^d} |c^0-c^{k,0}|\cdot|v^{k,0}|\cdot|f|dxds. \tag{3.60}$$

In view of (3.52) and the definition of $v^{k,0}$, we have

$$|v^{k,0}| \leq K \exp(\int_0^T \theta(t)dt) = K_1 \quad \text{(say)}$$

where K is an upper bound of g_0. Thus (3.60) yields

$$|\int f(t_0,x)g_0(x)dx|$$

$$\leq K_1 \int_{s_0}^{t_0}\int_{\mathbb{R}^d} |f|\cdot|c^0(s,x)|\{1_{\{|x|>k\}} + 1_{\{|c^0(s,x)|>k\}}\}dxds$$

(recall the choice of $c^{k,0}$). Since f satisfies (3.51), the dominated convergence theorem yields taking limit as $k \to \infty$

$$\int f(t_0,x)g_0(x)dx = 0.$$

Here, $g_0 \in C_0(\mathbb{R}^d)$ and $t_0 \in [s_0,T]$ is arbitrary and hence we can conclude that $f(t_0,x) = 0$ a.s. x for all $t_0 \in [s_0,T]$.

□

Let us return to the nonlinear filtering model (1.9)-(1.10), where we assume that the signal process (X_t) is an $\mathbb{R}^d$-valued diffusion process. In other words, (X_t) is an $\mathbb{R}^d$-valued continuous Markov process and the associated semi-group $\{T_t\}$ satisfies, for $f \in C_0^{1,2}([0,T]\times\mathbb{R}^d)$,

$$(T_tf)(s,x) = f(s,x) + \int_0^t[T_u(\frac{\partial}{\partial u} + L_u)f](s,x)du. \qquad (3.61)$$

For each u, L_u is the differential operator on $\mathbb{R}^d$ given by

$$(L_ug)(x) = \frac{1}{2} \sum_{i,j=1}^{d} a_{ij}(u,x)\frac{\partial^2}{\partial x^i\partial x^j}g(x) + \sum_{i=1}^{d} b_i(u,x)\frac{\partial}{\partial x^i}g(x) \qquad (3.62)$$

where a,b are the diffusion and drift coefficients of (X_t). The functions a,b are assumed to satisfy (II.3.2) and (II.3.3). Here

$$C_0^{1,2}([0,T]\times\mathbb{R}^d) \subseteq \mathcal{D}^{(e)}$$

and (3.63)

$$L^{(e)}f = (\frac{\partial}{\partial t} + L_t)f.$$

Moreover, if a,b are continuous, then $C_0^{1,2}([0,T]\times\mathbb{R}^d) \subseteq \mathcal{D}$ and

$$Lf = (\frac{\partial}{\partial t} + L_t)f. \qquad (3.64)$$

In view of this, we define a differential operator acting on $C^{1,2}([0,T]\S\mathbb{R}^d)$, also denoted by L, by

$$(Lf)(s,x) = (\frac{\partial}{\partial s} f)(s,x) + (L_s f)(s,x). \tag{3.65}$$

From now on, in the context of diffusion processes, L will always denote the differential operator defined by (3.65).

For $0 \leq s \leq T$, $\phi \in H$, let $\Gamma_s^\phi \in \mathcal{M}(\mathbb{R}^d)$ be defined by

$$\Gamma_s^\phi(B) = E_\Pi(1_B(X_s)q_s(\phi,\omega)) = \sigma_s(1_B,\phi) \tag{3.66}$$

for $B \in \mathcal{B}(\mathbb{R}^d)$, where q_s, σ_s are given by (2.7) and (2.9) respectively. That Γ_s^ϕ belongs to $\mathcal{M}(\mathbb{R}^d)$ follows easily from (3.66). Also, from (3.66) and (2.9) it is easy to see that for $f\colon \mathbb{R}^d \to \mathbb{R}$ such that $E|f(X_t)| < \infty$, we have

$$\sigma_t(f,\phi) = \int f(x)d\Gamma_t^\phi(x) \tag{3.67}$$

and hence from the Bayes formula (Theorem 1.1)

$$E_\alpha(f(X_t)|Q_t y) = \frac{1}{\Gamma_t^y(\mathbb{R}^d)} \cdot \int f(x)d\Gamma_t^y(x). \tag{3.68}$$

In view of (3.68), Γ_t^y is called the *unnormalized conditional distribution of* X_t *given* $Q_t y$. If the measure Γ_t^y is absolutely continuous with respect to the Lebesgue measure λ (on $\mathbb{R}^d$), then the density $p_t(x,y) = \frac{d\Gamma_t^y}{d\lambda}(x)$ is called the *unnormalized conditional density of* (X_t) *given* $Q_t y$. The next result shows that if the (unconditional) distribution of X_t admits a density with respect to λ, then $p_t(x,y)$ exists and then the Zakai equation (2.6) can be rewritten as a partial differential equation for $p_t(x,y)$.

Theorem 3.7: Suppose that for all t, the measure $\Pi \circ X_t^{-1}$ (on $\mathbb{R}^d$) admits a density $p_t(x)$ with respect to λ.

(i) Then, for all $t \in [0,T]$, $y \in H$, the measure Γ_t^y is absolutely continuous with respect to λ. Further, we can choose a version $p_t(x,y)$ of the density $\frac{d\Gamma_t^y}{dt}(x)$ such that $(t,x) \to p_t(x,y)$ is jointly measurable.

(ii) Suppose that h satisfies (2.5). Then for all $y \in H$, $p_t(x,y)$ is a generalized solution to the Cauchy problem

$$\frac{\partial p_t(x,y)}{\partial t} = L_t^* p_t(x,y) + [\sum_{i=1}^{N} h_t^i(x) y_t^i - \tfrac{1}{2}|h_t(x)|^2] p_t(x,y) \tag{3.69}$$

$$p_0(x,y) = p_0(x). \tag{3.70}$$

(iii) If $y_k \to y$ in H, then

$$\int_{\mathbb{R}^d} |p_t(x,y_k) - p_t(x,y)| dx \to 0. \tag{3.71}$$

Proof: (i) Fix $0 \leq t \leq T$, $y \in H$. Suppose $B \in \mathcal{B}(\mathbb{R}^d)$ is such that $\lambda(B) = 0$. Since $\Pi \circ X_t^{-1} << \lambda$, we have $\Pi(X_t \in B) = 0$. The relation (3.66) now gives $\Gamma_t^y(B) = 0$. Hence $\Gamma_t^y << \lambda$. The relation (3.66) and the joint measurability of $1_B(X_s(\omega)) q_s(y,\omega)$ in (s,ω) implies that for all $B \in \mathcal{B}(\mathbb{R}^d)$, $y \in H$, $s \to \Gamma_s^y(B)$ is measurable. Hence we can choose a jointly measurable (in (t,x)) function $p_t(x,y)$ such that for all $t \in [0,T]$, $y \in H$, $\frac{d\Gamma_t^y(x)}{d\lambda} = p_t(x,y)$ (see e.g., [61]).

(ii) Since h satisfies (2.5), the functional $\sigma_t(f,y)$

satisfies the Zakai equation (2.6) (see Remark 2.2). This immediately gives that $p_t(x,y)$ is a generalized solution to the Cauchy problem (3.69)-(3.70). Indeed, if $\phi \in C^\infty(\mathbb{R}^d)$, then $\phi \in \mathcal{D}_0$ and hence (2.6) (or (2.25)) and (3.67) imply

$$\int_{\mathbb{R}^d}\phi(x)p_t(x,y)dx = \int_{\mathbb{R}^d}\phi(x)p_0(x)dx + \int_0^t\int_{\mathbb{R}^d}(L_s\phi)(x)p_s(x,y)dxds \\ + \int_0^t\int_{\mathbb{R}^d}\phi(x)p_s(x,y)[\sum_{i=1}^N h_s^i(x)y_s^i - \tfrac{1}{2}|h_s(x)|^2]dxds. \tag{3.72}$$

Here we have also used the fact that $\Gamma_0^y = \Pi\circ X_0^{-1}$. The proof of Theorem 2.1 also shows that the integrals appearing in (3.72) are finite. This proves (ii).

For (iii), let us write

$$c_t^y(x) = \sum_{i=1}^N h_t^i(x)y_t^i - \tfrac{1}{2}|h_t(x)|^2. \tag{3.73}$$

Then note that

$$\Gamma_t^y(B) = \int 1_B(X_t(\omega))\exp(\int_0^t c_s^y(X_s(\omega))ds)d\Pi(\omega), \tag{3.74}$$

$$|c_t^y(x)| \le 2|h_t(x)|^2 + |y_t|^2 \tag{3.75}$$

and

$$c_t^y(x) \le \tfrac{1}{2}|y_t|^2. \tag{3.76}$$

Now we have

$$\int_{\mathbb{R}^d}|p_t(x,y_k)-p_t(x,y)|ds \le 2\sup_{B\in\mathcal{B}(\mathbb{R}^d)}|\Gamma_t^{y_k}(B)-\Gamma_t^y(B)| \tag{3.77}$$

$$\le 2\int|\exp(\int_0^t c_s^{y_k}(X_s(\omega))ds) - \exp(\int_0^t c_s^y(X_s(\omega))ds)|d\Pi(\omega).$$

As $y_k \to y$ in H, $c_s^{y_k}(x) \to c_s^y(x)$ and hence the integrand on the right hand side in (3.77) goes to zero pointwise. Also the same is bounded above by 2 $\exp(\frac{1}{2}|y|^2)$ in view of (3.76) and hence by the dominated convergence theorem, the right hand side in (3.77) goes to zero. This yields (3.71) □

Remark 3.1: Proceeding as in (iii) above and using continuity of paths of (X_t), it can be proved that if $t_k \to t$, then for all $y \in H$

$$\int |p_{t_k}(x,y) - p_t(x,y)|dx \to 0 \tag{3.78}$$

Remark 3.2: It is well known that under very general conditions on a,b, the distribution of (X_t) is absolutely continuous with respect to λ. In particular, when a,b satisfy (3.1), (3.2) and (3.18), then $\Pi \circ X_t^{-1} << \lambda$ for $t > 0$. This fact would also follow from Theorem 4.4 which is proved in the next section.

As a consequence of Theorem 3.4, we have the following result, which gives that for y belonging to a dense subset H_0 of H, $p_t(x,y)$ is the unique classical solution to the Cauchy problem (3.69)-(3.70). Thus, for $y \in H_0$, $p_t(x,y)$ can be computed via numerical methods and then in principle, $p_t(x,y)$ for any $y \in H$ can be obtained using (iii) of the previous theorem.

Let $H_0 = \{y \in H: y_t$ is Hölder continuous$\}$. Note that H_0 is dense in H.

Theorem 3.8: Suppose that the coefficients a,b appearing in the expression (3.62) for the generator L of (X_t) satisfy the conditions (3.1), (3.2) and (3.18). Suppose that

$p_0 = \dfrac{d\Pi \circ X_0^{-1}}{d\lambda}$ is a continuous function satisfying $|p_0(x)| \leq K \exp(|x|^{2-\epsilon})$ for some $\epsilon > 0$.

Suppose that h is a bounded Hölder continuous function.

Then for all $y \in H_0$, the unnormalized conditional density $p_t(x,y)$ is the unique classical solution to the Cauchy problem (3.69)-(3.70) in the class of $C^{1,2}((0,T]\times\mathbb{R}^d) \cap C([0,T]\times\mathbb{R}^d)$ functions f satisfying (3.24).

Proof: The conditions imposed on a,b, $\Pi \circ X_0^{-1}$ imply that $\Pi \circ X_t^{-1} \ll \lambda$ for all t (see Remark 3.2). Thus the unnormalized conditional density exists and is a generalized solution to the Cauchy problem (3.69)-(3.70). Note that $p_t(x,y)$ is the density of a positive mea-sure Γ_t^y and hence $p_t(x,y) \geq 0$. Thus

$$\sup_{0\leq t\leq T} \int |p_t(x,y)|dx = \sup_{0\leq t\leq T} \int p_t(x,y)dx \tag{3.79}$$

$$= \sup_{0\leq t\leq T} \Gamma_t^y(\mathbb{R}^d) \leq \exp(\|y\|^2).$$

Since h is Hölder continuous and bounded, for $y \in H_0$, i.e., y Hölder continuous, $c_s^y(x)$ defined by (3.74) is a bounded Hölder continuous function. Thus (3.79) shows that for $y \in H_0$, $p_t(x,y)$ satisfies (3.30) and (3.31) for $c^0 = c^y$. Let M_t^y be defined by

$$M_t^y g(x) = L_t g(x) + c_t^y(x)g(x)$$

for $g \in C^2(\mathbb{R}^d)$. Then for $y \in H_0$, M_t^y satisfies the conditions of Theorem 3.1. Hence by Theorem 3.1, the Cauchy problem (3.69)-(3.70) admits a unique classical solution $p'_t(x,y)$ which satisfies (3.24), (3.26) and hence (3.30), (3.31). Hence by Theorem 3.4

$$p_t(x,y) = p'_t(x,y) \quad \text{a.e. } x. \tag{3.80}$$

Thus, the classical solution $p'_t(x,y)$ is a version of the unnormalized conditional density. □

In the next result we show that if a, b, p_0 are as in the previous theorem, then for all $y \in H$, $p_t(x,y)$ is the unique generalized solution to the Cauchy problem (3.69)-(3.70).

Theorem 3.9: Suppose that a, b, p_0 satisfy the hypotheses of Theorem 3.8. Suppose that $h: [0,T]\times\mathbb{R}^d \to \mathbb{R}$ is a measurable function satisfying

$$E\int_0^T |h_t(X_t)|^2 dt < \infty. \tag{3.82}$$

Then for all $y \in H$, the unnormalized conditional density $p_t(x,y)$ exists and is the unique generalized solution to the Cauchy problem (3.69)-(3.70) in the class of measurable functions $f: [0,T] \to \mathbb{R}^d\times\mathbb{R}$ satisfying

$$\sup_{o\leq t\leq T} \int_{\mathbb{R}^d} |f(t,x)|dx < \infty \tag{3.83}$$

and

$$\int_0^T\int_{\mathbb{R}^d} |h_t(x)|^2 |f(t,x)|dxdt < \infty. \tag{3.84}$$

Proof: As noted in the proof of Theorem 3.8, $\Pi\circ X_t^{-1} << \lambda$ and hence by Theorem 3.7, the unnormalized conditional density

$p_t(x,y)$ exists and is a generalized solution to the Cauchy problem (3.69)-(3.70).

Let us note that if $p_t = \frac{d\Pi \circ X_t^{-1}}{d\lambda}$, then

$$p_t(x,y) \le \exp(\tfrac{1}{2}\|y\|^2) \cdot p_t(x) \quad \text{a.e.} \tag{3.85}$$

This follows from the fact that $q_s(y,\omega) \le \exp(\frac{1}{2}\|y\|^2)$ and hence $\frac{d\Gamma_t^y}{d\Pi \circ X_t^{-1}} \le \exp(\frac{1}{2}\|y\|^2)$. Hence

$$\sup_{0 \le t \le T} \int_{\mathbb{R}^d} p_t(x,y)dx \le \exp(\tfrac{1}{2}\|y\|^2) \sup_{0 \le t \le T} \int_{\mathbb{R}^d} p_t(x)dx$$

$$= \exp(\tfrac{1}{2}\|y\|^2) < \infty$$

and

$$\int_0^T \int_{\mathbb{R}^d} |h_t(x)|^2 p_t(x,y)dxdt \le \exp(\tfrac{1}{2}\|y\|^2) \int_0^T \int |h_t(x)|^2 p_t(x)dxdt$$

$$= \exp(\tfrac{1}{2}\|y\|^2) \int_0^T E|h_t(X_t)|^2 dt < \infty$$

in view of (3.82). Thus, $p_t(x,y)$ satisfies (3.83) and (3.84). Fix $y \in H$ and let f be a solution to the Cauchy problem (3.69)-(3.70) satisfying (3.83) and (3.84). Clearly, (3.83) implies (3.30) and further that

$$\int_0^T \int_{\mathbb{R}^d} |f(t,x)|dxds < \infty. \tag{3.86}$$

Also

$$\int_0^T \int_{\mathbb{R}^d} |y_t|^2 |f(t,x)|dxdt \le [\sup_{0 \le t \le T} \int |f(t,x)| \int_0^T |y_t|^2 dt] < \infty. \tag{3.87}$$

The inequalities (3.75), (3.84) and (3.87) yield

$$\int_0^T\int_{\mathbb{R}^d}[1+c_s^y(x)]|f(s,x)|dxds < \infty. \tag{3.88}$$

Thus, $p_t(x,y)$ and $f(t,x)$ are generalized solutions to the Cauchy problem (3.69)-(3.70), both satisfying (3.30) and (3.51) for $c^0 = c^y$. As noted earlier, c^y satisfies (3.50) with $\theta(t) = \frac{1}{2}|y_t|^2$. Hence by Theorem 3.6, we have, for all $t \in [0,T]$

$$f(t,x) = p_t(x,y) \quad \text{a.e. } x.$$

This proves the uniqueness part completing the proof. □

4. UNIQUENESS OF SOLUTION (UNBOUNDED COEFFICIENTS)

We will now extend the results on the unnormalized conditional density given in the previous section. In this section, we only consider classical solutions to the Cauchy problem but allow unbounded a,b and h. Our results rely crucially on the work of Besala and Bodanko on second order partial differential equations with unbounded coefficients.

The treatment given here is different from the one given in the previous section, where we used the generalized solution and its uniqueness in identifying the classical solution to the Cauchy problem as the unnormalized conditional density in Theorem 3.8. Here we use the Feynman-Kac formula for the identification part.

For an interval $I \subseteq [0,T]$, let $\mathscr{G}(I)$ be defined by

$$\mathscr{G}(I) = \{f \in C(\overline{I}\times\mathbb{R}^d) : f \in C^{1,2}(I\times\mathbb{R}^d) \text{ and } \sup_{t\in I}|f(t,x)| \le \exp(K(1+|x|^2)^{\frac{1}{2}}) \text{ for some } K\}.$$

Here $\overline{I}$ is the closure of I. Thus $\overline{[0,t_0)} = [0,t_0]$, $\overline{(s_0,T)} = [s_0,T]$.

Theorem 4.1: Let M_t be given by (3.3). Suppose that a^0 satisfies (3.1), (3.2) and that

$$a^0_{ij},\ \frac{\partial}{\partial x^i} a^0_{ij},\ \frac{\partial^2}{\partial x^i \partial x^j} a^0_{ij},\ b^0_i,\ \frac{\partial}{\partial x^i} b^0_i \text{ are locally} \tag{4.1}$$

Hölder continuous functions satisfying the growth condition

$$|g(t,x)| \leq K(1+|x|^2)^{\frac{1}{2}} \quad \text{for some } K < \infty. \tag{4.2}$$

Also, assume that c^0 is a locally Hölder continuous function and that for some constant K,

$$c^0(t,x) \leq K(1+|x|^2)^{\frac{1}{2}}. \tag{4.3}$$

Then the following assertions are true.

(i) Equations (3.4) and (3.13) admit fundamental solutions $G(t,z,s,x)$ and $G^*(s,x,t,z)$ $(0 \leq s < t < T,\ x,z \in \mathbb{R}^d)$ respectively.

(ii) The fundamental solutions are related by

$$G(t,z,s,x) = G^*(s,x,t,z). \tag{4.4}$$

(iii) There exist positive constants K_1, K_2 depending only on bounds on a^0, b^0, c^0 such that if

$$H_1(t,x) := \exp(K_1(1+|x|^2)^{\frac{1}{2}} \cdot \exp(K_2 t)), \tag{4.5a}$$

$$H_2(t,x) = \exp(-K_1(1+|x|^2)^{\frac{1}{2}} \cdot \exp(-K_2 t)) \tag{4.5b}$$

then for some constant K_3, we have for $i = 1,2$,

$$|G(t,z,s,x)| \leq K_3(t-s)^{-d/2} H_i(t,z) H_i(s,x)^{-1}, \tag{4.6}$$

$$\int |G(t,z,s,x)| H_i(s,x) dx \leq H_i(t,z) \tag{4.7}$$

and

$$\int |G(t,z,s,x)| H_i^{-1}(t,z) dz \leq H_i^{-1}(s,x). \tag{4.8}$$

(iv) Let $0 < t_0 \leq T$ and let $g_0 : \mathbb{R}^d \to \mathbb{R}$ be a continuous function such that for some $K < \infty$,

$$|g_0(x)| \leq K H_i^{-1}(t_0, x); \tag{4.9}$$

then v defined by (3.23) is the unique classical solution to the Cauchy problem (I) in the class $\mathcal{G}([0,t_0))$.

(v) Let $0 \leq s_0 < T$ and let $p_0 : \mathbb{R}^d \to \mathbb{R}$ be a continuous function such that for some $k < \infty$,

$$|p_0(x)| \leq K H_i(s_0, x); \tag{4.10}$$

then u defined by (3.25) is the unique classical solution to the Cauchy problem (II) in the class $\mathcal{G}((s_0,T])$.

The uniqueness of solutions in the class $\mathcal{G}$ to the Cauchy problems (I), (II) is proved in [12]. The existence of the fundamental solutions, the estimates given above on G and the other assertions follow from the results in [10] upon verifying conditions (4.11) and (4.12) given below.

There exist constants K_1, K_2 such that H_1, H_2 given by

(4.5) satisfy

$$(M_t^* - \frac{\partial}{\partial t})H_i(t,x) \leq 0 \tag{4.11}$$

and

$$(M_t + \frac{\partial}{\partial t})H_i^{-1}(t,x) \leq 0. \tag{4.12}$$

The verification is carried out in the next lemma.

Lemma 4.2: Let the conditions of Theorem 4.1 be satisfied. Then there exist constants K_1, K_2 depending only on bounds for a^0, b^0, c^0, such that H_i defined by (4.5) satisfies (4.9), (4.10).

Proof: Let K_1, K_2 be arbitrary positive numbers. Then

$$\frac{\partial}{\partial t} H_1(t,x) = H_1(t,x)e^{K_2 t} K_1(1+|x|^2)^{\frac{1}{2}} \cdot K_2$$

$$\frac{\partial}{\partial x^i} H_1(t,x) = H_1(t,x)e^{K_2 t} K_1 \cdot \tfrac{1}{2}(1+|x|^2)^{-\frac{1}{2}} \cdot 2x_i$$

$$\frac{\partial}{\partial x^i \partial x^j} H_1(t,x) = H_1(t,x)e^{K_2 t} K_1[(-\tfrac{3}{2})(1+|x|^2)^{-3/2} x_i 2x_j$$

$$+ (1+|x|^2)^{-\frac{1}{2}} \cdot 1_{\{i=j\}} + e^{K_2 t} K_1 (1+|x|^2)^{-1} x_i x_j].$$

Hence

$$|\frac{\partial}{\partial x^i} H_1(t,x)| \leq H_1(t,x)e^{K_2 t} K_1$$

and

$$|\frac{\partial^2}{\partial x^i \partial x^j} H_1(t,x)| \leq H_1(t,x)e^{K_2 t} K_1(4+e^{K_2 t} K_1).$$

Recall the expression (3.9) for M_t^*. From the assumed conditions on a^0, b^0, c^0, it follows that b^{0*} satisfies the condition (4.2) and c^{0*} satisfies (4.3). Hence using the estimates given above, it follows that there exists a constant K_3 such that

$$M_t^* H_1(t,x) \leq K_3 H_1(t,x) e^{K_2 t} K_1 (4 + e^{K_2 t} K_1)(1+|x|^2)^{\frac{1}{2}}$$

and hence

$$M_t^* H_1(t,x) - \frac{\partial}{\partial t} H_1(t,x) \tag{4.13}$$

$$\leq H_1(t,x) e^{K_2 t} K_1 (1+|x|^2)^{\frac{1}{2}} [K_3(4 + e^{K_2 t} K_1) - K_2].$$

Observe that

$$H_1^{-1}(t,x) = \exp\left[-K_1(1+|x|^2)^{\frac{1}{2}} e^{K_2 t}\right]$$

and hence the expressions for derivatives of H_1 remain valid with H_1, K_1 replaced by H_1^{-1}, $-K_1$ respectively. Thus the estimates on derivatives of H_1 are valid with H_1 replaced by H_1^{-1} and hence for some K_4

$$M_t H_1^{-1}(t,x) \leq K_4 H_1(t,x) e^{K_2 t} K_1 (4 + e^{K_2 T} K_1)(1+|x|^2)^{\frac{1}{2}}.$$

Thus

$$(M_t + \frac{\partial}{\partial t}) H_1^{-1}(t,x) \tag{4.14}$$

$$\leq H_1^{-1}(t,x) e^{K_2 t} K_1 (1+|x|^2)^{\frac{1}{2}} [K_4(4 + e^{K_2 t} K_1) - K_2].$$

Similarly, it can be proved that

$$(M_t^* - \frac{\partial}{\partial t})H_2(t,x) \tag{4.15}$$

$$\leq H_2(t,x)e^{-K_2 t}K_1(1+|x|^2)^{\frac{1}{2}}[K_5(4+e^{-K_2 t}K_1)-K_2]$$

and

$$(M_t + \frac{\partial}{\partial t})H_2^{-1}(t,x) \tag{4.16}$$

$$\leq H_2^{-1}(t,x)e^{-K_2 t}K_1(1+|x|^2)^{\frac{1}{2}}[K_6(4+e^{-K_2 t}K_1)-K_2]$$

where K_5, K_6 are positive constants. Thus if we can choose K_1, K_2 such that

$$K_j(4+e^{K_2 T}K_1) - K_2 \leq 0 \tag{4.17}$$

for $j = 3,4,5,6$, then (4.13)-(4.16) imply that for this choice, (4.11) and (4.12) are valid. Taking $K_2 = 5\,(K_3+K_4+K_5+K_6)$ and $K_1 = e^{-K_2 T}$, it is easy to see that (4.17) holds. Also note that K_3, K_4, K_5, K_6 and hence K_1, K_2 depend only on the bounds for a^0, b^0, c^0, or to be more specific on the constants appearing in (4.3) for c^0 and (4.2) for the functions appearing in (4.1). □

Remark 4.1: Suppose g_0 satisfies (4.12). Then the estimate (4.11) and the definition (3.23) of v implies that

$$|v(s,x)| \leq KH_i^{-1}(s,x) \tag{4.18}$$

for the same i for which g_0 satisfies (4.12).

This shows the advantage of having the estimates on G

in terms of both H_1 and H_2. Thus if $g_0(x)$ goes to zero as $x \to \infty$ fast enough (i.e., (4.9) holds for $i = 1$), then so does $v(s,x)$ and if $g_0(x)$ grows slower than a specified rate (i.e., (4.9) holds for $i = 2$), then the same is true of $v(s,x)$.

Similarly, if p_0 satisfies (4.10), then the solution u of (II) satisfies

$$|u(t,z)| \leq KH_i(t,z).$$

Remark 4.2: If for some $K_7 < \infty$, $\epsilon > 0$, g_0 satisfies

$$|g_0(x)| \leq \exp(K_7(1+|x|^2)^{\frac{1}{2}-\epsilon}), \tag{4.20}$$

then it follows that (4.9) holds for $i = 2$ with a suitable choice of K.

Similarly, if p_0 satisfies

$$|p_0(x)| \leq \exp(K_8(1+|x|^2)^{\frac{1}{2}-\epsilon}) \tag{4.21}$$

for some $K_8 < \infty$, $\epsilon > 0$, then (4.10) holds for $i = 1$ with a suitable choice of K.

We now return to the setup in the later part of Section 3. The signal process (X_t) in the filtering model (1.9)-(1.10) is an $\mathbb{R}^d$-valued diffusion process whose generator L is given by (3.61), (3.62). Recall that

$$H_0 = \{\phi \in H: t \to \phi_t \text{ is Hölder continuous}\}.$$

The next result is an extension of Theorem 3.8 which allows a,b,h to be unbounded. Let us impose the following conditions on the signal diffusion process (X_t) and on the observation model

(i) The diffusion and drift coefficients a,b of (X_t) satisfy conditions (3.1), (3.2) and (4.1).

(ii) The initial variable X_0 has a continuous density $p_0(x)$ which satisfies the growth condition

$$p_0(x) \leq \exp(K(1+|x|^2)^{\frac{1}{2}-\epsilon}) \tag{4.22}$$

for some constants K,ϵ $(K < \infty,\ \epsilon > 0)$.

(iii) The function h is locally Hölder continuous as a function of t and x.

<u>Theorem 4.3</u>: Under the conditions (i), (ii), (iii), for all $y \in H_0$, the unnormalized conditional density $p_t(x,y)$ of X_t given $Q_t y$ exists and is the unique classical solution to the Cauchy problem (3.69)-(3.70) in the class $\mathcal{G}((0,T])$.

<u>Proof</u>: It is easy to check that the product of two bounded Hölder continuous functions is itself Hölder continuous. From this it follows that the product of two locally Hölder continuous functions is locally Hölder continuous. Also, a finite sum of locally Hölder continuous functions is locally Hölder continuous. Hence for all $y \in H_0$, $c_t^y(x)$ given by (3.73) is a locally Hölder continuous function. For $y \in H_0$, $|y_t|^2$ is bounded and hence (3.76) implies

$$c_t^y(x) \leq K_y \tag{4.23}$$

where K_y is the bound for $|y_t|^2$. Thus, for $y \in H_0$, M_t^y given by

$$M_t^y f(x) = L_t f(x) + c_t^y(x) f(x)$$

satisfies the conditions imposed in Theorem 4.1, namely

(3.1), (3.2), (4.1), (4.3). From part (v) of Theorem 4.1, we conclude that the Cauchy problem (3.69)-(3.70) admits a unique classical solution in the class $\mathscr{G}$ for all $y \in H_0$. It remains only to show that this unique classical solution is the unnormalized conditional density of X_t given $Q_t y$. Fix $y \in H_0$ and denote the classical solution to (3.69)-(3.70) in the class $\mathscr{G}$ by $u(t,x)$.

To complete the proof, we need to show that for all bounded measurable functions $f\colon \mathbb{R}^d \to \mathbb{R}$, $0 \le t \le T$,

$$\int f(z)d\Gamma_t^y(z) = \int_{\mathbb{R}^d} f(z)u(t,z)dz. \tag{4.24}$$

Let $G(t,z,s,x)$, $G^*(s,x,t,z)$ be the fundamental solutions to the equations (3.4), (3.13) given by Theorem 4.1 (for $M_t = L_t + c_t^y$). Then by Theorem 4.1, u is given by

$$u(t,z) = \int p_0(x)G^*(0,x,t,z)dx, \quad 0 < t \le T. \tag{4.25}$$

Let $g_0 \in C_0(\mathbb{R}^d)$ and $0 < t_0 \le T$. Then again by Theorem 4.1, $v\colon [0,t_0]\times\mathbb{R}^d \to \mathbb{R}$ given by $v(t_0,z) = g_0(z)$,

$$v(s,x) = \int g_0(z)G(t_0,z,s,x)dz \tag{4.26}$$

is the classical solution to the Cauchy problem (I) (for $M_t = L_t + c_t^y$).

Let H_1 be as in the statement of Theorem 4.1 (given by (4.5) for some K_1,K_2). Since g_0 has compact support, for some constant K_0, we have

$$|g_0(z)| \le K_0 H_1^{-1}(t_0,z).$$

This and the estimate (4.8) on G yield the inequality

$$|v(s,x)| \le K_0 H_1^{-1}(s,x) \le K_0.$$

Thus, v is a bounded solution to the problem

$$\frac{\partial v}{\partial s}(s,x) + L_s v(s,x) + c_s^y(x)v(s,x) = 0, \ 0 \le s < t_0,$$

and $v(t_0,z) = g_0(z)$. Since c_s^y is bounded above, by the Feynman-Kac formula (Theorem II.4.1), we have

$$v(s,X_s) = E_\Pi[g(X_{t_0})\exp(\int_s^{t_0} c_t^y(X_t)dt)|\sigma(X_s)]. \tag{4.27}$$

Now, using (4.25) and the relation (4.4), we obtain

$$\int g_0(z)u(t_0,z)dz = \int g_0(z)\int G(t_0,z,0,x)p_0(x)dxdz. \tag{4.28}$$

Now the use of Fubini's theorem to interchange the integrals in the RHS of (4.28) is justified because we have $|p_0(x)| \le K_0 H_1(0,x)$ for some K_0 in view of (4.22) and

$$\begin{aligned}\int\int |g_0(z)|\cdot|G(t_0,z,0,x)|\cdot|p_0(x)|dxdz \\ \le \int\int |g_0(z)|\cdot|G(t_0,z,0,x)|\cdot K_0 H_1(0,x)dxdz \\ \le K_0\int|g_0(z)|H_1(t_0,z)dz < \infty\end{aligned}$$

by (4.7) and since $g_0 \in C_0(\mathbb{R}^d)$ and $H_1(t_0,z)$ is continuous. Hence using (4.26)

$$\begin{aligned}\int g_0(z)u(t_0,z)dz &= \int p_0(x)\cdot\int g_0(z)G(t_0,z,0,x)dzdx \\ &= \int p_0(x)v(0,x)dx\end{aligned} \tag{4.29}$$

from which it follows that

$$\int g_0(z)u(t_0,z)dz = E_\Pi[v(0,X_0)]$$

$$= E_\Pi[E_\Pi[g_0(X_{t_0})\cdot\exp(\int_0^{t_0} c_\tau^y(X_\tau)d\tau)\,|\sigma(X_0)]]$$

$$= E_\Pi[g_0(X_{t_0})\cdot\exp(\int_0^{t_0} c_\tau^y(X_\tau)d\tau)]$$

$$= \int g_0(z)d\Gamma_{t_0}^y(z).$$

Since g_0 is arbitrary we have proved that (4.24) holds for all t, $0 \le t \le T$ and for all $f \in C_0(\mathbb{R}^d)$. We will now show that this gives (4.24) for all bounded measurable functions f.

Let B be a bounded Borel subset of $\mathbb{R}^d$, $B \subseteq [-K,K]^d$ for some K. Fix $t \in [0,T]$ and let μ be the measure on $(\mathbb{R},\mathcal{B}(\mathbb{R}))$ given by $\mu = \lambda + \Gamma_t^y$ (here λ is the Lebesgue measure). Then $\mu([-K,K]^d) < \infty$. By Luzin's theorem we can find a sequence $\{f_j\}$ of continuous functions from $[-K,K]^d$ into $\mathbb{R}$ such that $0 < f_j \le 1$ and $f_j \to 1_B$ a.e. μ on $[-K,K]^d$. It is easy to see that for each $j \ge 1$, we can choose $g_j \in C_0(\mathbb{R}^d)$ such that $0 \le g_j \le 1$, $g_j = f_j$ on $[-K,K]^d$ and $g_j(x) = 0$ if $x \notin [-K - \frac{1}{j}, K + \frac{1}{j}]^d$. Then

$$g_j \to 1_B \text{ a.e. } \mu \text{ on } \mathbb{R}^d.$$

Since $u(t,\cdot)$ is a continuous function, it is bounded on $[-K-1,K+1]$ and hence by the dominated convergence theorem, we get

$$\int g_j(z)u(t,z)dz \rightarrow \int 1_B(z)u(t,z)dz \tag{4.30}$$

and

$$\int g_j(z)d\Gamma_t^y(z) \rightarrow \int 1_B(z)d\Gamma_t^y(z). \tag{4.31}$$

These relations and the fact that (4.24) holds for $f = g_j$ together imply that (4.24) is valid for $f = 1_B$. Thus

$$\int_B u(t,z)dz = \Gamma_t^y(B) \tag{4.32}$$

for all bounded Borel sets (and hence for all Borel sets). Hence

$$\frac{d\Gamma_t^y(\cdot)}{dx} = u(t,\cdot)$$

and (4.24) is proved. □

Remark 4.3: Since $u(t,\cdot)$ is the density of a positive measure Γ_t^y, it is nonnegative a.e. Here in addition $u(t,\cdot)$ is continuous and thus $u(t,x) \geq 0$ for all t. Indeed, we have proved that u given by (3.25) is nonnegative for all bounded continuous functions p_0. Thus $G^*(s_0,x,t,z) \geq 0$. This shows that the fundamental solutions G, G^* are both nonnegative.

We now make a brief digression from filtering theory to present a result on the existence and smoothness of transition probability densities. The proof is similar to that of the previous theorem. The result is well known and is available in the literature under different sets of conditions

Theorem 4.4: Suppose that the diffusion and drift coefficients a,b of (X_t) satisfy conditions (3.1), (3.2) and (4.1). Then we have the following:

(i) The diffusion process (X_t) admits a transition probability density $p(s,x,t,z)$, $0 \leq s < t \leq T$, i.e., for all bounded measurable functions $g: \mathbb{R}^d \to \mathbb{R}$, $0 \leq s < t \leq T$,

$$E_{\Pi}[g(X_t)\,|\sigma(X_s)] = \int g(z)p(s,X_s,t,z)dz. \tag{4.33}$$

(ii) For all $(s,x) \in [0,T)\times\mathbb{R}^d$, $p(s,x,\cdot,\cdot) \in C^{1,2}((s,T]\times\mathbb{R}^d)$ and

$$\frac{\partial}{\partial t} p(s,x,t,z) = [L_t^* p(s,x,t,\cdot)](z), \quad s<t\leq T, \quad z\in\mathbb{R}^d. \tag{4.34}$$

(iii) For all $(t,z) \in (0,T]\times\mathbb{R}^d$, $p(\cdot,\cdot,t,z) \in C^{1,2}([0,t)\times\mathbb{R}^d)$ and

$$\frac{\partial}{\partial t} p(s,x,t,z) + (L_s p(s,\cdot,t,z))(x) = 0, \quad 0\leq s<t, \quad x\in\mathbb{R}^d. \tag{4.35}$$

(iv) For all $0 < t \leq T$, $\Pi\circ X_t^{-1}$ is absolutely continuous with continuous with respect to the Lebesgue measure.

(v) Suppose that $\Pi\circ X_0^{-1}$ admits a continuous density $\hat{p}_0(x)$ with respect to the Lebesgue measure and that $\hat{p}_0$ satisfies the growth condition (4.22). Then $\Pi\circ X_t^{-1} << \lambda$ and $\frac{d\Pi\circ X_t^{-1}}{d\lambda} = p_t(x)$ is the unique classical solution in the class $\mathscr{G}((0,T])$ to the Cauchy problem

$$\frac{\partial}{\partial t} p_t(x) = [L_t^* p_t(\cdot)](x) \tag{4.36}$$

$$p_0(x) = \hat{p}_0(x).$$

<u>Proof</u>: The last part follows from Theorem 4.3 by taking $h = 0$. The proof of the remaining assertion is similar, the

only difference being that we have not made any assumptions on the initial distribution $\Pi\circ X_0^{-1}$.

Let $G(t,z,s,x)$ and $G^*(s,x,t,z)$ be the fundamental solutions to the equations (3.4) and (3.13) respectively with $M_t = L_t$. These exist in view of Theorem 4.1 and are related by (4.4). Take $p(s,x,t,z) = G^*(s,x,t,z)$. Then it follows from the definition of a fundamental solution and (4.4) that p satisfies (4.34) and (4.35) (and that p is smooth to the extent asserted in parts (ii) and (iii) above). For (i), it suffices to prove (4.33) for $g \in C_0(\mathbb{R})$. Then as in Theorem 4.3, we can conclude that (4.33) holds for all bounded measurable g. For $g \in C_0(\mathbb{R}^d)$, (4.33) follows from the Feynman-Kac formula (or take $c = 0$ in (4.27)). Part (iv) follows from (i). Take

$$p_t(z) = \int p(0,x,t,z)d\Pi\circ X_0^{-1}(x).$$

Then from (4.33), it follows that

$$E_\Pi[g(X_t)] = \int g(z)p_t(z)dz. \qquad \square$$

Equation (4.35) is the well known Kolmogorov forward equation (also known as the Fokker-Planck equations in physics) and equation (4.34) is the Kolmogorov backward equation.

A comparison of equations (3.69) for the unnormalized conditional density $p_t(x,y)$ with (4.35) for the density $p_t(x)$ reveals that the unnormalized conditional density is the unique solution to an equation which may be regarded as a perturbation of the Kolmogorov forward equation, the perturbation term being a potential depending on y.

We have proved in Theorem 4.3 that for all $y \in H_0$, $p_t(x,y)$ is the unique solution to the Zakai equation. It is natural to ask what happens when $y \notin H_0$. By Theorem 3.7, $p_t(x,y)$ exists for all $y \in H$ and is a generalized solution to the Cauchy problem. On the other hand, if $y \in H$ is not continuous, then $p_t(x,y)$ can not be a classical solution since $\frac{\partial}{\partial t} p_t(x,y)$ can not be continuous.

We will now show that if h is sufficiently smooth, equation (3.69) can be transformed into an equation where $y \in H$ appears only as $\int_0^s y_u du$ and that the transformed equation has a unique solution from which $p_t(x,y)$ can be recovered. The transformation is well known and has been used in the stochastic calculus approach to nonlinear filtering theory to reduce stochastic partial differential equations to (deterministic) partial differential equations.

Fix $y \in H$. Suppose that the unnormalized conditional density $p_t(x,y)$ exists and is a generalized solution to the Cauchy problem (3.69)-(3.70). Let

$$\psi_t(x,y) = p_t(x,y) \cdot \exp(-g_t^y(x)) \tag{4.38}$$

where

$$g_t^y(x) = \sum_{k=1}^{N} h_t^k(x) \cdot \int_0^t y_s^k ds. \tag{4.39}$$

Assuming that (3.69) holds for $p_t(x,y)$, formally, it follows that $\psi_t(x,y)$ satisfies the following equation.

$$\frac{\partial}{\partial t} \psi_t(x,y) = [\frac{\partial}{\partial t} p_t(x,y)]\exp(-g_t^y(x)) \tag{4.40}$$

$$+ p_t(x,y)[\frac{\partial}{\partial t} \exp(-g_t^y(x))]$$

$$= [L_t^* p_t(x,y) + c_t^y(x,y)p_t(x,y)]\exp(-g_t^y(x))$$

$$+ p_t(x,y)[-\frac{\partial}{\partial t} g_t^y(x)\exp(-g_t^y(x))]$$

$$= \hat{L}_t^y \psi_t(x,y) + \hat{c}_t^y(x)\psi_t(x,y)$$

where

$$[\hat{L}_t^y f](x) := [L_t^* f(\cdot)\exp(g_t^y(\cdot))](x)\exp(-g_t^y(x)) \tag{4.41}$$

and

$$\hat{c}_t^y(x) := c_t^y(x) - \frac{\partial}{\partial t} g_t^y(x).$$

Now

$$\frac{\partial}{\partial t} g_t^y(x) = \sum_{k=1}^{N} \frac{\partial}{\partial t} h_t^k(x)\int_0^t y_u^k du + \sum_{k=1}^{N} h_t^k(x)y_t^k \quad \text{a.e.},$$

and hence using (3.74), it follows that

$$\hat{c}_t^y(x) = -\tfrac{1}{2}|h_t(x)|^2 - \sum_{k=1}^{N} \frac{\partial}{\partial t} h_t^k(x)\int_0^t y_u^k du \quad \text{for a.e. } t. \tag{4.42}$$

Recall that (see (3.9))

$$L_t^* f(x) = \tfrac{1}{2} \sum_{i,j=1}^{d} a_{ij}(t,x)\frac{\partial^2}{\partial x^i \partial x^j} f(x) \tag{4.43}$$

$$+ \sum_{i=1}^{d} b_i^*(t,x)\frac{\partial}{\partial x^i} f(x) + c^*(t,x)f(x)$$

where b_i^*, c^* is given by

$$b_i^*(t,x) = -b_i(t,x) + \sum_{j=1}^{d} \frac{\partial}{\partial x^j} a_{ij}(t,x)$$

and

$$c^*(t,x) = -\sum_{i=1}^{d} \frac{\partial}{\partial x^i} b_i(t,x) + \frac{1}{2} \sum_{i,j=1}^{d} \frac{\partial^2 a_{ij}(t,x)}{\partial x^i \partial x^j} . \quad (4.45)$$

We will check that $\hat{L}_t^y$ is a second order differential operator and get an expression for its coefficients in terms of a,b,h,y. First, note that

$$\frac{\partial}{\partial x^j}[f(x)\exp(g_t^y(x))] = [\frac{\partial}{\partial x^j}f(x) + f(x)\frac{\partial}{\partial x^j}g_t^y(x)]\exp(g_t^y(x))$$

and

$$\frac{\partial}{\partial x^i \partial x^j}[f(x)\exp(g_t^y(x))] = [\frac{\partial^2}{\partial x^i \partial x^j} f(x) + \frac{\partial}{\partial x^i} f(x)\frac{\partial}{\partial x^j} g_t^y(x)$$

$$+ f(x)\frac{\partial}{\partial x^i \partial x^j} g_t^y(x)]\exp(g_t^y(x))$$

$$+ [\frac{\partial}{\partial x^j} f(x) + f(x)\frac{\partial}{\partial x^j} g_t^y(x)]\frac{\partial}{\partial x^i} g_t^y(x)\exp(g_t^y(x)).$$

Thus

$$\hat{L}_t^y f(x) = \frac{1}{2} \sum_{i,j=1}^{d} a_{ij}(t,x)\frac{\partial^2}{\partial x^i \partial x^j} \quad (4.46)$$

$$+ \sum_{j=1}^{d} \hat{b}_j(t,x,y)\frac{\partial}{\partial x^j} f(x) + \hat{c}(t,x,y)f(x)$$

where

$$\hat{b}_i(t,x,y) = b_i^*(t,x) + \sum_{k=1}^{N} [\sum_{j=1}^{d} a_{ij}(t,x)\frac{\partial}{\partial x^j} h_t^k(x)]\int_0^t y_s^k ds \quad (4.47)$$

and

$$\hat{c}(t,x,y) = c^*(t,x) + \sum_{k=1}^{N} [\sum_{i=1}^{d} b_i^*(t,x)\frac{\partial}{\partial x^i} h_t^k(x) \tag{4.48}$$

$$+ \frac{1}{2} \sum_{i,j=1}^{d} a_{ij}(t,x)\frac{\partial^2 h_t^k(x)}{\partial x^i \partial x^j}$$

$$+ \frac{1}{2} \sum_{i,j=1}^{d} a_{ij}(t,x)\frac{\partial h_t^k(x)}{\partial x^i} \frac{\partial h_t^k(x)}{\partial x^j} \int_0^t y_s^k ds]\int_0^t y_s^k ds.$$

The expression appearing on the right hand side of (4.42) is a continuous function of (t,x) (if h is smooth enough). Thus, we can define

$$\hat{c}_t^y(x) = - \tfrac{1}{2}|h_t(x)|^2 - \sum_{k=1}^{N} \frac{\partial}{\partial t} h_t^k(x)\int_0^t y_u^k du. \tag{4.49}$$

The equation (4.40) is thus a partial differential equation with continuous coefficients if $h \in C^{1,2}([0,T]\times\mathbb{R}^d)$ for all $y \in H$. In the equation (3.69) for $p_t(x,y)$, the potential term was discontinuous in t if $y \in H$ were not continuous. This singularity has been taken care of by the transformation (4.38). Thus, *formally*, $\psi_t(x,y)$ given by (4.38) satisfies the second order partial differential equation (4.40). It will now be proved that the PDE (4.40) has a unique solution ψ and that p given by (4.38) is the unnormalized conditional density of the filtering problem.

<u>Theorem 4.5</u>: Suppose that the diffusion and drift coefficients a,b of (X_t) satisfy (3.1), (3.2), (4.1) and further that a_{ij} is bounded (uniformly in (t,x)). Let the continuous density $p_0 := (d\Pi\circ X_0^{-1})/d\lambda$ exist, and satisfy (4.22). The following further conditions are imposed on $h = (h^1,\ldots,h^N)$

and the coefficients a_{ij} and b_i. We let $k = 1,\dots,N$ and $i,j = 1,\dots,d$.

The functions $h^k, \frac{\partial h^k}{\partial x^i}, \frac{\partial h}{\partial x^i \partial x^j}, \frac{\partial h^k}{\partial t}$ are (4.50)

locally Hölder continuous (in (t,x))

and

(4.51)

the functions $h^k, \frac{\partial h^k}{\partial t}, a_{ij}\frac{\partial h^k}{\partial x^j}, a_{ij}\frac{\partial^2 h^k}{\partial x^i \partial x^j}, a_{ij}\frac{\partial h^k}{\partial x^i} \cdot \frac{\partial h^k}{\partial x^j}$, $b_i\frac{\partial h^k}{\partial x^i}, \frac{\partial a_{ij}}{\partial x^j} \cdot \frac{\partial h^k}{\partial x^i}$ satisfy the growth condition (4.2).

Finally, assume that for some constant K_1,

$$E_{\Pi}(\exp(K_1 |X_0|^2)) < \infty. \tag{4.52}$$

Then

(i) For all $y \in H$, the Cauchy problem

$$\frac{\partial}{\partial t} \psi_t(x,y) = \hat{L}_t^y \psi_t(x,y) + \hat{c}_t^y(x)\psi_t(x,y) \tag{4.53}$$

$$\psi_0(x,y) = p_0(x) \tag{4.54}$$

has a unique classical solution in the class $\mathcal{G}((0,T])$, where $\hat{L}_t^y$ is given by (4.49), (4.50) and (4.51).

(ii) Let $\psi_t(x,y)$ be the unique solution in $\mathcal{G}((0,T])$ to (4.53), (4.54). Define

$$p_t(x,y) := \psi_t(x,y)\exp\left(\sum_{k=1}^{N} h_t^k(x)\int_0^t g_s^k ds\right). \tag{4.55}$$

Then $p_t(x,y)$ is the unnormalized conditional density of (X_t) given $Q_t y$.

Proof: Fix $y \in H$. Define, for $f \in C^2(\mathbb{R}^d)$,

$$(\overline{L}_t^y f)(x) = \exp(g_t^y(x))[L_t(f(\cdot)\exp(-g_t^y(\cdot)))](x). \tag{4.56}$$

Then it can be checked that $\overline{L}_t^y$ is a second order linear differential operator with coefficients $a, \overline{b}_i, \overline{c}$ where $\overline{b}_i, \overline{c}$ are given by

$$\overline{b}_i(t,x,y) = b_i(t,x) - \sum_{k=1}^{N} [\sum_{j=1}^{d} a_{ij}(t,x) \frac{\partial}{\partial x^j} h_t^k(x)] \int_0^t y_s^k ds \tag{4.57}$$

and

$$\overline{c}(t,x,y) = \sum_{k=1}^{N} [-\sum_{j=1}^{d} b_j(t,x) \frac{\partial}{\partial x^j} h_t^k(x) - \tfrac{1}{2} \sum_{i,j=1}^{d} a_{ij}(t,x) \frac{\partial^2 h_t^k(x)}{\partial x^i \partial x^j} + \tfrac{1}{2} \sum_{i,j=1}^{d} a_{ij}(t,x) \frac{\partial h_t^k(x)}{\partial x^i} \frac{\partial h_t^k(x)}{\partial x^i} \int_0^t y_s^k ds] \int_0^t y_s^k ds. \tag{4.58}$$

These computations are similar to the ones for $\hat{L}_t^y$. Also using the expressions (4.50), (4.51), (4.57) and (4.58), it can be checked that

$$\hat{b}_i(t,x,y) = -\overline{b}_i(t,x,y) + \sum_{j=1}^{d} \frac{\partial}{\partial x^j} a_{ij}(t,x) \tag{4.59}$$

and

$$\hat{c}(t,x,y) = \overline{c}(t,x,y) - \sum_{i=1}^{d} \frac{\partial}{\partial x^i} \overline{b}_i(t,x,y) + \tfrac{1}{2} \sum_{i,j=1}^{d} \frac{\partial^2 a_{ij}}{\partial x^i \partial x^j}. \tag{4.60}$$

Now, comparing these two relations between the coefficients of the differential operators $\hat{L}_t^y$ and $\vec{L}_t^y$ with differential operator M_t and its expressions (3.10), (3.11) connecting the coefficients of the formal adjoint M_t^*, it follows that

$$(\vec{L}_t^y)^* = \hat{L}_t^y. \tag{4.61}$$

Note that for all $y \in H$, $t \to \int_0^t y_s ds$ is a Hölder continuous function. In fact, for $t_1 < t_2$,

$$\left|\int_0^{t_1} y_s ds - \int_0^{t_2} y_s ds\right| = \left|\int_{t_1}^{t_2} y_s ds\right| \leq (t_2 - t_1)^{\frac{1}{2}} \|y\|_H.$$

Thus, $\hat{c}_t^y(x)$ (given by (4.43)) is locally Hölder continuous for all $y \in H$. In view of this observation and the assumptions on a,b and h, it follows that for all $y \in H$ (fixed), M_t given by

$$M_t f := \vec{L}_t^y f + \hat{c}_t^y f \tag{4.62}$$

satisfies the conditions of Theorem 4.1. The formal adjoint M_t^* for M_t given by (4.62) is the operator

$$M_t^* f = \hat{L}_t^y f + \hat{c}_t^y f$$

as can be easily checked using (4.61). Let $G^y(t,z,s,x)$, $G^{*y}(s,x,t,z)$ be the fundamental solutions to the equations

$$\left(\frac{\partial}{\partial t} + M_t\right)v = 0$$

and

$$\frac{\partial}{\partial t} u = M_t^* u.$$

The existence of G^y, G^{*y} is assured by Theorem 4.1, which also gives that

$$\psi_t(z,y) := \int p_0(x) G^{*y}(0,x,t,z)dx \tag{4.63}$$

is the unique solution to the Cauchy problem (4.53)-(4.54).

For part (ii), fox $0 < t_0 \leq T$ and $g_0 \in C_0(\mathbb{R}^d)$. Then

$$v_s(x,y) := \int g_0(z) G^y(t_0,z,s,x)dz \tag{4.64}$$

is the classical solution to the problem

$$\frac{\partial}{\partial s} v_s(x,y) + \overline{L}_s^y v_s(x,y) + \hat{c}_s^y(x) v_s(x,y) = 0 \tag{4.65}$$

and

$$v_{t_0}(x,y) = g_0(x). \tag{4.66}$$

Using the relation (4.4) between G^y and G^{*y}, we can prove that

$$\int p_0(x) v_0(x,y)dx = \int g_0(z) \psi_{t_0}(z,y)dz. \tag{4.67}$$

The proof of (4.67) is the same as that of (4.29), of course with appropriate modifications in the notation. Also, analogous to (4.27), we now have

$$|v_t(x,y)| \leq K_y \tag{4.68}$$

where the constant may depend on y. Define

$$v_1(s,x,y) := v_s(x,y)\exp(-g_s^y(x)). \tag{4.69}$$

Then

$$\frac{\partial}{\partial s} v_1(s,x,y) = [\frac{\partial}{\partial s} v_s(x,y)]\exp(-g_s^y(x)) \tag{4.70}$$

$$- \frac{\partial}{\partial s} g_s^y(x)\exp(-g_s^y(x))v_s(x,y) \quad \text{a.e. } s$$

and, by (4.56),

$$L_s v_1(s,x,y) = \exp(-g_s^y(x))[\bar{L}_s v_1(s,x,y)\exp(g_s^y(x))] \tag{4.71}$$

$$= \exp(-g_s^y(x))[\bar{L}_s v_s(x,y)].$$

Thus

$$(\frac{\partial}{\partial s} + L_s)v_1(s,x,y) \tag{4.72}$$

$$= \exp(-g_s^y(x))[(\frac{\partial}{\partial s} + \bar{L}_s)v_s(x,y) - (\frac{\partial}{\partial s} g_s^y(x))v_s(x,y)]$$

and as a consequence

$$(\frac{\partial}{\partial s} + L_s + c_s^y(x))v_1(s,x,y) \tag{4.73}$$

$$= \exp(-g_s^y(x))[(\frac{\partial}{\partial s} + \bar{L}_s + \hat{c}_s^y(x))v_s(x,y)] = 0.$$

The last two steps follow from the expression (4.49) for $\hat{c}^y$ and the relation (4.65).

Note that g^y satisfies the growth condition (4.2) in view of the assumption (4.51). Hence (4.68) gives

$$|v_1(s,x,y)| \le K_y \exp(K(1 + |x|^2)^{\frac{1}{2}}) \tag{4.74}$$

where K_y, K are constants.

We have assumed that a_{ij} is bounded above and that X_0 satisfies (4.52). Hence by the Feynman-Kac formula, Theorem II.4.4, and the relations (4.73), (4.74), we conclude that

$$v_1(0,X_0,y) = E_\Pi[v_1(t_0,X_{t_0},y)\exp(\int_0^{t_0} c_s^y(X_s)ds \,|\sigma(X_0)] \tag{4.75}$$

$$= E_\Pi[g_0(X_{t_0})\exp(-g_{t_0}^y(X_{t_0}))\exp(\int_0^{t_0} c_s^y(X_s)ds\,|\sigma(X_0)].$$

The last equality follows from (4.66) and (4.69). Note that $g_0^y(x) \equiv 0$ and thus, $v_1(0,x,y) = v_0(x,y)$. Hence

$$\int p_0(x)v_0(x,y)dx = \int p_0(x)v_1(0,x,y)dx = E_\Pi[v_1(0,X_0,y)] \tag{4.76}$$

$$= E_\Pi[g_0(X_{t_0})\exp(-g_{t_0}^y))\exp(\int_0^{t_0} c_s^y(X_s)ds)]$$

$$= \int g_0(z)\exp(-g_{t_0}^y(z))d\Gamma_{t_0}^y(z).$$

The relations (4.67) and (4.76) imply that

$$\int g_0(z)\psi_{t_0}(z,y)dz = \int g_0(z)\exp(-g_{t_0}^y(z))d\Gamma_{t_0}^y(z) \tag{4.77}$$

for all $g_0 \in C_0(\mathbb{R}^d)$. Given $f \in C_0(\mathbb{R}^d)$, taking

$$g_0(z) = f(z)\exp(g_{t_0}^y(z))$$

and noting that $g_0 \in C_0(\mathbb{R}^d)$, (4.77) for g_0 gives

$$\int f(z)\exp(g_{t_0}^y(z))\psi_{t_0}(z,y)dz = \int f(z)d\Gamma_{t_0}^y(z). \tag{4.78}$$

Thus, if p is defined by (4.55), then we have

$$\int f(z)p_{t_0}(z,y)dz = \int f(z)d\Gamma^y_{t_0}(z) \tag{4.79}$$

for all $f \in C_0(\mathbb{R}^d)$. In the last part of the proof of Theorem 4.3, we have proved that (4.79) for all $f \in C_0(\mathbb{R}^d)$ implies that

$$\frac{d\Gamma^y_{t_0}}{d\lambda} = p_{t_0}(\cdot,y). \tag{4.80}$$

Thus p defined by (4.55) is the unnormalized conditional density of X_t given $Q_t y$. □

CHAPTER VIII

MEASURE VALUED EQUATIONS OF FILTERING

In this chapter and the next, the finitely additive theory of nonlinear filtering, prediction and smoothing will be studied in a general framework that permits the observation and white noise processes to take values in an infinite dimensional Hilbert space and the signal process to be a Markov process having an arbitrary Polish (i.e., complete, separable metric) space S for its state space.

We shall first consider filtering. Recent attempts to develop the theory from the standpoint of stochastic calculus have not, to our knowledge, gone beyond the stage of deriving the stochastic differential equations for the optimal filter ([50]). In another direction, for S compact and separable, and with the observations and noise still assumed to be finite dimensional, it has been shown by Kunita that the conditional distribution of X_t, also known as the optimal filter, is the unique solution to a certain stochastic equation which, however, is *not* a stochastic differential equation [53]. (For purposes of identification, we refer to similar equations in this chapter as "Kunita-type" equations.) Kunita's work was completed by Szpirglas who showed that the equation derived by Kunita is equivalent to the FKK equation [73]. He further derived analogous results for the unnormalized conditional distribution.

The techniques on which the proofs of our main result depend, are based on properties of multiplicative functionals of Markov processes. In section 1, we consider measures induced by a Markov process transformed by a multiplicative functional and the corresponding normalized measures and show that these satisfy equations of the FKK type and Kunita type and that the equations of these two types are equivalent to each other. In this respect, what we do is related to the method in [73]. We prove, under two sets of conditions, that these equations have unique solutions.

In the second section, we deduce that the conditional distribution of X_t is the unique solution of the white noise versions of the Kunita and FKK type equations. One set of conditions under which this is proved imposes minimal conditions on the function h occurring in the filtering model which does not include any boundedness restrictions.

In the last section, we obtain an analogue of Kunita's result in the conventional theory that the conditional distribution of X_t is a Markov process. To do this, we first have to give an appropriate definition of a Markov process on a quasi-cylinder probability space $(E,\mathcal{E},\beta)$.

As will be seen from the proofs of the theorems in this chapter, some of the difficulties occurring in the stochastic calculus treatment of the filtering model with infinite dimensional noise do not arise in the finitely additive white noise formulation of the problem.

1. MULTIPLICATIVE FUNCTIONALS OF MARKOV PROCESSES AND CORRESPONDING INDUCED MEASURES

Let $(S,\mathcal{S})$ be a measurable space. Let $(X_t)_{0\leq t<\infty}$ be an S-valued Markov process on a (countably additive) probability space $(\Omega,\mathcal{A},\Pi)$. (See Sections II.1 and II.2 for the definition of a Markov process and related terminology and notation which appear in this section.) We assume throughout this chapter that

The Markov process (X_t) admits a transition probability function $P(s,x,t,B)$, $o \leq s,\ t < \infty$, $x \in S$, $B \in \mathcal{S}$ and that the paths of (X_t) are $\mathcal{F}_t^X$-progressively measurable. (1.1)

Let V_t^s be the two parameter semigroup acting on $J(S,\mathcal{S})$ defined by II.2.1. Let $\hat{S} = [0,\infty) \times S$ and $\hat{\mathcal{S}} = \mathcal{B}([0,\infty)) \otimes \mathcal{S}$. Let T_t be the one parameter semigroup associated with P, acting on $J(\hat{S},\hat{\mathcal{S}})$, given by II.2.3. Let

$$J_0 = \{f \in J(\hat{S},\hat{\mathcal{S}}): (T_t f)(s,x) \longrightarrow f(s,x) \text{ as } t\downarrow 0 \text{ for all } (s,x) \in \hat{S}\}.$$

We also make the following technical assumption:

$$J_0 \text{ is a measure determining class on } (\hat{S},\hat{\mathcal{S}}). \quad (1.2)$$

This means, if μ_1,μ_2 are measures on $(\hat{S},\hat{\mathcal{S}})$ such that $\int f d\mu_1 = \int f d\mu_2$ for all $f \in J_0$, then $\mu_1 = \mu_2$.

Let L be the weak generator of the Markov process (X_t) and let $\mathcal{D}$ be its domain.

Let $\mathcal{M}(S,\mathcal{S})$ denote the class of finite positive measures on $(S,\mathcal{S})$. For $g \in J(S,\mathcal{S})$ and $\mu \in \mathcal{M}(S,\mathcal{S})$, we will denote $\int g d\mu$ by the symbol $\langle g,\mu\rangle$. A function $f \in J(\hat{S},\hat{\mathcal{S}})$ will be written as $f(t,x)$ or $f_t(x)$ and the integral $\int f(t,x)d\mu(x)$ for $\mu \in \mathcal{M}(S,\mathcal{S})$ will be denoted by $\langle f(t,\cdot),\mu\rangle$ or $\langle f_t(\cdot),\mu\rangle$ or $\langle f_t,\mu\rangle$.

The following lemma has been used by Szpirglas in a similar context [73].

Lemma 1.1: Let $\mathscr{D}^{(2)} = \{f \in \mathscr{D}: Lf \in \mathscr{D}\}$. Let $\mu_1,\mu_2 \in \mathscr{M}(S,\mathscr{S})$ be such that for some t

$$\langle f(t,\cdot),\mu_1\rangle = \langle f(t,\cdot),\mu_2\rangle, \quad \text{for all } f \in \mathscr{D}^{(2)}. \tag{1.3}$$

Then $\mu_1 = \mu_2$.

Proof: It was shown in Theorem II.2.1 that under the assumption (1.2) on (X_t), $\mathscr{D}^{(2)}$ is a measure determining class on $(\hat{S},\hat{\mathscr{S}})$. Given μ_1,μ_2 satisfying (1.3), define μ_1',μ_2' on $(\hat{S},\hat{\mathscr{S}})$ by

$$\int_{\hat{S}} g d\mu_i' = \int_S g(t,\cdot)d\mu_i, \quad g \in J(\hat{S},\hat{\mathscr{S}}).$$

Then (1.3) implies that

$$\int f d\mu_1' = \int f d\mu_2'$$

for all $f \in \mathscr{D}^{(2)}$ and hence $\mu_1 = \mu_2$. □

Let $T < \infty$ be fixed for the rest of the section. Let $c: [0,T] \times S \to \mathbb{R}$ be a $\mathscr{B}([0,T]) \otimes \mathscr{S}$ measurable function such that

$$c_s(x) \leq \delta_s, \; 0 \leq s \leq T, \; x \in S \tag{1.4}$$

where $\delta: [0,T] \to \mathbb{R}$ is a positive measurable function with $K = \int_0^T \delta_s ds < \infty$ and

$$E_\Pi \int_0^T |c_s(X_s)|ds < \infty. \tag{1.5}$$

For $0 \leq t \leq T$, let $G_t,N_t: \mathscr{S} \to [0,\infty)$ be defined by

$$G_t(A) = E_\Pi 1_A(X_t) \exp(\int_0^t c_s(X_s)ds) \tag{1.6}$$

and

$$N_t(A) = \frac{1}{G_t(S)} \cdot G_t(A), \quad A \in \mathcal{S}. \tag{1.7}$$

In view of the condition (1.4), it is easy to check that G_t is a finite positive measure on $(S,\mathcal{S})$ and N_t is a probability measure. By Fubini's theorem it follows that for $A \in \mathcal{S}$, $t \to G_t(A)$ is a Borel measurable function. From the bound (1.4) on c, it follows that for all $A \in \mathcal{S}$, $0 \leq G_t(A) \leq \exp(K) \cdot \Pi(X_t \in A)$. Hence $G_t << \Pi \circ X_t^{-1}$ and

$$0 \leq \frac{dG_t}{d\Pi \circ X_t^{-1}} \leq \exp(K), \quad \text{a.s. } \Pi \circ X_t^{-1}.$$

It is easy to see using (1.4) that $G_t(S)$ is a continuous function on $[0,T]$ and $G_t(S) > 0$ for all t. Thus $G_t(S)$ is bounded below and hence

$$0 \leq \frac{dN_t}{d\Pi \circ X_t^{-1}} = \frac{1}{G_t(S)} \cdot \frac{dG_t}{d\Pi \circ X_t^{-1}} \leq K_1$$

for a suitable constant K_1.

Thus $\{G_t\}$ and $\{N_t\}$ satisfy the following conditions:

$\{K_t\} \subseteq \mathcal{M}(S,\mathcal{S})$; for $A \in \mathcal{S}$, $K_0(A) = E_\Pi 1_A(X_0)$ (1.8)
and $t \to K_t(A)$ is a bounded Borel measurable function and further K_t is absolutely continuous with respect to $\Pi \circ X_t^{-1}$ with

$$\left|\frac{dK_t}{d\Pi \circ X_t^{-1}}\right| \leq R, \quad 0 \leq t \leq T$$

for a suitable constant $R < \infty$.

If $\{K_t\}$ satisfies (1.8), then from (1.5) we have

$$\int_0^T \langle |c_s|, K_s\rangle ds \leq R\cdot\int_0^T \langle |c_s|, \Pi\circ X_s^{-1}\rangle ds \tag{1.9}$$

$$= R \int_0^T E_\Pi |c_s(X_s)| ds < \infty.$$

We will now obtain two equations for $\{G_t\}$. Later, we will prove that under additional assumptions on (X_t) and c, these equations admit a unique solution thus characterizing $\{G_t\}$. All the equations considered are on the interval [0,T], so that in what follows, s,t,r,u will all lie in [0,T].

<u>Theorem 1.2</u>: $\{G_t\}$ satisfies the equation

$$\langle g, G_t\rangle = \langle V_t^0 g, G_0\rangle + \int_0^t \langle c_s\cdot(V_t^s g), G_s\rangle ds, \tag{1.10}$$

$$g \in J(S,\mathcal{S}).$$

<u>Proof</u>: Note that for $g \in J(S,\mathcal{S})$, we have

$$g(X_t)\ \exp(\int_0^t c_s(X_s)ds)$$

$$= g(X_t)[1+\int_0^t c_s(X_s)\ \exp(\int_0^s c_r(X_r)dr)ds].$$

Thus

$$\langle g, G_t\rangle = E_\Pi g(X_t)\ \exp(\int_0^t c_s(X_s)ds) \tag{1.11}$$

$$= E_\Pi g(X_t) + E_\Pi g(X_t)\cdot\int_0^t c_s(X_s)\ \exp(\int_0^s c_r(X_r)dr)ds$$

$$= E_\Pi(V_t^0 g)(X_0) + \int_0^t E_\Pi g(X_t)c_s(X_s)\ \exp(\int_0^s c_r(X_r)dr)ds$$

by Fubini's theorem since (X_t) has been assumed to be

progressively measurable and c satisfies (1.4) and (1.5). This also implies that

$$c_s(X_s)\ \exp(\int_0^s c_r(X_r)dr)$$

is $\mathcal{F}_s^X$-measurable and hence

$$\begin{aligned} E_\Pi g(X_t)c_s(X_s)\ \exp(\int_0^s c_r(X_r)dr) \qquad (1.12) \\ = E_\Pi[E_\Pi[g(X_t)|\mathcal{F}_s^X]\cdot c_s(X_s)\cdot \exp(\int_0^s c_r(X_r)dr] \\ = E_\Pi(V_t^s g)(X_s)c_s(X_s)\ \exp(\int_0^s c_r(X_r)dr) \\ = \langle c_s\cdot(V_t^s g),G_s\rangle. \end{aligned}$$

The relations (1.11), (1.12) and the fact that $E_\Pi(V_t^0 g)(X_0) = \langle V_t^0 g,G_0\rangle$ imply that $\{G_t\}$ satisfies (1.10) □

The equation (1.10) is one of the two equations satisfied by $\{G_t\}$. The other equation is a differential equation and can be obtained proceeding as in the proof of Theorem VII.2.1. This equation takes the form

$$\frac{d}{dt}\langle f(t,\cdot),G_t\rangle = \langle (Lf)(t,\cdot) + c_t(\cdot)f(t,\cdot),G_t\rangle \qquad (1.13)$$
$$f \leq \mathcal{D}.$$

Instead of proving the validity of (1.13) directly, we will do so by proving that (1.13) is equivalent to the equation (1.10) in the sense made precise in the statement of the next result. The advantage is that when proving the uniqueness of solution for the equations (1.10) and (1.13), it would suffice to prove that *one* of these two equations admits a unique solution.

<u>Lemma 1.3</u>: Let $\{K_t\} \subseteq \mathcal{M}(S,\mathcal{S})$ satisfy (1.8). Then $\{K_t\}$ satisfies the equation

$$\langle g,K_t\rangle = \langle V_t^0 g,K_0\rangle + \int_0^t \langle c_s(V_t^s g),K_s\rangle ds,\ g \in J(S,\mathcal{S}) \quad (1.14)$$

if and only if it satisfies the equation

$$\langle f(t,\cdot),K_t\rangle \quad (1.15)$$
$$= \langle f(0,\cdot),K_0\rangle + \int_0^t \langle (Lf)(s,\cdot)+c_s(\cdot)f(s,\cdot),K_s\rangle ds,\ f \in \mathcal{D}.$$

<u>Proof</u>: Let $f \in J(\hat{S},\hat{\mathcal{S}})$, $t \in [0,T]$ be fixed and take $g(x) = f(t,x)$. Then

$$(V_t^s g)(x) = \int g(y)P(s,x,t-s,dy)$$
$$= \int f(t,y)P(s,x,t-s,dy) = (T_{t-s}f)(s,x).$$

Thus if $\{K_t\}$ satisfies (1.14), then we get

$$\langle f(t,\cdot),K_t\rangle \quad (1.16)$$
$$= \langle (T_t f)(0,\cdot),K_0\rangle + \int_0^t \langle c_s(\cdot)(T_{t-s}f)(s,\cdot),K_s\rangle ds.$$

On the other hand if (1.16) holds for all $f \in J(\hat{S},\hat{\mathcal{S}})$, then it is easy to see that (1.14) holds. Given $g \in J(S,\mathcal{S})$, take $f(t,x) = g(x)$ for all t. Thus, (1.14) holds if and only if (1.16) holds for all $f \in J(\hat{S},\hat{\mathcal{S}})$.

Let us introduce some temporary notation. For $f: \hat{S} \to \mathbb{R}$, $\hat{\mathcal{S}}$ measurable, such that $\int |f(t,\cdot)| dK_t < \infty$, let us write

$$\theta_t(f) := \langle f(t,\cdot),K_t\rangle$$

and for $f \in \mathcal{D}$, let $\delta_t(f)$ denote the difference between the right hand sides of (1.15) and (1.16), i.e.

$$\delta_t(f) := \theta_0(T_t f) + \int_0^t \theta_s(c\cdot(T_{t-s}f))ds - \int_0^t \theta_s(Lf+cf)ds - \theta_0(f).$$

Recall that for $f \in \mathcal{D}$ and $t \geq 0$ we have (see (II.2.8))

$$(T_t f)(s,x) = f(s,x) + \int_0^t (T_r Lf)(s,x)dr. \tag{1.17}$$

Thus we have using Fubini's theorem,

$$\theta_0(T_t f) = \theta_0(f) + \int_0^t \theta_0(T_r Lf)dr \tag{1.18}$$

and for all s such that $\theta_s(|c|) = \int |c_s| dK_s < \infty$, we have

$$\theta_s(c(T_{t-s}f)) = \theta_s(cf) + \int_0^{t-s} \theta_s(cT_r Lf)dr. \tag{1.19}$$

Since $\int_0^T \theta_s(|c|)ds = \int_0^T \langle |c_s|,K_s\rangle ds < \infty$ by (1.9), $\theta_s(|c|) < \infty$ for a.e. s and hence (1.19) holds for a.e. s. Using the relations (1.18), (1.19) we get the following expression for $\delta_t(f)$:

$$\begin{aligned}\delta_t(f) &= \int_0^t \theta_0(T_r Lf)dr + \int_0^t \int_0^{t-s} \theta_s(cT_r Lf)drds - \int_0^t \theta_s(Lf)ds \\ &= \int_0^t \theta_0(T_r Lf)dr + \int_0^t \int_0^{t-r} \theta_s(cT_r Lf)dsdr - \int_0^t \theta_s(Lf)ds\end{aligned} \tag{1.20}$$

by Fubini's theorem. Recall that in obtaining the expression (1.20) for $\delta_t(f)$ the only thing we have used is (1.17).

We now assume that $\{K_t\}$ satisfies (1.16). Let $f \in \mathcal{D}$ be fixed. We will show that $\delta_t(f) = 0$. This would imply that $\{K_t\}$ satisfies (1.15) as $\delta_t(f)$ is the difference between the right hand sides of (1.15) and (1.16) and their left hand sides are identical.

Using (1.16) for Lf, one obtains

$$\theta_u(Lf) = \theta_0(T_u Lf) + \int_0^u \theta_s(cT_{u-s}Lf)ds$$

and as a consequence,

$$\int_0^t \theta_u(Lf)du = \int_0^t \theta_0(T_u Lf)du + \int_0^t \int_0^u \theta_s(cT_{u-s}Lf)dsdu. \quad (1.21)$$

Changing the variable of integration from u to s in the first integral, from u to r in the second and using Fubini's theorem for interchanging the order of integration in the third integral in (1.21), we get

$$\int_0^t \theta_s(Lf)ds = \int_0^t \theta_0(T_r Lf)dr + \int_0^t \int_s^t \theta_s(cT_{u-s}Lf)duds. \quad (1.22)$$

Substituting u-s = r in the double integral above,

$$\int_0^t \theta_s(Lf)ds = \int_0^t \theta_0(T_r Lf)dr + \int_0^t \int_0^{t-s} \theta_s(cT_r Lf)drds. \quad (1.23)$$

The relation (1.23) and the expression (1.20) for $\delta_t(f)$ imply that $\delta_t(f) = 0$ and hence as observed earlier, this shows that $\{K_t\}$ satisfies the equation (1.15).

To prove the other half, we now assume that $\{K_t\}$ satisfies (1.15). First, fix f $\mathcal{D}^{(2)}$. Then Lf $\in \mathcal{D}$ and also $T_r Lf \in \mathcal{D}$. Using (1.15) for $T_r Lf$ with t replaced by t-r and rearranging the terms, we get

$$\int_0^{t-r} \theta_s(cT_r Lf)ds \quad (1.24)$$

$$= \theta_{t-r}(T_r Lf) - \theta_0(T_r Lf) - \int_0^{t-r} \theta_s(LT_r Lf)ds.$$

The relations (1.20) and (1.24) give

$$\delta_t(f) = \int_0^t \theta_{t-r}(T_r Lf)dr - \int_0^t \int_0^{t-r} \theta_s(LT_r Lf)dsdr - \int_0^t \theta_s(Lf)ds \tag{1.25}$$

$$= \int_0^t \theta_s(T_{t-s} Lf)ds - \int_0^t \int_0^{t-s} \theta_s(LT_r Lf)drds - \int_0^t \theta_s(Lf)ds$$

$$= \int_0^t \theta_s(T_{t-s} Lf)ds - \int_0^t \int_0^{t-s} \theta_s(T_r LLf)drds - \int_0^t \theta_s(Lf)ds$$

where we have used Fubini's theorem and also the fact that for $f_1 \in \mathscr{D}$, $LT_r f_1 = T_r Lf_1$. Finally, from the relation

$$T_{t-s}Lf = Lf + \int_0^{t-s} T_r LLf dr$$

it follows that

$$\theta_s(T_{t-s}Lf) = \theta_s(Lf) + \int_0^{t-s} \theta_s(T_r LLf)dr.$$

which implies that $\delta_t(f) = 0$ in view of (1.25).

As in the first part, this implies that $\{K_t\}$ satisfies (1.16) for $f \in \mathscr{D}^{(2)}$. Now Lemma 1.1 implies that $\{K_t\}$ satisfies (1.16) for all $f \in J(\hat{S},\hat{\mathscr{S}})$. To see this, let $0 \leq t \leq T$ and define $\mu \in \mathscr{M}(S,\mathscr{S})$ by

$$\langle g,\mu\rangle = \langle V_t^0 g, K_0\rangle + \int_0^t \langle c_s(V_t^s g), K_s\rangle ds \tag{1.26}$$

for $g \in J(S,\mathscr{S})$. Then for $f \in J(S,\mathscr{S})$, we have

$$\langle f(t,\cdot),\mu\rangle \tag{1.27}$$

$$= \langle (T_t f)(0,\cdot),K_0\rangle + \int_0^t \langle c_s(\cdot)(T_{t-s}f)(s,\cdot),K_s\rangle ds$$

We have proved that $\{K_t\}$ satisfies (1.16) for $f \in \mathscr{D}^{(2)}$. Thus

$$\langle f(t,\cdot),\mu\rangle = \langle f(t,\cdot),K_t\rangle$$

for all $f \in \mathscr{D}^{(2)}$. By Lemma 1.1, this shows $\mu = K_t$ and now (1.27) implies the required conclusion that (1.16) holds for all $f \in J(\hat{S},\hat{\mathscr{S}})$. This implies that $\{K_t\}$ satisfies (1.14) as noted at the beginning of the proof. □

The last two results, namely Theorem 1.2 and Lemma 1.3 yield the following theorem:

<u>Theorem 1.4</u>: $\{G_t\}$ satisfies

$$\langle f(t,\cdot),G_t\rangle \tag{1.28}$$
$$= \langle f(0,\cdot),G_0\rangle + \int_0^t \langle (Lf)(s,\cdot) + c_s(\cdot)f(s,\cdot),G_s\rangle ds$$

for all $f \in \mathscr{D}$.

We now turn to the normalized measures N_t and obtain equations analogous to (1.10) and (1.28). The first of these follows easily from (1.28)

<u>Theorem 1.5</u>: $\{N_t\}$ satisfies the following equation

$$\langle f(t,\cdot),N_t\rangle \tag{1.29}$$

$$= \langle f(0,\cdot),N_0\rangle + \int_0^t \langle (Lf)(s,\cdot) + c_s(\cdot)f(s,\cdot),N_s\rangle ds$$

$$- \int_0^t \langle f(s,\cdot),N_s\rangle\langle c_s(\cdot),N_s\rangle ds, \quad f \in \mathscr{D}.$$

Proof: Note that

$$\langle f(t,\cdot),N_t\rangle = \frac{\langle f(t,\cdot),G_t\rangle}{G_t(S)} = \frac{\langle f(t,\cdot),G_t\rangle}{\langle f_1(t,\cdot),G_t\rangle} \tag{1.30}$$

where $f_1(t,x) \equiv 1$. Then, $f_1 \in \mathfrak{D}$ and $Lf_1 = 0$ as is easy to see. For $f \in \mathfrak{D}$, $\langle f(t,\cdot),G_t\rangle$ and $\langle f_1(t,\cdot),G_t\rangle$ are absolutely continuous functions by Theorem 1.4. Thus (1.30) implies that $\langle f(t,\cdot),N_t\rangle$ is also an absolutely continuous function. Differentiating (1.30), we get

$$\frac{d}{dt}\langle f(t,\cdot),N_t\rangle \tag{1.31}$$

$$= \frac{\frac{d}{dt}\langle f(t,\cdot),G_t\rangle}{\langle f_1(t,\cdot),G_t\rangle} + \langle f(t,\cdot),G_t\rangle\left[\frac{-\frac{d}{dt}\langle f_1(t,\cdot),G_t\rangle}{\langle f_1(t,\cdot),G_t\rangle^2}\right]$$

$$= \frac{\langle (Lf)(t,\cdot)+c_t(\cdot)f(t,\cdot),G_t\rangle}{G_t(S)} - \frac{\langle f(t,\cdot),G_t\rangle}{G_t(S)}\ \frac{\langle c_t(\cdot),G_t\rangle}{G_t(S)}$$

$$= \langle (Lf)(t,\cdot) + c_t(\cdot)f(t,\cdot),N_t\rangle - \langle f(t,\cdot),N_t\rangle\langle c_t(\cdot),N_t\rangle.$$

Assertion (1.29) of the theorem now follows immediately.

□

The next result is analogous to Lemma 1.3.

Lemma 1.6: Let $\{K_t\}$ satisfy (1.8). Then $\{K_t\}$ satisfies

$$\langle f(t,\cdot),K_t\rangle \tag{1.32}$$

$$= \langle f(0,\cdot),K_0\rangle + \int_0^t\langle (Lf)(s,\cdot) + c_s(\cdot)f(s,\cdot),K_s\rangle ds$$

$$- \int_0^t\langle f(s,\cdot),K_s\rangle\ \langle c_s(\cdot),K_s\rangle ds, \quad f \in \mathfrak{D}$$

if and only if it satisfies

$$\langle g,K_t\rangle = \langle V_t^0 g,K_0\rangle + \int_0^t \langle c_s V_t^s g,K_s\rangle ds \tag{1.33}$$

$$- \int_0^t \langle c_s,K_s\rangle \langle V_t^s g,K_s\rangle ds, \quad g \in J(S,\mathcal{S}).$$

Proof: First, it can be noted that $\{K_t\}$ satisfies (1.33) if and only if it satisfies

$$\langle f(t,\cdot),K_t\rangle = \langle (T_t f)(0,\cdot),K_0\rangle \tag{1.34}$$

$$+ \int_0^t \langle c_s(\cdot)(T_{t-s}f)(s,\cdot),K_s\rangle ds$$

$$- \int_0^t \langle c_s(\cdot),K_s\rangle \langle (T_{t-s}f)(s,\cdot),K_s\rangle ds, \quad f \in J(\hat{S},\hat{\mathcal{S}}).$$

The arguments for this step are very similar to those given at the beginning of the proof of Lemma 1.3.

As in that lemma, let us write $\theta_t(f) = \langle f(t,\cdot),K_t\rangle$ for a real valued, $\hat{\mathcal{S}}$ measurable function f for which $\int |f(t,\cdot)| dK_t < \infty$. For $f \in \mathcal{D}$, let $\delta_t(f)$ denote the difference between the right hand sides of (1.34) and (1.32). Then

$$\delta_t(f) = \theta_0(T_t f) + \int_0^t \theta_s(cT_{t-s}f)ds - \int_0^t \theta_s(c)\theta_s(T_{t-s}f)ds \tag{1.35}$$

$$- \theta_0(f) - \int_0^t \theta_s(Lf+cf)ds + \int_0^t \theta_s(c)\theta_s(f)ds.$$

From the identity

$$T_t f = f + \int_0^t T_r Lf dr$$

for $f \in \mathcal{D}$ and Fubini's theorem, we have

$$\theta_0(T_t f) = \theta_0(f) + \int_0^t \theta_0(T_r Lf)dr, \tag{1.36}$$

$$\theta_s(T_{t-s}f) = \theta_s(f) + \int_0^{t-s} \theta_s(T_r Lf)dr \qquad (1.37)$$

and since $\theta_s(|c|) < \infty$ for a.e. s,

$$\theta_s(cT_{t-s}f) = \theta_s(cf) + \int_0^{t-s} \theta_s(cT_r Lf)dr \text{ for a.e. } s \qquad (1.38)$$

From (1.35) - (1.38), we get

$$\delta_t(f) = \int_0^t \theta_0(T_r Lf)dr + \int_0^t \int_0^{t-s} \theta_s(cT_r Lf)drds \qquad (1.39)$$

$$- \int_0^t \int_0^{t-s} \theta_s(c)\theta_s(T_r Lf)drds - \int_0^t \theta_s(Lf)ds.$$

Now suppose that $\{K_t\}$ satisfies (1.34). Fix $f \in \mathfrak{D}$. Then (1.34) for the function Lf gives

$$\theta_u(Lf) = \theta_0(T_u Lf) + \int_0^u \theta_s(cT_{u-s}f)ds - \int_0^u \theta_s(c)\theta_s(T_{u-s}f)ds.$$

Hence

$$\int_0^t \theta_u(Lf)du = \int_0^t \theta_0(T_u Lf)du + \int_0^t \int_0^u \theta_s(cT_{u-s}f)dsdu$$

$$- \int_0^t \int_0^u \theta_s(c)\theta_s(T_{u-s}f)dsdu$$

$$= \int_0^t \theta_0(T_u Lf)du + \int_0^t \int_s^t \theta_s(cT_{u-s}f)duds$$

$$- \int_0^t \int_s^t \theta_s(c)\theta_s(T_{u-s}f)duds$$

by Fubini's theorem. Substituting $u = r + s$ in the last two integrals, we get

$$\int_0^t \theta_u(Lf)du = \int_0^t \theta_0(T_u Lf)du + \int_0^t\int_0^{t-s} \theta_s(cT_r f)drds \qquad (1.40)$$

$$- \int_0^t\int_0^{t-s} \theta_s(c)\theta_s(T_r f)drds.$$

From (1.39) and (1.40) we have $\delta_t(f) = 0$. It follows that if $\{K_t\}$ satisfies (1.34), then it also satisfies (1.32), completing the proof of one half of the theorem.

For the other half, we will show that if $\{K_t\}$ satisfies (1.32), then for $f \in \mathcal{D}^{(2)}$, $\delta_t(f) = 0$.

Suppose that $\{K_t\}$ satisfies (1.32). Fix $f \in \mathcal{D}^{(2)}$. Then as noted in Lemma 1.3, $T_r Lf \in \mathcal{D}$. Use (1.32) for $T_r Lf$ to get

$$\theta_{t-r}(T_r Lf) = \theta_0(T_r Lf) + \int_0^{t-r} \theta_s(LT_r Lf + cT_r Lf)ds \qquad (1.41)$$

$$- \int_0^{t-r} \theta_s(T_r Lf)\theta_s(c)ds.$$

Integrating (1.41) with respect to r from 0 to t and interchanging the order of integration on the right hand side (justified by Fubini's theorem) yields the relation

$$(1.42)$$

$$\int_0^t \theta_{t-r}(T_r Lf)dr = \int_0^t \theta_0(T_r Lf)dr + \int_0^t\int_0^{t-s} \theta_s(LT_r Lf)drds$$

$$+ \int_0^t\int_0^{t-s} \theta_s(cT_r Lf)drds - \int_0^t\int_0^{t-s} \theta_s(T_r Lf)\theta_s(c)drds.$$

Using the identity $T_{t-s}Lf = Lf + \int_0^{t-s} LT_r Lf dr$ and Fubini's theorem, we get

$$\int_0^t \theta_{t-r}(T_r Lf)dr = \int_0^t \theta_s(T_{t-s}Lf)ds \qquad (1.43)$$

$$= \int_0^t \theta_s(Lf)ds + \int_0^t\int_0^{t-s} \theta_s(LT_r Lf)drds.$$

The two relations (1.42) and (1.43) imply that the right hand side of (1.39) is equal to zero and hence $\delta_t(f) = 0$.

This completes the proof of the lemma. □

Theorem 1.5 and Lemma 1.6 together imply the following result.

Theorem 1.7: The normalized measures $\{N_t\}$ satisfy

$$\langle g,N_t\rangle = \langle V_t^0 g,N_0\rangle + \int_0^t \langle c_s V_t^s g,N_s\rangle ds \qquad (1.44)$$

$$- \int_0^t \langle c_s,N_s\rangle \, \langle V_t^s g,N_s\rangle ds, \quad f \in J(S,\mathcal{S}).$$

Uniqueness of Solution

We now turn to the question of under what conditions on c and (X_t) do the equations that we have obtained admit a unique solution.

In view of Lemma 1.3, if we show that one of the equations (1.10) and (1.28) admits a unique solution, then it follows that the other equation also has a unique solution. The same is the case for equations (1.29) and (1.44) for $\{N_t\}$.

We will first prove that equations (1.10) and (1.44) admit unique solutions under a boundedness restriction on c. Later, we will suppose that the process (X_t) has some additional properties and show the uniqueness of solution for (1.28) and (1.29).

We will first prove a Gronwall type inequality.

Lemma 1.8: Let b: $[0,T] \to [0,\infty)$ be a measurable function such that

$$\int_0^T b(s)ds < \infty. \qquad (1.45)$$

(i) Suppose that $\{a_k: k\geq 1\}$ is a sequence of bounded measurable functions from $[0,T]$ into $[0,\infty)$ such that for all $k \geq 1$

$$a_{k+1}(t) \leq \int_0^t b(s)a_k(s)ds, \quad 0 \leq t \leq T. \tag{1.46}$$

Then

$$a_k(t) \to 0 \quad \text{as } k \to \infty \quad \text{for all } t \in [0,T].$$

(ii) Suppose that $a: [0,T] \to [0,\infty)$ is a bounded measurable function such that

$$a(t) \leq \int_0^t b(s)a(s)ds, \quad 0 \leq t \leq T. \tag{1.47}$$

Then $a(t) = 0$ for all $t \in [0,T]$.

Proof: Let $B(s) = \int_0^s b(u)du$, $\beta(s) = s + B(s)$, $0 \leq s \leq T$ and let $T^* = \beta(T)$. For $0 \leq t \leq T^*$, let

$$\tau(t) = \inf\{s\geq 0: \beta(s) \geq t). \tag{1.48}$$

Then β and τ are continuous strictly increasing functions and are inverses to each other, i.e., $\beta\circ\tau(t) = t$, $0 \leq t \leq T^*$ and $\tau\circ\beta(s) = s$, $0 \leq s \leq T$. Note that for $0 \leq t_2 \leq t_1 \leq T^*$

$$B(\tau(t_1)) - B(\tau(t_2)) \leq \beta(\tau(t_1)) - \beta(\tau(t_2)) = t_1 - t_2$$

and hence $\frac{dB\circ\tau(t)}{dt} \leq 1$. Let $A_k(t) = a_k(\tau(t))$, $0 < t \leq T^*$ for all $k \geq 1$, $t \in [0,T]$. Then we have

$$A_{k+1}(t) = a_{k+1}(\tau(t)) \leq \int_0^{\tau(t)} b(s)a_k(s)ds \tag{1.49}$$

$$= \int_0^{\tau(t)} a_k(s)dB(s) = \int_0^t A_k(u)dB\circ\tau(u)$$

$$= \int_0^t A_k(u) \frac{dB\circ\tau(u)}{du}\, du.$$

Thus

$$A_{k+1}(t) \leq \int_0^t A_k(u)du \tag{1.50}$$

for all $k \geq 1$, $0 \leq t \leq T^*$. From this inequality, it can be checked (by induction on k) that

$$A_{k+1}(t) \leq C \cdot \frac{t^k}{k!} \tag{1.51}$$

for all $k \geq 1$, $0 \leq t \leq T^*$, where C is an upper bound of $A_1(t)$. Hence for all t,

$$\lim_{k\to\infty} A_k(t) = 0. \tag{1.52}$$

This proves part (i). Part (ii) follows from this -- take $a_k(t) \equiv a(t)$. Then $\{a_k\}$ satisfies the hypothesis of part (i) and hence $a_k(t) \to 0$, which gives $a(t) \equiv 0$. □

Theorem 1.9: Suppose that c satisfies

there exists a positive measurable function f on $[0,T]$ with $\int_0^T b(s)ds < \infty$ such that (1.53)

$$|c_s(x)| \leq b(s), \quad x \in S,\ s \leq 0.$$

Then

(i) $\{G_t\}$ is the unique solution of the equation (1.10) in the class of $\{K_t\}$ satisfying (1.8).

(ii) $\{G_t\}$ is the unique solution of the equation (1.28) in the class of $\{K_t\}$ satisfying (1.8).

(iii) $\{N_t\}$ is the unique solution of the equation (1.44) in the class of $\{K_t\}$ satisfying (1.8).

(iv) $\{N_t\}$ is the unique solution of the equation (1.29) in the class of $\{K_t\}$ satisfying (1.8).

Proof: We need to prove only (i) and (iii). Then Lemma 1.3 and 1.6 would imply (ii) and (iv) respectively.

For (i), let $\{K_t\}$ satisfy (1.8) and (1.10). Then $\{G_t\}$ and $\{K_t\}$ both satisfy (1.10) and thus

$$\langle g, G_t-K_t\rangle = \int_0^t \langle c_s V_t^s g, G_s-K_s\rangle ds.$$

For any signed measure μ on $(S,\mathcal{S})$ (i.e., difference of two positive finite measures), let $|\mu|(A)$ be defined by

$$|\mu|(A) = \sup\{|\mu(B)| + |\mu(A-B)|: B \subseteq A, B \in \mathcal{S}\}.$$

It is well known that $|\mu|$ is a finite countably additive measure on $(S,\mathcal{S})$. Further, for any measurable function f on S, we have

$$\left|\int f(x)d\mu(x)\right| \leq \int |f(x)| \, d|\mu|(x).$$

Let $a(t) = \sup_{0\leq s\leq t} \sup_{B\in\mathcal{S}} |G_s(B)-K_s(B)|$. Since a(t) is an increasing function, it is Borel measurable. Note that

$$\int (V_t^s 1_A)(x) \, d|G_s-K_s|(x) \leq |G_s-K_s|(S) \leq a(s),$$

Thus, for any $B \in \mathcal{S}$

$$\begin{aligned}|G_t(B)-K_t(B)| &\leq \int_0^t \int |c_s(x)|(V_t^s 1_B)(x) \, d|G_s-K_s|(x)ds \\ &\leq \int_0^t b(s)a(s)ds\end{aligned}$$

so that

$$a(t) \leq \int_0^t b(s)a(s)ds. \tag{1.54}$$

Hence by Lemma 1.8, $a(t) \equiv 0$ which proves $G_t = K_t$.

For (iii), suppose $\{K_t\}$ satisfies (1.8) and (1.44). Then for $g \in J(S,\mathcal{S})$

$$\langle g, N_t - K_t\rangle = \int_0^t \langle c_s V_t^s g, N_s - K_s\rangle ds \tag{1.55}$$

$$- \int_0^t \langle c_s, N_s - K_s\rangle \langle V_t^s g, N_s\rangle ds$$

$$- \int_0^t \langle c_s, K_s\rangle \langle V_t^s g, N_s - K_s\rangle ds.$$

Let $a(t) = \sup_{0\leq s\leq t} \sup_{B\in\mathcal{S}} |N_s(B) - K_s(B)|$. As in the proof of (1.54), it can be proved using (1.55) that

$$a(t) \leq \int_0^t (2+C)b(s)a(s)ds$$

where $|K_s(S)| \leq C$, $0 \leq s \leq T$. By Lemma 1.8, this gives $a(t) \equiv 0$, so that $N_t = K_t$. □

We have proved that if c satisfies (1.53) then G_t (N_t) is the unique solution to (1.10) ((1.44) respectively). We will now show that G_t and N_t can be approximated using the successive approximation method and that the convergence is uniform in t in the total variation norm.

Theorem 1.10: Suppose that c satisfies (1.53).

(i) Let $\{G_t^k\}$ be defined inductively by

$$G_t^0(A) = E\, 1_A(X_t) \tag{1.56}$$

$$G_t^{k+1}(A) = G_t^0(A) + \int_0^t \langle c_s(V_t^s 1_A), G_s^k\rangle ds, \quad A\in\mathcal{S},\ 0\leq t\leq T,\ k\geq 1.$$

Let

$$a_k(t) = \sup_{0 \le s \le t} \sup_{A \in \mathcal{S}} |G_s^k(A) - G_s(A)|.$$

Then $a_k(T) \to 0$.

(ii) Let $\{N_t^k\}$ be defined inductively by

$$N_t^0(A) = E\, 1_A(X_t) \tag{1.57}$$

$$N_t^{k+1}(A) = N_t^0(A) + \int_0^t \langle c_s(V_t^s 1_A),\ N_s^k\rangle ds$$

$$- \int_0^t \langle c_s, N_s^k\rangle \langle V_t^s 1_A,\ N_s^k\rangle ds.$$

Let

$$a_k'(t) = \sup_{0 \le s \le t} \sup_{A \in \mathcal{S}} |N_s^k(A) - N_s(A)|.$$

Then $a_k'(T) \to 0$.

<u>Proof</u>: It is easy to see that for each k, t; G_t^k and N_t^k are signed measures on $(S,\mathcal{S})$. For this one needs to do induction on k and use (1.56), (1.57).

Also note that for $g \in J(S,\mathcal{S})$

$$\langle g, G_t^0\rangle = E\, g(X_t) = E(V_t^0 g)(X_0) = \langle V_t^0 g, G_0\rangle$$

and hence by (1.56)

$$\langle g, G_t^{k+1}\rangle = \langle V_t^0 g, G_0\rangle + \int_0^t \langle c_s V_t^s g,\ G_s^k\rangle ds, \tag{1.58}$$

for $0 \le t \le T$, $k \ge 1$, $g \in J(S,\mathcal{S})$. Hence

$$\langle g, G_t^{k+1} - G_t\rangle = \int_0^t \langle c_s V_t^s g,\ G_s^k - G_s\rangle ds. \tag{1.59}$$

Arguments similar to those which yielded (1.54) now give

$$a_{k+1}(t) \leq \int_0^t b(s)a_k(s)ds$$

which proves the assertion $a_k(T) \to 0$ in view of Lemma 1.8. The proof of the other part, namely $a_k'(T) \to 0$ is similar -- we need to write $\langle g, N_t^k - N_t \rangle$ in a form similar to (1.55) and get an estimate for $a_k'(T)$. □

As yet the only assumption on the signal process (X_t) is that it is a Markov process satisfying conditions (1.1) and (1.2). Condition (1.1) is a regularity condition on (X_t). If S is a topological space and $\mathcal{S}$ is the Baire σ-field on S (i.e., the smallest σ-field on S with respect to which all the real valued continuous functions are measurable), then (1.2) holds if $P(s,x,t,B)$ is stochastically continuous. Recall that the transition function is called stochastically continuous if for all $s \leq 0$, $x \in S$, $U \in \mathcal{S}$, U open, $x \in U$, we have

$$\lim_{t \downarrow 0} P(s,x,t,U) = 1.$$

In this setup, we have had to assume that c is bounded in x for all s (or more precisely (1.53)) in order to conclude that $\{G_t\}$ (respectively $\{N_t\}$) is the unique solution to the equation (1.10) or (1.28) (respectively (1.29) or (1.44)).

We will now show that if we impose more regularity on (X_t), then we can do away with (1.53). The idea is the following. Instead of proving uniqueness of solution of (1.10), let us concentrate on (1.28). Suppose that given $g \in J(S,\mathcal{S})$ and $t > 0$, the equation

$$(Lv + cv)(u,x) = 0, \quad 0 \le u < t, \ x \in S, \tag{1.60}$$

$$v(t,s) = g(x)$$

admits a solution $v \in \mathcal{D}$, then for any solution $\{K_t\}$ of (1.28), we would have

$$\langle g,K_t\rangle = \langle (T_t v)(0,\cdot),K_0\rangle \tag{1.61}$$

$$= \langle (T_t v)(0,\cdot),G_0\rangle = \langle g,G_t\rangle$$

Thus if the equation (1.60) admits a solution for a rich class of functions g, then (1.61) would imply $K_t = G_t$.

However, if c is not bounded, there is no hope of obtaining a solution to (1.60) as it forces cv to be bounded. We will get around this by proving that if c_k approximates c in an appropriate sense and if (1.60) has a solution for c_k, $(k \ge 1)$, then $G_t = K_t$.

Returning to the problem of proving existence of solution to (1.60), note that if for c bounded, v satisfies (1.60) then by the Feynman-Kac formula,

$$v(s,X_s) = E_{\Pi}[g(X_t)\ \exp(\textstyle\int_s^t c(u,X_u)du)\,|\mathcal{F}_s^X]. \tag{1.62}$$

Thus given c, one way to prove the existence of a solution to (1.60) is to define v using (1.62) and then prove that it satisfies (1.60). For this we need to make some more assumptions on the Markov Process (X_t). First, some notation. Let S be a separable metric space and let $D = D([0,\infty),S)$ be the space of all functions $\underline{X}$: $[0,\infty) \to S$ which are right continuous and admit left limits. We will denote by $\underline{X}_t$ the value of the function $\underline{X}$ at t and also treat $\underline{X}_t$ as a function from D into S, so that $(\underline{X}_t)$ are the coordinate mappings. For $0 \le s \le t \le \infty$, let $\mathcal{A}_t^s = \sigma(\underline{X}_u\colon s \le u \le t)$.

The following assumptions on (X_t) will be referred to as (1.63) in the sequel:

(i) Suppose that S is a separable metric space and that the paths of the process X_t are right continuous and admit left limits (i.e., $X(\omega) \in D$ for all ω).

(ii) Suppose that the Markov process (X_t) admits a transition probability function $P(\cdot,\cdot,\cdot,\cdot)$.

(iii) Suppose that for all $(s,x) \in [0,\infty)\times S$, there exists a countably additive probability measure $P_{s,x}$ on $(D,\mathcal{A}_\infty^s)$ such that for

$$0 \le s < t_1 < t_2 < \ldots < t_k, \quad A_i \in \mathcal{S}, \quad k \ge 1 \tag{1.64}$$

we have (with $t_0 = s$, $y_0 = x$)

$$P_{s,x}(\underline{X}_{t_i} \in A_i,\ 1 \le i \le k) \tag{1.65}$$
$$= \int \ldots \int \prod_{i=1}^{k} 1_{A_i}(y_i) P(t_{i-1}, y_{i-1}, t_i - t_{i-1}, dy_i).$$

The existence of $\{P_{s,x}\}$ satisfying (1.65) on D implies that the transition probability function $P(\cdot,\cdot,\cdot,\cdot)$ of (X_t) is stochastically continuous and hence that (1.2) holds.

For s, $\{t_i\}$, $\{A_i\}$ as in (1.64), the relation (1.65) and the fact that P is the transition probability function for (X_t) give

$$\Pi(X_{t_i} \in A_i;\ 1 \le i \le k \mid \mathcal{F}_s^X) = P_{s,X_s}(\underline{X}_{t_i} \in A_i;\ 1 \le i \le k) \text{ a.s. } \Pi. \tag{1.66}$$

We claim that (1.66) implies that for $B \in \mathcal{A}_\infty^s$,

$$\Pi(X \in B \mid \mathcal{F}_s^X) = P_{s,X_s}(B) \quad \text{a.s. } \Pi. \tag{1.67}$$

To verify the validity of this claim, let $\mathcal{A}'$ be the class of sets $B \in \mathcal{A}_\infty^s$ for which (1.67) is true (for a fixed s). Then it is easy to see that

(i) $B_1, B_2 \in \mathcal{A}'$, $B_1 \subseteq B_2$ implies $B_2 - B_1 \in \mathcal{A}'$

(ii) $B_1, B_2 \in \mathcal{A}'$, $B_1 \cap B_2 = \phi$ implies $B_1 \cup B_2 \in \mathcal{A}'$

(iii) $B_j \in \mathcal{A}'$, $B_j \subseteq B_{j+1}$, $j \geq 1$ implies $\bigcup_{j=1}^{\infty} B_j \in \mathcal{A}'$.

Also, by (1.66), $\mathcal{A}'$ contains sets of the form $\{\underline{X}_{t_i} \in A_i, 1 \leq i \leq k\}$ where $\{t_i\}$, $\{A_i\}$ are as in (1.64). These sets are closed under intersections and generate $\mathcal{A}_\infty^s$. Thus, we get $\mathcal{A}' = \mathcal{A}_\infty^s$. This proves (1.67).

Similarly, it can be proved that for $0 \leq s < t$, $B \in \mathcal{A}_\infty^t$, $x \in S$.

$$P_{s,x}(B \mid \mathcal{A}_t^s) = P_{t,\underline{X}_t}(B) \quad \text{a.s. } P_{s,x}. \tag{1.68}$$

We are now in a position to prove that the equation (1.60) admits a solution for a suitable c,v.

Let $S' = [0,T] \times S$ and $\mathcal{S}' = \mathcal{B}([0,T]) \otimes \mathcal{S}$. Let S' be equipped with the product topology, so that S' is itself a Polish space. Note that $\mathcal{S}' = \mathcal{B}(S')$.

Theorem 1.11: Let $c: S' \to \mathbb{R}$ be a bounded continuous function. Let $0 < t_0 < \infty$ be fixed and $f \in \mathcal{D}$. Let $v: \hat{S} \to \mathbb{R}$ be defined by

$$v(s,x) = E_{P_{s,x}}\left[f(t_0, \underline{X}_{t_0}) \exp\left(\int_0^{t_0} c_u(\underline{X}_u) du\right)\right], \tag{1.69}$$

$$s < t_0, \; x \in S$$

$$= f(s,x), \qquad s \geq t_0, \; x \in S.$$

Then $v \in \mathscr{D}$ and

$$Lv = f_1 \tag{1.70}$$

where $f_1 : \hat{S} \to \mathbb{R}$ is given by

$$f_1(s,x) = -c_s(x)v(s,x) \quad \text{for } s < t_0, \; x \in S \tag{1.71}$$

$$= (Lf)(s,x) \quad \text{for } s \geq t_0, \; x \in S.$$

<u>Proof</u>: First note that for $s+t \leq t_0$, $x \in S$,

$$v(s+t,\underline{X}_{s+t}) = E_{P_{s+t,\underline{X}_{s+t}}}\left[f(t_0,\underline{X}_{t_0}) \exp\left(\int_{s+t}^{t_0} c_u(\underline{X}_u)du\right)\right] \tag{1.72}$$

$$= E_{P_{s,x}}\left[f(t_0,\underline{X}_{t_0}) \exp\left(\int_{s+t}^{t_0} c_u(\underline{X}_u)du\right) \Big| \mathscr{A}_\infty^{s+t}\right]$$

and hence

$$(T_t v)(s,x) = \int v(s+t,y)\, P(s,x,t,dy) \tag{1.73}$$

$$= E_{P_{s,x}}[v(s+t, \underline{X}_{s+t})]$$

$$= E_{P_{s,x}}\left[f(t_0,\underline{X}_{t_0}) \exp\left(\int_{s+t}^{t_0} c_u(\underline{X}_u)du\right)\right].$$

Also, for $s+t < t_0$,

$$(T_t f_1)(s,x) = E_{P_{s,x}}[f_1(s+t, \underline{X}_{s+t})] \tag{1.74}$$

$$= E_{P_{s,x}}[-c_{s+t}(\underline{X}_{s+t})\, v(s+t,\underline{X}_{s+t})]$$

$$= E_{P_{s,x}}\left[-c_{s+t}(\underline{X}_{s+t})E_{P_{s+t,\underline{X}_{s+t}}}\left[f(t_0,\underline{X}_{t_0}) \times \exp\left(\int_{s+t}^{t_0} c_u(\underline{X}_u)du\right) \Big| \mathscr{A}_\infty^{s+t}\right]\right]$$

$$= -E_{P_{s,x}}\left[f(t_0,\underline{X}_{t_0})c_{s+t}(\underline{X}_{s+t}) \exp\left(\int_{s+t}^{t_0} c_u(\underline{X}_u)du\right)\right].$$

Since for all $\underline{X} \in \mathscr{D}$,

$$\exp\left(\int_{s+t}^{t_0} c_u(\underline{X}_u)du\right) = \exp\left(\int_s^{t_0} c_u(\underline{X}_u)du\right) - \int_0^t c_r(\underline{X}_r) \exp\left(\int_{s+r}^{t_0} c_u(\underline{X}_u)du\right)dr \tag{1.75}$$

it follows by multiplying both sides of (1.75) by $f(t_0,\underline{X}_{t_0})$ and taking expectation with respect to $P_{s,x}$ that for $s+t < t_0$

$$(T_t v)(s,x) = v(s,x) + \int_0^t (T_r f_1)(s,x)dr. \tag{1.76}$$

Of course, to get (1.76), we also need to use (1.73), (1.74) and Fubini's theorem which is justified as f, c (and as a consequence) v, f_1 are bounded. To verify (1.76) for $s+t \geq t_0$, note that for $s+t \geq t_0$, $s+r \geq t_0$, we have

$$(T_t v)(s,x) = \int v(s+t,y)\, P(s,x,t,dy) \tag{1.77}$$

$$= \int f(s+t,y)\, P(s,x,t,dy)$$

$$= (T_t f)(s,x)$$

and

$$(T_r f_1)(s,x) = (T_r(Lf))(s,x). \tag{1.78}$$

Thus for $s+t \geq t_0$, we have

$$(T_t v)(s,x) = (T_t f)(s,x) \tag{1.79}$$

$$= (T_{t_0-s} f)(s,x) + \int_{t_0-s}^{t} (T_r(Lf))(s,x)dr$$

$$= (T_{t_0-s} v)(s,x) + \int_{t_0-s}^{t} (T_r f_1)(s,x)dr$$

by (1.77), (1.78)

$$= v(s,x) + \int_0^{t_0-s} (T_r f_1)(s,x) + \int_{t_0-s}^{t} (T_r f_1)(s,x)dr$$

by (1.76)

$$= v(s,x) + \int_0^t (T_r f_1)(s,x)dr$$

This proves the validity of (1.76) for all s,t,x. To complete the proof, it remains to show that $f_1 \in J_0$. Then (1.76) would give $v \in \mathscr{D}$ and $Lv = f_1$.

For $s \geq t_0$, since $f \in \mathscr{D}$, we have

$$(T_t f_1)(s,x) = (T_t Lf)(s,x), \quad t \geq 0 \tag{1.80}$$

$$\to (Lf)(s,x) \quad \text{as} \quad t \downarrow 0$$

$$= f_1(s,x).$$

To prove (1.80) for $s < t_0$, note that since f_1 is bounded, (1.76) gives

$$(T_t v)(s,x) \to v(s,x) \quad \text{as } t \downarrow 0$$

i.e.,
$$E_{P_{s,x}} [v(s+t, \underline{X}_{s+t})] \to v(s,x) \quad \text{as } t \downarrow 0. \tag{1.81}$$

Since c is continuous, we also have for $\underline{X} \in D$

$$c_{s+t}(\underline{X}_{s+t}) \to c_s(\underline{X}_s) \quad \text{as } t \downarrow 0. \tag{1.82}$$

Now for $s+t < t_0$

$$\begin{aligned}
&|(T_t f_1)(s,x) - f_1(x,s)| \qquad (1.83)\\
&= |E_{P_{s,x}}[c_{s+t}(\underline{X}_{s+t})v(s+t,\ \underline{X}_{s+t}) - c_s(x)\ v(s,x)]|\\
&\le E_{P_{s,x}} |c_{s+t}(\underline{X}_{s+t}) - c_s(x)|\ |v(s+t,\ \underline{X}_{s+t})|\\
&\qquad + |c_s(x)|\ |E_{P_{s,x}} v(s+t,\ \underline{X}_{s+t}) - v(s,x)|\\
&\le K_1\Big[E_{P_{s,x}} |c_{s+t}(\underline{X}_{s+t}) - c_s(x)|\\
&\qquad + |E_{P_{s,x}} v(s+t,\ \underline{X}_{s+t}) - v(s,x)|\Big]
\end{aligned}$$

where K_1 is an upper bound for v,c. The first term in the previous expression goes to zero as $t \downarrow 0$ by (1.82) and the dominated convergence theorem and the second term goes to zero by (1.81). Hence we have proved that for all $(s,x) \in \hat{S}$,

$$(T_t f_1)(s,x) \to f_1(s,x) \quad \text{as } t \downarrow 0 \tag{1.84}$$

so that $f_1 \in J_0$. This completes the proof. □

<u>Theorem 1.12</u>: Suppose that $c\colon S' \to \mathbb{R}$ is a $\mathcal{S}'$-measurable function satisfying (1.4) and (1.5). Suppose that (X_t) is a Markov process satisfying (1.63). Then $\{G_t\}$ is the unique solution to the equation (1.28) in the class of $\{K_t\}$ satisfying (1.9).

<u>Proof</u>: We have seen that $\{G_t\}$ is a solution to (1.28). It suffices to show that if $\{K_t\}$ satisfies (1.28) and (1.8)

with the exception that $K_0 \equiv 0$ (instead of $K_0 = G_0$) and that each K_t is a signed measure, then $K_t \equiv 0$. Fix such a $\{K_t\}$. Then (1.28) reduces to

$$\langle f(t,\cdot),K_t\rangle = \int_0^t \langle (Lf+cf)(s,\cdot),K_s\rangle ds, \quad f \in \mathcal{D}. \tag{1.85}$$

For $k \geq 1$, let

$$c_s^k(x) = c_s(x)\, 1_{\{|c_s(x)|\leq k\}}. \tag{1.86}$$

Then $c^k \in J(S',\mathcal{S}')$ and $|c^k| \leq k$ and $c_s^k(x) \leq \delta_s$. Let $\mu\colon \mathcal{S}' \to [0,\infty)$ be defined by

$$\mu(B) = \int_0^T [E_\Pi\, 1_B(t,X_t)]dt, \quad B \in \mathcal{S}'.$$

Then μ is a countably additive finite measure on $(S',\mathcal{S}')$. Also, the condition (1.5) implies that

$$\int_{S'} |c|d\mu = \int_0^T E_\Pi |c_s(X_s)|ds < \infty. \tag{1.87}$$

Using Luzin's Theorem, for each $k \geq 1$, we can obtain a sequence $\{c^{kj}\colon j\geq 1\}$ of bounded continuous functions from $S' \to \mathbb{R}$, $|c^{kj}| \leq k$, such that $c^{kj} \to c^k$ as $j \to \infty$ a.e. μ.

Fix $f \in \mathcal{D}$, $0 < t_0 \leq T$ and let v^k, v^{kj} be defined by (1.69) with c^k, c^{kj} replacing c respectively. This implies that $v^{kj}(s,x) = v^k(s,x) = f(s,x)$, $s \geq t_0$ and

$$v^k(s,X_s) = E_\Pi[f(t_0,X_{t_0}) \exp(\int_s^{t_0} c_u^k(X_u)du)|\mathcal{F}_s^X], \quad s\leq t_0. \tag{1.88}$$

and

$$v^{kj}(s,X_s) = E_{\Pi}[f(t_0,X_{t_0}) \exp(\int_s^{t_0} c_u^{kj}(X_u)du) | \mathcal{F}_s^X], \quad s \le t_0. \tag{1.89}$$

Observe that

$$E\left[\left| \int_s^{t_0} c_u^{kj}(X_u)du - \int_s^{t_0} c_u^k(X_u)du \right| \right] \le \int_{S'} |c^{kj} - c^k| d\mu \to 0 \text{ as } j \to \infty$$

since $c^{kj} \to c^k$ a.e. μ and $|c^{kj}| \le k$. Thus as $j \to \infty$

$$f(t_0,X_{t_0}) \exp(\int_s^{t_0} c_u^{kj}(X_u)du) \to f(t_0,X_{t_0}) \exp(\int_s^{t_0} c_u^k(X_u)du)$$

in Π-probability and hence also in $L^1(\Omega,\mathcal{A},\Pi)$. This and (1.88) and (1.89) imply that for every s,

$$E_{\Pi} |v^{kj}(s,X_s) - v^k(s,X_s)| \to 0 \quad \text{as } j \to \infty$$

which also gives

$$\int |v^{kj} - v^k| d\mu = \int_0^T E_{\Pi} |v^{kj}(s,X_s) - v^k(s,X_s)| ds$$

$$\to 0 \qquad \text{as } j \to \infty$$

since for k fixed, $\{v^{kj}\}$ and v^k are uniformly bounded.

Now, recall that c^{kj} is bounded continuous and hence by the definition of v^{kj} and Theorem 1.11,

$$Lv^{kj} = -c^{kj} v^{kj} \quad \text{on } [0,t_0) \times S.$$

Then

$$|\langle f(t_0,\cdot), K_{t_0} \rangle| = |\langle v^{kj}(t_0,\cdot), K_{t_0} \rangle| \tag{1.90}$$

$$= |\int_0^{t_0} \langle (Lv^{kj})(s,\cdot)+c_s(\cdot)v^{kj}(s,\cdot),K_s\rangle ds|$$

$$= |\int_0^{t_0} \langle (c_s(\cdot)-c_s^{kj}(\cdot))v^{kj}(s,\cdot),K_s\rangle ds|$$

$$\le M_1 \int_0^{t_0} \int |c_s(x)-c_s^{kj}(x)|\ |v_s^{kj}(x)|d\Pi\circ X_s^{-1}(x)ds$$

$$\le M_1 \int_{S'} |c-c^{kj}|\cdot|v^{kj}|d\mu$$

where $M_1 < \infty$ is a constant such that $|\frac{dK_t}{d\Pi\circ X_t^{-1}}| \le M_1$. Such an M_1 exists as $\{K_t\}$ satisfies (1.8).

For k fixed, as $j \to \infty$, $|c-c^{kj}|\cdot|v^{kj}|$ converges in μ-probability to $|c-c^k|\cdot|v^k|$ and

$$|c-c^{kj}|\cdot|v^{kj}| \le e^{kT}(|c|+k).$$

Since $\int|c|d\mu < \infty$, the dominated convergence theorem now gives

$$\int|c-c^{kj}|\cdot|v^{kj}|d\mu \to \int|c-c^k|\cdot|v^k|d\mu$$

which in conjunction with (1.90) gives

$$|\langle f(t_0,\cdot),K_{t_0}\rangle| \le M_1\int|c-c^k|\cdot|v^k|d\mu. \tag{1.91}$$

Since $c_s^k(x) \le \delta_s$, $\int_0^T \delta_s ds = M_2 < \infty$, we have

$$|v^k| \le M_3 \exp(M_2) = M_4$$

where M_3 is a bound for f. Thus (1.91) gives

$$|\langle f(t_0,\cdot),K_{t_0}\rangle| \le M_4M_1 \int|c-c^k|d\mu$$

$$= M_4M_1 \int|c|\ 1_{\{|c|\ge k\}}d\mu$$

$$\to 0 \quad \text{as } k \to \infty$$

since $\int|c|d\mu < \infty$. Hence,

$$\langle f(t_0,\cdot),K_{t_0}\rangle = 0. \tag{1.92}$$

The relation (1.92) holds for all $f \in \mathscr{D}$ and hence by Lemma 1.1, $K_{t_0} = 0$. Recall that t_0 is also arbitrary. This proves $K_t = 0$, $0 \le t \le T$, proving the required uniqueness.

□

We now turn to the question of uniqueness of solution to the equation (1.29). As seen in Theorem 1.5, if $\{K_t\}$ is a solution to equation (1.28), then $\{M_t\}$, defined by

$$M_t(A) = \frac{1}{K_t(S)} K_t(A), \tag{1.93}$$

is a solution to (1.29). If we can express $K_t(S)$ in terms of $\{M_t\}$, then given a solution $\{M_t\}$ of (1.29), we can use this to produce a solution $\{K_t\}$ of (1.28) and conclude $K_t = G_t$ by Theorem 1.12. This will give $M_t = N_t$ (hopefully).

Note that $K_t(S) = \langle 1,K_t\rangle$, 1 denoting the function identically equal to one on $\hat{S}$, which belongs to $\mathscr{D}$ with $L1 = 0$. Thus if $K_t(S)$ is a solution to (1.28) (satisfying (1.8)), $K_t(S)$ is absolutely continuous and

$$\frac{d}{dt} K_t(S) = \langle c_t(\cdot),K_t\rangle$$

so that

$$\frac{d}{dt} \log K_t(S) = \frac{\frac{d}{dt} K_t(S)}{K_t(S)} = \frac{\langle c_t(\cdot),K_t\rangle}{K_t(S)} = \langle c_t(\cdot),M_t\rangle$$

where $\{M_t\}$ is defined by (1.93). Since $K_0(S) = 1$, this gives

$$K_t(S) = \exp(\int_0^t \langle c_s(\cdot),M_s\rangle ds)$$

so that

$$K_t(A) = M_t(A) \exp(\int_0^t \langle c_s(\cdot),M_s\rangle ds). \tag{1.94}$$

This expresses $\{K_t\}$ in terms of $\{M_t\}$. We will now prove uniqueness of solution to (1.29) following steps outlined above.

<u>Theorem 1.13</u>: Under the hypothesis of Theorem 1.12, $\{N_t\}$ is the unique solution to (1.29) in the class of $\{M_t\}$ satisfying (1.8).

<u>Proof</u>: Let $\{M_t\}$ be an arbitrary solution to (1.29) satisfying (1.8). First note that $M_t(S)$ is absolutely continuous and

$$\begin{aligned}\frac{d}{dt} M_t(S) &= \frac{d}{dt} \langle 1,M_t\rangle \\ &= \langle c_t(\cdot),M_t\rangle - \langle c_t(\cdot),M_t\rangle\langle 1,M_t\rangle.\end{aligned}$$

Hence

$$\frac{d}{dt} [M_t(S)-1) = -\langle c_t(\cdot),M_t\rangle [M_t(S)-1]$$

which gives, using $M_0(S) = 1$,

$$M_t(S) - 1 = -\int_0^t \langle c_s(\cdot),M_s\rangle [M_t(S)-1]ds.$$

Note that

$$|\langle c_s(\cdot),M_s\rangle| \le \langle |c_s(\cdot)|,M_s\rangle$$

$$\le C \langle |c_s(\cdot)|,\Pi\circ X_0^{-1}\rangle$$

$$= C\, E_\Pi |c_s(X_s)|.$$

Thus

$$|M_t(S)-1| \le C \int_0^t E_\Pi |c_s(X_s)|\cdot |M_s(S)-1|ds.$$

The assumption (1.5) and Lemma 1.8 now give $|M_t(S)-1| = 0$, so that

$$M_t(S) = 1, \quad 0 \le t \le T.$$

Define $\{K_t\}$ by (1.94). It is easy to see that $\{K_t\}$ satisfies (1.8) and that (1.93) holds (since $M_t(S) = 1$). Then for $f\in\mathscr{D}$

$$\frac{d}{dt}\langle f(t,\cdot),K_t\rangle = \frac{d}{dt}(K_t(S)\langle f(t,\cdot),M_t\rangle) \tag{1.95}$$

$$= K_t(S)[\langle (Lf)(t,\cdot),M_t + \langle c_t(\cdot)f(t,\cdot),M_t\rangle$$

$$- \langle f(t,\cdot),M_t\rangle \langle c_t,M_t\rangle] + \langle f(t,\cdot),M_t\rangle \frac{d}{dt}K_t(S)$$

and

$$\frac{d}{dt}K_t(S) = \frac{d}{dt}[\exp(\int_0^t \langle c_s(\cdot),M_s\rangle ds)] \tag{1.96}$$

$$= \langle c_t,M_t\rangle K_t(S).$$

The relations (1.95) and (1.96) give

$$\frac{d}{dt}\langle f(t,\cdot),K_t\rangle = K_t(S)[\langle (Lf)(t,\cdot),M_t\rangle + \langle c_t(\cdot)f(t,\cdot),M_t\rangle]$$

$$= \langle (Lf)(t,\cdot) + c_t(\cdot)f(t,\cdot),K_t\rangle.$$

Of course, $\langle f(t,\cdot),K_t\rangle$ is absolutely continuous for $f \in \mathcal{D}$ as $\langle f(t,\cdot),M_t\rangle$ and $K_t(S)$ are. Thus $\{K_t\}$ is a solution to (1.28). By Theorem 1.12, this gives $K_t = G_t$. The relation (1.93) (which holds as observed above) and the definition of $\{N_t\}$ (see (1.7)) give

$$M_t = N_t, \qquad 0 \leq t \leq T.$$

This completes the proof. □

As we have seen earlier, uniqueness of solution to equations (1.28), (1.29) implies the same for (1.10) and (1.44). We state this observation as a theorem for future reference.

Theorem 1.14: Suppose that the Markov process (X_t) satisfies (1.63) and that $c\colon S' \to \mathbb{R}$ is a $\mathcal{S}'$-measurable function satisfying (1.4) and (1.5).

Then the conclusions (i), (ii), (iii), and (iv) of Theorem 1.9 are valid.

The proof follows from Lemmas 1.3, 1.6 and Theorems 1.12 and 1.13. □

2. FILTERING WHEN SIGNAL AND NOISE ARE INFINITE DIMENSIONAL

The nonlinear filtering problem when the state space S of the signal process (X_t) is an arbitrary measurable space and the observations take values in an infinite dimensional Hilbert space has attracted considerable attention in recent years in view of the growing importance of prediction. In many cases, as we shall see, the latter can be recast as a problem involving infinite dimensional processes. We will apply the finitely additive white noise theory developed in the preceding chapters to obtain the optimal filter (which is now a measure valued process) as the unique solution of certain measure valued equations already introduced in Section 1. When the observation process is infinite dimensional, the natural model for white noise is an infinite dimensional Hilbert space. It should be noted, however, that the results derived here apply also to the finite dimensional situation and hence cover the case not covered by the Zakai equation for the conditional density.

Let $\mathcal{K}$ be a separable Hilbert space with inner product $(\cdot,\cdot)_{\mathcal{K}}$ and norm $\|\cdot\|_{\mathcal{K}}$. Let $H = L^2([0,T],\mathcal{K})$, i.e.

$$H = \{\phi:[0,T] \to \mathcal{K} \text{ such that } \|\phi\|^2 = \int_0^T \|\phi_s\|_{\mathcal{K}}^2 ds < \infty\}.$$

Then H is itself a separable Hilbert space with the inner product

$$(\phi,\phi') = \int_0^T (\phi_s,\phi_s')_{\mathcal{K}} ds. \tag{2.1}$$

Let the signal process (X_t) be as in the beginning of section 1, namely (X_t) is an $(S,\mathcal{S})$ valued Markov process on $(\Omega,\mathcal{A},\Pi)$. We assume that (X_t) satisfies (1.1). We will use the notation introduced in the previous section.

Let $h:[0,T] \times S \to \mathcal{K}$ be a measurable function such that

$$\int_0^T \|h_s(X_s(\omega))\|_{\mathcal{K}}^2 < \infty \quad \text{a.s. } \Pi. \tag{2.2}$$

As in Section VII.1, we assume that (2.2) holds for all ω, by redefining $X_t(\omega)$ on a null set if necessary. Let $\xi_s(\omega) = h_s(X_s(\omega))$, $o \le s \le T$. Since paths of (X_t) are measurable, it can be shown as in Section VII.1 that $\omega \to \xi_{\bullet}(\omega)$ is $\mathcal{A}/\mathcal{B}(H)$ measurable.

Let $(E,\mathcal{E},\alpha) = (\Omega,\mathcal{A},\Pi)\otimes(H,\mathcal{C},m)$. Regarded as random variables on $(E,\mathcal{E},\alpha)$, (X_t) and e are independent where, as before, e is the identity map from H to H. Now the abstract statistical model

$$y = \xi + e \tag{2.3}$$

takes the form

$$y_t = h_t(X_t) + e_t, \quad o \leq t \leq T. \tag{2.4}$$

This is our model of nonlinear filtering. Here (e_t) is $\mathcal{K}$ valued white noise. When $\mathcal{K} = \mathbb{R}^N$, this is the same as the model considered in Section VII.1. However, (2.4) allows the possibility of $\mathcal{K}$ being infinite dimensional, so that in this case, the noise (e_t) and observation (y_t) are infinite dimensional.

For $0 \leq t \leq T$, let $H_t = \{\phi \in H: \int_t^T \|\phi_s\|_{\mathcal{K}}^2 ds = 0\}$ and let Q_t be the orthogonal projection on H with range $Q_t = H_t$. Then

$$(Q_t\phi) = \phi \cdot 1_{[0,t]}, \quad \phi \in H, \tag{2.5}$$

so that $Q_t y$ represents the observations $\{y_s: 0 \leq s \leq t\}$ over $[0,t]$.

For $0 \leq t \leq T$, $\phi \in H$, let $\Gamma_t(\phi)$, $F_t(\phi) \in \mathcal{M}(S,\mathcal{S})$ be defined by

$$\Gamma_t(\phi)(A) = E_\Pi[1_A(X_t)\cdot\exp(\int_0^t (h_s(X_s),\phi_s)_{\mathcal{K}} ds - \tfrac{1}{2}\int_0^t \|h_s(X_s)\|_{\mathcal{K}}^2 ds)], \quad A \in \mathcal{S} \tag{2.6}$$

and

$$F_t(\phi)(A) = \frac{\Gamma_t(\phi)(A)}{\Gamma_t(\phi)(S)} . \tag{2.7}$$

Then by the Bayes formula, Theorem VI.3.4, for $Q = Q_t$, we have

$$E_\alpha(g(X_t)|Q_t y) = \int_S g(x)dF_t(y)(x) \tag{2.8}$$

for $g: S \to \mathbb{R}$ such that $E_\Pi|g(X_t)| < \infty$.

Thus $F_t(y)$ is the conditional distribution of X_t given $Q_t y$ and $\Gamma_t(y)$ is the unnormalized conditional distribution of X_t given $Q_t y$.

Note that $F_t(y)$, $\Gamma_t(y)$ are of the form of N_t, G_t for a suitable choice of c. Thus the results of the last section give us the following theorem.

Theorem 2.1: Suppose that one of the two sets of conditions (a), (b) given below is satisfied.

(a) (X_t) satisfies (1.2) and that $\|h_s(x)\|_{\mathcal{H}} \le a(s)$, where a is a measurable function such that $\int_0^T a^2(s)ds < \infty$.

(b) (X_t) satisfies (1.61) and that $E\int_0^T \|h_s(X_s)\|^2_{\mathcal{H}} ds < \infty$.

For $\phi \in H$, let $c_s^\phi(x)$: $[0,T] \to \mathbb{R}$ be defined by

$$c_s^\phi(x) = (h_s(x), \phi_s)_{\mathcal{H}} - \tfrac{1}{2}\|h_s(x)\|^2_{\mathcal{H}}. \tag{2.9}$$

Then we have the following conclusions, which are valid for all $y \in H$. All the uniqueness assertions stated below are in the class of $\{K_t\} \subseteq \mathcal{M}(S,\mathcal{S})$ satisfying (1.8). We denote $N_0 := \Pi \circ X_0^{-1}$

(i) $\{\Gamma_t(y)\}$ is the unique solution to

$$\langle g,\Gamma_t(y)\rangle = \langle V_t^0 g,N_0\rangle + \int_0^t \langle c_s^y(V_t^s g),\Gamma_s(y)\rangle ds, \quad g \in J(S,\mathcal{S}). \tag{2.10}$$

(ii) $\{\Gamma_t(y)\}$ is the unique solution to

$$\langle f(t,\cdot),\Gamma_t(y)\rangle = \langle f(0,\cdot),N_0\rangle + \int_0^t \langle (Lf)(s,\cdot)+c_s^y(\cdot)f(s,\cdot),\Gamma_s(y)\rangle ds, \quad f \in \mathfrak{D}. \tag{2.11}$$

(iii) $\{F_t(y)\}$ is the unique solution to

$$\langle g,F_t(y)\rangle = \langle V_t^0 g,N_0\rangle + \int_0^t \langle c_s^y\cdot(V_t^s g),F_s(y)\rangle ds - \int_0^t \langle c_s^y,F_s(y)\rangle\langle V_t^s g,F_s(y)\rangle ds, \quad \text{for } g \in J(S,\mathcal{S}). \tag{2.12}$$

(iv) $\{F_t(y)\}$ is the unique solution to

$$\langle f(t,\cdot),F_t(y)\rangle = \langle f(0,\cdot),N_0\rangle + \int_0^t \langle (Lf)(s,\cdot)+c_s^y(\cdot)f(s,\cdot),F_s(y)\rangle ds - \int_0^t \langle c_s^y,F_s(y)\rangle\langle f(s,\cdot),F_s(y)\rangle ds, \quad \text{for all } f \in \mathfrak{D}. \tag{2.13}$$

Proof: Note that for the choice of $c = c_s^y$, (given by (2.9)), $\Gamma_t(y) = G_t$, $F_t(y) = N_t$ (where the respective quantities are defined by (2.6), (1.6), (2.7), (1.7)) and the equations (1.10), (1.28), (1.29) and (1.44) in this notation reduce to (2.10), (2.11), (2.13) and (2.12) respectively. From the inequality

$$|(h_s(x),y_s)_{\mathcal{H}}| \le \tfrac{1}{2}\|h_s(x)\|_{\mathcal{H}}^2 + \tfrac{1}{2}\|y_s\|_{\mathcal{H}}^2, \tag{2.14}$$

it follows that

$$c_s^y(x) \leq \tfrac{1}{2}\|y_s\|_{\mathcal{H}}^2 \tag{2.15}$$

so that $c_s^y(x)$ satisfies (1.4) since $\int_0^T \|y_s\|_{\mathcal{H}}^2 ds < \infty$ for $y \in H$.

If (a) is satisfied, then

$$|c_s^y(x)| \leq |(h_s(x), y_s)_{\mathcal{H}}| + \tfrac{1}{2}\|h_s(x)\|_{\mathcal{H}}^2 \leq \|h_s(x)\|_{\mathcal{H}}^2 + \|y_s\|_{\mathcal{H}}^2$$

$$\leq a^2(s) + \|y_s\|_{\mathcal{H}}^2 \tag{2.16}$$

so that (1.53) holds with $b(s) = a^2(s) + \|y_s\|_{\mathcal{H}}^2$. Thus if (a) holds, the conditions of Theorem 1.9 are fulfilled.

On the other hand, if (b) holds, then

$$E_\Pi \int_0^T |c_s^y(X_s)| ds \leq E_\Pi \int_0^T \|h_s(X_s)\|_{\mathcal{H}}^2 ds + \int_0^T \|y_s\|_{\mathcal{H}}^2 ds < \infty \tag{2.17}$$

and hence the assumptions made in Theorem 1.14 are satisfied. Thus the required conclusions follow from Theorem 1.9 if (a) is satisfied and from Theorem 1.14 if (b) is true.

□

Theorem 1.10 yields the following result on approximation of $\{\Gamma_t^y\}$ and $\{F_t^y\}$.

<u>Theorem 2.2</u>: Suppose that the assumption (a) of Theorem 2.1 holds. Let $\{\Gamma_{t,k}^y\}$ and $\{F_{t,k}^y\}$ be defined inductively by

$$\Gamma_{t,0}^y(A) = E_\Pi 1_A(X_t). \tag{2.18}$$

$$\Gamma_{t,k+1}^y(A) = \Gamma_{t,0}^y(A) + \int_0^t \langle c_s^y \cdot (V_t^s 1_A), \Gamma_{s,k}^y \rangle ds, \tag{2.19}$$

$$k \geq 1,\ A \in \mathcal{S}$$

and

$$F^y_{t,0}(A) = E_\Pi 1_A(X_t) \tag{2.20}$$

$$F^y_{t,k+1}(A) = F^y_{t,0}(A) + \int_0^t \langle c^y_s \cdot (V^s_t 1_A), F^y_{s,k}\rangle ds \tag{2.21}$$

$$- \int_0^t \langle c^y_s, F^y_{s,k}\rangle \langle V^s_t 1_A, F^y_{s,k}\rangle ds, \quad k \geq 1, \ A \in S.$$

Then

$$\sup_{o \leq t \leq T} [\sup_{A \in \mathcal{S}} |\Gamma^y_{t,k}(A) - \Gamma_t(y)(A)|] \to 0 \quad \text{as } k \to \infty \tag{2.22}$$

and

$$\sup_{o \leq t \leq T} [\sup_{A \in \mathcal{S}} |F^y_{t,k}(A) - F_t(y)(A)|] \to 0 \quad \text{as } k \to \infty. \tag{2.23}$$

3. MARKOV PROPERTY OF THE OPTIMAL FILTER AS A MEASURE VALUED PROCESS

In the conventional theory of stochastic filtering, the optimal filter possesses the important property of being a Markov process (see [53]).

To prove that the same is true in the white noise setup, we have to begin by giving a definition of a Markov process on a quasi-cylinder probability space. We shall then show that $F_t(y)$, which is the conditional distribution of X_t given $Q_t y$, is a Markov process on $(E, \mathcal{E}, \alpha)$. It might be inter esting to observe that the proof does not make use of the fact that $F_t(y)$ satisfies a differential equation.

The proof is along the following lines.

(i) It is easy to observe that for $s < t$, $\Gamma_t(\phi)$ can be expressed as a function of $\Gamma_s(\phi)$ and $Q^s_t\phi$, where $Q^s_t\phi$ is the orthogonal projection on H with range

$$H_t^s = \{\phi \in H: \int_0^s + \int_0^T \|\phi_u\|_{\mathcal{H}}^2 du = 0\},\ 0 \leq s \leq t \leq T.$$

(ii) Under the canonical Gauss measure m on H, $Q_s^0\phi$ and $Q_t^s\phi$ are independent. This and (i) should give the Markov property of $\Gamma_t(\phi)$ on $(H,\mathcal{C},m)$.

(iii) Using that $n \ll m$ and the expression for $\frac{dn}{dm}$, we deduce the Markov property of $\Gamma_t(\phi)$ and $F_t(\phi)$ on $(H,\mathcal{C},n)$.

(iv) By proving that the functionals occuring in various expressions for conditional expectations are in appropriate $\mathcal{U}$ classes, we finally conclude that $\Gamma_t(y)$ and $F_t(y)$ are Markov processes on $(E,\mathcal{E},\alpha)$.

For (ii) we need a Fubini type theorem which we will prove first. Then we will define a Markov process and go to the other steps.

Let H be a real separable Hilbert space. Let Q_1 be an orthogonal projection with range H_1 and let H_2 be the orthogonal complement of H_1 in H. Let Q_2 be the orthogonal projection onto H_2. Then, $Q_2 = Q_1^{\perp}$, $Q_1Q_2 = Q_2Q_1 = 0$ and $I = Q_1 + Q_2$. Let m be the canonical Gauss measure on $(H,\mathcal{C})$.

<u>Theorem 3.1</u>: Suppose that

$$f \in \mathcal{L}^{1*}(H,\mathcal{C},m), \tag{3.1}$$

and let $f_1: H\times H \to \mathbb{R}$ be defined by

$$f_1(h_1,h_2) = f(Q_1h_1 + Q_2h_2),\ h_1,h_2 \in H. \tag{3.2}$$

Suppose that for all $h_1 \in H$ (fixed),

$$f_1(h_1,\cdot) \in \mathcal{L}^{1*}(H,\mathcal{C},m). \tag{3.3}$$

Let $f_2: H \to \mathbb{R}$ be defined by

$$f_2(h_1) = \int_H f_1(h_1,h_2)dm(h_2). \tag{3.4}$$

Suppose that

$$f_2 \in \mathcal{L}^*(H,\mathcal{C},m) \tag{3.5}$$

Then

$$E_m(f|Q_1) = f_2. \tag{3.6}$$

Proof: Let R_m be the m-lifting corresponding to a representative (L,Π_1) of m. We will give a proof in the case when both H_1 and H_2 are infinite dimensional. The proof in the other cases is similar (and simpler).

Let $\{e_i^1\}$ be a complete orthonormal system (CONS) in H_1 and $\{e_j^2\}$ be a CONS in H_2. Let P_i^1, P_j^2 be orthogonal projections on H with range span $\{e_1^1,\ldots,e_i^1\}$ and span $\{e_1^2,\ldots,e_j^2\}$ respectively, $i \geq 1$, $j \geq 1$. Let

$$P_{ij} = P_i^1 + P_j^2. \tag{3.7}$$

It is easy to see that P_{ij} is the orthogonal projection on H with range $\{e_1^1,e_2^1,\ldots,e_i^1,e_1^2,e_2^2,\ldots,e_j^2\}$. Thus, $P_{ij} \in \mathcal{P}$. If i_k,j_k are any subsequences of integers, $i_k\uparrow \infty$, $j_k\uparrow \infty$, then $P_{i_k,j_k}\uparrow I$, and then (3.1) implies that

$$R_m(f\circ P_{i_kj_k}) \to R_m(f) \quad \text{as } k \to \infty \text{ in } \mathcal{L}^1(\Omega_1,\mathcal{A}_1,\Pi_1). \tag{3.8}$$

Since (3.8) holds for all $i_k\uparrow \infty$, $j_k\uparrow \infty$, we can conclude that

$$R_m(f\circ P_{ij}) \to R_m(f) \text{ as } (i,j) \to \infty \text{ in } \mathscr{L}^1(\Omega_1,\mathscr{A}_1,\Pi_1). \tag{3.9}$$

For each (i,j), let $\psi_{ij}: \mathbb{R}^{i+j} \to \mathbb{R}$ be a Borel measurable function such that

$$f\circ P_{ij}(h) = \psi_{ij}((e_1^1,h),\ldots,(e_i^1,h),(e_1^2,h),\ldots,(e_j^2,h)). \tag{3.10}$$

It is easy to see that $Q_1P_{ij} = P_{ij}Q_1 = P_i^1$ and $Q_2P_{ij} = P_{ij}Q_2 = P_j^2$. Thus,

$$\begin{aligned} f_1(P_{ii}h_1,P_{jj}h_2) &= f(Q_1P_{ii}h_1 + Q_2P_{jj}h_2) \\ &= f(P_i^1h_1 + P_j^2h_2) \\ &= f(P_{ij}(Q_1h_1 + Q_2h_2)). \end{aligned} \tag{3.11}$$

From (3.10), (3.11), we get

$$\begin{aligned} &f_1(P_{ii}h_1,P_{jj}h_2) \\ &\quad = \psi_{ij}((e_1^1,h_1),\ldots,(e_i^1,h_1),(e_1^2,h_2),\ldots,(e_2^2,h_2)) \end{aligned} \tag{3.12}$$

For $j \geq 1$, define

$$f_2^j(h_1) = \int_H f_1(h_1,P_{jj}h_2)dm(h_2) \tag{3.13}$$

By the assumption (3.3) on f_1, we have

$$f_2^j(h_1) \to f_2(h_1) \quad \text{for all } h_1 \in H. \tag{3.14}$$

We now claim that

$$R_m(f_2^j\circ P_{ii}) = E_{\Pi_1}[R_m(f\circ P_{ij})|\mathscr{A}_{Q_1}] \tag{3.15}$$

where $\mathcal{A}_{Q_i} = \sigma(L(h): h \in H_i)$, $i = 1,2$. To verify (3.15), note that in view of (3.10)

$$R_m(f \circ P_{ij}) = \psi_{ij}(L(e_1^1),\ldots,L(e_i^1),L(e_1^2),\ldots,L(e_j^2))$$

and $L(e_k^1)$ is $\mathcal{A}_{Q_1}$-measurable for $1 \leq k \leq i$ and $L(e_k^2)$ is independent of $\mathcal{A}_{Q_1}$ for $1 \leq k \leq j$. Thus

$$E_{\Pi_1}[R_m(f \circ P_{ij}) | \mathcal{A}_{Q_1}] = \bar{\psi}_{ij}(L(e_1^1),\ldots,L(e_i^1)) \tag{3.16}$$

where for $a_1, a_2, \ldots, a_i \in \mathbb{R}$,

$$\bar{\psi}_{ij}(a_1,a_2,\ldots,a_i) \tag{3.17}$$

$$= \int_{\Omega_1} \psi_{ij}(a_1,a_2,\ldots,a_i,L(e_1^2),\ldots,L(e_j^2))d\Pi_1.$$

From (3.12), (3.13) it follows that

$$f_2^j(P_{ii}h_1) = \int f_1(P_{ii}h_1, P_{jj}h_2)dm(h_2)$$

$$= \int \psi_{ij}((e_i^1,h_1),\ldots,(e_i^1,h_1),(e_1^2,h_2),\ldots,(e_j^2,h_2))dm(h_2)$$

$$= \int \psi_{ij}((e_1^1,h_1),\ldots,(e_i^1,h_1),L(e_1^2),\ldots,L(e_j^2))d\Pi_1$$

$$= \bar{\psi}_{ij}((e_i^1,h_1),\ldots,(e_i^1,h_1))$$

so that

$$R_m(f_2^j \circ P_{ii}) = \bar{\psi}_{ij}(L(e_1^1),\ldots,L(e_i^1)). \tag{3.18}$$

The assertion (3.15) follows from (3.17) and (3.18). The

convergence of $R_m(f \circ P_{ij})$ to $R_m(f)$ in $\mathscr{L}^1(\Omega_1, \mathscr{A}_1, \Pi_1)$ (see (3.9)) implies

$$E_{\Pi_1}[R_m(f \circ P_{ij}) | \mathscr{A}_{Q_1}] \to E_{\Pi_1}[R_m(f) | \mathscr{A}_{Q_1}]$$

and thus from (3.15) we have

$$R_m(f_2^j \circ P_{ii}) \to E_{\Pi_1}[R_m(f) | \mathscr{A}_{Q_1}] \text{ as } (i,j) \to \infty \tag{3.19}$$

in $\mathscr{L}^1(\Omega_1, \mathscr{A}_1, \Pi_1)$.

Let $\tilde{\psi}_i$ be Borel measurable function from $\mathbb{R}^i \to \mathbb{R}$ such that

$$f_2 \circ P_{ii}(h_1) = \tilde{\psi}_i((e_1^1, h_1), \ldots, (e_i^1, h_1)).$$

The relation (3.14) for $P_{ii}h_1$ implies

$$f_2^j(P_{ii}h_1) \to f_2(P_{ii}h_1) \quad \text{for all } h_1 \in H$$

and hence (3.20)

$$\tilde{\psi}_{ij}(a_1, a_2, \ldots, a_i) \to \tilde{\psi}_i(a_1, a_2, \ldots, a_i), \quad \forall (a_1, \ldots, a_i) \in \mathbb{R}^i.$$

Now

$$\begin{aligned} R_m(f_2^j \circ P_{ii}) &= \tilde{\psi}_{ij}(L(e_1^1), \ldots, L(e_i^1)) \\ &\to \tilde{\psi}_i(L(e_1^1), \ldots, L(e_i^1)) \quad \text{as } j \to \infty \text{ (pointwise)} \\ &= R_m(f_2^j). \end{aligned}$$

Thus

$$R_m(f_2^j \circ P_{ii}) \to R_m(f_2 \circ P_{ii}) \text{ as } j \to \infty \text{ in } \Pi_1\text{-probability.} \tag{3.21}$$

The assumption (3.4) on f_2 implies

$$R_m(f_2 \circ P_{ii}) \to R_m(f_2) \text{ as } i \to \infty \text{ in } \Pi_1\text{-probability.} \tag{3.22}$$

The convergence relations (3.19), (3.21), (3.22) give us

$$R_m(f_2) = E_{\Pi_1}[R_m(f)|\mathcal{A}_{Q_1}]. \tag{3.23}$$

Note that the assumption $f_2 \in \mathcal{L}^*(H,\mathcal{C},m)$ and Proposition IV.2.3 imply that

$$f_2' \in \mathcal{U}(H,\mathcal{C},m;Q_1) \tag{3.24}$$

where f_2' is the restriction of f_2 to H_1.

The conclusion follows from (3.23) and (3.24) and part (iv) of Theorem IV.3.2. We need to check that $\overline{\mathcal{D}}_{Q_1}$ (as defined just before the statement of Theorem IV.3.2) is equal to $\overline{\mathcal{A}}_{Q_1}$. This is essentially done in the proof of the Theorem referred to and can be verified directly. □

<u>Remark 3.1</u>: If m_1,m_2 are the canonical Gauss measures on H_1,H_2 respectively then it is natural to expect that m is the 'product' of m_1 and m_2 and the Fubini theorem will be a statement of the form

$$\int\int g(h_1',h_2')dm_2(h_2')dm_1(h_1') = \int g(h_1',h_2')dm(h_1',h_2'). \tag{3.25}$$

Theorem 3.1 gives some conditions for this to be valid. If $f(h) = g(Q_1h,Q_2h)$ satisfies the conditions of Theorem 3.1, then (3.6) gives

$$\int f_2(h)dm(h) = \int f(h)dm(h).$$

But $f_2(h_1) = \int g(Q_1h_1, Q_2h_2)dm(h_2) = \int g(Q_1h_1, h_2')dm_2(h_2')$ (by (3.3)) and also by (3.5)

$$\int f_2(h_1)dm(h_1) = \int f_2(h_1')dm_1(h_1').$$

Thus, a set of sufficient conditions for (3.25) to hold is that $f(h) := g(Q_1h, Q_2h)$ satisfy the conditions of Theorem 3.1. For this reason, we have referred to Theorem 3.1 as a Fubini type theorem. It should be noted that (3.25) is only one half of Fubini's theorem -- we also need the equality with the other iterated integral for which also the previous theorem will give a sufficient condition.

Definition of a Markov Process on $(E, \mathcal{E}, \beta)$

Let Φ: $(E, \mathcal{E}, \beta) \to (H', \mathcal{C}')$ be a QCM where $(E, \mathcal{E}, \beta)$ is a quasi-cylinder probability space and H' is a separable Hilbert space, $\mathcal{C}' = \mathcal{C}(H')$. For each t, $0 \le t \le T$, let Q_t be an orthogonal projection on h' such that

$$t_1 \le t_2 \Rightarrow (\text{range } Q_{t_1}) \subseteq (\text{range } Q_{t_2}).$$

The family of QCM's $\{Q_t\Phi\}$ here plays the role played by an increasing family of σ-fields $(\mathcal{F}_t)$ in the usual theory.

Definition 3.1: For $0 \le t \le T$, let $Z_t \in \mathcal{L}(E, \mathcal{E}, \beta; S)$. The S-valued process $\{Z_t\}$ is said to be adapted to $\{Q_t\Phi\}$ if for all t,

$$Z_t = g_t(Q_t\Phi)$$

for some $g_t \in \mathcal{U}(E, \mathcal{E}, \beta; Q_t\Phi, S)$.

<u>Definition 3.2</u>: Let $\{Z_t\}$ be an S-valued process adapted to $\{Q_t\Phi\}$. $\{Z_t\}$ is called a Markov process w.r.t. $\{Q_t\phi\}$ if for all $0 \leq s < t \leq T$, $f \in C_b(S)$, we have

$$E_\beta(f(Z_t)|Q_s\Phi) = g(Z_s)$$

for a suitable function g.

We now return to the setup of the previous section. We will freely use the notation established there. In addition, we assume that the signal process satisfies the condition (1.63) and that S is a Polish space. Denote by $\mathcal{M}(S)$ the collection of all finite positive measures on $(S, \mathcal{B}(S))$.

Let (L_0,Π_0) be a fixed measurable representation of m with representation space $(\Omega_0,\mathcal{A}_0)$ and R_m be the corresponding lifting. Let $(\rho,L,\tilde{\Pi})$ be defined by (V.2.3-V.2.5). As seen in Section V.2, $(\rho,L,\tilde{\Pi})$ is a representation of $\alpha = \Pi\odot m$ with representation space $(\tilde{\Omega},\tilde{\mathcal{A}})$. Here

$$(\tilde{\Omega},\tilde{\mathcal{A}},\tilde{\Pi}) = (\Omega,\mathcal{A},\Pi) \otimes (\Omega_0,\mathcal{A}_0,\Pi_0)$$

and for $\tilde{\omega} = (\omega,\omega_0) \in \tilde{\Omega}$, $h \in H$,

$$\rho(\tilde{\omega}) := \omega$$

$$L(h)(\tilde{\omega}) := L_0(h)(\omega_0).$$

Let R_α be the α-lifting corresponding to $(\rho,L,\tilde{\Pi})$.

We assume that for all $\underline{X} \in D$, $\int_0^T|h_u(\underline{X}_u)|^2du < \infty$ (so that analogue of (2.2) holds for all $P_{s,z}$).

For $0 \leq s < t \leq T$, $\phi \in H$, $\underline{X} \in D$ let

$$M_t^s(\underline{X},\phi) = \exp(\int_s^t(h_u(\underline{X}_u),u)_{\mathcal{H}}du - \tfrac{1}{2}\int_s^t|h_u(\underline{X}_u)|_{\mathcal{H}}^2du). \quad (3.26)$$

For $0 \leq s < t \leq T$, $\phi \in H$, $z \in S$, let $\overline{\Gamma}^z(\phi)(\cdot) \in \mathcal{M}(S)$ be defined by

$$\overline{\Gamma}^z(\phi)(A) := E_{P_{s,z}}[1_A(\underline{X}_t)M_t^s(\underline{X},\phi)], \quad A \in \mathcal{S} \tag{3.27}$$

and for $\mu \in \mathcal{M}(S)$,

$$\overline{\Gamma}^\mu(\phi)(A) := \int \overline{\Gamma}^z(\phi)(A)d\mu(z), \quad A \in \mathcal{S}. \tag{3.28}$$

Then $\overline{\Gamma}^\mu(\phi)$ also belongs to $\mathcal{M}(S)$. Note that

$$\Gamma_t(\phi)(A) = E_\Pi[1_A(X_t)M_t^0(X(\cdot),\phi)]. \tag{3.29}$$

Let $\overline{F}^z(\phi)$, $\overline{F}^\mu(\phi)$ be defined by

$$\overline{F}^z(\phi)(A) = \frac{1}{\overline{\Gamma}^z(\phi)(S)} \cdot \overline{\Gamma}^z(\phi)(A) \tag{3.30}$$

and

$$\overline{F}^\mu(\phi)(A) = \frac{1}{\overline{\Gamma}^\mu(\phi)(S)} \cdot \overline{\Gamma}^\mu(\phi)(A), \ A \in \mathcal{S} \tag{3.31}$$

so that $\overline{F}^z(\phi)$, $\overline{F}^\mu(\phi) \in \mathcal{M}(S)$.

We have suppressed the variables s,t in the notation $\overline{\Gamma}^z$, $\overline{\Gamma}^\mu$, $\overline{F}^z$, $\overline{F}^\mu$, but it will be clear from the context. Indeed, we will fix $0 \leq s < t \leq T$ in the arguments leading to the proof of Markov property, where these quantities will make their appearance.

Before we can prove that $\{F_t\}$, $\{\Gamma_t\}$ are Markov processes, we have to prove that these are random variables on $(H,\mathcal{C},m)$ and $\{F_t(y)\}$, $\{\Gamma_t(y)\}$ are random variables on $(E,\mathcal{E},\alpha)$. Now Γ_t, F_t are $\mathcal{M}(S)$ valued functions. We can define weak convergence topology on $\mathcal{M}(S)$ as follows.

$$\mu_k \to \mu \text{ in } \mathcal{M}(S) \text{ if } \int f d\mu_k \to \int f d\mu \text{ for all } f \in C_b(S). \tag{3.32}$$

It is known that under this topology, $\mathcal{M}(S)$ is itself a Polish space. As in the case of $\mathcal{M}_0(S)$, we can define a metric d_0 on $\mathcal{M}(S)$ by

$$d_0(\mu_1,\mu_2) = \sum_{j=0}^{\infty} \frac{1}{2^{j+1}} [|\int f_j d\mu_1 - \int f_j d\mu_2| \wedge 1],$$

where $f_0 = 1$ and $\{f_j: j\geq 1\}$ is as in the definition of d_0 on $\mathcal{M}_0(S)$ in section I.2. Here again, $\mathcal{M}(S)$ is not complete under d_0.

The Borel σ-field on $\mathcal{M}(S)$ (for d_0-topology) is the smallest σ-field with respect to which the mappings $\mu \to \int f d\mu$ are measurable for all $f \in C_b(S)$. Thus, a $\mathcal{M}(S)$ valued mapping V (on some measurable space) is measurable if and only if the mappings $\langle f,V\rangle$ are measurable for all $f \in C_b(S)$.

Before we proceed, note that if $|f| \leq 1$

$$|\int f d\mu_1 - \int f d\mu_2| \leq 2 \sup_{A\in\mathcal{S}} |\mu_1(A) - \mu_2(A)|$$

and hence

$$d_0(\mu_1,\mu_2) \leq 2 \sup_{A\in\mathcal{S}} |\mu_1(A) - \mu_2(A)|.$$

<u>Theorem 3.2</u>:

(i) Γ_t, $F_t \in \mathcal{L}^*(H,\mathcal{C},m;\mathcal{M}(S))$ and $R_m(\Gamma_t)(S) > 0$ a.s.

(ii) $\Gamma_t,F_t \in \mathcal{L}^*(H,\mathcal{C},n;\mathcal{M}(S))$

(iii) $\Gamma_t(y)$, $F_t(y) \in \mathcal{L}^*(E,\mathcal{E},\alpha;\mathcal{M}(S))$

(iv) $\Gamma_t,F_t \in \mathcal{U}(E,\mathcal{E},\alpha;y,\mathcal{M}(S))$.

<u>Proof</u>: (i) Fix $P_i \underset{s}{\to} I$. Since

$$\Gamma_t \circ P_i(\phi)(A) = \int 1_A(X_t(\omega)) \exp((Q_t\xi(\omega),P_i\phi)-\tfrac{1}{2}|Q_t\xi(\omega)|^2)d\Pi(\omega) \tag{3.33}$$

$$= \int 1_A(X_t(\omega)) \exp((P_iQ_t\xi(\omega),\phi)-\tfrac{1}{2}|Q_t\xi(\omega)|^2 d\Pi(\omega)$$

it follows from the definition of lifting for cylinder functions that

$$R_m(\Gamma_t \circ P_i)(\omega_0) = V_i(\omega_0) \tag{3.34}$$

where for $\omega_0 \in \Omega_0$, $V_i(\cdot)(\omega_0) \in \mathcal{M}(S)$ is defined by

$$V_i(\omega_0)(A) := \int 1_A(X_t(\omega)) \exp(L_0(P_iQ_t\xi(\omega))(\omega_0) - \tfrac{1}{2}|Q_t\xi(\omega)|^2)d\Pi(\omega) \quad \text{for } A \in \mathcal{S}. \tag{3.35}$$

See the beginning of the proof of Theorem VI.2.2 for details of a similar assertion, namely (VI.2.12).

Let $V(\omega_0) \in \mathcal{M}(S)$ be defined by, for $A \in \mathcal{S}$

$$V(\omega_0)(A) := \int 1_A(X_t(\omega) \exp(L_0(Q_t\xi(\omega))(\omega_0) - \tfrac{1}{2}|Q_t\xi(\omega)|^2)d\Pi(\omega). \tag{3.36}$$

It is easy to see (using Fubini's theorem) that $\omega_0 \to \langle f,V_i(\omega_0)\rangle$, $\omega_0 \to \langle f,V(\omega_0)\rangle$ are measurable mappings on $(\Omega_0,\mathcal{A}_0)$ and hence it follows that V, V_i (for $i \geq 1$) are $\mathcal{M}(S)$ valued measuable mappings on $(\Omega_0,\mathcal{A}_0)$ and that $\omega_0 \to d_0(V_i(\omega_0),V(\omega_0))$ is also $\mathcal{A}_0$-measurable.

$$d_0(V_i(\omega_0),V(\omega_0)) \leq 2 \sup_{A\in\mathcal{S}} |V_i(\omega_0)(A) - V(\omega_0)(A)| \tag{3.37}$$

$$\leq 2\int \Big|\exp(L_0(P_iQ_t\xi(\omega))(\omega_0)-\tfrac{1}{2}|Q_t\xi(\omega)|^2) - \exp(L_0(Q_t\xi(\omega))(\omega_0)-\tfrac{1}{2}|Q_t\xi(\omega)|^2)\Big| d\Pi(\omega)$$

$$= 2\int \Big|\exp(L_0(P_ik)(\omega_0)-\tfrac{1}{2}|k|^2) - \exp(L_0(k)-\tfrac{1}{2}|k|^2)\Big| d\nu(k)$$

$$= U_i(\omega_0) \quad \text{(say)}$$

where $\nu = \Pi \circ (Q_t \xi(\cdot))^{-1} \in \mathcal{M}(H)$. If $Z_i(k,\omega_0)$, $Z(k,\omega_0)$ are defined by (VI.2.13) and (VI.2.14) respectively, then

$$U_i(\omega_0) = 2\int |Z_i(k,\omega_0) - Z(k,\omega_0)| d\nu(k). \tag{3.38}$$

As seen in the proof of Theorem VI.2.2, Z_i, Z satisfy the condition of Proposition VI.2.1 with $(X_1, \mathcal{D}_1, \lambda_1) = (H, \mathcal{B}(H), \nu)$ and $(X_2, \mathcal{D}_2, \lambda_2) = (\Omega_0, \mathcal{A}_0, \Pi_0)$. Hence by Proposition VI.2.1, we have

$$U_i \to 0 \quad \text{in } \mathcal{L}^1(\Omega_0, \mathcal{A}_0, \Pi_0). \tag{3.39}$$

The relations (3.34), (3.37) and (3.39) give

$$d_0(R_m(\Gamma_t \circ P_i), V) \to 0 \quad \text{in } \Pi_0\text{-probability}. \tag{3.40}$$

Since $\{P_i\} \subseteq \mathcal{P}$, $P_i \underset{s}{\to} I$ is arbitrary, it follows that

$$\Gamma_t \in \mathcal{L}^*(H, \mathcal{C}, m; \mathcal{M}(S)) \tag{3.41}$$

and

$$R_m(\Gamma_t) = V. \tag{3.42}$$

Note that $V(\omega_0)(S) > 0$ for all $w_0 \in \Omega_0$ and hence that $\Pi_0(R_m(\Gamma_t)(\omega_0) \in \mathcal{M}^*(S)) = 1$ where $\mathcal{M}^*(S)$ is the class of all $\mu \in \mathcal{M}(S)$, with $\mu(S) > 0$. Since

$$\mathcal{M}^*(S) = \{\mu: \langle 1,\mu \rangle > 0\}$$

and the mapping $\mu \to \langle 1,\mu \rangle$ is continuous, it follows that $\mathcal{M}^*(S)$ is a Borel subset of $\mathcal{M}(S)$. Clearly, $\Gamma_t(\phi) \in \mathcal{M}^*(S)$. Let

$$\theta: \mathcal{M}^*(S) \to \mathcal{M}(S) \tag{3.43}$$

be defined by

$$\theta(\mu) = \frac{1}{\mu(S)} \cdot \mu, \quad \mu \in \mathcal{M}^*(S). \tag{3.44}$$

Then θ is continuous and $F_t = \theta(\Gamma_t)$. Hence by Theorem V.3.4, and (3.41), (3.42), we conclude

$$F_t \in \mathcal{L}^*(H, \mathcal{C}, m; \mathcal{M}(S)) \tag{3.45}$$

and

$$R_m(F_t) = \theta(V) = \frac{1}{V(S)} \cdot V. \tag{3.46}$$

This proves (i)

For (ii), first recall that $n \ll m$ (and $m \ll n$) (see Theorem VI.3.1). Observe that though in Theorem IV.1.1, the statement

$$g \in \mathcal{L}^*(H, \mathcal{C}, m; S_1) \text{ implies } g \in \mathcal{L}^*(H, \mathcal{C}, n; S_1) \tag{3.47}$$

has been proved only for the case $S_1 = \mathbb{R}$, the case of a general Polish space S_1 can be proved along the same lines. Thus Part (i) gives

$$\Gamma_t, F_t \in \mathcal{L}^*(H, \mathcal{C}, m; \mathcal{M}(S)). \tag{3.48}$$

Further, if Π_0' on $(\Omega_0, \mathcal{A}_0)$ is defined by $d\Pi_0' = R_m(\frac{dn}{dm}) \cdot d\Pi_0$, then (L_0, Π_0') is a representation of n, and if R_n' is the corresponding lifting, then

$$R_n'(\Gamma_t) = R_m(\Gamma_t) \tag{3.49}$$

and

$$R_n'(F_t) = R_m'(F_t) = \frac{1}{V(S)} \cdot V. \tag{3.50}$$

Let $(L_1,\tilde{\Pi})$ be the representation of n induced by the QCM y (corresponding to the representation $(\rho,L,\tilde{\Pi})$ of α). As seen in the proof of Theorem VI.3.2,

$$L_1(k)(\tilde{\omega}) = (\xi(\omega),k) + L_0(k)(\omega_0); \ \tilde{\omega} = (\omega,\omega_0)\in\tilde{\Omega}. \tag{3.51}$$

If R_n is the n-lifting corresponding to $(L_1,\tilde{\Pi})$, then (3.49) and arguments given in Theorem VI.2.5 yield

$$R_n(\Gamma_t) = W \tag{3.52}$$

where (3.53)

$$W(A)(\tilde{\omega}) = \int 1_A(X_t(\omega'))\exp(L_1(Q_t\xi(\omega'))(\tilde{\omega})-\tfrac{1}{2}|Q_t\xi(\omega')|^2)d\Pi(\omega'),$$

for $A \in \mathcal{S}$, $\tilde{\omega} \in \tilde{\Omega}$. Also

$$R_n(F_t) = \frac{1}{W(S)} \cdot W. \tag{3.54}$$

Coming to the proof of part (iv), let $f(\omega,h) = [\Gamma_t \circ y](\omega,h)$. Then for $\{P_i\} \subseteq \mathcal{P}$, $P_i \xrightarrow{s} I$,

$$\begin{aligned} f_{P_i}(\omega,h) &= \int 1_A(X_t(\omega'))\exp((Q_t\xi(\omega'),y(\omega,P_ih)) \\ &\qquad - \tfrac{1}{2}|Q_t\xi(\omega')|^2)d\Pi(\omega') \\ &= \int 1_A(X_t(\omega'))\exp((Q_t\xi(\omega'),\xi(\omega)) + (Q_t\xi(\omega'),P_ih) \\ &\qquad - \tfrac{1}{2}|Q_t\xi(\omega')|^2)d\Pi(\omega'). \end{aligned}$$

Hence

$$R_\alpha([\Gamma_t \circ y]_{P_i}) = W_i \tag{3.55}$$

where

$$W_i(\tilde{\omega})(A) = \int 1_A(X_t(\omega'))\exp((Q_t\xi(\omega'),\xi(\omega)) + L_0(P_iQ_t\xi(\omega'))(\omega_0) - \tfrac{1}{2}|Q_t\xi(\omega')|^2)d\Pi(\omega') \tag{3.56}$$

for $\tilde{\omega} = (\omega,\omega_0) \in \tilde{\Omega} = \Omega\times\Omega_0$. Using (3.51), W can be expressed as

$$W(\tilde{\omega})(A) = \int 1_A(X_t(\omega'))\exp((Q_t\xi(\omega'),\xi(\omega)) + L_0(Q_t\xi(\omega'))(\omega_0) - \tfrac{1}{2}|Q_t\xi(\omega')|^2)d\Pi(\omega'). \tag{3.57}$$

From (3.56), (3.57), it follows that, for $\tilde{\omega} = (\omega,\omega_0) \in \tilde{\Omega}$,

$$\begin{aligned} d_0(W_i(\tilde{\omega}),W(\tilde{\omega})) &\le 2\int|\exp((\xi(\omega),k) + L_0(P_ik)(\omega_0) - \tfrac{1}{2}|k|^2) \\ &\quad - \exp((\xi(\omega),k) + L_0(k)(\omega_0) - \tfrac{1}{2}|k|^2)|d\upsilon(k) \\ &= 2\int|Z_i(k,\tilde{\omega})-Z(k,\tilde{\omega})|d\upsilon(k) = U_i'(\tilde{\omega}) \quad \text{(say)} \end{aligned} \tag{3.58}$$

where $\upsilon = \Pi\circ[Q_t\xi]^{-1}$ and Z_i, Z are defined by (VI.3.11), (VI.3.12). We have seen in the course of proof of Theorem VI.3.2 that $\{Z_i\}$, Z satisfy the conditions of proposition VI.2.1, where $(X_1,\mathcal{D}_1,\lambda_1) = (H,\mathcal{B}(H),\upsilon)$ and $(X_2,\mathcal{D}_2,\lambda_2) = (\tilde{\Omega},\tilde{\mathcal{A}},\Pi')$, Π' being a probability measure such that $\Pi' \ll \tilde{\Pi}$. Thus, we have

$$U_i' \to 0 \quad \text{in } \Pi'\text{-probability}$$

which gives

$$U_i' \to 0 \quad \text{in } \tilde{\Pi}\text{-probability.} \tag{3.59}$$

Hence,

$$d_0([R_\alpha(\Gamma_t \circ y)]_{P_i}, W) \to 0 \quad \text{in } \tilde{\Pi}\text{-probability} \tag{3.60}$$

for all $\{P_i\} \subseteq \mathcal{P}$, $P_i \underset{s}{\to} I$. Thus

$$\Gamma_t \circ y \in \mathcal{L}^*(E, \mathcal{E}, \alpha; \mathcal{M}(S)) \tag{3.61}$$

and

$$R_\alpha(\Gamma_t \circ y) = W. \tag{3.62}$$

Also, as seen earlier in the proof of part (i),

$$F_t \circ y = \theta(\Gamma_t \circ y) \in \mathcal{L}^*(E, \mathcal{E}, \alpha; \mathcal{M}(S)) \tag{3.63}$$

with

$$\begin{aligned} R_\alpha(F_t \circ y) &= \theta(W) \\ &= \frac{1}{W(S)} \cdot W. \end{aligned} \tag{3.64}$$

The relations (3.52), (3.54), (3.63) and (3.64) imply that $R_\alpha(\Gamma_t \circ y) = R_n(\Gamma_t)$; $R_\alpha(F_t \circ y) = R_n(F_t)$ completing the proof of part (iv) as well. □

Proceeding exactly as in Theorem 3.2, we can prove analogous statements for $\overline{\Gamma}^\mu$, $\overline{F}^\mu$ (for any $\mu \in \mathcal{M}^*(S)$), i.e., $\overline{\Gamma}^\mu$, $\overline{F}^\mu$ are accessible random variables on $(H, \mathcal{C}, m)$ and $\overline{\Gamma}^\mu(y)$, $\overline{F}^\mu(y)$ are accessible random variables on $(E, \mathcal{E}, \alpha)$. Later we need only the analogue of (i) for $\overline{\Gamma}^\mu$, which we state as a Theorem for future reference. Let Q_t^s be the orthogonal projection onto H_t^s (as defined at the beginning of this section).

Theorem 3.3: For $\mu \in \mathcal{M}^*(S)$, $\bar{\Gamma}^\mu \in \mathcal{L}^*(H,\mathcal{C},m;\mathcal{M}(S))$ and

$$R_m(\bar{\Gamma}^\mu)(\omega_0)(A) = \int_S[\int_D 1_A(\underline{X}_t) \exp(L_0(Q_t^s\bar{\xi}) - \tfrac{1}{2}|Q_t^s\bar{\xi}|^2)d\Pi_{s,z}]d\mu(z) \tag{3.65}$$

where $\bar{\xi}$: $D \to H$ is given by $\bar{\xi}(\underline{X}) = (h_u(\underline{X}_u)\colon 0 \leq u \leq T)$.

As a consequence of Theorem 3.2, it follows that if g is a continuous function from $\mathcal{M}(S)$ into $\mathbb{R}$, then $g(\Gamma_t) \in \mathcal{L}^*(H,\mathcal{C},m)$; $g(\Gamma_t) \in \mathcal{L}^*(H,\mathcal{C},n)$ and $g(\Gamma_t) \in \mathcal{U}(E,\mathcal{E},\alpha;y)$. However we will later need these properties for a larger class of functions g, which are continuous for a stronger topology on $\mathcal{M}(S)$ - the norm topology - which is induced by the metric

$$d_1(\mu_1,\mu_2) = |\mu_1-\mu_2|(S)$$

where for any signed measure μ, $|\mu|$ denotes the 'total variation measure' defined by

$$|\mu|(B) = \sup_{\substack{A\in\mathcal{S} \\ A\subseteq B}} \{|\mu(A)| + |\mu(B-A)|\}.$$

For any measurable function f on S, we have

$$|\int f d\mu| \leq \int |f| d|\mu|$$

and also

$$d_1(\mu_1,\mu_2) \leq 2 \sup_{\substack{A\in\mathcal{S} \\ A\subseteq S}} |\mu_1(A) - \mu_2(A)|.$$

Under the metric d_1, $\mathcal{M}(S)$ is not separable except when S is countable. Thus when dealing with $\mathcal{M}(S)$ valued random variables, we consider only the d_0-topology on $\mathcal{M}(S)$. Given $\mathcal{M}(S)$ valued random variables, say V_i, V,

$$\omega_0 \to d_1(V_i(\omega_0), V(\omega_0)$$

may not be measurable (as d_1 involves a supremum over possibly uncountably many variables). This poses some problems as we cannot talk of convergence of V_i to V in probability in d_1-topology in $\mathcal{M}(S)$: the sets

$$\{\omega_0 : d_1(V_i(\omega_0), V(\omega_0) > \epsilon\}$$

may not belong to $\mathcal{A}_0$. We will get over this difficulty, as will be seen in the proof below.

In the sequel, unless otherwise mentioned, the terms continuous, Borel measurable for functions on S *will always refer to the weak convergence topology (induced by the metric* d_0*).*

Theorem 3.3: Let $g: \mathcal{M}(S) \to \mathbb{R}$ be a Borel measurable function. Suppose that g is continuous in the d_1-topology. Then we have

(i) $g(\Gamma_t) \in \mathcal{L}^*(H, \mathcal{C}, m)$

(ii) $g(\Gamma_t) \in \mathcal{L}^*(H, \mathcal{C}, n)$

(iii) $g(\Gamma_t(y)) \in \mathcal{L}^*(E, \mathcal{E}, \alpha)$

(iv) $g(\Gamma_t) \in \mathcal{U}(E, \mathcal{E}, \alpha; y)$.

Proof: Fix $P_i \xrightarrow{s} I$. Let V_i, V be given by (3.35), (3.36). Since g is Borel measurable, we have

$$R_m(g(\Gamma_t \circ P_i) = g(R_m(\Gamma_t \circ P_i)) = g(V_i).$$

Further, we have seen in (3.37) that

$$d_1(V_i(\omega_0), V(\omega_0) \leq U_i(\omega_0) \tag{3.66}$$

where $U_i(\omega_0) \to 0$ in Π_0-probability. If $d_1(V_i, V)$ were measurable, this would give that $V_i \to V$ (d_1 topology) in probability and since g is d_1-continuous, $g(V_i) \to g(V)$ in probability. We will prove this latter statement.

Suppose $g(V_i)$ does not converge to $g(V)$ in Π_0-probability. Then for some $\epsilon > 0$, $\delta > 0$,

$$\Pi_0(|g(V_i)-g(V)| > \epsilon) > \delta \quad \text{for infinitely many } i\geq 1.$$

Thus choose a subsequence $\{i_k)$ such that

$$\Pi_0(|g(V_{i_k})-g(V)| > \epsilon) > \delta, \quad k\geq 1. \tag{3.67}$$

Now $U_{i_k} \to 0$ in Π_0-probability and hence there exists a subsequence $\{i_{k_j}\}$ of $\{i_k\}$ such that $U_{i_{k_j}} \to 0$ a.s. Π_0. The inequality (3.65) gives

$$d_1(V_{i_{k_j}}, V) \to 0 \quad \text{a.s. } \Pi_0$$

Since g is d_1-continuous, this implies

$$g(V_{i_{k_j}}) \to g(V) \quad \text{a.s. } \Pi_0$$

which contradicts (3.67) as $\{i_{k_j}\}$ is a subsequence of $(i_k\}$. Hence

$$g(V_i) \to g(V) \quad \text{in } \Pi_0\text{-probability.}$$

We have shown that for all $P_i \xrightarrow{s} I$,

$$R_m(g\circ\Gamma_t\circ P_i) \to g(V) \quad \text{in } \Pi_0\text{-probability.}$$

Thus, $g(\Gamma_t) \in \mathcal{L}^*(H,\mathcal{C},m)$ and

$$R_m(g(\Gamma_t) = g(V).$$

This proves (i). The second part follows from (i) as in Theorem 3.2 and also that

$$R_n(g(\Gamma_t\circ y)) = g(W) \tag{3.68}$$

where W is given by (3.53). For (iii), note that

$$R_\alpha([g(\Gamma_t\circ y)]_{P_i}) = g(W_i)$$

(W_i defined by (3.56)) and, as in (3.58),

$$d_1(W_i(\tilde{\omega}),W(\tilde{\omega})) \le U_i'(\tilde{\omega})$$

with $U_i' \to 0$ in $\tilde{\Pi}$-probability. Arguments given in part (i) above now yield

$$g(W_i) \to g(W) \quad \text{in } \tilde{\Pi}\text{-probability}$$

and hence that $g(\Gamma_t\circ y) \in \mathcal{L}^*(E,\mathcal{E},\alpha)$,

$$R_\alpha(g(\Gamma_t\circ y)) = g(W). \tag{3.69}$$

Finally (3.68), (3.69) imply the remaining assertion, namely

$$g(\Gamma_t) \in \mathcal{U}(E,\mathcal{E},\alpha;y). \qquad \square$$

The following lemma accomplishes what was promised in step (i) at the beginning of this section. From now on we fix $0 \le s < t \le T$.

Lemma 3.5: For all $\phi \in H$, we have

$$\Gamma_t(\phi) = \bar{\Gamma}^{\mu}(\phi) \quad \text{for } \mu = \Gamma_s(\phi).$$

Proof: We have, for $A \in \mathcal{S}$ arbitrary

$$\Gamma_t(\phi)(A) = E_{\Pi}[1_A(X_t)\cdot M_t^0(X(\cdot),\phi)]$$

$$= E_{\Pi}[1_A(X_t)\cdot M_t^s(X(\cdot),\phi)\cdot M_s^0(X(\cdot),\phi)]$$

$$= E_{\Pi}[1_A(X_t)M_t^s(X(\cdot),\phi)\cdot M_s^0(X(\cdot),\phi)\,|\mathcal{F}_s^X)]$$

where we have used $M_t^0 = M_s^0\cdot M_t^s$. Note that $M_s^0(X(\cdot),\phi)$ is $\mathcal{F}_s^X$ measurable and hence

$$\Gamma_t(\phi)(A) = E_{\Pi}[M_s^0(X(\cdot),\phi)\cdot E_{\Pi}[1_A(X_t)M_t^s(X(\cdot),\phi)\,|\mathcal{F}_s^X]]$$

$$= E_{\Pi}[M_s^0(X(\cdot),\phi)\cdot E_{P_{s,X_s}}[1_A(\underline{X}_t)\cdot M_t^s(\underline{X},\phi)]]$$

by (1.67) since $1_A(\underline{X}_t)\cdot M_t^s(\underline{X},\phi)$ is $\mathcal{A}_\infty^s$ measurable. Thus

$$\Gamma_t(\phi)(A) = E_{\Pi}[g(X_s)\cdot M_s^0(X(\cdot),\phi)] \tag{3.70}$$

where

$$g(z) = E_{P_{s,z}}[1_A(\underline{X}_t)M_t^s(\underline{X},\phi)] \tag{3.71}$$

$$= \bar{\Gamma}^z(\phi)(A).$$

The definition of $\Gamma_s(\phi)$ yields

$$E_{\Pi}[g(X_s)M_s^0(X(\cdot),\phi)] = \langle g,\Gamma_s(\phi)\rangle = \langle g,\mu\rangle, \tag{3.72}$$

for $\mu = \Gamma_s(\phi)$. Thus we have from (3.70), (3.71), (3.72),

$$\Gamma_t(\phi)(A) = \int \overline{\Gamma}^z(A)d\mu(z) \quad \text{for } \mu = \Gamma_s(\phi).$$

This proves the required assertion as $A \in \mathcal{S}$ is arbitrary.

□

In order to apply Theorem 3.1 to get Markov property of $\{\Gamma_t\}$ as outlined in step (ii) (at the beginning of this section), we need one more result, which is,

<u>Theorem 3.6</u>: Let $g: \mathcal{M}^*(S) \to \mathbb{R}$ be a continuous function such that for some constant $C < \infty$,

$$|g(\mu)| \leq C(1 + |\mu(S)|). \tag{3.73}$$

Let $g_1: \mathcal{M}^*(S) \to \mathbb{R}$ be defined by

$$g_1(\mu) = \int_H g(\overline{\Gamma}^\mu(\phi))dm(\phi). \tag{3.74}$$

Then

(i) g_1 is a Borel measurable function;

(ii) g_1 is continuous in d_1-topology.

<u>Proof</u>: Using (3.65), we can conclude that

$$R_m(\overline{\Gamma}^\mu)(\omega_0)(A) = \int_S R_m(\overline{\Gamma}^z)(\omega_0)(A)d\mu(z). \tag{3.75}$$

From the fact that L_0 is a measurable representation and the observation that $z \to \Pi_{s,z}(B)$ (for $B \in \mathcal{A}^s_\infty$) is $\mathcal{S}$ measurable, it follows that for $f \in C_b(S)$

$$(z,\omega_0) \to \langle f, R_m(\overline{\Gamma}^z)(\omega_0)\rangle$$

is $\mathcal{S}\otimes\mathcal{A}_0$ measurable. This implies

$$(\mu,\omega_0) \to \int_S \langle f, R_m(\vec{\Gamma}^z)(\omega_0)\rangle d\mu(z) \quad = \langle f, R_m(\vec{\Gamma}^{\mu})(\omega_0)\rangle$$

is $\mathcal{B}(\mathcal{M}(S)) \otimes \mathcal{A}_0$ measurable. Hence

$$(\mu,\omega_0) \to R_m(\vec{\Gamma}^{\mu})(\omega_0)$$

is $\mathcal{B}(\mathcal{M}(S))\otimes\mathcal{A}_0/\mathcal{B}(\mathcal{M}(S))$ measurable. Continuity of g gives

$$(\mu,\omega_0) \to g(R_m(\vec{\Gamma}^{\mu})(\omega_0)) = R_m(g(\vec{\Gamma}^{\mu}))(\omega_0) \tag{3.76}$$

is $\mathcal{B}(\mathcal{M}(S)) \otimes \mathcal{A}_0$ measurable. We will later prove that

$$E_{\Pi_0}|R_m(g(\vec{\Gamma}^{\mu}))| < \infty. \tag{3.77}$$

Then the measurability of the mapping in (3.76) and Fubini's theorem yield

$$\mu \to E_{\Pi_0}[R_m(g(\vec{\Gamma}^{\mu}))] = \int g(\vec{\Gamma}^{\mu})dm = g_1(\mu)$$

is $\mathcal{B}(\mathcal{M}(S))$ measurable. This proves (i).

For (ii), note that for $\mu_i \to \mu$ in d_1-topology,

$$d_0(R_m(\vec{\Gamma}^{\mu_i})(\omega_0), R_m(\vec{\Gamma}^{\mu})(\omega_0)) \tag{3.78}$$

$$\leq 2 \sup_{A\in\mathcal{S}} |R_m(\vec{\Gamma}^{\mu_i})(\omega_0)(A) - R_m(\vec{\Gamma}^{\mu})(\omega_0)(A)|$$

$$\leq 2 \sup_{A\in\mathcal{S}} |\int_S R_m(\vec{\Gamma}^z)(\omega_0)(A)[d\mu_i(z) - d\mu(z)]|$$

$$\leq 2 \int R_m(\vec{\Gamma}^z)(\omega_0)(S)d|\mu_i-\mu|(z).$$

Note that

$$E_{\Pi_0} R_m(\bar{\Gamma}^z)(\cdot)(S) = E_{\Pi_0} \int \exp(L_0(Q_t^s\bar{\xi}) - \tfrac{1}{2}|Q_t^s\bar{\xi}|^2) d\Pi_{s,z}$$

$$= \int E_{\Pi_0} \exp(L_0(Q_t^s\bar{\xi}) - \tfrac{1}{2}|Q_t^s\bar{\xi}|^2) d\Pi_{s,z}$$

$$= \int d\Pi_{s,z} = 1$$

and hence

$$E_{\Pi_0} |R_m(\bar{\Gamma}^\mu)(S)| \le \int E_{\Pi_0}[R_m(\bar{\Gamma}^z)(S)] d|\mu|(z) = |\mu|(S)$$

and

$$\int_{\Omega_0} \int_S R_m(\bar{\Gamma}^z)(\omega_0)(S) d|\mu_i - \mu|(z) d\Pi_0(\omega_0) \tag{3.79}$$

$$= |\mu_i - \mu|(S) \qquad \to 0 \text{ as } i \to \infty.$$

The relations (3.78) and (3.79) imply

$$d_0(R_m(\bar{\Gamma}^{\mu_i}), R_m(\bar{\Gamma}^\mu)) \to 0 \quad \text{in } \Pi_0\text{-probability}$$

and continuity of g gives

$$g(R_m(\bar{\Gamma}^{\mu_i})) \longrightarrow g(R_m(\bar{\Gamma}^\mu)) \quad \text{in } \Pi_0\text{-probability.}$$

which is the same as

$$R_m(g(\bar{\Gamma}^{\mu_i})) \longrightarrow R_m(g(\bar{\Gamma}^\mu)) \quad \text{in } \Pi_0\text{-probability.} \tag{3.80}$$

Also

$$|R_m(\bar{\Gamma}^{\mu_i})(\omega_0)(S) - R_m(\bar{\Gamma}^\mu)(\omega_0)(S)|$$

$$\leq \int R_m(\overline{\Gamma}^z)(\omega_0)(S)d|\mu_i-\mu|(z)$$

and hence in view of (3.79), we have

$$\{R_m(\overline{\Gamma}^{\mu_i})(\cdot)(S): i\geq 1\} \text{ is uniformly integrable.} \tag{3.81}$$

The observation (3.81) and the assumption (3.73) imply that

$$\{R_m(g(\overline{\Gamma}^{\mu_i})): i\geq 1\} \text{ is uniformly integrable.}$$

This and (3.80) finally yield

$$g_1(\mu_i) = \int g(\overline{\Gamma}^{\mu_i})dm = \int R_m(g(\overline{\Gamma}^{\mu_i}))d\Pi_0$$

$$\to \int R_m(g(\overline{\Gamma}^{\mu}))d\Pi_0 = g_1(\mu).$$

This proves the continuity of g_1 in the metric d_1. □

We are in a position now to prove that $\{\Gamma_t\}$ is a Markov process on $(H,\mathscr{C},m)$.

<u>Theorem 3.7</u>: Let Φ be the identity mapping on H. Then

(i) $\{\Gamma_t\}$ is an $\mathscr{M}(S)$ valued Markov process with respect to $\{Q_t\Phi\}$.

(ii) If g is a continuous function from $\mathscr{M}^*(S)$ into $\mathbb{R}$ satisfying (3.73), then for $0 \leq s < t \leq T$, we have

$$E_m[g(\Gamma_t)|Q_s\Phi] = g_1(\Gamma_s) \tag{3.82}$$

where for $\mu \in \mathscr{M}^*(S)$,

$$g_1(\mu) = \int_H g((\overline{\Gamma}^{\mu})(\phi))dm(\phi). \tag{3.83}$$

[In (3.83), $\bar{\Gamma}^{\mu}$ is defined by (3.27), (3.28) for the pair $s < t$ appearing in (3.82).]

Proof: We have seen in Theorem 3.2 that

$$\Gamma_t \in \mathcal{L}^*(H,\mathcal{C},m;\mathcal{M}(S)).$$

It is easy to check that $\Gamma_t(\phi) = \Gamma_t(Q_t\phi)$ and hence if Γ'_t is the restriction of Γ_t to H_t, then

$$\Gamma_t(\phi) = \Gamma'_t(Q_t\phi) \quad \text{for all } \phi \in H.$$

Hence by Proposition IV.2.3, $\Gamma'_t \in \mathcal{U}(Q_t)$. Thus $\{\Gamma_t\}$ is an $\mathcal{M}(S)$ valued process on $(H,\mathcal{C},m)$ adapted to $Q_t\Phi$.

Now (i) will follow once we prove (ii), indeed for (i) we only need the validity of (3.82) for bounded continuous functions g.

Thus fix g as in (ii) in the statement of the theorem. Define $f: H \to \mathbb{R}$ by

$$f(\phi) = g(\Gamma_t(\phi)).$$

Let $f_1: H\times H \to \mathbb{R}$ be defined by, for $\phi_1,\phi_2 \in H$

$$f_1(\phi_1,\phi_2) = f(Q_s^0\phi_1 + Q_T^s\phi_2).$$

Then using that $\bar{\Gamma}^{\mu}(Q_s^0\phi_1 + Q_T^s\phi_2) = \bar{\Gamma}^{\mu}(Q_T^s\phi_2) = \bar{\Gamma}^{\mu}(\phi_2)$, $\Gamma_s(Q_s^0\phi_1 + Q_T^s\phi_2) = \Gamma_s(Q_s^0\phi_1) = \Gamma_s(\phi_1)$ and Lemma 3.5, it follows that

$$f_1(\phi_1,\phi_2) = g(\bar{\Gamma}^{\mu}(\phi_2)) \quad \text{for } \mu = \Gamma_s(\phi_1).$$

Thus the required relation (3.82) would follow from Theorem

3.1 once we check that the conditions of the same are satisfied for the function f defined above. So we will now verify that for f defined above, the conditions (3.1), (3.3) and (3.5) hold.

The continuity of g and part (i) of Theorem 3.2 give

$$f = g(\Gamma_t) \in \mathcal{L}^*(H,\mathcal{C},m).$$

To conclude that $f \in \mathcal{L}^{1*}(H,\mathcal{C},m)$ as required, note that

$$|f(\phi)| \le C(1 + \Gamma_t(\phi)(S)). \tag{3.84}$$

For $P_i \underset{s}{\rightarrow} I$, we have $R_m(f \circ P_i) \rightarrow R_m(f)$ in Π_0-probability (as $f \in \mathcal{L}^*(H,\mathcal{C},m)$). Further, as seen in Theorem 3.2,

$$R_m(\Gamma_t \circ P_i)(S) = V_i(S) \longrightarrow R_m(\Gamma_t)(S) \text{ in } \mathcal{L}^1(\Omega_0,\mathcal{A}_0,\Pi_0).$$

Hence $\{R_m(\Gamma_t \circ P_i)\}$ is uniformly integrable. The estimate (3.84) gives the uniform integrability of the sequence $\{R_m(f \circ P_i)\}$. Thus

$$R_m(f \circ P_i) \longrightarrow R_m(f) \quad \text{in } \mathcal{L}^1(\Omega_0,\mathcal{A}_0,\Pi_0).$$

This proves $f \in \mathcal{L}^{1*}(H,\mathcal{C},m)$.

Similarly, using Theorem 3.3, it follows that for $\mu \in \mathcal{M}^*(S)$, $g((\overline{\Gamma}^{\mu})) \in \mathcal{L}^{1*}(H,\mathcal{C},m)$. Hence for all $\phi_1 \in H$

$$f_1(\phi_1,\cdot) = g(\overline{\Gamma}^{\mu})(\cdot) \quad (\text{for } \mu = \Gamma_s(\phi_1))$$

$$\in \mathcal{L}^{1*}(H,\mathcal{C},m).$$

Thus we have verified (3.1) and (3.3). Finally,

$$f_2(\phi_1) := \int f(\phi_1,\phi_2)dm(\phi_2)$$

$$= \int g(\overline{\Gamma}^{\mu})(\phi_2)dm(\phi_2) \quad (\text{for } \mu = \Gamma_s(\phi_1))$$

$$= g_1(\Gamma_s(\phi_1))$$

where g_1 is defined by (3.83). By Theorem 3.6, g_1 is Borel measurable and continuous in d_1 topology. Consequently, by Theorem 3.4, $f_2 \in \mathscr{L}^*(H,\mathscr{C},m)$.

Thus we have verified all the conditions of Theorem 3.1 and this proves (ii). □

Remark 3.1: In Theorem 3.6, we cannot conclude that g_1 is continuous in d_0-topology without making further assumptions on the family $\{\Pi_{s,x}\}$ of conditional distributions of (X_t). Thus it becomes necessary to have Theorem 3.4 for d_1-continuous functions g. The reason why we have proved part (ii) above for g satisfying (3.73) instead of only bounded functions will become clear in the next result when we deduce the Markov property of $\{\Gamma_t\}$ on $(H,\mathscr{C},n)$.

We will deduce the Markov property of $\{\Gamma_t\}$ on $(H,\mathscr{C},n)$ from the same property on $(H,\mathscr{C},m)$ by using the form of the R-N derivative of n with respect to m. The following lemma will be useful.

Lemma 3.8: Let $0 < t_0 \leq T$ and let $f \in \mathscr{L}^*(H,\mathscr{C},n)$ and $f \in \mathscr{L}^1(H,\mathscr{C},n)$. Further suppose that $f(\phi) = f(Q_{t_0}\phi)$ for all $\phi \in H$. Then for all $C' \in \mathscr{C}_{t_0} = \mathscr{C}(H_{t_0})$, we have

$$\int 1_{C'}(Q_{t_0}\phi)f(\phi)dn(\phi) = \int 1_{C'}(Q_{t_0}\phi)f(\phi)\Gamma_{t_0}(\phi)(S)dm(\phi). \quad (3.85)$$

<u>Proof</u>: Since $f \in \mathcal{L}^*(H,\mathcal{C},n)$, $f' = f|_{H_{t_0}} \in \mathcal{U}(H,\mathcal{C},n;Q_{t_0})$ by Proposition IV.2.3. Thus, if $n' = n\circ[Q_t]^{-1}$

$$\int_H 1_{C'}(Q_{t_0}\phi)f(\phi)dn(\phi) = \int_{H_{t_0}} 1_{C'}(\phi')f'(\phi')dn'(\phi'). \quad (3.86)$$

Let Γ'_{t_0} be the restriction of Γ_{t_0} to H_{t_0}. Then from the definition of Γ_{t_0}, the fact that $n' = \alpha\circ(Q_{t_0}y)^{-1}$ and Theorem VI.2.4, it follows that $n' \ll m'$ with $\frac{dn'}{dm'} = \Gamma'_{t_0}$, where m' is the canonical Gauss measure on H_{t_0}. Thus, by Theorem IV.1.1, $f'\cdot\Gamma'_{t_0} \in \mathcal{L}^1(H_{t_0},\mathcal{C}_{t_0},m')$ and

$$\int_{H_{t_0}} 1_{C'}(\phi')f'(\phi')dn'(\phi') \quad (3.87)$$
$$= \int_{H_{t_0}} 1_{C'}(\phi')f'(\phi')\Gamma'_{t_0}(\phi')(S)dm'(\phi').$$

By Theorem VI.2.6, $m' = m\circ[Q_{t_0}]^{-1}$. Note that $\mathcal{L}^*(H,\mathcal{C},m) = \mathcal{L}^*(H,\mathcal{C},n)$ as $n \ll m$ and $m \ll n$. Thus $f \in \mathcal{L}^*(H,\mathcal{C},m)$. By Theorem 3.2, $\Gamma_{t_0}(S) = \langle 1,\Gamma_{t_0}\rangle \in \mathcal{L}^*(H,\mathcal{C},m)$. Hence $f\cdot\Gamma_{t_0} \in \mathcal{L}^*(H,\mathcal{C},m)$, so that its restriction $f'\cdot\Gamma'_{t_0} \in \mathcal{U}(H,\mathcal{C},m;Q_{t_0})$ by Proposition IV.2.3. This gives

$$\int_{H_{t_0}} 1_{C'}(\phi')f'(\phi')\Gamma'_{t_0}(\phi')(S)dm'(\phi') \quad (3.88)$$

$$= \int_H 1_C(Q_{t_0}\phi)f(\phi)\Gamma_{t_0}(\phi)(S)dm(\phi).$$

The assertion (3.85) follows from (3.86), (3.87) and (3.88).

□

<u>Theorem 3.9</u>: $\{\Gamma_t\}$, $\{F_t\}$ are $\mathcal{M}(S)$ valued Markov processes on $(H,\mathcal{C},n)$ with respect to $\{Q_t\Phi\}$.

<u>Proof</u>: We have seen in Theorem 3.8 that $\{\Gamma_t\}$ is adapted to $\{Q_t\Phi\}$. Similarly, it can be proved that $\{F_t\}$ is adapted to $\{Q_t\Phi\}$.

Let g_0 be a bounded continuous function on $\mathcal{M}^*(S)$, let $0 \leq s < t \leq T$. Let $C \in \mathcal{C}(H_s)$. Then noting that $1_C(Q_s\phi) = 1_{C_1}(Q_t\phi)$ for a suitable choice of $C_1 \in \mathcal{C}(H_t)$ and that $f'(\phi) = g_0(\Gamma_t(\phi))$ satisfies the conditions of Lemma 3.8, we have

$$\int 1_C(Q_s\phi)g_0(\Gamma_t(\phi))dn(\phi) = \int 1_C(Q_s\phi)g_0(\Gamma_t(\phi))\Gamma_t(\phi)dm(\phi)$$

$$= \int 1_C(Q_s\phi)g(\Gamma_t(\phi))dm(\phi) \quad (3.89)$$

where $g(\mu) = g_0(\mu)\cdot\mu(S)$, $\mu \in \mathcal{M}^*(S)$. Clearly, g is a continuous function from $\mathcal{M}^*(S)$ into $\mathbb{R}$ and it satisfies the bound (3.73). Thus by Theorem 3.7,

$$\int 1_C(Q_s\phi)g(\Gamma_t(\phi))dm(\phi) = \int 1_C(Q_s\phi)g_1(\Gamma_s(\phi))dm(\phi) \quad (3.90)$$

where g_1 is defined by (3.83). Let g_2: $\mathcal{M}^*(S) \to \mathbb{R}$ be defined by

$$g_2(\mu) = \frac{g_1(\mu)}{\mu(S)}, \quad \mu \in \mathcal{M}^*(S). \quad (3.91)$$

Then by Theorem 3.6, g_1 (and hence g_2) satisfies the conditions of Theorem 3.4. As a consequence,

$$g_2(\Gamma_s) \in \mathscr{L}^*(H,\mathscr{C},m) = \mathscr{L}^*(H,\mathscr{C},n). \tag{3.92}$$

Besides (3.90), Theorem 3.7 also implies

$$g_1(\Gamma_s(\phi)) = g_2(\Gamma_s(\phi))\cdot\Gamma_s(\phi) \in \mathscr{L}^1(H,\mathscr{C},m).$$

Arguments given in the proof of Lemma 3.8 would imply that

$$g_2(\Gamma_s(\phi)) \in \mathscr{L}^1(H,\mathscr{C},n).$$

Thus by Lemma 3.8,

$$\begin{aligned}\int 1_C(Q_s\phi)g_1(\Gamma_s(\phi))dm(\phi) &= \int 1_C(Q_s\phi)g_2(\Gamma_s(\phi))\Gamma_s(\phi)(S)dm(\phi)\\ &= \int 1_C(Q_s\phi)g_2(\Gamma_s(\phi))dm(\phi) \qquad (3.93)\end{aligned}$$

From (3.89), (3.90) and (3.93) we get the equality

$$\int 1_C(Q_s\phi)g_0(\Gamma_s(\phi))dn(\phi) = \int 1_C(Q_s\phi)g_2(\Gamma_s(\phi))dn(\phi). \tag{3.94}$$

If Γ'_s is the restriction of Γ_s to H_s, then $\Gamma_s(\phi) = \Gamma'_s(Q_s\phi)$ and hence (3.92) and Proposition IV.2.3 imply that

$$g_2(\Gamma'_s) \in \mathscr{U}(H,\mathscr{C},n;Q_s).$$

This and (3.94) (which is true for all $C \in \mathscr{C}(H_s)$) yield

$$E_n(g_0(\Gamma_t)\,|Q_s\Phi) = g_2(\Gamma_s). \tag{3.95}$$

Thus $\{\Gamma_t\}$ is a Markov process on $(H,\mathscr{C},n)$, since (3.95) holds for all $g_0 \in C_b(\mathscr{M}^*(S))$, and hence in particular for all $g_0 \in C_b(\mathscr{M}(S))$.

We will now prove that $\{F_t\}$ is a Markov process. Let $f \in C_b(\mathscr{M}(S))$ and let $g_0\colon \mathscr{M}^*(S) \to \mathbb{R}$ be defined by

$$g_0(\mu) = f(\theta(\mu)), \quad \mu \in \mathcal{M}^*(S) \tag{3.96}$$

where θ is defined by (3.44) ($\theta(\mu)$ is the normalization of μ such that $\theta(\mu)(S) = 1$). Thus we have

$$g_0(\Gamma_t) = f(F_t)$$

so that in view of (3.95), we get

$$E_n(f(F_t)|Q_s\Phi) = g_2(\Gamma_s)$$

where g_2 is given by (3.91), g_1 is given by (3.83) and $g(\mu) = g_0(\mu)\cdot\mu(S)$. We will prove that

$$g_2(\mu) = g_2(\theta(\mu)) \tag{3.97}$$

which would imply $g_2(\Gamma_s) = g_2(F_s)$ and hence that

$$E_n(f(F_t)|Q_s\Phi) = g_2(F_s). \tag{3.98}$$

For (3.97), note that

$$\overline{\Gamma}^{\mu} = \mu(S)\cdot\overline{\Gamma}^{\theta(\mu)}.$$

Thus

$$\begin{aligned} g_1(\mu) &= \int g(\overline{\Gamma}^{\mu})dm \\ &= \int g_0(\overline{\Gamma}^{\mu})\cdot\overline{\Gamma}^{\mu}(S)dm \\ &= \int f(\theta(\overline{\Gamma}^{\mu}))\cdot\overline{\Gamma}^{\mu}(S)dm. \end{aligned}$$

For any $\mu \in \mathcal{M}^*(S)$, $\overline{F}^{\mu}$ is a probability measure. Hence

$$\begin{aligned} \theta(\overline{\Gamma}^{\mu}) &= \theta(\overline{\Gamma}^{\theta(\mu)}) \\ &= \overline{F}^{\theta(\mu)}. \end{aligned}$$

Hence

$$g_1(\mu) = \int f(\overline{F}^{\theta(\mu)})\cdot\mu(S)\cdot\overline{\Gamma}^{\theta(\mu)}dm$$

$$= \mu(S)\cdot\int f(\overline{F}^{\theta(\mu)})\overline{\Gamma}^{\theta(\mu)}dm.$$

The definition (3.91) of g_2 now gives

$$g_2(\mu) = \frac{1}{\mu(S)}\cdot g_1(\mu)$$

$$= \int f(\overline{F}^{\theta(\mu)})\overline{\Gamma}^{\theta(\mu)}dm.$$

Since $\theta(\theta(\mu)) = \theta(\mu)$, this gives

$$g_2(\theta(\mu)) = g_2(\mu).$$

As noted earlier, this proves (3.98). Thus $\{F_t\}$ is a Markov process on $(H,\mathscr{C},n)$. □

Finally, we are in a position to prove that the conditional distribution $\{F_t(y)\}$ of X_t gives $Q_t y$ is a Markov process on $(E,\mathscr{E},\alpha)$.

<u>Theorem 3.10</u>: $\{\Gamma_t(y)\}$, $\{F_t(y)\}$ are $\mathscr{M}(S)$ valued Markov processes on $(E,\mathscr{E},\alpha)$ with respect to $\{Q_t y\}$.

<u>Proof</u>: We have seen in Theorem 3.2 that

$$\Gamma_t,\ F_t \in \mathscr{L}^*(H,\mathscr{C},n;\mathscr{M}(S)).$$

Thus by Proposition IV.2.3,

$$\Gamma_t',\ F_t' \in \mathscr{U}(H,\mathscr{C},n;Q_t,\mathscr{M}(S))$$

where Γ_t', F_t' are restrictions of Γ_t, F_t to H_t. Also as seen in Theorem 3.2,

$$\Gamma_t = \Gamma_t'\circ Q_t,\ F_t = F_t'\circ Q_t \in \mathscr{U}(E,\mathscr{E},\alpha;y,\mathscr{M}(S)).$$

Hence by Theorem V.4.4

$$\Gamma'_t,\ F'_t \in \mathcal{U}(E,\mathcal{E},\alpha;Q_t y,\mathcal{M}(S)).$$

This shows that $\{\Gamma_t(y)\}$, $\{F_t(y)\}$ are adapted to $\{Q_t(y)\}$.

Now for $g_1 \in C_b(\mathcal{M}(S))$, $C \in \mathcal{C}(H_s)$, we have by Theorem 3.9

$$\int_H 1_C(Q_s\phi)g_0(\Gamma_t(\phi))dn(\phi) = \int_H 1_C(Q_s\phi)g_2(\Gamma_s(\phi))dn(\phi) \quad (3.99)$$

where g_2 satisfies the conditions of Theorem 3.4. Thus, $g_0(\Gamma_t)$, $g_2(\Gamma_s) \in \mathcal{U}(E,\mathcal{E},\alpha;y)$ and hence we have

$$\int_E 1_C(Q_s y)g_0(\Gamma_t(y))d\alpha = \int_E 1_C(Q_s y)g_2(\Gamma_s(y))d\alpha. \quad (3.100)$$

Further, $g_2(\Gamma_s) \in \mathcal{U}(y)$ (Theorem 3.4) and $\Gamma'_s \in \mathcal{U}(Q_t)$ implies $g_2(\Gamma'_s) \in \mathcal{U}(Q_t y)$, by Theorem V.4.4. Thus (3.100) gives

$$E_\alpha(g_0(\Gamma_t(y))|Q_t y) = g_2(\Gamma_s(y)). \quad (3.101)$$

So that $\{\Gamma_t(y)\}$ is a Markov process on $(E,\mathcal{E},\alpha)$.

Given $f \in C_b(\mathcal{M}(S))$, define $g_0 = f\circ\theta$ as in the proof of Theorem 3.9 and use (3.101) and the observation $g_2 = g_2\circ\theta$ to conclude

$$E_\alpha(f(F_t(y))|Q_t y) = g_2(F_t(y)). \quad (3.102)$$

This proves the Markov property of $\{F_t(y)\}$. □

Remark 3.2: In [39], a proof of Theorem 3.9 was given. At that time, the notions of a QCM, the $\mathcal{U}$ class associated with it were not developed and thus the Theorem 3.10 was not proved - indeed it could not be formulated then.

Remark 3.3: As in the proof of Theorem 3.9, it can be shown that $\{F_t\}$ is an $\mathcal{M}(S)$ valued Markov process with respect to

$\{Q_t\Phi\}$. What we need to check is that if the function g in (3.82) has the property $g = g\circ\theta$, then so does g_1, defined by (3.83).

4. A SEMIGROUP DESCRIPTION OF THE WHITE NOISE FILTERING THEORY

The solution to the general filtering problem can be alternatively given using a semigroup, or rather, a family of semigroups parametrized by the observations. A semigroup approach to the pathwise (conventional) theory of filtering has been previously considered by M.H.A. Davis [16, 17].

Suppose that in the filtering model (2.4), (X_t) is a Markov process satisfying (1.63), S is a Polish space and h is such that (2.2) holds and

$$E_{P_{s,x}} \int^T |h_u(\underline{X}_u)|^2 du < \infty. \tag{4.1}$$

For $y \in H$, $s \geq 0$, $x \in S$, define

$$\begin{aligned} c^y(s,x) &= (h_s(x), y_s)_{\mathcal{H}} - \tfrac{1}{2}\|h_s(x)\|^2_{\mathcal{H}}, \quad s \leq T \\ &= 0 \qquad\qquad\qquad\qquad\qquad\;\; , \quad s > T. \end{aligned} \tag{4.2}$$

Also for $s \geq 0$, $t \geq 0$, $x \in S$ and $y \in H$, define $P^y(s,x,t,\cdot) \in \mathcal{M}(S)$ by

$$P^y(s,x,t,A) = E_{P_{s,x}}[1_A(\underline{X}_{s+t})\exp(\int_s^{s+t} c^y(u,\underline{X}_u)du)]. \tag{4.3}$$

Then it is easy to see that for each $y \in H$, P^y is an "unnormalized" transition function - i.e., it has all the properties of a transition probability function except that

$P^y(s,x,t,\cdot)$ is a measure instead of a probability measure. Thus, we have

(i) $(s,x) \to P^y(s,x,t,A)$ is Borel measurable, for each $t \geq 0$, $A \in \mathcal{S}$ and $y \in H$,

(ii) $P^y(s,x,t,\cdot) \in \mathcal{M}(S)$ for all $s \geq 0$, $t \geq 0$, $x \in S$, $y \in H$, and

(iii) for all $y \in H$, $s \geq 0$, $t \geq 0$, $u \geq 0$, $A \in \mathcal{S}$, we have

$$\int P^y(s,x,t,dz)P^y(s+t,z,u,A) = P^y(s,x,t+u,A). \qquad (4.4)$$

Equation (4.4) is the Chapman-Kolmogorov equation for P^y. If we define T_t^y: $J \to J$ by

$$(T_t^y f)(s,x) = \int f(s+t,z)P^y(s,x,t,dz), \ (t \geq o), \qquad (4.5)$$

then (4.4) shows that T_t^y is a semigroup. Since $c^y(s,x) \leq \frac{1}{2}|y_s|^2$ for $s \leq T$ and $c^y(s,x) = 0$ for $s > T$, it follows that

$$P^y(s,x,t,A) \leq \exp(\tfrac{1}{2}\|y\|^2)P(s,x,t,A)$$

and hence

$$(T_t^y f)(s,x) \leq \exp(\tfrac{1}{2}\|y\|^2)(T_t f)(s,x) \qquad (4.6)$$

for all $f \in J(\hat{S},\hat{\mathcal{S}})$, $f \geq 0$, $s \geq 0$, $x \in S$, $t \geq 0$. Moreover, we also have

$$(T_t^y f)(s,x) = E_{P_{s,x}}[f(s+t,\underline{X}_{s+t}) \exp(\int_s^{s+t} c^y(u,\underline{X}_u)du)].$$

Thus, (1.68) yields

$$E_{\Pi}[(T_t^y f)(0,X_0) = E_{\Pi}[f(t,X_t)\exp(\int_0^t c^y(u,X_u)du)] \tag{4.7}$$

$$= \langle f(t,\cdot),\Gamma_t(y)(\cdot)\rangle.$$

So the semigroup $\{T_t^y\}$ and the initial distribution $\Pi\circ X_0^{-1}$ completely determines $\Gamma_t(y)$ and hence also $E_\alpha(X_t\in\cdot\,|Q_t y)$. We have

$$E_\alpha[f(t,X_t)\,|Q_t y] = \frac{\int(T_t^y f)(0,x)dN_0(x)}{\int(T_t^y 1)(0,x)dN_0(x)}\ . \tag{4.8}$$

So one way to solve the filtering problem is to obtain the semigroup $\{T_t^y\}$, or equivalently, the generator of $\{T_t^y\}$. It is easy to see that if $f \in C_b(\hat{S})$, then

$$(T_t^y f)(s,x) \to f(s,x)$$

weakly as $t \downarrow 0$. Thus, the weak generator of T_t^y will completely determine $\{T_t^y\}$.

We are unable to directly describe the generator of $\{T_t^y\}$ in terms of the generator L of (X_t), h and y. However, for a rich class of functions f belonging to the domain $\mathscr{D}_{(e)}^y$ of the extended generator $L_{(e)}^y$ of $\{T_t^y\}$, we will obtain an expression for $L_{(e)}^y f$. Recall that a measurable function f: $\hat{S} \to \mathbb{R}$ belongs to $\mathscr{D}_{(e)}^y$ if there exists g: $\hat{S} \to \mathbb{R}$, such that

$$(T_t|f|)(s,x) < \infty, \tag{4.9}$$

$$\int_0^t(T_u|g|)(s,x)du < \infty \quad \text{for all } (s,x) \in \hat{S}, \tag{4.10}$$

and

$$(T_t f)(x,s) = f(s,x) + \int_0^t (T_u g)(s,x)du \qquad (4.11)$$

for all $(s,x) \in \hat{S}$.

Here $T_t|f|$ is defined by (4.5) and if $T_t|f|(s,x) < \infty$, then $(T_t f)(s,x)$ is also defined by (4.5).

With this notation, we have the following result.

Theorem 4.1: Fix $y \in H$. Then we have

(i) $f \in \mathscr{D} \Rightarrow f \in \mathscr{D}^y_{(e)}$ and

$$L^y_{(e)} f = Lf + c^y f. \qquad (4.12)$$

(ii) Suppose T'_t is any semigroup of linear operators on $J(\hat{S},\hat{\mathscr{S}})$ such that for a finite constant K,

$$(T'_t f)(s,x) \leq K(T_t f)(s,x) \qquad (4.13)$$

for all non-negative $f \in J(\hat{S},\hat{\mathscr{S}})$, $(s,x) \in \hat{S}$, $t \geq 0$.

Let $L'_{(e)}$ be the extended generator of T'_t with domain $\mathscr{D}'_{(e)}$. Suppose that $\mathscr{D} \subseteq \mathscr{D}'_{(e)}$ and for $f \in \mathscr{D}'_{(e)}$ let

$$L'_{(e)} f = Lf + c^y f = L^y_{(e)} f. \qquad (4.14)$$

Then $T'_t \equiv T^y_t$. In other words, (4.12) uniquely determines T^y_t in the class of semigroups satisfying (4.13).

Proof: This is a consequence of Theorem 1.12. Fix $x \in S$ and consider the process $\underline{X}_t$ on $(D, \mathscr{A}^0_\infty, P_{0,x})$. For this process $\underline{X}_t$ and $c = c^y$, the measure G_t defined by (1.6) is equal to $P^y(0,x,t,\cdot)$. Thus Theorem 1.12 implies that (for fixed x) $P^y(0,x,t,\cdot)$ is the unique solution to

$$\langle f(t,\cdot),P^y(0,x,t,\cdot)\rangle = \langle f(0,\cdot),P^y(0,x,0,\cdot)\rangle + \int_0^t \langle (Lf+c^y f)(u,\cdot),P^y(0,x,u,\cdot)\rangle du \tag{4.15}$$

for $f \in \mathcal{D}$ in the class of measures $\{K_t\}$ satisfying (1.8). From the definition of $T_t^y f$, we can rewrite this as

$$(T_t^y f)(0,x) = f(0,x) + \int_0^t T_u^y (Lf+c^y f)(0,x)du, \quad f \in \mathcal{D}. \tag{4.16}$$

Instead, taking $X_t = \underline{X}_{r+t}$ on $(D,\mathcal{A}_\infty^0,P_{r,x})$ and $c(s,x) = c^y(s+r,x)$, Theorem 1.12 yields

$$(T_t^y f)(r,x) = f(r,x) + \int_0^t T_u^y (Lf+c^y f)(r,x)du, \quad f \in \mathcal{D}. \tag{4.17}$$

For $f \in \mathcal{D}$, f and Lf are bounded. Assumption (4.2) implies that c^y and hence $g = Lf + c^y f$ satisfy the integrability condition (4.10). Thus (4.17) implies that if $f \in \mathcal{D}$, then $f \in \mathcal{D}_{(e)}^y$ and $L_{(e)}^y f = Lf + c^y f$. For the uniqueness part, let $P'(s,x,t,A)$ be defined by

$$P'(s,x,t,A) = (T_t' 1_A)(s,x)$$

for $s \geq 0$, $t \geq 0$, $x \in S$, $A \in \mathcal{S}$ (considered as a subset of $[0,\infty)\times S$). Then condition (4.13) implies

$$P'(s,x,t,A) \leq K\cdot P(s,x,t,A) \tag{4.18}$$

where P is the transition probability function of (X_t). The expression (4.14) for $L_{(e)}' f$ implies, in particular, that $P'(0,x,t,\cdot)$ satisfies (4.15). The condition (4.18) implies that $P'(0,x,t,\cdot) \ll P(0,x,t,\cdot)$ and the density is bounded by K. Thus from the uniqueness part of Theorem 1.12, we conclude that

$$P'(0,x,t,\cdot) = P(0,x,t,\cdot), \quad t \geq 0.$$

Similarly, we can show that $P'(s,x,t,\cdot) = P(s,x,t,\cdot)$ for $s \geq 0$. This shows that $T'_t = T^y_t$. □

Though we have not been able to describe the generator L^y of T^y_t completely, we have an expression for $L^y_{(e)}$ for a large enough subset of $\mathscr{D}^y_{(e)}$, i.e., large enough to characterize T^y_t. Hence (4.8) can be considered as a solution to the filtering problem.

CHAPTER IX

PREDICTION AND SMOOTHING

This chapter is concerned with the recursive calculation of the quantity $E_{\alpha}[f(X_s)|Q_t y]$ when $s \neq t$. The case $s = t$ is the filtering problem which has been dealt with at great length in the preceding chapters. Here we consider briefly two other problems of interest: prediction (when $s > t$) and smoothing (when $s < t$), t being regarded as the present instant of time.

In the first section, the signal process is d-dimensional, in particular, a diffusion. The second section is devoted to the general case when the observation process y_t is Hilbert-space valued.

1. PREDICTION AND SMOOTHING FOR THE FINITE DIMENSIONAL CASE

Let us return to the model considered in Chapter VII -- namely (VII.1.9) where ξ,e are given by (VII.1.7) and (VII.1.5) repsectively. We will use the notation introduced in Chapter VII.

Consider the problem of predicting the behavior of the signal at a time s, $s > t$, on the basis of the observations $\{y_u: 0 \leq u \leq t\}$. This will be achieved if we can calculate

$$E_{\alpha}[f(X_s)|Q_t y] := \pi_{st}(f,y) \tag{1.1}$$

for $f \in J(\mathbb{R}^d, \mathcal{B}(\mathbb{R}^d))$. The prediction problem thus consists of obtaining $\pi_{st}(f,y)$ for $s > t$.

On the other hand, let s denote a time point in the past, i.e., $s < t$. As t increases, i.e., as we have more observations on y, our estimate of X_s becomes better or smoother. Finding a procedure for calculating $\pi_{st}(f,y)$ for $s < t$ is known as the smoothing problem.

Of course $\pi_{st}(f,y)$ for $s = t$ is nothing but $\pi_t(f,y)$, the quantity we have obtained in Chapter VII while studying the filtering problem.

The following is an immediate consequence of the Bayes formula (Theorem VII.1.1).

Theorem 1.1: For $0 \leq s \leq T$, $0 \leq t \leq T$, $f \in J(\mathbb{R}^d, \mathscr{B}(\mathbb{R}^d))$, we have

$$E_\alpha[f(X_s)|Q_t y] = \frac{1}{\Gamma_{st}(y)(\mathbb{R}^d)} \int f(x) d\Gamma_{st}(y)(x) \tag{1.2}$$

where $\Gamma_{st}(\phi) \in \mathscr{M}(\mathbb{R}^d)$ is given by

$$\Gamma_{st}(\phi)(B) = E_\Pi[1_B(X_s) q_t(\phi,\cdot)] \tag{1.3}$$

for $0 \leq s,t \leq T$, $\phi \in H$, $B \in \mathscr{B}(\mathbb{R}^d)$.

Proof: Recall that $q_t(\phi,\cdot)$, which is given by (VII.2.7) is bounded for each ϕ and hence $q_t(\phi,\cdot)$ is Π-integrable. Hence it follows that $\Gamma_{st}(\phi) \in \mathscr{M}(\mathbb{R}^d)$. It is easy to see that

$$\begin{aligned} \int f(x) d\Gamma_{st}(\phi)(x) &= E_\Pi[f(X_s) q_t(\phi,\cdot)] \\ &= \tilde{\sigma}_t(f(X_s),\phi). \end{aligned} \tag{1.4}$$

The relation (1.2) follows from (1.4) and (VII.1.13).

Remark 1.1: Since $q_t(\phi,\omega) = q_t(Q_t\phi,\omega)$ for all ω,ϕ, it follows that

$$\Gamma_{st}(\phi) = \Gamma_{st}(Q_t\phi).$$

Thus $\Gamma_{st}(y)$, which is called the unnormalized conditional distribution of X_s given $Q_t y$, is a functional of $Q_t y$, as one would expect.

Thus the Bayes formula gives a partial solution to the prediction and the smoothing problems. It remains to find ways of obtaining $\Gamma_{st}(\phi)$ without having to evaluate the function space integral (appearing in (1.3)).

When (X_t) is a diffusion process, we can study the density of $\Gamma_{st}(\phi)$ with respect to the Lebesgue measure and, as in Chapter VII, obtain it as a solution to a partial differential equation. From now on, (X_t) will denote a diffusion process.

The first result in this direction is common to both the prediction and smoothing problems.

Theorem 1.2: Suppose that for all t, the measure $\Pi \circ X_t^{-1}$ admits a density $p_t(x)$ with respect to the Lebesgue measure λ on $\mathbb{R}^d$.

(i) Then for all $s,t,\phi \in H$, $\Gamma_{st}(\phi) \ll \lambda$. Further, we can choose a jointly measurable version $(s,t,x) \to p_{st}(x,\phi)$ of the density $p_{st}(x,\phi) = d\Gamma_{st}(\phi)/d\lambda$.

(ii) If $y_k \to y$ in H, then (writing dx for $d\lambda(x)$)

$$\lim_{k\to\infty} \int_{\mathbb{R}^d} |p_{st}(x,y_k) - p_{st}(x,y)|dx = 0. \tag{1.5}$$

But for a change of notation, the proof of these assertions is identical to the proof of parts (i) and (iii) in Theorem

VIII.3.7 and is omitted.

The Prediction Problem

We will now concentrate on the prediction problem, so that $0 \leq t \leq s \leq T$ and find ways of obtaining $p_{st}(x,y)$. The first result, which is an analogue of Theorem VII.3.9, gives conditions under which $p_{st}(x,y)$ can be characterized as the unique generalized solution to a Cauchy problem.

Theorem 1.3: Suppose that a, b, p_0 satisfy the hypothesis of Theorem VII.3.8. Suppose that h is a measurable function from $[0,T]\times\mathbb{R}^d$ into $\mathbb{R}$ satisfying (VII.3.82).

(i) Then for all $y \in H$ fixed, the Cauchy problem

$$\frac{\partial v(s,x)}{\partial s} = L_s^* v(s,x) + [\sum_{i=1}^{N} h_s^i(x)y_s^i - \tfrac{1}{2}|h_s(x)|^2]1_{\{s\leq t\}}v(s,x), \tag{1.6}$$

$$v(0,x) = p_0(x) \tag{1.7}$$

has a unique generalized solution in the class of functions f satisfying (VII.3.83) and

$$\int_0^t \int_{\mathbb{R}^d} |h_u(x)|^2 |f(u,x)| dxdu < \infty \tag{1.8}$$

(ii) For all $y \in H$, $\Gamma_{st}(y)$ is absolutely continuous with respect to λ and the density $p_{st}(x,y) = d\Gamma_{st}(y)/d\lambda$ satisfies

$$p_{st}(x,y) = v(s,x) \text{ for a.e. } x \text{ if } s \geq t \tag{1.9}$$

where v is the unique generalized solution in (i).

Proof: Fix $t \in [0,T]$. Let $h'_u(x) = h_u(x)\cdot 1_{\{u\leq t\}}$. Then the Cauchy problem (1.6)-(1.7) is the same as the Cauchy problem (VII.3.69)-(VII.3.70) for h'. Hence by Theorem VII.3.9, it follows that the said Cauchy problem admits a unique generalized solution in the class of functions f satisfying (VII.3.83) and (1.14). Note that (1.14) is the same as (VII.3.84) for h'. This proves (i).

For (ii), note that by Theorem VII.3.9, we have for $s \leq t$, $B \in \mathscr{B}(\mathbb{R}^d)$,

$$\int_B v(s,x)dx = E_\Pi 1_B(X_s)\exp(\int_0^s[\sum_{i=1}^N h'^i_u(X_u)y^i_u du - \tfrac{1}{2}|h'_u(X_u)|^2]du)$$

$$= E_\Pi 1_B(X_s)\exp(\int_0^t[\sum_{i=1}^N h^i_u(X_u)y^i_u - \tfrac{1}{2}|h'_u(X_u)|^2]du)$$

$$= \Gamma_{st}(y)(B)$$

because $h'_u = h_u\cdot 1_{\{u\leq t\}}$. Hence

$$\frac{d\Gamma_{st}(y)}{d\lambda} = v(s,x). \qquad \square$$

Suppose that the diffusion and drift coefficients a,b of (X_t) satisfy the conditions of Theorem VII.4.4 and $p(s,x,t,z)$ is its transition probability density. Suppose that the density $p_0(x)$ of X_0 satisfies the growth condition (VII.4.22). Recall that in this case, the density $p_t(x)$ of X_t exists and is the unique classical solution to the Cauchy problem (VII.4.36)-(VII.4.37). By Theorem VII.3.7, the unnormalized conditional density $p_t(x,y)$ of X_t given $Q_t y$ exists for all $y \in H$ and is a positive measurable function.

Theorem 1.4:

(i) Fix $t \in [0,T]$. For $s > t$, define $p_{st}(x,y)$ by

$$p_{st}(x,y) = \int_{\mathbb{R}^d} p_t(z,y)p(t,z,s,x)dz, \tag{1.10}$$

$y \in H$, $x \in \mathbb{R}^d$. Then $p_{st}(x,y)$ is the unnormalized conditional density of X_s given $Q_t y$, for all $y \in H$.

(ii) If for a certain $y \in H$, $0 \leq t \leq T$, $x \to p_t(x,y)$ is a continuous function, then $p_{st}(x,y)$ is the unique classical solution in the class $\mathcal{G}((t,T])$ to the Cauchy problem

$$\frac{\partial}{\partial s} p_{st}(x,y) = L_s^*(p_{st}(\cdot,y))(x) \qquad s > t \tag{1.11}$$

$$p_{tt}(x,y) = p_t(x,y). \tag{1.12}$$

Proof: Since $p_t(x,y) \geq 0$ a.e., the integral (1.10) is well defined. For $B \in \mathcal{B}(\mathbb{R}^d)$, $0 \leq t \leq T$, $y \in H$ fixed, we have

$$\int_B p_{st}(x,y)dx \tag{1.13}$$

$$= \int\int_B p_t(z,y)p(t,z,s,x)dzdx = \int[\int_B p(t,z,s,x)dx]p_t(z,y)dz$$

$$= E_\Pi[q_t(y,\cdot)\int_B p(t,X_t,s,x)dx] = E_\Pi[q_t(y,\cdot)E_\Pi[1_{(X_s \in B)}|\mathcal{F}_t^X]]$$

$$= E_\Pi[1_{\{X_s \in B\}}q_t(y,\cdot)] = \Gamma_{st}(y)(B).$$

The use of Fubini's theorem in the above steps is justified as the integrand is positive. Since (1.13) holds for all $B \in \mathcal{B}(\mathbb{R}^d)$, it follows that $p_{st}(x,y)$ defined by (1.10) is the unnormalized conditional density.

For (ii), note that for suitable constants K_1, K_2,

$H_1(t,x)$ defined by (VII.4.5) satisfies (VII.4.11)-(VII.4.12) with $M_t = L_t^*$. Hence for this choice, the transition density $p(s,x,t,z) = G^*(t,z,s,x)$ satisfies the estimates (VII.4.6)-(VII.4.8). (See Theorems VII.4.1 and VII.4.4.) As noted in Remarks VII.4.1 and VII.4.2, the assumption that $p_0(x)$ satisfies (VII.4.22) implies

$$p_0(x) \leq K_3 H_1(0,x)$$

and hence that

$$p_t(x) \leq K_3 H_1(t,x).$$

As seen in Chapter VII (see VII.3.85), we have

$$p_t(x,y) \leq \exp(\tfrac{1}{2}\|y\|^2) p_t(x) \quad \text{for a.e. } x \tag{1.14}$$

for all $y \in H$, $t \in [0,T]$ and hence

$$p_t(x,y) \leq K_3 \exp(\tfrac{1}{2}\|y\|^2) H_1(t,x) \quad \text{for a.e. } x. \tag{1.15}$$

Thus if for a certain $y \in H$, $t \in [0,T]$, $x \to p_t(x,y)$ is a continuous function, then (1.15) holds for all x, i.e.,

$$p_t(x,y) \leq K_3 \exp(\tfrac{1}{2}\|y\|^2) H_1(t,x) \tag{1.16}$$

and then by Theorem VII.4.1, $p_{st}(x,y)$ defined by (1.10) is the unique classical solution in the class $\mathcal{G}((t,T])$ to the Cauchy problem (1.11)-(1.12). □

Remark 1.2: Theorem VII.4.3 gives conditions under which, for $y \in H_0$, $p_t(x,y)$ is a unique classical solution to a PDE, and in particular is continuous so that (ii) of the Theorem proved above is applicable for all $y \in H_0$ in this case. (See

Chapter VII for the definition of H_0.) Further, if conditions of Theorem VII.4.5 are fulfilled, then it implies that $p_t(x,y)$ is a continuous function for each $t \in [0,T]$, $y \in H$ and then (ii) is applicable for all $y \in H$.

Thus, as in the filtering case, the unnormalized conditional density $p_{st}(x,y)$ of X_s given $Q_t y$ (for $s > t$) is the unique classical solution to the Cauchy problem (1.11)-(1.12), for all $y \in H_0$ if the conditions of Theorem VII.4.3 are satisfied, and for all $y \in H$ under the conditions of Theorem VII.4.5.

The Smoothing Problem

We now focus our attention on $\Gamma_{st}(y)$ for $0 \le s \le t \le T$. The first result in this direction is the following.

Theorem 1.5: For all $\phi \in H$, $0 \le s < t \le T$, we have

$$\Gamma_{st}(\phi) \ll \Gamma_s^{\phi} \tag{1.17}$$

and

$$\frac{d\Gamma_{st}(\phi)}{d\Gamma_s^{\phi}}(x) = v_t(s,x,\phi) \tag{1.18}$$

where v satisfies

$$v_t(s,X_s,\phi) = E_{\Pi}\left[\exp\left(\int_s^t\left[\sum_{i=1}^N h_u^i(X_u)y_u^i - \tfrac{1}{2}|h_u(X_u)|^2\right]du\,\Big|\,\sigma(X_s)\right)\right]. \tag{1.19}$$

Proof: Note that $\dfrac{q_t(\phi,\omega)}{q_s(\phi,\omega)}$ is $\sigma(X_u: s \le u \le t)$ measurable.

Hence we have for $B \in \mathcal{B}(\mathbb{R}^d)$,

$$\Gamma_{st}(\phi)(B) = E_{\Pi}1_B(X_s)q_t(\phi,\cdot) \tag{1.20}$$

$$= E_{\Pi}\left[1_B(X_s)q_s(\phi,\cdot)E_{\Pi}\{\frac{q_t(\phi,\cdot)}{q_s(\phi,\cdot)}|\sigma(X_u:\ u\leq s)\}\right]$$

$$= E_{\Pi}\left[1_B(X_s)q_s(\phi,\cdot)E_{\Pi}\{\frac{q_t(\phi,\cdot)}{q_s(\phi,\cdot)}|\sigma(X_s)\}\right]$$

$$= E_{\Pi}[1_B(X_s)q_s(\phi,\cdot)v_t(s,X_s,\phi)] = \int 1_B(x)v_t(s,x,\phi)d\Gamma_s^{\phi}(x).$$

We have used the Markov property of (X_t). The relation (1.20) implies (1.17) and (1.18). □

<u>Corollary 1.6</u>: Suppose that $p_s(x,\phi) = \frac{d\Gamma_s^{\phi}}{d\lambda}(x)$ exists for all s,ϕ. Then $p_{st}(x,\phi) = \frac{d\Gamma_{st}(\phi)}{d\lambda}(x)$ also exists and we have, for all $\phi \in H$,

$$p_{st}(x,\phi) = p_s(x,\phi)v_t(s,x,\phi). \tag{1.21}$$

This is an immediate consequence of (1.17) and (1.18).

The relation (1.21) gives a method of obtaining $p_{st}(x,\phi)$ -- first obtain $p_s(x,\phi)$ by methods given in Chapter VII, obtain the function $v_t(s,x,\phi)$ and multiply the two. Thus we now need to look at ways to get the functional v without performing any function space integration. The next result gives a method of obtaining $v_t(s,x,\phi)$ for $\phi \in H$.

<u>Theorem 1.7</u>: Suppose that a,h,p_0,h satisfy the conditions of Theorem VII.4.3.

(i) Then for $y \in H_0$, the Cauchy problem

$$\frac{\partial}{\partial s}v_t(s,x,y) + L_s v_t(s,x,y) + c_s^y(x)v_t(s,x,y) = 0,\ s<t \tag{1.22}$$

$$v_t(t,x,y) = 1 \tag{1.23}$$

admits a unique solution in the class $\mathscr{G}([0,t))$.

(ii) Let $p_s(x,y)$, $y \in H_0$, be the unnormalized conditional density of X_s given $Q_s y$. (It is the unique solution to a Cauchy problem, by Theorem VII.4.3.) Then we have for $y \in H_0$,

$$p_{st}(x,y) = p_s(x,y)v_t(s,x,y). \tag{1.24}$$

Proof: Part (i) follows from Theorem VII.4.1 because $M_t = L_t+c_t^y$ (for $y \in H_0$) satisfies the conditions of Theorem VII.4.1 as already noted in the proof of Theorem VII.4.3 and

$$1 \leq K\cdot H_2^{-1}(t,x)$$

for a suitable choice of $K < \infty$. The second part will follow from Corollary 1.6 once we show that the solution $v_t(s,x,y)$ of (1.22)-(1.23) satisfies (1.19). This in turn will follow from the Feynman-Kac formula if it can be shown that v is bounded. However, we cannot deduce that v is bounded using the estimates on the fundamental solution to (1.22). Instead, we proceed as follows. Let g_k be a sequence of continuous func-tions from $\mathbb{R}^d$ into $\mathbb{R}$ such that

$$1_{\{|x|\leq k\}} \leq g_k(x) \leq 1_{\{|x|\leq k+1\}}.$$

Then $|g_k| \leq 1$ and $g_k(x)$ increases to one as $k \to \infty$. Let $G(t,z,s,x)$ be the fundamental solution to the equation (1.22). Then by Theorem VII.4.1, v_t^k defined by

$$v_t^k(s,x,y) = \int g_k(z)G(t,z,s,x)dz \tag{1.25}$$

is a solution to the Cauchy problem (1.22)-(1.26), where

$$v_t^k(t,x,y) = g_k(x). \tag{1.26}$$

Also, $v_t(s,x,y)$ is given by

$$v_t(s,x,y) = \int G(t,z,s,x)dz. \tag{1.27}$$

As noted in Remark VII.4.3, $G(t,z,s,x) \geq 0$ and hence by the monotone convergence theorem, using (1.25), (1.27) and $g_k \uparrow 1$,

$$v_t^k(s,x,y) \to v_t(s,x,y). \tag{1.28}$$

Since g_k has compact support, v_t^k is bounded (as noted in the proof of Theorem VII.4.3) and hence by the Feynman-Kac formula, we have

$$v_t^k(s,x,y) = E_{\Pi}[g_k(X_t)\exp(\int_s^t c_u^y(X_u)du)\,|\sigma(X_s)] \tag{1.29}$$

Since c_u^y is bounded above (as $y \in H_0$) and $g_k(X_t) \to 1$, it follows upon taking limits as $k \to \infty$ that

$$v_t(s,X_s,y) = E_{\Pi}[\exp(\int_s^t c_u^y(X_u)du)\,|\sigma(X_s)]. \tag{1.30}$$

This and Corollary 1.6 now imply (1.24), completing the proof. □

The result given above provides a method of obtaining $p_{st}(x,y)$ for $y \in H_0$ in terms of classical solutions to Cauchy problems. Since H_0 is dense in H and $y \to p_{st}(x,y)$ is continuous (in the sense of (1.5)), this is sufficient to solve the smoothing problem.

We have seen in Chapter VII that under additional conditions on a,b,h, $p_t(x,y)$ can be obtained for all $y \in H$ via a solution to a certain PDE. Now we impose the same conditions on a,b,h and transform the equation (1.22) so that it admits a solution for all $y \in H$.

Fix $0 \leq t \leq T$, $y \in H$ and let

$$g_s^{\prime y}(x) = \sum_{i=1}^{N} h_s^i(x) \int_s^t y_u^i du. \tag{1.31}$$

If $v_t(s,x,y)$ is given to satisfy (1.22), then defining

$$v_t'(s,x,y) = v_t(s,x,y)\exp(-g_s^{\prime y}(x)) \tag{1.32}$$

formally, it follows that v' satisfies

$$\frac{\partial}{\partial s} v_t'(s,x,y) + L_s^{\prime y} v_t'(s,x,y) + c_s^{\prime y} v_t'(s,x,y) = 0 \tag{1.33}$$

where

$$(L_s^{\prime y} f)(x) = [L_s f(\cdot)\exp(g_s^{\prime y}(\cdot))](x)\exp(-g_s^{\prime y}(x)) \tag{1.34}$$

and

$$c_s^{\prime y}(x) = -\tfrac{1}{2}|h_s(x)|^2 + \sum_{i=1}^{N} \frac{\partial}{\partial s} h_s^i(x) \cdot \int_s^t y_u^i du. \tag{1.35}$$

The calculations (and the ones that follow) are similar to the ones given before the proof of Theorem VII.4.5, and hence the details are omitted. In (1.34) above, $L_s^{\prime y}$ can be shown to be a differential operator, given by

$$(L_s^{\prime y}f)(x) = \frac{1}{2}\sum_{i,j=1}^{d} a_{ij}(s,x)\frac{\partial^2}{\partial x^i \partial x^j} f(x) \tag{1.36}$$

$$+ \sum_{i=1}^{d} b_i^{\prime}(s,x,y)\frac{\partial}{\partial x^i} f(x) + c^{\prime}(s,x,y)f(x)$$

where

$$b_i^{\prime}(s,x,y) = b_i(s,x) + \sum_{k=1}^{N}\left[\sum_{j=1}^{d} a_{ij}(s,x)\frac{\partial h_s^k(x)}{\partial x^j}\right]\int_s^t y_u^k du \tag{1.37}$$

and

$$c^{\prime}(s,x,y) = \sum_{k=1}^{N}\left[\sum_{i=1}^{d} b_i(s,x)\frac{\partial h_s^k(x)}{\partial x^i} + \frac{1}{2}\sum_{i,j=1}^{d} a_{ij}(s,x)\frac{\partial^2 h_s^k(x)}{\partial x^i \partial x^j}\right.$$

$$\left. + \frac{1}{2}\sum_{i,j=1}^{d} a_{ij}(s,x)\frac{\partial h_s^k(x)}{\partial x^i}\,\frac{\partial h_s^k(x)}{\partial x^j}\int_s^t y_u^k du\right]\int_s^t y_u^k du. \tag{1.38}$$

We will now show that the equation (1.33) admits a solution and v defined by (1.32) is the required function.

Theorem 1.8: Suppose that a, b, p_0, h satisfy the conditions of Theorem VII.4.5. Then

(i) For all $y \in H$, the Cauchy problem (1.33) and $v_t^{\prime}(t,x,y) = 1$ admits a unique classical solution in the class $\mathcal{G}([0,t))$.

(ii) Let $v_t^{\prime}$ be as in (i) above and let $\psi_s(x,y)$ be the unique classical solution to the Cauchy problem (VII.4.53), (VII.4.54). Then p_{st} defined by

$$p_{st}(x,y) = \psi_t(x,y)v_t^{\prime}(s,x,y)\exp\left(\sum_{k=1}^{N} h_s^k(x)\int_0^t y_u^k du\right) \tag{1.39}$$

is the unnormalized conditional distribution of X_s given $Q_t y$.

Proof: Part (i) is a consequence of Theorem VII.4.1. If v' is as in (i) and v is defined by (1.32), then it follows that v satisfies (1.22) for a.e. s. Hence by the Feynman-Kac formula Theorem II.4.4, we have that v so defined satisfies (1.30). Thus by Corollary 1.6, we have

$$p_{st}(x,y) = p_s(x,y)v_t(s,x,y). \tag{1.40}$$

Noting the relation (VII.4.55) between p_s and ψ_s and (1.32) between v and v', it is easy to see that (1.40) is the same as (1.39). □

2. THE GENERAL CASE

We will now consider the model of Chapter VIII where the signal process takes values in a general space S while the observations and the noise take values in a separable infinite dimensional Hilbert space. For this model, given by (VIII.2.3) and (VIII.2.4), solutions to the predictions and smoothing problems can be obtained in a manner analagous to the finite dimensional case.

The notation established in Sections VIII.1 and VIII.2 will be used throughout this section. For $0 \leq s \leq T$, $0 \leq t \leq T$, $\phi \in H$, define $\Gamma_{st}(\phi)$, and $F_{st}(\phi) \in \mathcal{M}(S,\mathcal{S})$ by

$$\Gamma_{st}(\phi)(A) = E_{\Pi}[1_A(X_s)\exp(\int_0^t[(h_u(X_u),y_u)_{\mathcal{H}} - \tfrac{1}{2}\|h_u(X_u)\|^2_{\mathcal{H}}]du)] \tag{2.1}$$

and

$$F_{st}(\phi)(A) = \frac{\Gamma_{st}(\phi)(A)}{\Gamma_{st}(\phi)(S)}, \qquad A \in \mathcal{S}. \tag{2.2}$$

Then from the Bayes formula, Theorem VI.3.4 for $Q = Q_t$, we have

$$E_\alpha(g(X_s)|Q_t y) = \int_S g(x) dF_{st}(y)(x) \tag{2.3}$$

for any $g: S \to \mathbb{R}$, measurable such that $E_\Pi|g(X_s)| < \infty$. Thus $F_{st}(y)$ is the conditional distribution of X_s given $Q_t y$. In view of this and (2.2), $\Gamma_{st}(\phi)$ will be called the unnormalized conditional distribution of X_s given $Q_t y$.

The prediction problem is to obtain $\Gamma_{st}(\phi)$ (or $F_{st}(\phi)$) for $s > t$ and the smoothing problem is to obtain $\Gamma_{st}(\phi)$, (or $F_{st}(\phi)$) for $s < t$. The case $s = t$ is the filtering case for which $\Gamma_{tt}(\phi) = \Gamma_t(\phi)$, the quantity considered and studied in Section VIII.2. Recall that Theorem VIII.2.1 gives conditions under which $\Gamma_t(\phi)$, $F_t(\phi)$ can be obtained as the unique solution of certain measure valued equations.

The Prediction Problem

The following result gives a method of obtaining $\Gamma_{st}(\phi)$, $F_{st}(\phi)$ for $s > t$.

Theorem 2.1: Under the conditions stated in Theorem VIII.2.1 we obtain the following: Let $s > t$.

(i) Then for $A \in \mathcal{S}$ and $y \in H$,

$$\Gamma_{st}(y)(A) = \int P(t,x,s,A) d\Gamma_t(y)(x) \tag{2.4}$$

and

$$F_{st}(y)(A) = \int P(t,x,s,A) dF_t(y)(x). \tag{2.5}$$

Here $P(t,x,s,A)$ denotes the transition probability function of (X_t).

(ii) The equation

$$\langle f(s,\cdot),K_s\rangle = \langle f(0,\cdot),K_0\rangle + \int_0^s\langle (Lf)(u,\cdot) + c_u^y(\cdot)1_{\{u\le t\}}f(u,\cdot),K_u\rangle du, \tag{2.6}$$

where $f \in \mathcal{D}$, has a unique solution in the class of $\{K_t\} \subseteq \mathcal{M}(S,\mathcal{S})$ satisfying VIII.1.8.

(iii) If $\{K_s\}$ is the unique solution to (2.6) above satisfying VIII.1.8, (for t,y fixed), then we have

$$\Gamma_{st}(y) = K_s \qquad \text{for } s \ge t.$$

<u>Proof</u>: The proof of (i) is along the same lines as that of (1.12) and is omitted. Parts (ii) and (iii) follow from Theorems VIII.1.9 and VIII.1.14 for the choice

$$c_u = c_u^y \cdot 1_{\{u\le t\}}.$$

It has already been seen in the proof of Theorem VIII.2.1 that (for a similar choice of c), (i) the conditions of Theorem VIII.1.9 are satisfied if condition (a) of Theorem VIII.2.1 holds and (ii) the conditions of Theorem VIII.1.14 are satisfied if, instead, (b) is true. □

<u>The Smoothing Problem</u>

The following result gives a relation between $\Gamma_{st}(\phi)$ and $\Gamma_s(\phi)$.

<u>Theorem 2.2</u>: Let $0 \le s < t$. Then for $\phi \in H$,

$$\Gamma_{st}(\phi) \ll \Gamma_s(\phi) \tag{2.7}$$

and

$$\frac{d\Gamma_{st}(\phi)}{d\Gamma_s(\phi)} = v_t(s,x,\phi) \tag{2.8}$$

where $v_t(s,x,\phi)$ is given by

$$v_t(s,X_s,\phi) = E_\Pi[\exp(\int_s^t\{(h_u(X_u),\phi_u)_{\mathcal{H}} - \tfrac{1}{2}\|h_u(X_u)\|_{\mathcal{H}}^2\}du)\,|\sigma(X_s)]. \tag{2.9}$$

<u>Proof</u>: The proof of this result is the same as that of Theorem 1.5. Indeed, if we temporarily denote by $q_t(\phi,\omega)$, the quantity

$$\exp(\int_0^t\{(h_u(X_u(\omega)),\phi_u)_{\mathcal{H}} - \tfrac{1}{2}\|h_u(X_u(\omega))\|_{\mathcal{H}}^2\}du),$$

then the proof is exactly the same. □

CHAPTER X

CONSISTENCY AND ROBUSTNESS OF THE WHITE NOISE THEORY

The general aim of this chapter is to reconcile the results of the white noise theory with the mainstream of research in this field which is based on the stochastic calculus of semimartingales and stochastic partial differential equations. The first three sections discuss these questions. The last two sections are devoted to studying robustness properties of the optimal filter (predictor or smoother) as a function of the observations.

The term "consistency" is used here in an informal sense. In saying that our theory is consistent with the conventional theory, we simply mean that the optimal filter (predictor or smoother) given by the stochastic calculus approach can be approximated in a suitable manner by the corresponding optimal quantity provided by the white noise theory.

1. GENERAL CONSISTENCY RESULTS FOR FILTERING, PREDICTION AND SMOOTHING

The white noise model

$$y = \xi + e \tag{1.1}$$

or

$$y_t = h_t(X_t) + e_t, \qquad 0 \leq t \leq T \tag{1.1'}$$

where $H = L^2[0,T]$, $\xi(\omega) := (h_u(X_u(\omega)))_{o\leq u\leq T} \in H$ and $X = (X_u)$

is an $\mathbb{R}^d$-valued process has been described in detail in Chapter VII. Throughout what follows we will be using the notation and terminology of that chapter.

The corresponding model in the conventional theory may be set up in the following manner. Let

$$(\tilde{\Omega},\tilde{A},\tilde{\Pi}) = (\Omega,A,\Pi) \otimes (\Omega_0,A_0,\Pi_0)$$

where $\Omega_0 = C_0([0,T],\mathbb{R}^N)$, A_0 is the σ-field of Borel sets of Ω_0 and Π_0 is the Wiener measure. Let $Z_t(\omega_0) = \omega_0(t)$, $\omega_0 \in \Omega_0$, $0 \le t \le T$ be the coordinate process on Ω_0. Under Π_0, Z_t is an $\mathbb{R}^N$-valued standard Brownian motion. The observation process (Y_t) in the conventional approach is given by

$$Y_t = \int_0^t h_u(X_u)du + Z_t \tag{1.2}$$

i.e.

$$Y_t(\tilde{\omega}) = \int_0^t h_u(X_u(\omega))du + Z_t(\omega_0), \quad \tilde{\omega} = (\omega,\omega_0) \in \tilde{\Omega}. \tag{1.2)'$$

To get a representation of α with representation space $(\tilde{\Omega},\tilde{A},\tilde{\Pi})$, recall that

$$L_0(\phi) = \int_0^T \phi dZ, \qquad \phi \in H \tag{1.3}$$

is a representation of m, where $\int_0^T \phi dZ$ is the Wiener integral. See Section III.4 for the definition of the Wiener integral and the proof of the fact that (L_0,Π_0) is a representation of m. For $\tilde{\omega} = (\omega,\omega_0) \in \tilde{\Omega}$, $\phi \in H$, define $\rho(\tilde{\omega}) = \omega$ and $L(\phi)(\tilde{\omega}) = L_0(\phi)(\omega_0)$. Then as seen in Section V.2, $(\rho,L,\tilde{\Pi})$ is a representation of $\alpha = \Pi\otimes m$. Let R_α be the corresponding α-lifting. The representation $(\rho, L, \tilde{\Pi})$ and R_α will be fixed throughout this section.

We have seen in Chapter VI that y defined by (1.1) is a QCM. Let $n = \alpha \circ y^{-1}$ and let L_1 be the representation of n induced by y, i.e.,

$$L_1(\phi) = R_\alpha((y,\phi)). \tag{1.4}$$

and let R_n be the corresponding n-lifting.

We now prove that $L_1(\phi)$ can be expressed as $\int_0^T \phi dY$.

<u>Lemma 1.1</u>: For all $\phi \in H$,

$$L_1(\phi) = \int_0^T \phi dY \qquad \tilde{\Pi} \text{ a.s.} \tag{1.5}$$

where $\int_0^T \phi dY$ is defined by

$$[\int_0^T \phi dY](\omega,\omega_0) = \sum_{j=1}^N \int_0^T \phi_u^j h_u^j(X_u(\omega))du + (\int_0^T \phi dZ)(\omega_0), \tag{1.6}$$

<u>Proof</u>: Since $(y,\phi) = (\xi,\phi) + (e,\phi)$, we have

$$R_\alpha((y,\phi)) = (\xi,\phi) + R_m((e,\phi)) = (\xi,\phi) + L(\phi)$$

$$= (\xi,\phi) + \int_0^T \phi dZ. \qquad \square$$

The following result is an important link in our consistency results. For $0 \le t \le T$ fixed, let Q_t be the orthogonal projection on H with range H_t given by $H_t = \{\phi \in H: \phi_u = 0 \text{ for a.e. } u \in [t,T]\}$. Let

$$\mathcal{G}_t = \sigma(R_\alpha((Q_t y,\phi)): \phi \in H) \tag{1.7}$$

and

$$\mathcal{F}_t^Y = \sigma(Y_u: 0 \leq u \leq t). \tag{1.8}$$

Then we have

Theorem 1.2: $\overline{\mathcal{G}}_t = \mathcal{F}_t^Y$, $0 \leq t \leq T$, where the bar denotes the completion of a σ-field.

Proof: For $\phi \in H$, $Q_t\phi$ is given by

$$\begin{aligned}(Q_t\phi)(u) &= \phi(u), \quad u \leq t \\ &= 0, \quad\quad u > t.\end{aligned}$$

If ϕ is a step function ϕ: $[0,T] \to \mathbb{R}^N$ of the form

$$\phi(u) = \sum_{i=1}^{p} a_i 1_{(t_{i-1}, t_i]}(u) \tag{1.9}$$

where $0 = t_0 < .. < t_p = T$, $a_i = (a_i^j) \in \mathbb{R}^N$, $1 \leq i \leq p$, we have

$$(Q_t\phi)(u) = \sum_{i=1}^{p} a_i 1_{(t_{i-1} \wedge t, t_i \wedge t]}(u), \tag{1.10}$$

so that from (1.6) and the definition of the Wiener integral $\int \phi dZ$ for a step function (III.4.36), we have for ϕ given by (1.9)

$$\begin{aligned}[\int_0^T (Q_t\phi)dY](\tilde{\omega}) &= \sum_{j=1}^{N} \int_0^T (Q_t\phi)_u^j h_u^j(X_u(\omega))du + [\int_0^T \phi dZ](\omega_0) \\ &= \sum_{j=1}^{N} \sum_{i=1}^{p} a_i^j \int_0^T 1_{(t_{i-1} \wedge t, t_i \wedge t)}(u) h_u^j(X_u(\omega))du \\ &\quad + \sum_{j=1}^{N} \sum_{i=1}^{p} a_i^j (Z_{t_i \wedge t}^j(\omega_0) - Z_{t_{i-1} \wedge t}^j(\omega_0))\end{aligned} \tag{1.11}$$

$$= \sum_{j=1}^{N} \sum_{i=1}^{p} a_i^j \{Y^j_{t_i \wedge t}(\tilde{\omega}) - Y^j_{t_{i-1} \wedge t}(\tilde{\omega})\}.$$

Relation (1.11) shows that for a step function ϕ, $\int_0^T (Q_t\phi)dY$ is $\mathcal{F}_t^Y$ measurable. In (1.9), if we make $p = 2$, $t_1 = s$, $t_2 = T$, $a_1^j = 1$, $a_1^k = 0$, $k \neq j$, $a_2 = 0$, then (1.11) yields

$$\int_0^T (Q_t\phi)dY = Y^j_{s\wedge t}.$$

Hence $\{Y_s: s \leq t\}$ is $\mathcal{G}_t'$ measurable where $\mathcal{G}_t'$ is the σ-field, $\mathcal{G}_t' := \sigma(\int_0^T (Q_t\phi)dY: \phi \in H$, step function). These observations can be summarized as

$$\mathcal{G}_t' = \mathcal{F}_t^Y. \tag{1.12}$$

Remains to prove

$$\overline{\mathcal{G}}_t' = \overline{\mathcal{G}}_t. \tag{1.13}$$

Clearly, $\mathcal{G}_t' \subseteq \mathcal{G}_t$. Fix $\phi \in H$. We will prove that $\int \phi dY$ is $\overline{\mathcal{G}}_t'$ measurable. This would imply $\overline{\mathcal{G}}_t \subseteq \overline{\mathcal{G}}_t'$, giving the relation (1.13). Since step functions are dense in H, we can get a sequence ϕ_k converging to ϕ in H. Then $Q_t\phi_k \to Q_t\phi$ in H. By the construction of $\int \phi dZ$, we have

$$\int_0^T (Q_t\phi_k)dZ \longrightarrow \int_0^T Q_t\phi dZ \quad \text{in } \Pi_0\text{-probability}. \tag{1.14}$$

Also

$$(Q_t\phi_k, \xi(\omega)) \longrightarrow (Q_t\phi, \xi(\omega)) \quad \text{for all } \omega \in \Omega. \tag{1.15}$$

Thus

$$\int_0^T (Q_t\phi_k)dY = (Q_t\phi_k, \xi) + \int_0^T (Q_t\phi_k)dZ$$

$$\longrightarrow (Q_t\phi,\xi) + \int_0^T(Q_t\phi)dZ = \int_0^T(Q_t\phi)dY$$

in $\tilde{\Pi}$-probability. Since $\int_0^T(Q_t\phi_k)dY$ is $\mathcal{G}'_t$ measurable, this shows that $\int_0^T(Q_t\phi)dY$ is $\overline{\mathcal{G}}'_t$ measurable. □

Note that for each ϕ, $\int_0^t\phi_u dY_u$, which we write in place of $\int_0^T(Q_t\phi)dY$, is defined $\tilde{\Pi}$-a.s. Since $L_0(\phi) = \int\phi dZ$ is a representation of m, by Theorem VI.1.1, we can choose a version of $\int\phi dZ$ which is $\mathcal{B}(H)\otimes\mathcal{A}_0$ measurable. Thus, in view of (1.16), we can choose a version $(\phi,(\omega,\omega_0)) \to [\int_0^T\phi dY](\omega,\omega_0)$ of $\int_0^T\phi dY$ which is $\mathcal{B}(H)\otimes\tilde{\mathcal{A}}$ measurable. From now on $\int_0^T\phi dY$ will always denote this measurable version. In view of Theorem 1.2, $\int_0^T\phi dY$ may be thought of as a function of Y and it will be convenient for us to do so.

Theorem 1.2 enables us to get a relation between the conditional expectation of an integrable function g on $(E,\mathcal{E},\alpha)$ given Q_ty, $E_\alpha(g|Q_ty)$ and the conditional expectation of $R_\alpha g$ given $\mathcal{F}_t^Y$. Theorem V.4.5 identifies the latter as the α-lifting of the former. The case when g: $E \to \mathbb{R}$ is a function of ω alone is important for the filtering problem, for then $R_\alpha(g) = g$ (regarded as a function on $\tilde{\Omega}$). The result will be used in the sequel.

<u>Theorem 1.3</u>: Let $g \in L^1(\Omega,A,\Pi)$. Then we have

$$R_\alpha[E_\alpha(g|Q_ty)] = E_{\tilde{\Pi}}[g|\mathcal{F}_t^Y]. \tag{1.16}$$

<u>Proof</u>: From the first part of Theorem V.4.5, it follows that

$\overline{\mathscr{D}}_{Q_t y} = \overline{\mathscr{G}}_t$ (= $\mathscr{F}_t^Y$ by Theorem 1.2). Also, we have $R_\alpha(g) = g \circ \rho = g$. Thus (1.16) is a consequence of (iii) in Theorem V.4.5.

Let us recall the Bayes formula for $E_\alpha(g|Q_t y)$ and the related notation: (from Sections VII.2 and IX.1).

$$E_\alpha[g|Q_t y] = \tilde{\pi}_t(g,y), \tag{1.17}$$

$$\tilde{\pi}_t(g,y) = \frac{\tilde{\sigma}_t(g,y)}{\tilde{\sigma}_t(1,y)}, \tag{1.18}$$

$$\tilde{\sigma}_t(g,y) = \int g(\omega)\exp((Q_t\xi(\omega),y) - \tfrac{1}{2}\|Q_t\xi(\omega)\|^2)d\Pi(\omega). \tag{1.19}$$

Also, observe that for $f: \mathbb{R}^d \to \mathbb{R}$, we have been using the notation

$$\pi_t(f,y) = \tilde{\pi}_t(f(X_t),y) \tag{1.20}$$

$$\pi_{st}(f,y) = \tilde{\pi}_t(f(X_s),y) \tag{1.21}$$

$$\sigma_t(f,y) = \tilde{\sigma}_t(f(X_t),y) \tag{1.22}$$

$$\sigma_{st}(f,y) = \tilde{\sigma}_t(f(X_s),y). \tag{1.23}$$

If $g \geq 0$ is such that $\int g d\Pi = 1$, then $\tilde{\sigma}_t(g,\cdot) \in \mathscr{I}(H)$ and we have from Theorem VI.3.2 and VI.2.5 that

$$\begin{aligned} R_\alpha(\tilde{\sigma}_t(g,y)) &= R_n(\tilde{\sigma}_t(g,\cdot)) \qquad (1.24) \\ &= \int g(\omega)\exp(L_1(Q_t\xi(\omega)) - \tfrac{1}{2}\|Q_t\xi(\omega)\|^2)d\Pi(\omega) \\ &= \int g(\omega)\exp(\textstyle\int_0^t h_u(X_u(\omega))dY_u - \tfrac{1}{2}\int_0^t |h_u(X_u(\omega))|^2 du)d\Pi(\omega). \end{aligned}$$

In this integral, Y is treated as a constant. In other words, the above integral is to be interpreted as

$$\int_H \exp\left(\int_0^t \phi_u dY_u - \tfrac{1}{2}|\phi_u|^2 du\right) d\nu(\phi)$$

where ν on H is given by $\nu(B) = \int g(\omega) 1_{(\xi(\omega)\in B)} d\Pi(\omega)$. The same comments applies to the two integrals appearing below. From the linearity of R_α and the integral, it follows that (1.24) is valid for all $g\colon \Omega \to \mathbb{R}$ such that $\int g d\Pi < \infty$. Define

$$\hat{\sigma}_s(f,Y) := E_\Pi\left[f(X_s)\exp\left(\int_0^s h_u(X_u)dY_u - \tfrac{1}{2}\int_0^s |h_u(X_u)|^2 du\right)\right] \tag{1.25}$$

and

$$\hat{\sigma}_{st}(f,Y) := E_\Pi\left[f(X_s)\exp\left(\int_0^t h_u(X_u)dY_u - \tfrac{1}{2}\int_0^t |h_u(X_u)|^2 du\right)\right] \tag{1.26}$$

for $0 \le s \le T$, $0 \le t \le T$, $f\colon \mathbb{R}^d \to \mathbb{R}$ such that $E_\Pi|f(X_s)| < \infty$. For s,t,f as above, write

$$\hat{\pi}_s(f,Y) := \frac{\hat{\sigma}_s(f,Y)}{\hat{\sigma}_s(1,Y)} \tag{1.27}$$

and

$$\hat{\pi}_{st}(f,Y) := \frac{\hat{\sigma}_{st}(f,Y)}{\hat{\sigma}_{st}(1,Y)}\,. \tag{1.28}$$

The functional $\hat{\sigma}_s(f,Y)$ will be treated as a function of $Y \in C_0([0,T],\mathbb{R}^N)$ as well as a function of $\tilde{\omega} \in \tilde{\Omega}$, i.e., $\tilde{\omega} \to \hat{\sigma}_s(f,Y(\tilde{\omega}))$. The same applies to the other functionals defined above.

Theorem 1.3 and (1.24) have an important consequence which is our next result. Recall the notation

$$E_\alpha(f(X_t)|Q_t y) = \pi_t(f,y) \tag{1.29}$$

and

$$E_\alpha(f(X_s)|Q_t y) = \pi_{st}(f,y). \tag{1.30}$$

Theorem 1.4: For any f such that $E_\Pi|f(X_s)| < \infty$, we have

$$R_\alpha(\sigma_s(f,y)) = \hat{\sigma}_s(f,Y) \tag{1.31}$$

$$R_\alpha(\sigma_{st}(f,y)) = \hat{\sigma}_{st}(f,Y) \tag{1.32}$$

$$R_\alpha(\pi_s(f,y)) = \hat{\pi}_s(f,Y) \tag{1.33}$$

$$R_\alpha(\pi_{st}(f,y)) = \hat{\pi}_{st}(f,Y). \tag{1.34}$$

Further we also have

$$E_{\underset{\sim}{\Pi}}(f(X_s)|\mathcal{F}_s^Y) = \hat{\pi}_s(f,Y) \tag{1.35}$$

and

$$E_{\underset{\sim}{\Pi}}(f(X_s)|\mathcal{F}_t^Y) = \hat{\pi}_{st}(f,Y). \tag{1.36}$$

Proof: Relations (1.31) and (1.32) are direct consequences of (1.24). The next two relations follow from the properties of the lifting map and the fact that $\hat{\sigma}_t(1,Y) > 0$ a.s. and $\hat{\sigma}_{st}(1,Y) > 0$ a.s. The last two assertions follow from Theorem 1.2, (1.29) and (1.30). □

The above result (together with the previous one) shows that the "estimate" $\pi_t(f,y)$ of $f(X_t)$ based on $Q_t y$ in the white noise approach is consistent with the estimate $\hat{\pi}_t(f,Y)$ of $f(X_t)$ based on $\{Y_u, u \le t\}$ in the conventional approach.

We have obtained formulas (1.35) and (1.36) as a

consequence of the corresponding formulas for the white noise model and Theorem V.4.5 by identifying the α-lifting of $\pi_t(f,y)$. These formulas can be (and have been) derived directly without any reference to the white noise theory.

We will now show that the identification made in (1.33) and (1.34) yield methods of approximating the conditional distributions $\hat{\pi}_t(f,Y)$ for the filtering problem and $\hat{\pi}_{st}(f,Y)$ for prediction $(s > t)$ and smoothing problems $(s < t)$ for the model (1.2).

<u>Theorem 1.5</u>: Let $\{P_k\} \subseteq \mathscr{P}$, $P_k \xrightarrow{s} I$. Then

$$R_\alpha(\pi_s(f,P_k y)) \longrightarrow \hat{\pi}_s(f,Y) \tag{1.37}$$

and

$$R_\alpha(\pi_{st}(f,P_k y)) \longrightarrow \hat{\pi}_{st}(f,Y) \tag{1.38}$$

in $\tilde{\Pi}$-probability, for any $f\colon \mathbb{R}^d \to \mathbb{R}$ such that $E_\Pi|f(X_s)| < \infty$.

<u>Proof</u>: Note that $\sigma_s(f,\cdot) \in \mathscr{P}(H)$ if $f \geq 0$, $E_\Pi f(X_s) = 1$. Hence, $\sigma_s(f,\cdot) \in \mathscr{L}^*(H,\mathscr{C},n)$ and $\sigma_s(f,\cdot) \in \mathscr{U}(y)$. By linearity of $\sigma_s(f,\cdot)$ in f, it follows that these relations continue to hold for any f such that $E_\Pi|f(X_s)| < \infty$. It follows that

$$R_n(\sigma_s(f,P_k\cdot)) \longrightarrow R_n(\sigma_s(f,\cdot)) = R_\alpha(\sigma_s(f,y)) \tag{1.39}$$

in $\tilde{\Pi}$-probability. Also $\phi \to \sigma_s(f,P_k\phi)$ is a cylinder function and so

$$R_n(\sigma_s(f,P_k\cdot)) = R_\alpha(\sigma_s(f,P_k y)). \tag{1.40}$$

Therefore, we have

$$R_\alpha(\sigma_s(f,P_k y)) \longrightarrow R_\alpha(\sigma_s(f,y)) = \hat{\sigma}_s(f,Y) \tag{1.41}$$

in $\tilde{\Pi}$-probability. Since $\hat{\sigma}_s(1,Y) > 0$ a.s. $\tilde{\Pi}$, (1.41) implies (1.37) since

$$R_\alpha(\pi_s(f,P_k y)) = \frac{R_\alpha(\sigma_s(f,P_k y))}{R_\alpha(\sigma_s(1,P_k y))}$$

by Theorem V.2.7 and $\hat{\pi}_s(f,Y) = \hat{\sigma}_s(f,Y)/\hat{\sigma}_s(1,Y)$ by (1.27). The other relation (1.38) follows similarly. □

For some specific choices of P_k, it is possible, as we will now show, to express $R_\alpha(\pi_s(f,P_k y))$ as a function of Y and obtain an explicit approximation result for $\hat{\pi}_t(f,Y)$. We need to recall some definitions and notations from Section III.4. Let $\mathcal{H}$ be the Hilbert space given by (III.4.23), i.e.,

$$\mathcal{H} = \{\eta\colon [0,T] \to \mathbb{R}^N\colon \|\eta\|^2 := \sum_{j=1}^{N} \int_0^T (D\eta_s^j)ds < \infty\},$$

where $D\eta_s^j = \frac{d}{ds}\eta_s^j$ (and is assumed to exist almost everywhere for $\eta \in \mathcal{H}$). Let γ be the injection mapping from $\mathcal{H} \to \Omega_0$. The dual Ω_0^* of Ω is given by

$$\Omega_0^* = \{\theta = (\theta^1,\ldots,\theta^N)\colon \theta^i \text{ a signed measure on } (0,T]\}$$

and the action $\theta[\omega_0]$ of $\theta \in \Omega_0^*$ on $\omega_0 \in \Omega_0$ is given by

$$\theta[\omega_0] = \sum_{j=1}^{N} \int_0^T Z_u^j(\omega_0)d\theta(u).$$

We will also write $\theta[Z]$ for $\theta[\omega_0]$, where $Z = (Z_t)$ represents the coordinate process. Let γ^* be the adjoint of γ. γ^* is explicitly given by (III.4.27). The range of γ^* is the class of right continuous functions η of bounded variation, continuous and vanishing at T.

Let $\hat{P} \in \mathcal{P}(\mathcal{H})$ be an orthogonal projection with range $\hat{P} \subseteq$ range γ^*. Then we can extend $\hat{P}$ to Ω_0 in a canonical way as follows. Choose an orthonormal basis $(\eta_1, \eta_2, \ldots, \eta_k)$ in range $\hat{P}$. Then for all $\eta \in \mathcal{H}$, we have

$$\hat{P}\eta = \sum_{j=1}^{k} (\eta, \eta_j)\eta_j = \sum_{j=1}^{k} (\gamma^{*-1}\eta_j)[\gamma(\eta)]\eta_j$$

by the property of the adjoint map γ^*. Here, γ^{*-1}: range $\gamma^* \to \Omega_0^*$ denotes the inverse of γ^*. Thus we can extend $\hat{P}$ to Ω_0 (as a continuous mapping from Ω_0 into $\mathcal{H}$) by defining

$$\hat{P}Z = \sum_{j=1}^{k} (\gamma^{*-1}\eta_j)[Z]\eta_j. \tag{1.42}$$

Since $\mathcal{H}$ is dense in Ω_0, the continuous extension is unique. Recall that J: $\mathcal{H} \to H$ defined by $J\eta = D\eta$ is an isometry so that if $P \in \mathcal{P}(H)$, then $\hat{P} = J^{-1}PJ \in \mathcal{P}(\mathcal{H})$. Our aim is to prove that if $P_k \in \mathcal{P}(H)$ is such that range $\hat{P}_k (= J^{-1}P_kJ) \subseteq$ range γ^*, then

$$R_\alpha(\pi_s(f, P_k y)) = \pi_s(f, J(\hat{P}_k Y)). \tag{1.43}$$

Substituting from (1.43) in (1.37) we obtain an explicit approximation for $\hat{\pi}_s(f, Y)$. To prove (1.43) we need the following lemma.

Let $\int\phi dZ$ be the Wiener integral and $\int\phi dY$ be as defined in (1.6).

Lemma 1.6: Let $\eta \in$ range γ^*. Then

$$\int_0^T (J\eta)dZ = (\gamma^{*-1}\eta)[Z] \quad \text{a.s. } \Pi_0 \tag{1.44}$$

and

$$\int_0^T (J\eta)dY = (\gamma^{*-1}\eta)[Y] \quad \text{a.s. } \tilde{\Pi}. \tag{1.45}$$

Proof: First, suppose that $\theta = \gamma^{*-1}\eta$ is given by

$$\theta^j(A) = \sum_{i=1}^{r} b_i^j 1_A(t_{i+1}), \quad A \in \mathcal{B}((0,T]) \tag{1.46}$$

where $b_i = (b_i^j) \in \mathbb{R}^N$, $i = 1,2,\dots,r$ and $0 = t_1 < t_2 < \dots < t_{r+1} = T$. Then as seen in Section III.4,

$$\eta_s = \gamma^*(\theta)_s = \int_0^s \{\sum_{i=1}^{r} a_i 1_{(t_i,t_{i+1}]}(t)\}dt$$

where $a_i = \Sigma_{k=i}^r b_k$. Then

$$(J\eta)_t = \sum_{i=1}^{r} a_i 1_{(t_i,t_{i+1}]}(t)$$

and hence

$$\int_0^T (J\eta)dZ = \sum_{j=1}^{N} \sum_{i=1}^{r} a_i^j (Z_{t_{i+1}}^j - Z_{t_i}^j).$$

It is easy to check (and was proved in Section III.4) that

$$(\gamma^{*-1}\eta)[Z] = \theta[Z] = \sum_{j=1}^{N} \sum_{i=1}^{r} a_i^j (Z^j_{t_{i+1}} - Z^j_{t_i}).$$

Hence, (1.44) holds if $\theta = \gamma^{*-1}\eta$ is of the form (1.46). In general, let $\theta = \gamma^{*-1}\eta \in \Omega_0^*$. We can find a sequence $\theta_k \in \Omega_0^*$ such that each θ_k is of the form (1.46) and

$$\theta_k[Z] \longrightarrow \theta[Z] \quad \text{for all } Z \in \Omega_0. \tag{1.47}$$

This is possible because finite linear combinations of Dirac measures (measures concentrated at one point) are dense in $\mathcal{M}((0,T])$ in the topology of weak convergence. Let $\eta(k) = \gamma^*(\theta_k)$. Then

$$\eta_s(k) = \left(\int_0^s \theta_k^j((t,T])dt\right)_{1\le j\le N}, \quad 0 \le s \le T \tag{1.48}$$

(see (III.4.27)). Since $\theta_k^j \to \theta^j$ in the topology of weak convergence, we have

$$\theta_k^j((t,T]) \longrightarrow \theta^j((t,T]) \quad \text{for a.e. } t$$

and

$$\theta_k^j((0,T]) \longrightarrow \theta^j((0,T]).$$

Hence by the dominated convergence theorem,

$$\int_0^T |\theta_k^j((t,T]) - \theta^j((t,T])|^2 dt \longrightarrow 0. \tag{1.49}$$

(1.49) implies that $\eta(k) \to \eta$ in $\mathcal{H}$ and hence

$$J\eta(k) \longrightarrow J\eta \qquad \text{in } H.$$

Since $\phi \to \int_0^T \phi dZ$ is an isometry from H into $L^2(\Omega_0, \mathcal{A}_0, \Pi_0)$,

$$\int_0^T (J\eta(k))dZ \longrightarrow \int_0^T (J\eta)dZ \quad \text{in } \Pi_0\text{-probability.} \tag{1.50}$$

Now (1.47), (1.50) and the fact that (1.44) holds for $\eta = \eta(k)$ implies that (1.44) holds for the given η as well.

To prove (1.45), note that for $\eta \in$ range γ^*

$$\sum_{j=1}^{N} \int_0^T (J\eta)_t^j (J\eta')_t^j dt = (\eta,\eta') = (\gamma^{*-1}\eta)[\gamma(\eta')]. \tag{1.51}$$

by the definition of the adjoint map γ^*. Let $\hat{\xi}(\omega) \in \mathcal{H}$ be given by

$$\hat{\xi}_t(\omega) = \int_0^t \xi_u(\omega)du = \int_0^t h_u(X_u(\omega))du,$$

so that $Y = \gamma(\hat{\xi}) + Z$. Using (1.51) for $\hat{\xi}$, we get

$$\sum_{j=1}^{N} \int_0^T (J\eta)_t^j (J\hat{\xi})_t^j dt = (\gamma^{*-1}\eta)[\gamma(\hat{\xi})]. \tag{1.52}$$

Since $(\gamma^{*-1}\eta)[Y] = (\gamma^{*-1}\eta)[Z] + (\gamma^{*-1}\eta)[\gamma(\hat{\xi})]$, the required relation (1.45) follows from (1.6), (1.44) and (1.52). □

<u>Lemma 1.7</u>: Let $P \in \mathcal{P}(H)$ be such that $\hat{P} = J^{-1}PJ \in \mathcal{P}(\mathcal{H})$ satisfies

$$\text{range } \hat{P} \subseteq \text{range } \gamma^*.$$

Then

$$R_\alpha(\pi_s(f,Py)) = \pi_s(f,J(\hat{P}Y)) \tag{1.53}$$

and

$$R_\alpha(\pi_{st}(f,Py)) = \pi_{st}(f,J(\hat{P}Y)) \tag{1.54}$$

for all f such that $E|f(X_s)| < \infty$.

Proof: Choose a orthonormal basis $(\eta_1, \eta_2, \ldots, \eta_k)$ of range $\hat{P}$ and let $\phi_i = J\eta_i$. Define $g: \mathbb{R}^k \to \mathbb{R}$ by

$$g(a_1, a_2, \ldots, a_k) = \pi_s(f, \sum_{i=1}^{k} a_i\phi_i). \tag{1.55}$$

Then, since $Py = \Sigma_{i=1}^{k}(y, \phi_i)$, we have

$$g((y, \phi_1), \ldots, (y, \phi_k)) = \pi_s(f, Py). \tag{1.56}$$

Hence

$$R_\alpha(\pi_s(f, Py)) = g(L_1(\phi_1), \ldots, L_1(\phi_k)). \tag{1.57}$$

Now $L_1(\phi_i) = \int_0^T \phi_i dY = \int_0^T (J\eta_i) dY = (\gamma^{*-1}\eta_i)[Y]$

and hence

$$\sum_{i=1}^{k} L_1(\phi_i)\phi_i = J\left[\sum_{i=1}^{k} L_1(\phi_i)\eta_i\right] = J\left[\sum_{i=1}^{k} (\gamma^{*-1}\eta_i)[Y]\eta_i\right]$$

$$= J(\hat{P}Y).$$

This relation along with (1.57) and (1.55) imply (1.53). The other assertion can be proved similarly. □

The above result is also valid if π is replaced by σ -- the same proof holds.

We are now in a position to show that the conditional expectations $\hat{\pi}_s(f, Y)$ and $\hat{\pi}_{st}(f, Y)$ occurring in the conventional theory of filtering, smoothing, prediction can be approximated by the corresponding conditional expectations $\pi_s(f, y)$, $\pi_{st}(f, y)$ in the white noise theory.

Theorem 1.8: Let $\hat{P}_k \in \mathcal{P}(\mathcal{H})$ be such that $\hat{P}_k \overset{s}{\rightarrow} I$ and range $\hat{P}_k \subseteq$ range γ^*. Then for all $f: \mathbb{R}^d \rightarrow \mathbb{R}$ such that $E|f(X_s)| < \infty$, we have

$$\pi_s(f, J(\hat{P}_k Y)) \longrightarrow \hat{\pi}_s(f, Y)$$

and

$$\pi_{st}(f, J(\hat{P}_k Y)) \longrightarrow \pi_{st}(f, Y)$$

in $\tilde{\Pi}$-probability. This is a direct consequence of Theorem 1.5 and Lemma 1.7.

An interesting application of the above theory occurs when $\{\hat{P}_k Y\}$ is a polygonal approximation to Y. We have to show in this case that the function obtained by linearly interpolating Y over suitable intervals can be represented as $\hat{P}Y$.

Fix $0 = t_0 < t_1 < \ldots < t_r = T$. Let $\psi_\cdot(i)$: $[0,T] \rightarrow \mathbb{R}$ be defined by

$$\psi_s(i) = (t_i - t_{i-1})^{-\frac{1}{2}} \int_0^s 1_{[t_{i-1}, t_i]}(\omega) du.$$

Let $\eta(j,i) \in \mathcal{H}$ be defined by

$$\eta(j,i) = a_j \psi(i)$$

where $a_j = (a_j^k)$ is the unit vector in $\mathbb{R}^N$ given by $a_j^j = 1$, $a_j^k = 0$ for $k \neq j$. For any $\bar{\eta} \in \mathcal{H}$,

$$(\bar{\eta}, \eta(j,i)) = \sum_{k=1}^{N} \int_0^T (D\bar{\eta}_s^k) D\eta_s^k(j,i) ds$$

$$= (t_j - t_{j-1})^{-\frac{1}{2}} \int_0^T (D\overline{\eta}_s^j) 1_{(t_{i-1}, t_i]}(s) ds$$

$$= (t_j - t_{j-1})^{-\frac{1}{2}} (\overline{\eta}_{t_i}^j - \overline{\eta}_{t_{i-1}}^j)$$

and hence

$$\sum_{j=1}^{N} (\overline{\eta}, \eta(j,i)) \eta_s(j,i)$$
$$= (\overline{\eta}_{t_i} - \overline{\eta}_{t_{i-1}})(t_i - t_{i-1})^{-1} \int_0^s 1_{(t_{i-1}, t_i]}(t) dt.$$

Let $\hat{P}$ be the orthogonal projection onto the linear span of $\{\eta(j,i): 1 \leq j \leq N, 1 \leq i \leq r\}$. It is easy to see that $\eta(j,i) \in$ range γ^* and hence range $\hat{P} \subseteq$ range γ^*. Also

$$(\hat{P}\overline{\eta}) = \sum_{i=1}^{r} \sum_{j=1}^{N} (\overline{\eta}, \eta(j,i)) \eta_s(j,i)$$

$$= \sum_{i=1}^{r} (\overline{\eta}_{t_i} - \overline{\eta}_{t_{i-1}})(t_i - t_{i-1})^{-1} \int_0^s 1_{(t_{i-1}, t_i]}(t) dt$$

so that

$$(\hat{P}\overline{\eta})_s = \overline{\eta}_{t_{i-1}} + (\overline{\eta}_{t_i} - \overline{\eta}_{t_{i-1}})(t_i - t_{i-1})^{-1}(s - t_{i-1}), \qquad (1.58)$$
$$t_{i-1} \leq s < t_i.$$

The continuous extension of $\hat{P}$ to Ω_0 is therefore given by

$$(\hat{P}Z)_s = Z_{t_{i-1}} + (Z_{t_i} - Z_{t_{i-1}})(t_i - t_{i-1})^{-1}(s - t_{i-1}); \qquad (1.59)$$
$$t_{i-1} \leq s \leq t_i.$$

Indeed $Z \to \hat{P}Z$ (defined above) is continuous and agrees with $\hat{P}\eta$ if $Z = \eta \in \mathcal{H}$.

For each $k \geq 1$, let $0 = t_0^k < \ldots < t_{r_k}^k = T$ be given such that

$$\lim_{k\to\infty} \left[\max_{1\leq i\leq r_k} |t_i^k - t_{i-1}^k| \right] = 0. \tag{1.60}$$

Let $\hat{P}_k \in \mathcal{P}(\mathcal{H})$ be defined by the right hand side of (1.58) with $(t_0^k, \ldots, t_{r_k}^k)$ in place of $(t_0, t_1, \ldots, t_r)$. We will check that $\hat{P}_k \xrightarrow{s} I$. Fix $\eta \in \mathcal{H}$ such that η is continuously differentiable. Let M be the upper bound of $|\dot{\eta}|$ where $\dot{\eta} = \frac{d}{ds}\eta$. Then

$$\begin{aligned} \frac{d}{ds}(\hat{P}_k\eta)_s &= \left[\eta_{t_i^k} - \eta_{t_{i-1}^k}\right]\left[t_i^k - t_{i-1}^k\right]^{-1} \quad \text{if } t_{i-1}^k < s < t_i^k, \\ &= \dot{\eta}(s_{i,k}^*) \end{aligned}$$

for some $s_{i,k}^* \in (t_{i-1}^k, t_i^k)$. Hence, $\frac{d}{ds}(\hat{P}_k\eta)_s$ is bounded by M for almost all $s \in [0,T]$ and if $s \notin \{t_i^k\colon i \leq r_k, k \geq 1\}$, then

$$\frac{d}{ds}(\hat{P}_k\eta)_s \longrightarrow \dot{\eta}_s.$$

Hence

$$\|\hat{P}_k\eta - \eta\| = \sum_{j=1}^{N} \int_0^T \left|\frac{d}{ds}(\hat{P}_k\eta)_s^j - \dot{\eta}_s^j\right|^2 ds \longrightarrow 0 \quad \text{as } k \to \infty \tag{1.61}$$

by the dominated convergence theorem.

Now let $\bar{\eta}$ be any element in $\mathcal{H}$. For an arbitrary positive ϵ we can find a continuously differentiable $\eta \in \mathcal{H}$ such that $\|\eta-\bar{\eta}\| < \epsilon$. Then $\|\hat{P}_k\eta - \hat{P}_k\bar{\eta}\| < \epsilon$ and in view of (1.61),

$$\limsup_{k\to\infty} \|\hat{P}_k\bar{\eta}-\bar{\eta}\| \le \limsup_{k\to\infty}\{\|\hat{P}_k\bar{\eta}-\hat{P}_k\eta\| + \|\hat{P}_k\eta-\eta\| + \|\eta-\bar{\eta}\|\}$$

$$\le 2\epsilon + \limsup_{k\to\infty} \|P_k\eta-\eta\| = 2\epsilon.$$

$\|\hat{P}_k\bar{\eta}-\bar{\eta}\| \to 0$ for all $\bar{\eta} \in \mathcal{H}$. Thus, $\hat{P}_k$ satisfy conditions of Theorem 1.8. We will state the conclusion of Theorem 1.8 for this choice of $\hat{P}_k$ explicitly, writing Y^k for $\hat{P}_kY$.

<u>Corollary 1.9</u>: For $k \ge 1$, let $0 = t_0^k < t_1^k < \ldots < t_{r_k}^k = T$ satisfy (1.60). Let $Y^k(\tilde{\omega})$ be defined by

$$Y_s^k(\tilde{\omega}) = Y_{t_{i-1}^k}(\tilde{\omega}) + \left[Y_{t_i^k}(\tilde{\omega})-Y_{t_{i-1}^k}(\tilde{\omega})\right]\left[t_i^k-t_{i-1}^k\right]^{-1}(s-t_{i-1}^k); \quad t_{i-1}^k \le s < t_i^k. \tag{1.62}$$

Then

$$\pi_t(f,\dot{Y}^k) \longrightarrow \hat{\pi}_t(f,Y)$$

and

$$\pi_{st}(f,\dot{Y}^k) \longrightarrow \hat{\pi}_{st}(f,Y)$$

in $\tilde{\Pi}$-probability. Similar results are valid if we replace Π by σ.

<u>Remark 1.1</u>: Throughout this section, the signal process has

been assumed to be $\mathbb{R}^d$-valued. It is to be noted that this did not play any special role and (X_t) could as well have been S-valued, for any measurable space $(S,\mathcal{S})$. However, the finite dimensionality of the noise and observation processes has played an important part.

2. CONSISTENCY OF THE UNNORMALIZED CONDITIONAL DENSITIES FOR DIFFUSION SIGNAL PROCESSES

In this section, we assume that the signal process (X_t) is an $\mathbb{R}^d$-valued diffusion. Under suitable conditions on the diffusion and drift coefficients a,b and on the function h of the filtering model, it was shown in Chapter VII that the unnormalized conditional distribution of X_t given Q_ty admits a density $p_t(x,y)$ with respect to d-dimensional Lebesgue measure. If $Y_t = \int_0^t y_s ds$, then $Y \in \mathcal{H}$ and $p_t(x,y)$ may be denoted by $p_t(x,\dot{Y})$ (the dot denoting differentiation with respect to t). It will now be shown that the map $Y \to p_t(x,\dot{Y})$ from $\mathcal{H}$ to $C([0,T]\times\mathbb{R}^d)$ has a continuous extension to Ω^* (defined below) denoted by $\hat{p}_t(s,Y)$. Moreover, we will prove that $\hat{p}_t(x,Y)$ is the unnormalized conditional density of X_t given $\mathcal{F}_t^Y$ for the model (1.2).

The main result on which the above consistency property is based concerns the continuous dependence of solutions of PDE's on their coefficients. We will give a probabilistic proof of the latter fact using the Feynman-Kac formula and a result on the continuous dependence of diffusion measures on the coefficients.

Similar consistency properties for the smoothing and

prediction conditional densities are proved in the last two subsections.

Theorem 2.1: Consider the sequence $\{M_t^{(k)}\}_{k\geq 0}$ of differential operators on $\mathbb{R}^d$ given by

$$M_t^{(k)}f(x) := \tfrac{1}{2}\sum_{i,j=1}^{d} a_{ij}^k(t,x)\frac{\partial^2 f(t,x)}{\partial x^i \partial x^j} + \sum_{i=1}^{d} b_i^k(t,x)\frac{\partial f(t,x)}{\partial x^i} \tag{2.1}$$

for $f \in C^2(\mathbb{R}^d)$, $t \in [0,T]$, $k \geq 0$. Suppose that a_{ij}^k, b_i^k for $k \geq 0$, $1 \leq i$, $j \leq d$ satisfy, for some constants K_1, K_2, the following conditions:

$$a_{ij}^k,\ b_i^k \text{ are continuous functions of } (t,x); \tag{2.2}$$

$$\sup_{x\in\mathbb{R}^d}\ \sup_{0\leq t\leq T} |a_{ij}^k(t,x)| \leq K_1; \tag{2.3}$$

$$\sup_{0\leq t\leq T} |b_i^k(t,x)| \leq K_2(1+|x|). \tag{2.4}$$

Further, suppose that for each k, there exists $\delta_k > 0$ such that

$$\sum_{i,j=1}^{d} a_{ij}^k(t,x)\lambda_i\lambda_j \geq \delta_k \sum_{i=1}^{d} \lambda_i^2 \tag{2.5}$$

for all $0 \leq t \leq T$, $x \in \mathbb{R}^d$, $\lambda_i \in \mathbb{R}^d$, $k \geq 0$.

Let $\{c^k\}_{k\geq 0}$ be a sequence of continuous real-valued functions on $[0,T]\times\mathbb{R}^d$ such that for a constant $K_3 < \infty$, $\epsilon_1 > 0$,

$$c^k(t,x) \leq K_3(1+|x|^2)^{1-\epsilon}, \tag{2.6}$$

for $0 \leq t \leq T$, $x \in \mathbb{R}^d$, $k \geq 0$.

Let $\{g^k\}_{k\geq 0}$ be a sequence of continuous functions satisfying

$$|g^k(x)| \leq \exp(K_4(1+|x|^2)^{1-\epsilon_2}) \tag{2.7}$$

for some constants $K_4 < \infty$, $\epsilon_2 > 0$. Let $v^k(t,x)$, $u^k(t,x)$ be classical solutions to

$$\frac{\partial v^k(t,x)}{\partial t} + M_t^k v^k(t,x) + c^k(t,x)v^k(t,x) = 0 \tag{2.8}$$

$$v^k(T,x) = g^k(x), \tag{2.9}$$

and

$$\frac{\partial u^k(t,x)}{\partial t} = M_t^k u^k(t,x) + c^k(t,x)u^k(t,x) \tag{2.10}$$

$$u^k(0,x) = g^k(x), \tag{2.11}$$

respectively, satisfying, in addition, the growth condition,

$$|u^k(t,x)| \leq \exp\left[C_k(1+|x|^2)^{1-\epsilon^k}\right] \tag{2.12}$$

and

$$|v^k(t,x)| \leq \exp\left[C_k(1+|x|^2)^{1-\epsilon^k}\right] \tag{2.13}$$

for some constants $C_k < \infty$, $\epsilon^k > 0$.

Finally, suppose that $a^k \to a^0$, $b^k \to b^0$, $c^k \to c^0$, $g^k \to g^0$, in the following sense: for every $R > 0$,

$$\lim_{k\to\infty} \int_0^T \sup_{|x|\le R} \{ \sum_{i,j=1}^{d} |a_{ij}^k(t,x) - a_{ij}^0(t,x)| \tag{2.14}$$

$$+ \sum_{i=1}^{d} |b_i^k(t,x) - b_i^0(t,x)| + |c^k(t,x) - c^0(t,x)|\}dt = 0$$

$$\lim_{k\to\infty} \sup_{|x|\le R} |g^k(x) - g^0(x)| = 0. \tag{2.15}$$

Then, $u^k(t,x) \to u^0(t,x)$ and $v^k(t,x) \to v^0(t,x)$ as $k \to \infty$ uniformly on compact subsets on $[0,T]\times\mathbb{R}^d$.

<u>Proof</u>: For $k \ge 0$, $(s,x) \in [0,T)\times\mathbb{R}^d$, let $Q_{s,x}^k$ be the solution to the martingale problem for a^k,b^k starting at (s,x). (See Theorem II.3.4.) $Q_{s,x}^k$ is a probability measure on $(\Omega_d = C([0,T],\mathbb{R}^d), \mathcal{G})$ where $\mathcal{G} = \sigma(Z_u: u \ge 0)$ and $Z_u(\omega') = \omega'(u)$, $\omega' \in \Omega_d$, $u \ge 0$ is the coordinate process on Ω^d. The assumptions (2.2) - (2.5) ensure the existence of $\{Q_{s,x}^k\}$.

From the conditions (2.3), (2.4) and (2.14) we have $Q_{s,x}^k \to Q_{s,x}^0$ in the topology of weak convergence uniformly over compact subsets of $[0,T]\times\mathbb{R}^d$, i.e., for any $s_k \to s$ and $x_k \to x$ (see Theorem II.3.5),

$$Q_{s_k,x_k}^k \longrightarrow Q_{s,x}^0 \quad \text{as } k \to \infty. \tag{2.16}$$

The conditions (2.3) and (2.4) imply that for a positive constant δ and a finite constant K_5, (see Theorem II.3.3),

$$E_{Q^k_{s,x}}[\exp(\delta|Z|^{*2}_T)] \leq K_5. \tag{2.17}$$

where $|Z|^*_t = \left[\sup_{0\leq u\leq t} \sum_{j=1}^{d} |Z^j_u|^2\right]^{\frac{1}{2}}$.

Since for each $k \geq 0$, a^k, b^k, c^k, v^k satisfy the conditions of the Theorem II.4.2, we have the following representation of v^k given by the Feynman-Kac formula:

$$v^k(s,x) = E_{Q^k_{s,x}}[g^k(Z_T)\exp(\int_s^T c^k(u,Z_u)du)] \tag{2.18}$$

for $k \geq 0$, $s \in [0,T]$, $x \in \mathbb{R}^d$.

It should be noted that our assumptions are precisely the ones that ensure the existence of $\{Q^k_{s,x}\}$ and the validity of (2.16), (2.17) and (2.18).

We now show that $v^k(s_k,x_k) \to v^0(s_0,x_0)$ as $s_k \to s_0$ and $x_k \to x_0$. Let $f_k\colon \Omega_d \to \mathbb{R}$ be defined by

$$f_k(\omega') = g^k(Z_t(\omega'))\exp(\int_{s_k}^t c^k(u,Z_u(\omega'))du) \tag{2.19}$$

so that

$$v^k(s_k,x_k) = E_{Q^k_{s_k,x_k}} f_k. \tag{2.20}$$

The convergence of c^k to c^0 and g^k to g^0 as required by the assumptions (2.14), (2.15) implies that for each $R > 1$,

$$\lim_{k\to\infty} \sup_{|Z|^*_T(\omega') \leq R} |f_k(\omega')-f_0(\omega')| = 0. \tag{2.21}$$

Also, the assumptions (2.6), (2.7) imply that for some constants $K_6 < \infty$, $\epsilon_4 > 0$,

$$|f_k(\omega')| \leq \exp[K_6(1+|Z|_T^{*2}(\omega'))^{1-\epsilon_4}]. \tag{2.22}$$

For each $R > 1$, let $G_R\colon \Omega_d \to \mathbb{R}$ be a continuous function such that

$$1_{\{|Z|_T^* \leq R\}} \leq G_R \leq 1_{\{|Z|_T^* \leq R+1\}}.$$

Then (2.21), (2.22) imply that for each $R > 1$, the family of continuous function $\{f_k G_R\colon k \geq 1\}$ is uniformly bounded and converges uniformly to $f_0 G_R$. Thus the weak convergence of $Q^k_{s_k,x_k}$ to $Q^0_{s_0,x_0}$ gives

$$E_{Q^k_{s_k,x_k}}[f_k G_R] \longrightarrow E_{Q^0_{s_0,x_0}}[f_k G_R] \quad \text{as } k \to \infty \tag{2.23}$$

for every $R > 1$. We also have

$$E_{Q^k_{s_k,x_k}}|f_k(1-G_R)| \leq E_{Q^k_{s_k,x_k}} 1_{\{|Z|_T^* > R\}} \exp(K_6(1+|Z|_T^{*2})^{1-\epsilon_4})$$

in view of (2.22). The estimate (2.17) now yields

$$\lim_{R\to\infty} \sup_{k\geq 0} E_{Q^k_{s_k,x_k}}|f_k(1-G_R)| = 0.$$

This and (2.23) imply that

$$\lim_{k\to\infty} E_{Q^k_{s_k,x_k}} f_k = E_{Q^0_{s_0,x_0}} f_0$$

which is the same as

$$\lim_{k\to\infty} v^k(s_k,x_k) = v^0(s_0,x_0).$$

Thus, $v^k(s,x) \to v^0(s,x)$ uniformly on compact subsets of $[0,T]\times\mathbb{R}^d$.

The equations (2.10), (2.11) can be transformed to (2.8) - (2.9) by the substitution $s = T-t$ and since the assumptions are invariant under this substitution, we can conclude that $u^k \to u^0$ from what we have proved above. To be precise, define $\bar{u}^k(s,x) = u^k(T-s,x)$, $\bar{c}^k(s,x) = c^k(T-s,x)$, and $(\bar{M}_s^k f)(x) = (M_{T-s}^k f)(x)$. Then, it is easy to see that $\bar{u}^k$ satisfies

$$\frac{\partial\bar{u}^k(s,x)}{\partial s} + \bar{M}_s^k\bar{u}^k(s,x) + \bar{c}_s^k\bar{u}^k(s,x) = 0$$

$$\bar{u}(T,x) = g^k(x).$$

The collection $\{\bar{M}^k,\bar{c}^k,\bar{u}^k,g^k\}$ satisfies the assumptions made in the statement of the theorem for $\{M^k,c^k,v^k,g^k\}$ and hence $\bar{u}^k$ converges to $\bar{u}^0$ uniformly on compact subsets of $[0,T]\times\mathbb{R}^d$. The same conclusion, therefore, holds with $\bar{u}^k,\bar{u}^0$ replaced by u^k and u^0. This completes the proof. □

Remark 2.1: From the Feynman-Kac representation, it follows that if $g^0 \geq 0$, then u^0 and v^0 are also positive.

Let us return to the setup of Section VII.4 so that (X_t) is a diffusion process with drift coefficient b and diffusion coefficient a. For the remaining part of this section we will assume the conditions of Theorem VII.4.5 on a,b,h, their derivatives and the density p_0 of X_0. Recall that under these conditions, the unnormalized conditional density $p_t(x,y)$ of X_t given $Q_t y$ (for the model (1.1)) exists and is given by

$$p_t(x,y) = \psi_t(x,y) \exp\left[\sum_{k=1}^{N} h_t^k(x) \int_0^t y_s^k ds\right] \tag{2.24}$$

where $\psi_t(x,y)$ is the solution to (VII.4.53) - (VII.4.54).

For each $Y \in \Omega_0 = C([0,T],\mathbb{R}^N)$, consider the Cauchy problem

$$\frac{\partial\hat{\psi}_t(x,Y)}{\partial t} = \hat{L}_t^Y\hat{\psi}_t(x,Y) + \hat{c}_t^Y(x)\hat{\psi}(s,Y) \tag{2.25}$$

$$\hat{\psi}_0(x,Y) = p_0(x) \tag{2.26}$$

where

$$(\hat{L}_t^Y f)(x) = \tfrac{1}{2}\sum_{i,j=1}^{d} a_{ij}(t,x)\frac{\partial^2 f(x)}{\partial x^i \partial x^j} + \sum_{i=1}^{d} \hat{b}_i(t,x,Y)\frac{\partial f(x)}{\partial x^i}$$
$$+ \hat{c}(t,x,Y)f(x),$$

$$\hat{b}_i(t,x,Y) = b_i^*(t,x) + \sum_{k=1}^{N}\left[\sum_{j=1}^{d} a_{ij}(t,x)\frac{\partial h_t^k(x)}{\partial x^j}\right]Y_t^k,$$

$$\hat{c}(t,x,Y) = c^*(t,x) + \sum_{k=1}^{N}\left[\sum_{i=1}^{d} b_i^*(t,x)\frac{\partial h_t^k(x)}{\partial x^i}\right.$$
$$\left. + \tfrac{1}{2}\sum_{i,j=1}^{d} a_{ij}(t,x)\frac{\partial^2 h_t^k(x)}{\partial x^i \partial x^j}\right.$$

$$+ \tfrac{1}{2} \sum_{i,j=1}^{d} a_{ij}(t,x) \frac{\partial h_t^k(x)}{\partial x^i} \frac{\partial h_t^k(x)}{\partial x^j} Y_t^k \Big] Y_t^k$$

and

$$\hat{c}_t^Y(x) = -\tfrac{1}{2}|h_t(x)|^2 - \sum_{k=1}^{N} \frac{\partial}{\partial t} h_t^k(x) Y_t^k.$$

Note that if $Y \in \mathscr{H}$ and $y = JY$ (i.e., $Y_t = \int_0^t y_s ds$), then $\hat{L}_t^Y$, c_t^Y, coincide with $\hat{L}_t^y$, $\hat{c}_t^y$ defined by (VII.4.46), (VII.4.48), and hence (2.25), (2.26) have a unique solution given by

$$\hat{\psi}_t(x,Y) = \psi_t(x,JY), \qquad Y \in \mathscr{H}. \tag{2.27}$$

Let $\Omega^* = \{\omega_0 \in \Omega_0 : t \to \omega_0(t)$ is Hölder continuous$\}$.

Theorem 2.2: Suppose that the conditions of Theorem VII.4.5 are satisfied.

(i) For all $Y \in \Omega^*$, the Cauchy problem (2.25), (2.26) admits a unique solution $\hat{\psi}_t(x,Y)$ in the class $\mathscr{G}((0,T])$ and, furthermore, $\hat{\psi}_t(x,Y) \geq 0$ for all $t \geq 0$, $x \in \mathbb{R}^d$, $Y \in \Omega^*$.

(ii) The mapping $Y \to \hat{\psi}_t(x,Y)$ from Ω^* into $C([0,T]\times\mathbb{R}^d)$ is continuous (in the topology of uniform convergence on compact subsets of $[0,T]\times\mathbb{R}^d$).

Proof: For each $Y \in \Omega^*$, the differential operators $\{M_t\}$ defined by

$$M_t f = \hat{L}_t^Y f + \hat{c}_t^Y f$$

satisfy the conditions of Theorem VII.4.1 and $p_0(x) \geq 0$. Thus part (i) of the Theorem follows from Theorem VII.4.1

and Remark 2.1 above. For (ii) note that if $Y^k \to Y^0$ in Ω_0 (i.e., uniformly), then taking $a^k \equiv a$, $b^k \equiv b(\cdot,\cdot,Y^k)$, $c^k \equiv \hat{c}(\cdot,\cdot,Y^k) + \hat{c}_{\bullet}^{Y^k}(\cdot)$, $g^k \equiv p_0$ and $u^k \equiv \hat{\psi}(\cdot,Y^k)$, it is easy to see that the conditions of Theorem 2.1 are satisfied. Hence the convergence of $\psi_t(x,Y^k)$ to $\psi_t(x,Y^0)$ in the required sense follows from Theorem 2.1. □

We are now in a position to establish the consistency result for the unnormalized conditional density for the filtering problem.

Theorem 2.3: For $Y \in \mathcal{H}$, define

$$\hat{p}_t(x,Y) = p_t(x,JY). \tag{2.28}$$

Then the mapping $Y \to \hat{p}_t(x,Y)$ from $\mathcal{H}$ into $C([0,T]\times\mathbb{R}^d)$ has a continuous extension to Ω^* (also denoted by $\hat{p}_t(x,Y)$). The extension is given by

$$\hat{p}_t(x,Y) = \hat{\psi}_t(x,Y)\ \exp\Big(\sum_{k=1}^{N} h_t^k(x)Y_t^k\Big) \tag{2.29}$$

for $Y \in \Omega^*$, $\hat{\psi}_t(x,Y)$ being the unique solution (in $\mathcal{G}((0,T]))$ to the Cauchy problem (2.25) - (2.26). Further, $\hat{p}_t(x,\cdot)$ is the unnormalized conditional density of X_t given $\mathcal{F}_t^Y$ (for the model (1.2)).

Proof: Define $\hat{p}_t(x,Y)$ by the relation (2.29). Then by Theorem 2.2, it follows that $Y \to \hat{p}_t(x,Y)$ is continuous in the topology of uniform convergence on compact subsets of

$[0,T]\times\mathbb{R}^d$. The relations (2.24) and (2.27) imply that $\hat{p}_t(x,Y)$ so defined also satisfies (2.28). Note that $\hat{\psi}_t(x,Y)$ and hence $\hat{p}_t(x,Y)$ are nonnegative.

It remains to show that $\hat{p}_t(x,Y)$ is the unnormalized conditional density of X_t given $\mathcal{F}_t^Y$. To see this, fix $f \in C_0(\mathbb{R}^d)$ and let

$$\sigma_t'(f,Y) = \int f(x)\hat{p}_t(x,Y)dx, \quad Y \in \Omega^*, \ t \in [0,T]. \tag{2.30}$$

Since f has compact support, $Y \to \sigma_t'(f,Y)$ is a continuous mapping from Ω^* into $\mathbb{R}$. Also, (2.28) implies

$$\sigma_t'(f,Y) = \sigma_t(f,JY), \quad \text{for } Y \in \mathcal{H}.$$

Recall that $\tilde{\Pi}\{\tilde{\omega}: Y(\tilde{\omega}) \in \Omega^*\} = 1$. Define $\sigma_t'(f,Y) = 0$ if $Y \in \Omega_0-\Omega^*$.

Let $Y^k(\tilde{\omega})$ be the polygonal approximation to $Y(\tilde{\omega})$ given by (1.62). Then $Y^k(\tilde{\omega}) \in \mathcal{H} \subseteq \Omega^*$ and $Y^k(\tilde{\omega})$ converges uniformly to $Y(\tilde{\omega})$. Hence for $Y(\tilde{\omega}) \in \Omega^*$,

$$\sigma_t'(f,Y^k(\tilde{\omega})) := \sigma_t(f,JY^k(\tilde{\omega})) \longrightarrow \sigma_t'(f,Y(\tilde{\omega})).$$

Since $\tilde{\Pi}\{\tilde{\omega}: Y(\tilde{\omega}) \in \Omega^*\} = 1$, this gives

$$\sigma_t(f,JY^k(\tilde{\omega})) \longrightarrow \sigma_t'(f,Y(\tilde{\omega})) \quad \text{a.s. } \tilde{\Pi}.$$

As seen in Theorem 1.8,

$$\sigma_t(f,JY^k) \longrightarrow \hat{\sigma}_t(f,Y) \quad \text{in } \tilde{\Pi}\text{-probability}.$$

Hence

$$\sigma_t'(f,Y) = \hat{\sigma}_t(f,Y) \qquad \text{a.s. } \tilde{\Pi}. \tag{2.31}$$

Usual arguments now give the validity of (2.31) for all bounded continuous functions f. Thus

$$E_{\tilde{\Pi}}[f(X_t)|\mathcal{F}_t^Y] = \frac{\int f(x)\hat{p}_t(x,Y)dx}{\int \hat{p}_t(x,Y)dx} \qquad \tilde{\Pi}\text{-a.s.}$$

for all $f \in C_b(\mathbb{R}^d)$. □

Consistency for the Smoothing Problem

We have seen in Section IX.2 that the unnormalized conditional density $p_{st}(x,y)$ of X_s given $Q_t y$ for the smoothing case (i.e., $s < t$) is given by

$$p_{st}(x,y) = \psi_s(x,y)v_t'(s,x,y)\exp\left(\sum_{i=1}^{N} h_s^i(x) \int_0^t y_u^i du\right) \tag{2.32}$$

where $\psi_s(x,y)$ is the solution to (VII.4.53) - (VII.4.54) and $v_t'(s,x,y)$ is the solution to (IX.1.33) and $v_t'(t,x,y) = 1$.

As shown in Theorem 2.3, the function $\psi_s(x,y)$, considered as a function of $Y(\cdot) = \int_0^{(\cdot)} y_u du$ has a continuous extension $\hat{\psi}_s(x,Y)$ which provides a continuous version of the unnormalized conditional density $\hat{p}(s,Y)$ of the filtering problem. In order to obtain a consistent solution to the smoothing problem, it remains to show that $v_t'(s,x,y)$ also has a continuous extension.

For $s < t$ (t fixed) and for each $Y \in \Omega^*$, let L'^Y_s be defined by

$$L'^{Y}_{s}f(x) = \tfrac{1}{2}\sum_{i,j=1}^{d} a_{ij}(s,x)\frac{\partial^2 f(x)}{\partial x^i \partial x^j} \tag{2.33}$$

$$+ \sum_{i=1}^{d} b'_i(s,x,Y)\frac{\partial f(x)}{\partial x^i} + c'(s,x,Y)f(x)$$

for $f \in C^2(\mathbb{R}^d)$, where

$$b'_i(s,x,Y) = b_i(s,x) + \sum_{k=1}^{N}\left[\sum_{j=1}^{d} a_{ij}(s,x)\frac{\partial h^k_s(x)}{\partial x^j}\right](Y^k_t - Y^k_s)$$

and

$$c'(s,x,Y) = \sum_{k=1}^{N}\left[\sum_{i=1}^{d} b_i(s,x)\frac{\partial h^k_s(x)}{\partial x^i} + \tfrac{1}{2}\sum_{i,j=1}^{d} a_{ij}(s,x)\frac{\partial^2 h^k_s(x)}{\partial x^i \partial x^j}\right.$$

$$\left. + \tfrac{1}{2}\sum_{i,j=1}^{d} a_{ij}(s,x)\frac{\partial h^k_s(x)}{\partial x^i}\,\frac{\partial h^k_s(x)}{\partial x^j}(Y^k_t - Y^k_s)\right](Y^k_t - Y^k_s).$$

If $Y \in \mathcal{H}$ and $y = JY$, then L'^{Y}_{s} coincides with the operator L'^{y}_{s} defined by (IX.1.36). Consider the equation

$$\frac{\partial \hat{v}'_t(s,x,Y)}{\partial s} + L'^{Y}_{s}\hat{v}'_t(s,x,Y) + c'^{Y}_{s}(x)\hat{v}'_t(s,x,Y) = 0, \quad s > t, \tag{2.34}$$

with

$$\hat{v}'_t(t,x,Y) = 1 \tag{2.35}$$

where

$$c'^{Y}_{s}(x) = -\tfrac{1}{2}|h_s(x)|^2 + \sum_{i=1}^{N}\frac{\partial}{\partial s}h^i_s(x)(Y^i_t - Y^i_s).$$

By Theorem VII.4.1, it follows that the Cauchy problem (2.34) - (2.35) admits a unique classical solution in the class $\mathcal{G}([0,t))$. By Theorem 2.1, it follows that $Y \to v'_t(\cdot,\cdot,Y)$ is a continuous mapping. The proof is similar to that of Theorem 2.2. Here again, c'^{Y} coincides with c'^{y} defined by (IX.1.35) if $Y \in \mathcal{H}$ and $y = JY$. Hence

$$v'_t(s,x,Y) = v'_t(s,x,y) \quad \text{if } Y \in \mathcal{H},\ y = JY. \tag{2.36}$$

The consistency result for the smoothing problem stated below follows from these observations.

Theorem 2.4: Fix $t \le T$. For $Y \in \Omega^*$, let $\hat{p}_{st}(x,Y)$, $s \le t$, $x \in \mathbb{R}^d$, be defined by

$$\hat{p}_{st}(x,Y) = \hat{\psi}_s(s,Y)\cdot\hat{v}'_t(s,x,Y) \exp\Big(\sum_{k=1}^{N} h^k_s(x)Y^k_t\Big) \tag{2.37}$$

where $\hat{\psi}_s(x,Y)$ is the unique solution to (2.25) - (2.26) and $\hat{v}'_t(s,x,Y)$ is the unique solution to (2.34) - (2.35).

(i) Then, for $Y \in \mathcal{H}$, with $y = JY$, we have

$$\hat{p}_{st}(x,Y) = p_{st}(x,y). \tag{2.38}$$

(ii) $Y \to \hat{p}_{st}(x,Y)$ is a continuous mapping from Ω^* into $C([0,T]\times\mathbb{R}^d)$.

(iii) $\hat{p}_{st}(x,Y)$ is a version of the unnormalized conditional density of X_s given $\mathcal{F}^Y_t$.

Proof: The relation (2.38) follows from (2.27), (2.32) and (2.36). The continuity of $Y \to \hat{p}_{st}(s,Y)$ follows from the continuity of $\hat{\psi}_s(x,Y)$, that of $\hat{v}'_t(s,x,Y)$ and the definition of

$\hat{p}_{st}(x,Y)$. The identification of $\hat{p}_{st}(x,Y)$ as the unnormalized conditional density is similar to the proof of Theorem 2.3. We will need to use the consistency (part (i)), continuity (part (ii)) and Corollary 1.9.

Consistency for the Prediction Problem

The unnormalized conditional density $p_{st}(s,y)$ in the prediction case $(s > t)$ in the white noise approach is related to $p_t(x,y)$ by the formula

$$p_{st}(s,y) = \int p_t(z,y)p(t,z,s,x)dz$$

as seen in Theorem IX.1.4. We will prove a similar formula $\hat{p}_{st}(x,Y)$ for the conventional model.

Theorem 2.5:

(i) Fix $t \in [0,T]$. For $Y \in \Omega^*$, $s \geq t$, define

$$\hat{p}_{st}(x,Y) := \int \hat{p}_t(z,Y)p(t,z,s,x)dz \tag{2.40}$$

where $p(t,z,s,x)$ is the transition probability function for (X_s) and $\hat{p}_t(z,Y)$ is the unnormalized conditional density of X_t given $\mathcal{F}_t^Y$. Then for $Y \in \mathcal{H}$ with $JY = y$, we have

$$\hat{p}_{st}(x,Y) = p_{st}(x,y) \tag{2.41}$$

and further, $\hat{p}_{st}(x,Y)$ is the unnormalized conditional density of X_s given $\mathcal{F}_t^Y$.

(ii) In addition to the conditions of Theorem VII.4.5, assume that h is bounded. Then

$Y \to \hat{p}_{st}(x,Y)$ is a continuous mapping from Ω^* into $C([t,T]\times\mathbb{R}^d)$.

Proof: For $B \in \mathscr{B}(\mathbb{R}^d)$, note that

$$\int_B \hat{p}_{st}(x,Y)dx = \int_B \int \hat{p}_t(z,Y)p(t,z,s,x)dzdx \tag{2.42}$$

$$= \int \hat{p}_t(z,Y)[\int_B p(t,z,s,x)dx]dz$$

$$= \int \hat{p}_t(z,Y)f(z)dz = E_{\underset{\sim}{\Pi}}[f(X_t)|\mathscr{F}_t^Y] \quad \text{a.s.}$$

where

$$f(z) = \int_B p(t,z,s,x)dx.$$

Then

$$f(X_t) = E_{\underset{\sim}{\Pi}}[1_{(X_s \in B)}|\mathscr{F}_t^X].$$

Since $\{Z_t\}$ is independent of $\{X_t\}$ (see the beginning of Section 1), we also have

$$f(X_t) = E_{\underset{\sim}{\Pi}}[1_{(X_s \in B)}|\mathscr{F}_t^{X,Z}]. \tag{2.43}$$

Since Y is related to X,Z via (1.2), it follows that

$$\mathscr{F}_t^Y \subseteq \mathscr{F}_t^{X,Z}$$

and hence for $s > t$,

$$E_{\underset{\sim}{\Pi}}[1_{(X_s \in B)}|\mathscr{F}_t^Y] = E_{\underset{\sim}{\Pi}}[E_{\underset{\sim}{\Pi}}[1_{(X_s \in B)}|\mathscr{F}_t^{X,Z}]|\mathscr{F}_t^Y]$$

$$= E_{\underset{\sim}{\Pi}}[f(X_t)|\mathscr{F}_t^Y] = \int_B \hat{p}_{st}(x,Y)dx \quad \text{a.e.}$$

by (2.42) and (2.43). This shows that $\hat{p}_{st}(x,Y)$ is a version of the unnormalized conditional density of X_s given Q_tY. The consistency, namely (2.41), follows from the consistency in the filtering case (2.28), and the defining equations (2.39) and (2.40).

For the continuity part, observe that for a fixed integer $R \geq 1$, we can choose constants K_1, K_2 by Lemma VII.4.2, such that $H_1(t,x)$ defined by VII.4.5 satisfies the following inequality for all $Y \in \Omega^*$ with $|Y|_T^* \leq R$

$$(\hat{L}_s^Y + \hat{c}_s^Y - \frac{\partial}{\partial s})H_1(s,x) \leq 0, \tag{2.44}$$

$$((\hat{L}_s^Y)^* + \hat{c}_s^Y + \frac{\partial}{\partial s})H_1^{-1}(s,x) \leq 0, \tag{2.45}$$

$$(L^*{}_s - \frac{\partial}{\partial s})H_1(s,x) \leq 0, \tag{2.46}$$

and

$$(L_s + \frac{\partial}{\partial s})H_1^{-1}(s,x) \leq 0. \tag{2.47}$$

Hence by Remarks VII.4.1, VII.4.2, we have

$$|\hat{\psi}_t(x,Y)| \leq K_3H_1(t,x), \tag{2.48}$$

for all $Y \in \Omega^*$ with $|Y|_T^* \leq R$, K_3 being a fixed constant. Since h is assumed to be bounded, (2.29) and (2.48) yield

$$|\hat{p}_t(x,Y)| \leq K_4H_1(t,x) \tag{2.49}$$

for all $Y \in \Omega^*$, $|Y|_T^* \leq R$, for a suitable constant K_4.

By Theorem VII.4.1, inequalities (2.46), (2.47) and (2.49) imply that $\hat{p}_{st}(x,Y)$ defined by (2.40) is the unique classical solution in $\mathcal{G}((t,T])$ to the Cauchy problem

$$\frac{\partial}{\partial s} \hat{p}_{st}(x,Y) = L_s^* \hat{p}_{st}(x,Y), \quad s > t \tag{2.50}$$

$$\hat{p}_{tt}(x,Y) = \hat{p}_t(x,Y). \tag{2.51}$$

If $Y^k \to Y$ in Ω^*, then note that $|Y^k|_T^* \leq R$ for some R. Then uniform convergence of $\hat{p}_t(x,Y^k)$ to $\hat{p}_t(x,Y)$ on compact subsets of $\mathbb{R}^d$ (as proved in Theorem 2.3) and the bound (2.49), which is now valid for all Y^k, yield the convergence of $\hat{p}_{st}(x,Y^k)$ to $\hat{p}_{st}(x,Y)$ uniformly in (s,x) belonging to any compact subset of $[s,T]\times\mathbb{R}^d$. This is a straightforward consequence of Theorem 2.1. □

3. CONSISTENCY OF THE MEASURE VALUED OPTIMAL FILTER

In this section we derive consistency results analogous to those obtained in Section 1 for the filtering, prediction and smoothing problems where the noise is allowed to be infinite dimensional. The white noise model for this context has been discussed in Chapter VIII for filtering and in Section IX.2 for prediction and smoothing. The corresponding theory in the conventional approach has been been considered, but not much progress seems to have been made.

Choice of Model for the Conventional Theory

Let $\mathcal{H}$ be a separable Hilbert space. The white noise model is now given by

$$y = \xi + e \tag{3.1}$$

or

$$y_t = h_t(X_t) + e_t \qquad 0 \le t \le T \tag{3.1)'$$

where $e = (e_t)$ is the identity mapping from $H = L^2([0,T],\mathcal{H})$ into itself, X_t is an S-valued process on $(\Omega,\mathcal{A},\Pi)$ and h: $[0,T]\times S \to \mathcal{H}$ is such that

$$\int_0^T \|h_u(X_u(\omega))\|_{\mathcal{H}}^2 du < \infty \quad \text{for all } \omega \in \Omega. \tag{3.2}$$

and $\xi(\omega) = (\xi_u(\omega)) = (h_u(X_u(\omega)))$.

The model (3.1) is considered on $(E,\mathcal{E},\alpha) = (\Omega,\mathcal{A},\Pi) \otimes (H,\mathcal{C},m)$, m being the canonical Gauss measure on H.

As in the finite dimensional case, one may be tempted to get the corresponding model in the conventional theory by integrating both sides in (3.1) from 0 to t, replacing $\int_0^t e_s ds$ by a $\mathcal{H}$-valued Brownian motion W_t and $\int_0^t y_s ds$ by Y_t. We would want the finite dimensional distributions of W_t to be the same as those of $\int_0^t e_s ds$. This demands that

$$E(W_t,\eta)_{\mathcal{H}}(W_s,\eta')_{\mathcal{H}} = (s\wedge t)(\eta,\eta')_{\mathcal{H}} \tag{3.3}$$

for $s,t \in [0,T]$ and $\eta,\eta' \in \mathcal{H}$. If $\mathcal{H}$ is infinite dimensional, such a process cannot be realized so as to have paths in $C_0([0,T],\mathcal{H})$ -- for if that were so, then we would have

$$E \exp(i(W_t,\eta)_{\mathcal{H}}) = \exp(-\tfrac{t}{2}\|\eta\|_{\mathcal{H}}^2) \tag{3.4}$$

and thus the distribution μ_t of W_t would be a countably additive extension of the Gauss measure m_t on H with variance parameter t. As we have seen in Chapter III, m_t does not admit such an extension.

Let $(\gamma,\mathcal{H},B)$ be an abstract Wiener space, so that B is a Banach space, η: $\mathcal{H} \to B$ is the injection and B is obtained as

R

the completion of $\mathcal{H}$ under a measurable seminorm. Denote the dual of B by B^*. It was remarked in Section III.4 (see Example 4.5) that there exists a Gaussian measure μ on $(\mathcal{X},\mathcal{B}(\mathcal{X}))$, where $\mathcal{X} = C_0([0,T],B)$ such that if $(W_t)_{t\geq 0}$ denotes the coordinate mappings on $\mathcal{X}$ (into B), then

$$E_\mu f_1[W_t] = 0 \tag{3.5}$$

and

$$E_\mu f_1[W_t]f_2[W_s] = (s\wedge t)(\gamma^*(f_1),\gamma^*(f_2))_{\mathcal{H}} \tag{3.6}$$

for $s,t \in [0,T]$, $f_1,f_2 \in B^*$ and $\gamma^*: B^* \to \mathcal{H}^* = \mathcal{H}$ is the adjoint of γ and $f[\eta]$, $f \in B^*$, $\eta \in B$ is the duality map. The existence of the measure μ can also be proved directly.

Let $(\tilde{\Omega},\tilde{\mathcal{A}},\tilde{\Pi}) = (\Omega,\mathcal{A},\Pi) \otimes (\mathcal{X},\mathcal{B}(\mathcal{X}),\mu)$ and consider the model

$$Y_t = \int_0^t \gamma(h_u(X_u))du + W_t, \qquad 0 \leq T \leq T. \tag{3.7}$$

Here the observation process Y_t takes values in B and is defined on $\tilde{\Omega}$: $Y_t(\omega,\omega') = \int_0^t \gamma(h_u(X_u(\omega)))du + W_t(\omega')$ for $(\omega,\omega') \in \tilde{\Omega}$. To establish a connection between the conditional distributions for the respective models (3.1) and (3.7) in the sense of Section 1, we first need to obtain a representation of m, the canonical Gauss measure on $H = L^2([0,T],\mathcal{H})$ with the representation space $(\mathcal{X},\mathcal{B}(\mathcal{X}),\mu)$. In order to do this, we need to introduce the Wiener integral $\int_0^t \phi dW$ for $\phi \in H$.

<u>The Wiener Integral</u>

Since $(\gamma,\mathcal{H},B)$ is an abstract Wiener space, $\mathcal{H}$ is dense in B and $\gamma^*(B^*)$ is dense in $\mathcal{H}$. This allows us to choose $\{f^j\} \subseteq B^*$ such that $\eta^j = \gamma^*(f^j)$ is a CONS in $\mathcal{H}$. Fix $\{f^j\}$. Define

$$W_t^j = f^j[W_t].$$

Then $\{W_t^j\colon t \geq 0\}$ is a sequence of independent real-valued standard Brownian motions. Note that for $\phi \in H = L^2([0,T],\mathcal{H})$,

$$\|\phi\|_H^2 = \int_0^T \|\phi_t\|_{\mathcal{H}}^2 dt = \int_0^T \sum_{j=1}^{\infty} (\phi_t, \eta^j)_{\mathcal{H}}^2 dt \tag{3.8}$$

$$= \sum_{j=1}^{\infty} \int_0^T (\phi_t, \eta^j)_{\mathcal{H}}^2 dt.$$

Let $I^j(\phi)$ $(j \geq 1)$ be defined by

$$I^j(\phi) = \sum_{i=1}^{j} \int_0^T (\phi_t, \eta^i)_{\mathcal{H}} dW_t^i$$

for $\phi \in H$. Then from the properties of a finite dimensional Wiener integral, we have

$$E_\mu |I^{j+r}(\phi) - I^j(\phi)|^2 = E_\mu \left| \sum_{i=j+1}^{j+r} \int_0^T (\phi_t, \eta^i)_{\mathcal{H}} dW_t^i \right|^2 \tag{3.9}$$

$$= \sum_{i=j+1}^{j+r} \int_0^T (\phi_t, \eta^i)_{\mathcal{H}}^2 dt.$$

Thus $\{I^j(\phi)\}$ is a Cauchy sequence in $L^2(\mathcal{X},\mathcal{B}(\mathcal{X}),\mu) = L^2(\mu)$ in view of (3.8), (3.9). This allows us to define

$$\int_0^T \phi dW = \lim_{r\to\infty} \sum_{i=1}^{r} \int_0^T (\phi_t, \eta^i) dW_t^i \tag{3.10}$$

where the limit is taken in $L^2(\mu)$. We then have

$$E_\mu \exp(i\int_0^T \phi dW) = \lim_{r\to\infty} E_\mu \exp(iI^r(\phi)) \tag{3.11}$$

$$= \lim_{r\to\infty} \exp(-\tfrac{1}{2}\sum_{j=1}^{r} \int_0^T (\phi_t, \eta^j)^2 dt)$$

$$= \exp(-\tfrac{1}{2}\|\phi\|_H^2).$$

As in the finite dimensional case, we define

$$\int_0^T \phi dY = (\phi, \xi)_H + \int_0^T \phi dW, \tag{3.12}$$

i.e.,

$$(\int_0^T \phi dY)(\omega, \omega') = (\phi, \xi(\omega))_H + (\int_0^T \phi dW)(\omega'), \tag{3.12'}$$

$$(\omega, \omega') \in \tilde{\Omega}.$$

For $\phi \in H$, $\int_0^T \phi dY \in \mathscr{L}(\tilde{\Omega}, \tilde{\mathscr{A}}, \tilde{\Pi})$. It is easy to check that

$$\int_0^T \phi dY = \lim_{r\to\infty} \sum_{j=1}^{r} \int_0^T (\phi_t, \eta^j)_{\mathscr{K}} dY_t^j \tag{3.13}$$

where $Y_t^j = f^j[Y_t]$ and the limit is in $\tilde{\Pi}$-probability. The quantity $\int_0^T (\phi_t, \eta^j)_{\mathscr{K}} dY_t^j$ is defined in Section 1 and is equal to

$$\int_0^T (\phi_t, \eta^j)_{\mathscr{K}} dY_t^j = \int_0^T (\phi_t, \eta^j)_{\mathscr{K}} (h_t(X_t), \eta^j)_{\mathscr{K}} dt + \int_0^T (\phi_t, \eta^j)_{\mathscr{K}} dW_t^j. \tag{3.14}$$

For $t \geq 0$ if Q_t is the orthogonal projection onto

$$H_t = \{\phi \in H: \int_t^T \|\phi_u\|_{\mathscr{K}}^2 du = 0\},$$

then as in the proof of Lemma 1.2, we have for every $r \geq 1$

$$\overline{\sigma}\left(\sum_{j=1}^{r}\int_0^T(Q_t\phi_u,\eta^j)_{\mathcal{H}}dY_u^j:\ \phi\in H\right)=\overline{\sigma}(Y_s^j:\ j\leq r,\ s\leq t)$$

and hence from (3.13),

$$\overline{\sigma}\left(\int_0^T(Q_t\phi)dY,\ \phi\in H\right)=\overline{\sigma}(Y_s^j:\ s\leq t,\ j\geq 1) \tag{3.15}$$

$$=\overline{\sigma}(Y_s:\ s\leq t)=\mathcal{F}_t^Y.$$

Here the bar over σ in (3.14) denotes completion with respect to μ.

Let $L_0: H\to\mathcal{L}(\mathcal{X},\mathcal{B}(\mathcal{X}),\mu)$ be defined by

$$L_0(\phi)=\int_0^T\phi dW.$$

Then in view of (3.11), L_0 is a representation of m. Defining $\rho((\omega,\omega'))=\omega$ and $L(\phi)(\omega,\omega')=L_0(\phi)(\omega')$ for $(\omega,\omega')\in\tilde{\Omega}$, it follows that $(\rho,L,\tilde{\Pi})$ is a representation of $\alpha=\Pi\otimes m$. Let L_1 be the representation of $n=\alpha\circ[y]^{-1}$ induced by y. Then

$$L_1(\phi)=(\phi,\xi)+L_0(\phi)=\int_0^T\phi dY. \tag{3.16}$$

from (3.12). Let R_α and R_n be the corresponding α and n liftings. Then

$$R_\alpha((y,\phi))=L_1(\phi)=\int_0^T\phi dY$$

and

$$R_\alpha((Q_ty,\phi))=R_\alpha((y,Q_t\phi))=\int_0^T(Q_t\phi)dY.$$

Thus (3.15) can be written as

$$\overline{\sigma}(R_\alpha((Q_ty,\phi)),\ \phi\in H)=\mathcal{F}_t^Y. \tag{3.17}$$

It is convenient from now on to write $\int_0^t \phi dY$ in place of $\int_0^T (Q_t\phi)dY$.

The following result is an immediate consequence of the identification of the σ-fields (3.17) and Theorem V.4.5.

<u>Theorem 3.1</u>: Let $g \in \mathscr{L}^1(\Omega,\mathscr{A},\Pi)$. Then for all $t \in [0,T]$

$$R_\alpha[E_\alpha(g|Q_t y)] = E_{\tilde{\Pi}}[g|\mathscr{F}_t^Y]. \tag{3.18}$$

Taking $g(\omega) = f(X_t(\omega))$, (3.18) yields the consistency of the solution to the filtering problem for the model (3.1) with that for the conventional model (3.7). Similarly, substituting $g(\omega) = f(X_s(\omega))$ gives consistency results for the prediction (if s>t) and smoothing (s<t) problems as well.

Once we have chosen a model in the convention theory such that (3.18) holds, the other results obtained in Section 1 for the finite dimensional case follow without much additional work. We will deal with these briefly.

Let $\Gamma_s(y)$, $\Gamma_{st}(y)$ be the unnormalized conditional distribution of X_s given $Q_s y$ and given $Q_t y$ respectively. $\Gamma_s(\phi)$ is defined by (VIII.2.6) and $\Gamma_{st}(\phi)$ is defined by (IX.2.1). Let $F_s(y)$ and $F_{st}(y)$ be the corresponding (normalized) conditional distributions, defined by (VIII.2.7) and (IX.2.2) respectively.

Define $\hat{\Gamma}_s(Y)$, $\hat{\Gamma}_{st}(Y)$ as elements of $\mathscr{M}(S,\mathscr{S})$ by

$$\hat{\Gamma}_{st}(Y)(B) = E_{\Pi}[1_B(X_s)\exp(\int_0^t h_u(X_u)dY_u - \tfrac{1}{2}\int_0^t \|h_u(X_u)\|_{\mathscr{H}}^2 du)]. \tag{3.19}$$

and

$$\hat{\Gamma}_s(Y)(B) = \hat{\Gamma}_{ss}(Y)(B). \tag{3.20}$$

Also, let $\hat{F}_s(Y)$ and $\hat{F}_{st}(Y)$ be the corresponding normalized measures.

Theorem 3.2: For any $f: S \to \mathbb{R}$ such that $E_\Pi|f(X_s)| < \infty$, we have

$$R_\alpha(\langle f, \Gamma_{st}(y)\rangle) = \langle f, \hat{\Gamma}_{st}(Y)\rangle \tag{3.21}$$

$$R_\alpha(\langle f, F_{st}(y)\rangle) = \langle f, \hat{F}_{st}(Y)\rangle \tag{3.22}$$

and

$$E_{\tilde{\Pi}}[f(X_s)|\mathcal{F}_t^Y] = \langle f, \hat{F}_{st}(Y)\rangle. \tag{3.23}$$

Proof: The first relation follows from Theorems VI.3.2 and VI.2.5 since $\phi \to \langle f, \Gamma_{st}(\phi)\rangle$ belongs to $\mathcal{S}(H)$ (for f such that $f \geq 0$, $E_\Pi f(X_s) = 1$). Indeed, $\langle f, \Gamma_{st}(y)\rangle \in \mathcal{L}^*(E, \mathcal{E}, \alpha)$. The second follows from the first and properties of the lifting map, since $\hat{\Gamma}_{st}(Y)(S) > 0$ a.s. The last assertion follows from the previous one and Theorem 3.1, since $F_{st}(y)$ is the conditional distribution of X_s given $Q_t y$. □

Assertion (3.23) in the above theorem provides the solution to the filtering, prediction and smoothing problems (by taking $s = t$, $s > t$ and $s < t$ respectively) for the model (3.7).

When S is a Polish space, $\Gamma_{st}(y)$ is actually an S-valued random variabled on $(E, \mathcal{E}, \alpha)$ (this was proved in Section VIII.3 for the case $s = t$, and the case $s \neq t$ can be treated similarly). Then (3.21) and (3.22) imply

$$R_\alpha(\Gamma_{st}(y)) = \hat{\Gamma}_{st}(Y) \tag{3.24}$$

and

$$R_\alpha(F_{st}(y)) = \hat{F}_{st}(Y). \tag{3.25}$$

It should be noted that we are not stating the results for filtering separately -- the latter is covered by taking s = t in the above relations.

If $\{P_k\} \subseteq \mathcal{P}$, $P_k \overset{s}{\rightarrow} I$, Theorem 3.2 and the fact noted in its proof that $\langle f, \Gamma_{st}(y)\rangle \in \mathcal{L}^*(E,\mathcal{E},\alpha)$ show that

$$R_\alpha([\langle f, \Gamma_{st}(y)\rangle]_{P_k}) \longrightarrow \langle f, \hat{\Gamma}_{st}(Y)\rangle \text{ in } \tilde{\Pi}\text{-probability.}$$

Since $\langle f, \Gamma_{st}(\cdot)\rangle \in \mathcal{J}(H)$, it also belongs to $\mathcal{U}(y)$ and $\mathcal{L}^*(H,\mathcal{C},n)$ (see Theorems VI.2.5 and VI.3.2). Hence we also have

$$R_\alpha(\langle f, \Gamma_{st}(P_k y)\rangle) \longrightarrow \langle f, \hat{\Gamma}_{st}(Y)\rangle \text{ in } \tilde{\Pi}\text{-probability.} \tag{3.26}$$

We state these conclusions in the theorem given below.

<u>Theorem 3.3</u>: Let $\{P_k\} \subseteq \mathcal{P}$, $P_k \xrightarrow{s} I$. Let S be a Polish space. Then

$$R_\alpha(\Gamma_{st}(P_k y)) \longrightarrow \hat{\Gamma}_{st}(Y) \quad \text{in } \tilde{\Pi}\text{-probability} \tag{3.27}$$

in the topology of weak convergency. Equivalently,

$$d_0(R_\alpha(\Gamma_{st}(P_k y)), \hat{\Gamma}_{st}(Y)) \longrightarrow 0 \text{ in } \tilde{\Pi}\text{-probability} \quad (3.27)'$$

and

$$R_\alpha(F_{st}(P_k y)) \longrightarrow \hat{F}_{st}(Y) \quad \text{in } \tilde{\Pi}\text{-probability.} \tag{3.28}$$

<u>Remark 3.1</u>: In the course of proving Theorem 3.2, it has been shown that for $P_k \xrightarrow{s} I$, $\{P_k\} \subseteq \mathcal{P}$,

$$\sup_{A\in\mathcal{S}} |R_m(\Gamma_t\circ P_k)(A) - R_m(\Gamma_t)(A)| \le U_k$$

where $U_k \to 0$ in $\mathcal{L}^1(\Omega_0,\mathcal{A}_0,\Pi_0)$. Thus

$$d_1(R_m(\Gamma_t\circ P_k),\ R_m(\Gamma_t)) \longrightarrow 0 \quad \text{in } \Pi_0\text{-probability}, \tag{3.29}$$

d_1 being the metric corresponding to the total variation norm on $\mathcal{M}(S)$. (3.29) is to be understood as

$$\Pi_0^*(d_1(R_m(\Gamma_t\circ P_k),\ R_m(\Gamma_t)) > \epsilon) \longrightarrow 0 \quad \text{as } k\to\infty \tag{3.30}$$

where Π_0^* is the outer measure: $\Pi_0^*(A) = \inf\{\Pi_0(B) : A \subseteq B,\ B \in \mathcal{S}\}$. Similar arguments will yield

$$d_1(R_\alpha(\Gamma_{st}(P_k7)),\hat{\Gamma}_{st}(Y)) \longrightarrow 0 \text{ in } \tilde{\Pi}\text{-probability}, \tag{3.31}$$

as $k\to\infty$. Thus (3.22) is valid even if we take the total variation norm topology on $\mathcal{M}(S)$. For this it is not necessary to assume that S is a Polish space.

For some choices of $\{P_k\}$, it is possible to describe $R_\alpha(\Gamma_{st}(P_k y))$ explicitly in terms of Y, as in the finite dimensional case. We show this for a particular sequence $\{P_k\}$ -- corresponding to a finite dimensional polygonal approximation.

Fix $0 = t_0^k < t_1^k < \ldots < t_{t_k}^k = T$, where $m_k \ge 1$, and

$$\lim_{k\to\infty} \sup_{1\le i\le m_k} |t_1^k - t_{i-1}^k| = 0, \tag{3.32}$$

and

$$\{t_i^k : 0 \le i \le m_k\} \subseteq \{t_j^{k+1} : 0 \le j \le m_{k+1}\}. \tag{3.33}$$

Let $\psi^{k,i} \in L^2([0,T])$ be defined by

$$\psi_s^{k,i} = (t_i^k - t_{i-1}^k)^{-\frac{1}{2}} 1_{(t_{i-1}^k, t_i^k)}(s)$$

for $1 \le i \le m_k$, $k \ge 1$, $0 \le s \le T$. Choose $\{f^j\} \subseteq B^*$ such that $\eta^j = \gamma^*(f^j)$ is a CONS in $\mathcal{H}$. For each k, let

$$H^{(k)} = \text{linear span of } \{\psi^{k,i}\eta_j\colon 1 \le i \le m_k,\ 1 \le j \le K\}.$$

Here $\psi^{k,i}\eta_j$ is considered as an element of H. It is easy to see that $\{\psi^{k,i}\eta_j\colon 1 \le i \le m_k,\ 1 \le j \le k\}$ is an ONB for $H^{(k)}$ and that $H^{(k)} \subseteq H^{(k+1)}$ in view of (3.33). Let P_k be the orthogonal projection onto $H^{(k)}$. The arguments given at the end of Section 1 show that $L^2([0,T],\mathcal{H})$ is the closed linear span of $\{\psi^{k,i}\colon 1 \le i \le m_k,\ k \ge 1\}$. From this, one can check that $P_k \uparrow I$. Now we have the Wiener integral

$$\int_0^T \left[\psi^{k,i}\eta_j\right] dY = \int_0^T \psi_s^{k,i} dY_s^j = \left[t_i^k - t_{i-1}^k\right]^{-\frac{1}{2}} \left[Y_{t_i^k}^j - Y_{t_{i-1}^k}^j\right] \tag{3.34}$$

where $Y^j = f^j[Y]$.

Define

$$Y_t^{(k)}(\tilde{\omega}) = \sum_{j=1}^{k} \left[Y_{t_{i-1}^k}^j + (Y_{t_i^k}^j - Y_{t_{i-1}^k}^j)(t - t_{i-1}^k)^{-1}\right]\eta^j \tag{3.35}$$

$$\text{for } t_{i-1}^k \le t \le t_i^k.$$

Writing $\dot{Y}_t^{(k)}$ for $\frac{d}{dt} Y_t^{(k)}$, we have

$$\dot{Y}^{(k)} = \sum_{j=1}^{k} \sum_{i=1}^{m_k} (\int_0^T (\psi_s^{k,i}\eta^j) dY) \psi^{k,i}\eta^j \tag{3.36}$$

$$= \sum_{j=1}^{k} \sum_{j=1}^{m_k} L_1(\psi^{k,i}\eta^j)\psi^{k,i}\eta^j.$$

Thus, as in the finite dimensional case, it follows that

$$R_\alpha(\Gamma_{st}(P_k y)) = \Gamma_{st}(\dot{Y}^{(k)}). \tag{3.37}$$

The identification in (3.37) yields the following corollary to Theorem 3.3.

Corollary 3.4: Let S be a Polish space and let $Y^{(k)}(\tilde{\omega})$ be defined by (3.35), where $\{t_i^k\}$ satisfy (3.32) and (3.33). Then we have

$$\Gamma_{st}(\dot{Y}^{(k)}) \longrightarrow \hat{\Gamma}_{st}(Y) \quad \text{in } \tilde{\Pi}\text{-probability} \tag{3.38}$$

in the topology of weak convergence. Observe that (3.38) also holds for convergence in total variation norm.

Remark 3.2: When $\mathcal{H} = L^2(\mathbb{R}^d)$, instead of the Banach space B and its dual B^* we can take $\mathcal{S}^*(\mathbb{R}^d)$ and $\mathcal{S}(\mathbb{R}^d)$ respectively, $\mathcal{S}^*(\mathbb{R}^d)$ being the space of Schwartz distributions on $\mathbb{R}^d$. In fact, we have now the Gilfand triplet $\mathcal{S}(\mathbb{R}^d) \subset L^2(\mathbb{R}^d) \subset \mathcal{S}^*(\mathbb{R}^d)$. The existence of a Gaussian measure μ on $(\mathcal{X},\mathcal{B}(\mathcal{X}))$ satisfying (3.5) and (3.6) has been outlined in Example III.4.3.

4. ROBUSTNESS -- PATHWISE AND STATISTICAL

Two notions of robustness will be discussed in this section. The term robustness as customarily used in the context of filtering theory refers to the continuous dependence of the solution on the observations. In the stochastic calculus version of the theory, it has also been used to emphasize the existence of a pathwise solution.

As we have seen in the previous chapters, the solution to the estimation problem is always obtained pathwise, so that this type of robustness is inherent in the finitely additive white noise approach both in the finite and infinite dimensional cases. The continuous dependence of the optimal filter on the observations has been noted earlier for the finite dimensional problem. The same is true in the infinite dimensional case as well. Indeed, we will prove in Theorems 4.1 and 4.4 that the optimal estimator (i.e., filter, predictor or smoother) is a Lipschitz continuous function of the observations.

The other notion of robustness pertains to the continuous dependence of the optimal filter on the underlying distribution of the signal process. We call it 'statistical robustness' since it is in this sense that the word 'robustness' has gained currency in the statistical literature. Let us briefly examine what this property means. Suppose Π is the probability measure on the appropriate space of paths of the signal process $X = (X_t)$ in the filtering model of Section VIII.2. An important question is to know how the optimal filter given by the conditional distribution $\Pi[X_t \in \cdot|Q_t y]$ changes if the 'true' distribution of X is not the hypothesized Π but deviates slightly from it. What statistical robustness means is that small changes in Π produce small changes in the optimal filter ('small' being understood in the sense of the topology of weak convergence). The main results on statistical robustness are Theorems 4.3, 4.5, 4.6 and 4.7. We consider the general setup of Chapter VIII which includes the finite dimensional model as a special case. The robustness properties for the latter are formulated in terms of densities. Our methods also yield some robustness results in the conventional theory which are included at the end of the section.

Let us return to the white noise model (2.1), used in Section VIII.2 for the filtering problem and in Section IX.2 for prediction and smoothing. We refer the reader to these sections for unexplained notation.

The unnormalized conditional distribution $\Gamma_t(y)$ of X_t given $Q_t y$ and $\Gamma_{st}(y)$ of X_s given $Q_t y$ are given respectively by (VIII.2.6) and (IX.2.1). It is worth observing that these formulas do not require the assumption that the signal (X_t) be a Markov process.

Here is the first result on Lipschitz continuity of $\Gamma_t(y), \Gamma_{st}(y)$.

<u>Theorem 4.1</u>: Suppose that $E_{\Pi}\int_0^T \|h_s(X_s)\|_{\mathcal{H}}^2 = K < \infty$. Then for

$$\sup_{A \in \mathcal{S}} |\Gamma_t(y^1)(A) - \Gamma_t(y^2)(A)| \tag{4.1}$$

$$\leq K\|Q_t y^1 - Q_t y^2\| \exp(\tfrac{1}{2}\|Q_t y^1\|^2 + \tfrac{1}{2}\|Q_t y^2\|^2).$$

$$\sup_{0 \leq s \leq T} \sup_{A \in \mathcal{S}} |\Gamma_{st}(y^1)(A) - \Gamma_{st}(y^2)(A)| \tag{4.2}$$

$$\leq K\|Q_t y^1 - Q_t y^2\| \cdot \exp(\tfrac{1}{2}\|Q_t y^1\|^2 + \tfrac{1}{2}\|Q_t y^2\|^2).$$

Proof: Since $\Gamma_{tt} = \Gamma_t$, (4.1) is a special case of (4.2). Recall that

$$\Gamma_{st}(y)(A) = \int 1_A(X_s(\omega))\exp((\xi(\omega),Q_t y) - \tfrac{1}{2}\|Q_t\xi(\omega)\|^2)d\Pi(\omega)$$

where $\xi(\omega) = (h_s(X_s(\omega))$. Thus

$$\sup_{0\leq s\leq T}\sup_{A\in\mathcal{S}} |\Gamma_{st}(y^1)(A) - \Gamma_{st}(y^2)(A)| \tag{4.3}$$

$$\leq \int|\exp((\xi(\omega),Q_t y^1) - \tfrac{1}{2}\|Q_t\xi(\omega)\|^2 - \exp((\xi(\omega),Q_t y^2)$$

$$- \tfrac{1}{2}\|Q_t\xi(\omega)\|^2)|d\Pi(\omega).$$

For any two real numbers a,b, it is easy to check that

$$|\exp(a)-\exp(b)| \leq |a-b|\cdot\max\{\exp(|a|),\exp(|b|)\}. \tag{4.4}$$

Also, as we have seen several times earlier,

$$(\xi(\omega),Q_t y^i) - \tfrac{1}{2}\|Q_t\xi(\omega)\|^2 = (Q_t\xi(\omega),Q_t y^i) - \tfrac{1}{2}\|Q_t\xi(\omega)\|^2 \tag{4.5}$$

$$\leq \tfrac{1}{2}\|Q_t y^i\|^2$$

for $i = 1,2$. Hence the right hand side in (4.3) is less than or equal to

$$\int|(\xi(\omega),Q_t y^1-Q_t y^2)|\ \exp(\tfrac{1}{2}\|Q_t y^1\|^2 + \tfrac{1}{2}\|Q_t y^2\|^2)d\Pi(\omega).$$

Since $|(\xi(\omega),Q_t y^1-Q_t y^2)| \leq \|\xi(\omega)\|\cdot\|Q_t y^1-Q_t y^2\|$, we can conclude that the right hand side of (4.3) is dominated by

$$\|Q_t y^1-Q_t y^2\|\ \exp(\tfrac{1}{2}\|Q_t y^1\|^2 + \tfrac{1}{2}\|Q_t y^2\|^2)\cdot\int\|\xi(\omega)\|d\Pi(\omega)$$

which implies the required inequality (4.2). □

<u>Remark 4.1</u>: The estimate (4.2) implies that the functional $y \to \Gamma_{st}(y)$ restricted to $\{y \in H: \|Q_t y\|^2 \leq a\}$ is Lipschitz continuous in the total variation norm, uniformly in $s \in [0,T]$.

<u>Remark 4.2</u>: If $y^1, y^2 \in H$ are such that $Q_t y^1 = Q_t y^2$, then (4.2) yields $\Gamma_{st}(y^1)(A) = \Gamma_{st}(y^2)(A)$ for all $A \in \mathcal{S}$, $0 \leq s \leq T$. Of course, this follows directly from the definition of $\Gamma_{st}(y)$ itself. This property is expressed in the theory of stochastic processes by saying that $\Gamma_{s\bullet}(y)$ is a non-anticipative functional of y.

Now we turn to statistical robustness. To consider this, assume the following setup.

Let S be a complete separable metric space and $D = D([0,T],S)$ the space of right continuous functions $\underline{X}$: $[0,T] \to S$ which admit left limits. We denote the value of $\underline{X} \in D$ at $t \in [0,T]$ by $\underline{X}_t$ and equip D with the Skorokhod topology.

Let X and $\{X^k: k > 1\}$ be processes on $(\Omega,\mathcal{A},\Pi)$ with paths in D. Assume that

$$X^k \text{ converges in distribution to } X \text{ (on D)} \tag{4.6}$$

which means that for all bounded, real continuous functions G on D,

$$E_{\Pi}G(X^k) \to E_{\Pi}G(X). \tag{4.7}$$

Let $\Gamma^k_{st}(y)$ be the conditional distribution of X^k_s given $Q_t y$ (and $\Gamma^k_{tt}(y) = \Gamma^k_t(y)$) where

$$y_t = h_t(X^k_t) + e_t.$$

$\Gamma^k_{st}(y)$ is given by a formula analogous to (IX.2.1) with X^k in place of X. Indeed, if

$$G^y_t: D \to \mathbb{R}$$

is defined by

$$G^y_t(\underline{X}) = \exp(\int_0^t (h_u(\underline{X}_u), y_u)_{\mathcal{H}} du - \tfrac{1}{2}\int_0^t \|h_u(\underline{X}_u)\|^2_{\mathcal{H}} du) \qquad (4.8)$$

then

$$\Gamma^k_{st}(y)(A) = E_\Pi 1_A(X^k_s) G^y_t(X^k) \qquad (4.9)$$

and

$$\Gamma_{st}(y)(A) = E_\Pi 1_A(X_s) G^y_t(X). \qquad (4.10)$$

Our aim is to prove that Γ^k_{st} converges to Γ_{st} in an appropriate sense. For this we need the following Lemma.

Lemma 4.2: Suppose that h: $[0,T]\times S \to \mathcal{H}$ is a continuous function. Then G^y_t is a continuous function from D into $\mathbb{R}$, for every $s,t \in [0,T]$ and $y \in H$.

Proof: Let $\underline{X}^j \to \underline{X}$ in D. Convergence in the Skorokhod topology has the following implications:

There exists a compact set S_0 in S such that $\underline{X}^j_u \in S_0$ for all $u \in [0,T]$, $j \geq 1$. (4.11)

For all u such that $\underline{X}_u = \underline{X}_{u-}$; $\underline{X}^j_u \to \underline{X}_u$. (4.12)

Continuity of h and (4.11) imply that

$$\|h_u(\underline{X}^k_u)\|_{\mathcal{H}} \leq K_1 \quad \text{for } j \geq 1,\ u \in [0,T] \tag{4.13}$$

for a constant $K_1 < \infty$ and that

$$h_u(\underline{X}^j_u) \to h_u(\underline{X}_u) \tag{4.14}$$

for all u such that $\underline{X}_u = \underline{X}_{u-}$. Since $\{u: \underline{X}_u \neq \underline{X}_{u-}\}$ is at most countable, (4.14) holds for almost all $u \in [0,T]$. Thus in view of the bound (4.13), we have

$$\int_0^t \|h_u(\underline{X}^j_u)\|^2_{\mathcal{H}} du \longrightarrow \int_0^t \|h_u(\underline{X}_u)\|^2_{\mathcal{H}} du \tag{4.15}$$

and for all $y \in H$,

$$\int_0^t (h_u(\underline{X}^j_u), y_u) du \longrightarrow \int_0^t (h_u(\underline{X}_u), y_u) du. \tag{4.16}$$

Hence $G^y_t(\underline{X}^j) \to G^y_t(\underline{X})$ and the lemma is proved. □

<u>Theorem 4.3</u>: Suppose h: $[0,T]\times S \to \mathcal{H}$ is continuous. Suppose X^k converge in distribution to X (as D-valued random variables). Suppose $\Pi(X_t \neq X_{t-}) = 0$ for all t. Then for all $y \in H$,

$$\Gamma^k_{st}(y) \longrightarrow \Gamma_{st}(y) \text{ in the weak topology on } \mathcal{M}(S). \tag{4.17}$$

$$\Gamma^k_t(y) \longrightarrow \Gamma_t(y) \quad \text{in the weak topology on } \mathcal{M}(S). \tag{4.18}$$

<u>Proof</u>: (4.18) is a special case of (4.17). To prove (4.17), we have to prove that for all $f \in C_b(S)$,

$$\langle f, \Gamma^k_{st}(y)\rangle \longrightarrow \langle f, \Gamma_{st}(y)\rangle. \tag{4.19}$$

In view of (4.9), (4.10), we have

$$\langle f, \Gamma^k_{st}(y)\rangle = E_\Pi f(X^k_s)G^y_t(X^k)$$

and

$$\langle f, \Gamma_{st}(y)\rangle = E_\Pi f(X_s)G^y_t(X).$$

Fix s,t $\in$ [0,T] and y $\in$ H. Let

$$G'(\underline{X}) = f(\underline{X}_s)G^y_t(\underline{X}).$$

Since G^y_t is continuous (as seen in Lemma 4.2), we have

$$\{\underline{X}: G' \text{ is discontinuous at } \underline{X}\} \subseteq \{\underline{X}: \underline{X}_s \neq \underline{X}_{s-}\}$$

and hence the set of discontinuities of G' has $\Pi \circ X^{-1}$ measure zero. Thus from the weak convergence of X^k to X, we conclude

$$E_\Pi G'(X^k) \longrightarrow E_\Pi G'(X).$$

This proves (4.19) and hence the Theorem. □

Remark 4.3: If the signal process has continuous sample paths, we may substitute C = C([0,T,S) with the topology of uniform convergence instead of D in the statement of the Theorem. For this case, Lemma 4.2 is easy to prove and the argument involving the set of discontinuity is unnecessary because on C, the map $\underline{X} \to \underline{X}_t$ is continuous.

The Finite Dimensional Case

Let us now specialize to the finite dimensional model of Chapter VII where the signal is $\mathbb{R}^d$-valued and the noise is $\mathbb{R}^N$-valued. The following refinement of Theorems VII.3.7 and

IX.1.2 is a consequence of Theorem 4.1.

We will use below the notation from Sections VII.3, VII.4 and IX.1. In particular, we are dealing with the model (VII.1.9) - (VII.1.10).

Theorem 4.4: Suppose that for all t, the measure $\Pi \circ X_t^{-1}$ admits a density with respect to λ and that h satisfies the condition

$$E\int_0^T |h_t(X_t)|^2 dt = K < \infty. \tag{4.20}$$

Let $p_t(x,y)$, $p_{st}(x,y)$ be the unnormalized conditional densities of X_t, X_s (respectively) given $Q_t y$. Then we have

$$\int |p_t(x,y^1) - p_t(x,y^2)|dx \tag{4.21}$$

$$\leq 2K\|Q_t y^1 - Q_t y^2\| \exp(\tfrac{1}{2}\|Q_t y^1\|^2 + \tfrac{1}{2}\|Q_t y^2\|^2)$$

and

$$\sup_{0\leq s\leq T} \int |p_{st}(x,y^1) - p_{st}(x,y^2)|dx \tag{4.22}$$

$$\leq 2K\|Q_t y^1 - Q_t y^2\| \exp(\tfrac{1}{2}\|Q_t y^1\|^2 + \tfrac{1}{2}\|Q_t y^2\|^2)$$

for all $s,t \in [0,T]$, $y^1, y^2 \in H$.

Proof: Existence of $p_t(x,y)$ was proved in Theorem 3.7, where we also proved continuity in y. The corresponding result for $p_{st}(x,y)$ is Theorem IX.1.

The assertions made above follow from (4.1) and (4.2). Upon noting that

$$p_t(x,y) = \frac{d\Gamma_t^y}{d\lambda}(x), \quad p_{st}(x,y) = \frac{d\Gamma_{st}^y}{d\lambda}(x)$$

and the fact that for any integrable functions f_1, f_2 on $\mathbb{R}^d$,

$$\int|f_1-f_2|d\lambda = \int(f_1-f_2)1_{\{f_1>f_2\}}d\lambda + \int(f_2-f_1)1_{\{f_2>f_1\}}d\lambda$$
$$\leq 2 \sup_{A\in\mathcal{B}(\mathbb{R}^d)} |\int_A f_1 d\lambda - \int_A f_2 d\lambda|.$$

□

Theorem 4.3 has an analogue for the unnormalized conditional densities. We assume conditions on the signal processes $\{X^k\}$ and X which ensure $X^k \to X$ in distribution and further that the densities $p_t^k(x)$. $p_t(x)$ of X_t^k, X_t respectively exist and $p_t^k(x) \to p_t(x)$ uniformly on compact subsets of $[0,T]\times\mathbb{R}^d$. We show then that a similar conclusion can be drawn about the convergence of unnormalized conditional densities $p_{st}^k(x,y)$ to $p_{st}(x,y)$. This is much stronger than the weak convergence of the unnormalized conditional distributions implied by Theorem 4.3.

We will be working with the following setup. Let $\{X^k: k \geq 1\}$ and X be diffusion processes on $\mathbb{R}^d$ with diffusion and drift coefficients (a^k,b^k), (a,b) respectively. We will impose the following conditions on $\{X^k\}$ and X.

(i) The coefficients (a,b), (a^k,b^k) for each $k \geq 1$ satisfy the conditions of Theorem VII.4.1 and further, the constant appearing in the growth condition (VII.4.2) for (a^k,b^k) and their derivatives can be chosen independently of k.

(ii) For all $1 \leq i,j \leq d$; $a_{ij}^k \to a_{ij}$, $\frac{\partial}{\partial x^i} a_{ij}^k \to \frac{\partial}{\partial x^i} a_{ij}$,

$\frac{\partial^2}{\partial x^i \partial x^j} a^k_{ij} \to \frac{\partial^2}{\partial x^i \partial x^j} a_{ij}$, $b^k_i \to b_i$, $\frac{\partial}{\partial x^i} b^k_i \to \frac{\partial}{\partial x^i} b_i$ as $k \to \infty$ in the following sense: for all $R > 0$

$$\lim_{k\to\infty} \int_0^T \sup_{|x|\le R} |\alpha^k_{ij}(t,x) - \alpha_{ij}(t,x)| dt = 0 \qquad (4.23)$$

for all i,j where α^k_{ij} stands for a^k_{ij}, $\frac{\partial a^k_{ij}}{\partial x^i}$, $\frac{\partial^2 a^k_{ij}}{\partial x^i \partial x^j}$ and α_{ij} stands for a_{ij} and its derivatives. Similar conditions hold for (b^k_i, b_i) and their first order derivatives.

(iii) For some constant $K < \infty$, we have

$$|a^k_{ij}(t,x)| \le K$$

and

$$|a_{ij}(t,x)| \le K.$$

for all $k \ge 1$, $1 \le i,j \le d$, $t \in [0,T]$, $x \in \mathbb{R}^d$.

(iv) The distribution of X^k_0, X_0 admit continuous densities p^k_0 and p_0 respectively with respect to the Lebesgue measure satisfying the growth condition (VII.4.22) with constants K, ϵ not depending on $k \ge 1$ and such that for all $R > 1$,

$$\lim_{k\to\infty} \sup_{|x|\le R} |p^k_0(x) - p_0(x)| = 0.$$

For each $k \ge 1$, we consider the model

$$y_t = h_t(X^k_t) + e_t, \quad 0 \le t \le T \qquad (4.24)$$

where (e_t) is white noise (as in Chapter VII) and $h: [0,T]\times\mathbb{R}^d \to \mathbb{R}^N$ is a locally Hölder continuous function.

Under these conditions, Theorems VII.4.3, IX.1.4 and IX.1.7 imply the existence of the unnormalized conditional density $p^k_{st}(x,y)$ of X_s given $Q_t y$, for $y \in H_0$ for the cases $s = t$, $s > t$ and $s < t$ respectively.

Under these assumptions, we have the following result.

Theorem 4.5: Let conditions (i) - (iv) be satisfied and suppose h is a locally Hölder continuous function. Then for each $y \in H_0$,

$p^k_t(x,y)$ converges to $p_t(x,y)$ uniformly in $(t,x) \in [0,T] \times \{x: |x| \le R\}$ for every $R < \infty$, (4.25)

and

for all $t \in [0,T]$ fixed, $p^k_{st}(x,y)$ converges to $p_{st}(x,y)$ uniformly in $(s,x) \in I \times \{x: |x| \le R\}$ for every $R < \infty$ where $I = [t,T]$ in the case of prediction and $I = [0,t]$ for the smoothing problem. (4.26)

Proof: Thoughout this proof, y will denote a fixed element of H_0. Let L^k_t be the differential operator corresponding to $a^k(t,\cdot)$, $b^k(t,\cdot)$ (given by VII.3.64 with a^k,b^k in place of a,b), and let L^{k*}_t be its adjoint. Let b^{k*} and c^{k*} be defined by (VII.3.10) - (VII.3.11) with a^k,b^k in place of a^0,b^0 and $c^0 = 0$. Then L^{k*}_t is the differential operator corresponding to a^k ,b^{k*},c^{k*}-- expressed by (VII.3.9).

By Theorem VII.4.3, $p^k_t(x,y)$ is the unique classical solution in the class $\mathcal{G}((0,T])$ to the Cauchy problem

$$\frac{\partial p_t^k(x,y)}{\partial t} = L_t^{k*} p_t^k(x,y) + c_t^y(x) p_t^k(x,y) \tag{4.26}$$

$$p_0^k(x,y) = p_0^k(x). \tag{4.27}$$

Take $M_t^k = L_t^{k*} + c_t^y$, $M_t^0 = L_t^* + c_t^y$, $g^k = p_0^k$, $g^0 = p_0$ and $u^k(t,x) = p_t^k(x,y)$, $u^0(t,x) = p_t(x,y)$. The various assumptions on the coefficients a, b, a^k, b^k and initial densities p_0^k, p_0 imply (easy to check) that M_t^k, M_t^0, g^k, g^0 satisfy conditions of Theorem 2.1. The fact that $p_t^k(x,y)$, $p_t^0(x,y)$ belong to $\mathcal{G}((0,T])$ (a consequence of Theorem VII.4.3) now allows us to apply Theorem 2.1 to conclude (4.25).

Coming to (4.26), fix $t \in [0,T]$. We will prove the uniform convergence of $p_{st}^k(x,y)$ to $p_t(x,y)$ first in $(s,x) \in [t,T] \times \{x: |x| \leq R\}$ and then in $(s,x) \in [0,t] \times \{x: |x| \leq R\}$. First is the prediction case and second is the smoothing case.

Let $s \geq t$. By Theorem IX.1.4, $p_{st}^k(x,y)$ is the unique classical solution in the class $\mathcal{G}((t,T])$ to

$$\frac{\partial}{\partial s} p_{st}^k(x,y) = L_s^{k*} p_{st}^k(x,y), \quad s > t, \tag{4.28}$$

$$p_{tt}^k(x,y) = p_t^k(x,y). \tag{4.29}$$

As in the proof of (4.25), we can apply Theorem 2.1 to get the desired conclusion if we can verify that $g^k(x) = p_t^k(x,y)$ and $g^0(x) = p_t(x,y)$ satisfy (2.7) and (2.15). The uniform convergence of $p_t^k(x,y)$ to $p_t(x,y)$ in $x \in \{x: |x| \leq R\}$ follows from (4.25), so this gives (2.15).

In view of assumption (i), we can choose constants K_1, K_2 such that the function $H_1(t,x)$ defined by (VII.4.5) satisfies (VII.4.11), (VII.4.12),

$$(L_t^{k*} - \frac{\partial}{\partial t})H_1(t,x) \le 0$$

and

$$(L_t^k + \frac{\partial}{\partial t})H_1^{-1}(t,x) \le 0.$$

This follows from Lemma VII.4.2. The assumption on $\{p_0^k(x)\}$, $p_0(x)$ implies that for a constant K_3,

$$|p_0^k(x)| \le K_3, \quad |p_0(x)| \le K_3.$$

Hence from Theorem VII.4.1 (see Remarks 4.1, 4.2 following the Theorem also), it follows that the densities $p_t^k(x)$, $p_t(x)$ of X_t^k, X_t respectively satisfy $|p_t^k(x)| \le K_3H_1(t,x)$ and $|p_t(x)| \le K_3H_1(t,x)$. Then the arguments leading to the proof of (IX.1.16) yield

$$|p_t^k(x,y)| \le K_3H_1(t,x)\cdot\exp(\tfrac{1}{2}\|y\|^2) \tag{4.30}$$

and

$$|p_t(x,y)| \le K_3H_1(t,x)\cdot\exp(\tfrac{1}{2}\|y\|^2). \tag{4.31}$$

These bounds imply that (2.7) holds for $g^k(x) = p_t^k(x,y)$, $g^0(x) = p_t(x,y)$. Thus we can apply Theorem 2.1 as noted earlier to conclude that $p_{st}^k(x,y)$ converges to $p_{st}(x,y)$ uniformly in $(s,x) \in [t,T] \times \{x: |x| \le R\}$.

To get the corresponding result for $s \le t$ (smoothing

problem), note that by Theorem IX.1.7, $p^k_{st}(x,y)$ can be expressed as

$$p^k_{st}(x,y) = p^k_s(x,y)v^k_t(s,x,y) \tag{4.32}$$

where $v^k_t(s,x,y)$ is the unique classical solution in $\mathcal{G}([0,t))$ to

$$\left(\frac{\partial}{\partial s} + L^k_s + c^y_s\right)v^k_t(s,x,y) = 0, \quad s < t, \tag{4.33}$$

$$v^k_t(t,x,y) = 1, \tag{4.34}$$

and a similar representation holds for $p_{st}(x,y)$, i.e.,

$$p_{st}(x,y) = p_s(x,y)v_t(s,x,y) \tag{4.35}$$

where $v_t(s,x,y)$ is the unique classical solution in $\mathcal{G}([0,t))$ to

$$\left(\frac{\partial}{\partial s} + L_s + c^y_s\right)v_t(s,x,y) = 0, \quad s < t, \tag{4.36}$$

$$v_t(t,x,y) = 1. \tag{4.37}$$

We can apply Theorem 2.1 as our assumptions imply that its conditions are fulfilled with $M^k_s = L^k_s + c^y_s$, $M^0_s = L_s + c^y_s$, $T = t$, $g^k = 1 = g^0$. Thus we have

$$v^k_t(s,x,y) \longrightarrow v_t(s,x,y) \tag{4.38}$$

uniformly in $(s,x) \in [0,t] \times \{x: |x| \le R\}$

for every R. (4.38), (4.25), (4.32) and (4.35) yield the other half of (4.26), namely uniform convergence of $p^k_{st}(x,y)$

to $p_{st}(x,y)$ in $(s,x) \in [0,t] \times \{x: |x| \leq R\}$. This completes the proof. □

Remark 4.3: Under conditions (i) - (iv), we can prove that for each t, X_t^k, X_t admit densities $p_t^k(x)$, $p_t(x)$ with respect to Lebesgue measure, that $p_t^k(x)$, $p_t(x)$ are continuous functions of (t,x) and further that

$$p_t^k(x) \longrightarrow p_t(x) \tag{4.39}$$

uniformly over compact subsets of $[0,T] \times \mathbb{R}^d$. This can be shown following the steps of the proof of (4.25) above, or it can be deduced from (4.25) by taking $h = 0$, so that $p_t^k(x,y) \equiv p_t^k(x)$ and $p_t(x,y) \equiv p_t(x)$ for all y.

We have thus established the continuous dependence of the unnormalized conditional density $p_{st}(x,y)$ of X_s given $Q_t y$ on the signal process for each $y \in H_0$. With more careful analysis, it should be possible to prove that for all $R > 0$, $y \in H_0$,

$$p_{st}^k(x,y) \longrightarrow p_{st}(x,y) \qquad \text{uniformly in}$$

$$(s,t,x) \in [0,T] \times [0,T] \times \{x: |x| \leq R\}.$$

We do not attempt to prove this. Instead, we will show how the result proved above for $y \in H_0$ can be extended to all $y \in H$ without much difficulty.

Theorem 4.6: Suppose that the conditions (i) - (iv) stated just prior to the statement of Theorem 4.5 are satisfied. Further, suppose that (h,a,b) and (h,a^r,b^r) for $r \geq 1$

satisfy the conditions (VII.4.50) - (VII.4.51) in Theorem VII.4.5, and that the constants appearing in the growth conditions can be chosen independently of r. Then the conclusions (4.25), (4.26) in Theorem 4.5 are valid for all $y \in H$.

Proof: (i) The existence of $p_t(x,y)$ and $p_t^k(x,y)$ for all $y \in H$ is assured by Theorem VII.4.5: Indeed,

$$p_t(x,y) = \psi_t(x,y)\cdot\exp\Big(\sum_{r=1}^{N} h_t^r(x) \int_0^t y_u^r du\Big) \qquad (4.40)$$

$$p_t^k(x,y) = \psi_t^k(x,y)\cdot\exp\Big(\sum_{r=1}^{N} h_t^r(x) \int_0^t y_u^r du\Big) \qquad (4.41)$$

where ψ_t is the solution to (VII.4.53) - (VII.4.54) and ψ_t^k is the solution to the corresponding differential equation with (a,b,p_0) replaced by (a^k,b^k,p_0^k).

As in the proof of Theorem 4.5, we can verify that the coefficients in the differential equations for ψ and ψ^k satisby the conditions of Theorem 2.1. The coefficients in the PDE for ψ are given by (VII.4.46) - (VII.4.48) and those for ψ^k are obtained by substituting a^k for a and b^k for b.

Thus, by Theorem 2.1, $\psi_t^k(x,y)$ converges to $\psi_t(x,y)$ uniformly over compact subsets of $[0,T]\times\mathbb{R}^d$ for all $y \in H$ and hence (4.25) follows from (4.40) and (4.41).

The proof of convergence of $p_{st}^k(x,y)$ to $p_{st}(x,y)$ uniformly in $(s,x) \in [t,T] \times \{x: |x|\leq R\}$ given in Theorem 4.5 for $y \in H_0$ goes through for all $y \in H$.

For the last part, $s \leq t$, note that by Theorem IX.1.8, p_{st}^k and p_{st} continue to satisfy (4.32) and (4.35) above

where

$$v_t(s,x,y) = v_t'(s,x,y) \exp\Big(\sum_{i=1}^{N} h_s^i(x)\cdot\int_s^t y_u^i du\Big) \tag{4.42}$$

$$v_t^k(s,x,y) = v_t^{k'}(s,x,y) \exp\Big(\sum_{i=1}^{N} h_s^i(x)\cdot\int_s^t y_u^i du\Big), \tag{4.43}$$

and $v_t'(s,x,y)$ is the classical solution to (IX.1.33) and $v_t'(t,x,y) = 1$; for each k, $v_t^{k'}(s,x,y)$ is the classical solution to an analogous equation, with (a,b) replaced by (a^k, b^k). Again Theorem 2.1 yields uniform convergence of $v_t^{k'}(s,x,y)$ over compact subsets of $[0,t]\times\mathbb{R}^d$ and then the relations (4.42), (4.43); (4.32), (4.35) and (4.25) (proved above) yield the convergence of $p_{st}^k(x,y)$ to $p_{st}(x,y)$ uniformly in $(s,x) \in [0,t] \times \{x: |x|\le R\}$ for every $R > 0$. □

The results on statistical robustness proved above can be applied to more specific problems in which the transition function of the signal process (X_t) (assumed to be Markov) is known but the initial distribution $\mu_0 := \Pi\circ X_0^{-1}$ may not be completely known. We may have only a rough idea of μ_0 in many practical problems and it is important to find out whether the optimal filter depends continuously on μ_0.

Suppose that the signal (X_t) is a diffusion process whose diffusion and drift coefficients a,b satisfy the conditions of Theorem II.4.4. Without loss of generality, assume that (X_t) is the coordinate process on $\Omega_d = C([0,T],\mathbb{R}^d)$. Let $Q_{s,x}$ be the solution to the martingale problem for (a,b) starting from (s,x), $0 \le s \le T$, $x \in \mathbb{R}^d$.

Let μ^k, $k \geq 0$, be probability measures on $\mathbb{R}^d$ and let $\{\Pi^k\}$ be probability measures on $(\Omega_d, \mathcal{B}(\Omega_d))$ defined by

$$\Pi^k(A) = \int_{\mathbb{R}^d} Q_{0,z}(A) d\mu_0^k(z), \tag{4.44}$$

$A \in \mathcal{B}(\Omega_d)$. Then under Π^k, (X_t) is a diffusion process with drift and diffusion coefficients a,b and $\Pi_0^k \circ X_0^{-1} = \mu_0^k$.

Let $F_t^k(y)$, $\Gamma_t^k(y)$, $k \geq 0$, be, respectively, the conditional and unnormalized conditional distribution of X_t given $Q_t y$ when the process (X_t) is considered on $(\Omega_d, \mathcal{B}(\Omega_d), \Pi^k)$.

It may be remarked that (X_t) satisfies (VIII.1.63) and hence $F_t^k(y)$, $\Gamma_t^k(y)$ are the unique solutions of measure valued equations given in Section VIII.2 provided

$$E_{\Pi^k}[\int_0^T |h_t(X_t)|^2 dt] < \infty. \tag{4.45}$$

Theorems 4.3, 4.5 and 4.6 yield the following result.

<u>Theorem 4.7</u>:

(i) Suppose the function h appearing in the filtering model is continuous. Let $\mu_0^k \xrightarrow{d} \mu_0$ as $k \to \infty$. Then for all $y \in H$, $t \in [0,T]$, we have

$$\Gamma_t^k(y) \xrightarrow{d} \Gamma_t(y); \quad F_t^k(y) \xrightarrow{d} F_t(y). \tag{4.46}$$

(ii) Suppose (a,b,h) satisfy the conditions of Theorem VII.4.3 and suppose that each μ^k admits a density ϕ^k which is continuous and satisfies the growth condition (VII.4.22). Suppose further that for all $R < \infty$,

$$\lim_{k\to\infty} \sup_{|x|\leq R} |\phi^k(x) - \phi^0(x)| = 0. \tag{4.47}$$

Then for each Hölder continuous y, the (unnormalized conditional) density $p_t^k(x,y)$ of $\Gamma_t^k(y)$ satisfies

$$\lim_{k\to\infty} \sup_{|x|\leq R} \sup_{0\leq t\leq T} |p_t^k(x,y) - p_t^0(x,y)| = 0 \tag{4.48}$$

for all $R < \infty$. If in addition, a,b,h ϕ^k satisfy the conditions of Theorem VII.4.5 for each $k \geq 0$, then (4.48) is true for all $y \in H$.

Proof: As noted just after Theorem II.3.5, the mapping $x \to Q_{0,x}$ is continuous and hence the assumption $\mu^k \xrightarrow{d} \mu^0$ implies that $\Pi^k \xrightarrow{d} \Pi^0$ as $k \to \infty$. Thus (4.45) follows from Theorem 4.3. The last part about convergence of densities follows from Theorems 4.5 and 4.6. □

Robustness in the Conventional Theory

We have already proved in Section 2 that for the conventional model (1.2), the unnormalized conditional density $\hat{p}_{st}(x,Y)$ exists for each $Y \in \Omega^*$, and can be obtained in terms of the classical solution to PDE's in which Y occurs as a parameter and, further, $Y \to \hat{p}_{st}(x,Y)$ is a continuous mapping. (See Theorems 2.3, 2.4, 2.5.) This is described usually by saying that $\{\hat{p}_{st}(x,Y)\}$ is a robust solution.

We will now note that as in the white noise approach, we also have the other kind of robustness -- namely if $X^k \to X$, then $\hat{p}^k_{st}(x,Y) \to \hat{p}_{st}(x,Y)$. Suppose that $\{X^k\}$, $\{X\}$ satisfy the conditions of Theorem 4.5. Consider the model

$$Y_t = h_t(X_t^k) + Z_t, \qquad 0 \le t \le T, \tag{4.49}$$

where $\{Z_t\}$ is a $\mathbb{R}^N$-valued Wiener process, as in (1.2). Suppose that h is such that the conditions of Theorem 4.6 are satisfied.

Then, Theorems 2.3, 2.4, 2.5 imply that the unnormalized conditional densities $\hat{p}_{st}^k(x,Y)$ of X_s^k given $\mathscr{F}_t^Y$ and $\hat{p}_{st}(x,Y)$ of X_s given $\mathscr{F}_t^Y$ exist for all $Y \in \Omega^*$ and are continuous functionals of $Y \in \Omega^*$ (for $s > t$, we need to assume that h is bounded). As before, let us write

$$\hat{p}_t^k(x,Y) = \hat{p}_{tt}^k(x,Y).$$

Then we have the following result.

<u>Theorem 4.8</u>: Suppose that the conditions of Theorem 4.6 are satisfied. Then we have for all $Y \in \Omega^*$ fixed, for all $R \ge 1$,

$$\hat{p}_t^k(x,Y) \longrightarrow \hat{p}_t(x,Y) \text{ uniformly in } (t,x) \in [0,T]\times\{x: |x|\le R\} \tag{4.50}$$

and for $0 \le t \le T$ fixed, we further have

$$\hat{p}_{st}^k(x,Y) \longrightarrow \hat{p}_{st}(x,Y) \tag{4.51}$$

uniformly in $(s,x) \in [0,t]\times\{x: |x|\le R\}$.

If h is assumed to be bounded

$$\hat{p}_{st}^k(x,Y) \longrightarrow \hat{p}_{st}(x,Y) \tag{4.52}$$

uniformly in $(s,x) \in [t,T]\times\{x: |x|\le R\}$.

Proof: The proofs of (4.50) and (4.51) are the same as the proof of the corresponding parts in Theorem 4.6. We will need to use the representations (2.29) and (2.37) with X^k in place of X:

$$\hat{p}^k_t(x,Y) = \hat{\psi}^k_t(x,Y) \exp\left[\sum_{j=1}^{N} h^j_t(x)Y^j_t\right] \tag{4.53}$$

$$\hat{p}^k_{st}(x,Y) = \hat{\psi}^k_s(x,Y)\ \hat{v}^{k'}_t(s,x,Y) \exp\left[\sum_{j=1}^{N} h^j_t(x)Y^j_t\right] \tag{4.54}$$

where $\hat{\psi}^k_t(x,Y)$, $\hat{v}^{k'}_t(s,x,Y)$ are solutions to appropriate Cauchy problems. As in Theorem 4.6, we can prove uniform convergence of $\hat{\psi}^k_t(x,Y)$ to $\hat{\psi}_t(x,Y)$ and $\hat{v}^{k'}_t(s,x,Y)$ to $\hat{v}_t(s,x,Y)$ over the compact subsets of $[0,T]\times\mathbb{R}^d$ and $[0,t]\times\mathbb{R}^d$ respectively. For (4.52), let us note that as in the proof of Theorem 2.5, we can get constants K_1, K_2 such that H_1 defined by VII.4.5 satisfies (2.44) - (2.47) with (a^k,b^k) instead of (a,b), for all $k \geq 1$ (here $Y \in \Omega^*$ is fixed).

Thus, we can conclude

$$|\hat{p}^k_t(x,Y)| \leq K_4 H_1(t,x). \tag{4.55}$$

for a fixed constant K_4. As seen in the proof of Theorem 2.5, $\hat{p}^k_{st}(x,Y)$ satisfies

$$\frac{\partial}{\partial s}\hat{p}^k_{st}(x,Y) = L^{k*}_s\ \hat{p}_{st}(x,Y), \qquad s > t \tag{4.56}$$

$$\hat{p}^k_{tt}(x,Y) = \hat{p}^k_t(x,Y). \tag{4.57}$$

The required conclusion (4.52) now follows from Theorem 2.1, since we have assumed conditions (i) - (iv) and also (4.55), (4.56), (4.57) are satisfied. We need to take in Theorem 2.1, $M_r^k = L_{t+r}^{k*}$, $M_r = L_{t+r}^{*}$, $T = T-t$, $u^k(r,x) = \hat{p}_{t+r}^k(x,Y)$, $u(r,x) = \hat{p}_{t+r}(x,Y)$, $g^k(x) = \hat{p}_t^k(x,Y)$ and $g^0(x) = \hat{p}_t(x,Y)$.' □

5. SMOOTHNESS PROPERTIES OF THE CONDITIONAL EXPECTATION

It has been shown recently that the conditional expectation in the nonlinear filtering problem in the conventional approach is a C^∞-functional in Malliavin's sense [ ].

A similar result in the white noise setup turns out to be surprisingly easy to establish as we shall show in this section. The reasons seem to be the independence of signal and noise, which is a basic assumption in our model, and the fact that in contrast to the Malliavin theory, all directional derivatives are admissible. The derivatives are taken in the Fréchet sense.

Our treatment includes infinite dimensional signal and noise and applies to prediction and smoothing problems as well as filtering. We do not even need to assume the Markovian nature of the signal. In fact, we will deal with the abstract statistical model of Section VI.3 which includes all the models considered in the later chapters.

An interesting point to note is that Fréchet differentiability follows from straightforward calculations. The part that requires attention is the one related to proving that the derivatives are accessible random variables in our sense, i.e., belong to the appropriate $\mathcal{L}$ spaces.

We begin by recalling the definition of Fréchet derivatives. For two Banach spaces B_1, B_2 (with norms $\|\cdot\|_1$, $\|\cdot\|_2$ respectively), let $L(B_1,B_2)$ denote the class of all linear transformations $A: B_1 \to B_2$. $L(B_1,B_2)$ is itself a Banach space with the operator norm

$$\|A\| = \sup\{\|Ab_1\|_2: b_1 \in B_1, \|b_1\| \leq 1\}. \tag{5.1}$$

Let H be a real separable Hilbert space with norm $|\cdot|$, and $f: H \to B_1$ be a mapping. Then f is said to be *Fréchet* (F) *differentiable* if for every $h \in H$, there exists $f_1(h) \in L(H,B_1)$ such that

$$\|f(h+h') - f(h) - f_1(h)[h']\|_1 = o(|h'|), \text{ as } |h'| \to o$$

(where $o(|h'|)$ denotes a quantity which when divided by $|h'|$ converges to 0 as $|h'| \to o$), and then $f_1(h)$ is called the F-derivative of f at h and is written as $(Df)(h)$.

Let $L^0(H) = \mathbb{R}$, $L^1(H) = L(H,\mathbb{R})$ and for $r \geq 1$, $L^{r+1}(H) = L(H,L^r(H))$. Let us denote the operator norm (defined by (5.1)) on $L^r(H)$ by $\|\cdot\|_r$. The Banach space $L^r(H)$ can be identified with the class of all linear mappings from the r-fold product $H\times\ldots\times H$ into $\mathbb{R}$, and under this identification, the norm $\|\cdot\|_r$ is defined by

$$\|g\|_r = \sup\{|g[h_1,\ldots,h_r]|: h_i \in H, |h_i| \leq 1\}.$$

Thus, an element $g \in L^r(H)$ will be thought of as a linear operator from $H \to L^{r-1}(H)$ and as a linear map from $H\times\ldots\times H$ (r times) into $\mathbb{R}$.

If a function $f: H \to \mathbb{R}$ is F-differentiable (take $B_1 = \mathbb{R}$ in the definition given above), then Df is a mapping from H into $L^1(H)$. If Df is F-differentiable, then we say that f is twice differentiable and then $D^2 f := D(Df)$, which in turn is a mapping from H in $L^2(H)$. Similarly, f is defined to be (r+1) times F-differentiable if $D^r f$ is F-differentiable and $D^{r+1} f := D(D^r f)$.

Let $L^r_{(2)}(H)$ be the subclass of $L^r(H)$ consisting of $g \in L^r(H)$ for which

$$\|g\|^2_{r,2} := \sum_{j_1,\ldots,j_r} |g[\phi_{j_1},\ldots,\phi_{j_r}]|^2 < \infty$$

where $\{\phi_j\}$ is any CONS in H. It is well known that the right hand side in the above expression does not depend on the choice of CONS $\{\phi_j\}$. Now $L^r_{(2)}(H)$ is a Hilbert space with norm $\|g\|_{r,2}$. It is easy to see that for $f \in L^r_{(2)}(H)$

$$\|g\|_r \leq \|g\|_{r,2}. \tag{5.2}$$

Lemma 5.1: Let $\nu \in \mu_0(H)$. Let g_0: $H \to \mathbb{R}$ and for $r \geq 1$, g_r: $H \to L^r(H)$ be defined by

$$g_0(h) = \int \exp\{(h,k) - \tfrac{1}{2}|k|^2\} d\nu(k) \tag{5.3}$$

and

$$g_r(h)[h_1,\ldots,h_r] \tag{5.4}$$

$$= \int \exp\{(h,k) - \tfrac{1}{2}|k|^2\}(h_1,k)(h_2,k)\ldots(h_r,k) d\nu(k).$$

Then $g_r(h) \in L^r_{(2)}(H)$ for all $h \in H$, $r \geq 1$, and $Dg_r = g_{r+1}$, $r \geq 0$. Thus, g_0 is r-times F-differentiable and $D^r g_0 = g_r$.

<u>Proof</u>: Note that the integral appearing is finite since

$$|(h,k)| \le \tfrac{1}{2}\{|2h|^2 + |\tfrac{1}{2}k|^2\} = 2|h|^2 + \tfrac{1}{8}|k|^2 \tag{5.5}$$

and

$$|(h,k)| \le |h|\ |k|. \tag{5.6}$$

Fix a CONS $\{\phi_j\}$ in H. Then

$$\begin{aligned}
\|g_r\|^2_{r,2} &= \sum_{j_1\cdots j_r} \Big[\int \exp\{(h,k) - \tfrac{1}{2}|k|^2\} \\
&\qquad \times (\phi_{j_1},k)(\phi_{j_2},k)\ldots(\phi_{j_r},k)\,d\nu(k)\Big]^2 \\
&\le \sum_{j_1\cdots j_r} \int \exp\{2(h,k) - |k|^2\} \\
&\qquad \times (\phi_{j_1},k)^2(\phi_{j_2},k)^2\ldots(\phi_{r_r},k)^2\,d\nu(k) \\
&= \int \exp\{2(h,k) - |k|^2\}\cdot|k|^{2r}\,d\nu(k) \\
&< \infty
\end{aligned}$$

and hence $g_r \in L^r_{(2)}(H)$.

For $h,h' \in H$ let

$$v(h,h') := \|g_r(h+h') - g_r(h) - g_{r+1}(h)(h')\|_{r,2}. \tag{5.7}$$

Recalling that $g_{r+1}(h)(h')[h_1,\ldots,h_r] = g_{r+1}(h)[h_1,\ldots,h_r,h']$,

$$\begin{aligned}
v^2(h,h') = \sum_{j_1\cdots j_r} &|\{g_r(h+h') - g_r(h)\}[\phi_{j_1},\ldots,\phi_{j_r}] \\
&- g_{r+1}(h)[\phi_{j_1},\ldots,\phi_{j_r},h']|^2
\end{aligned}$$

$$= \sum_{j_1,\dots,j_r} \Big| \int \exp\{(h,k) - \tfrac{1}{2}|k|^2\} \times (\phi_{j_1},k)\dots(\phi_{j_r},k) v_1(h',k) d\nu(k) \Big|^2$$

$$\leq \int \exp\{2(h,k) - |k|^2\} |k|^{2r} v_1^2(h',k) d\nu(k)$$

$$\leq \exp(4|h|^2) \int \exp(-\tfrac{3}{4}|k|^2) |k|^{2r} v_1^2(h',k) d\nu(k)$$

using (5.5), (5.6) where

$$\begin{aligned} v_1(h',k) &= |\exp((h',k)) - 1 - (h',k)| \\ &\leq |(h',k)|^2 \exp\{|(h',k)|\} \\ &\leq |h'|^2 |k|^2 \exp\{2|h'|^2 + \tfrac{1}{8}|k|^2\}. \end{aligned}$$

Thus

$$v^2(h,h') \leq |h'|^4 \exp\{4|h|^2 + 4|h'|^2\} \int \exp\{-\tfrac{1}{2}|k|^2\} |k|^{2r+2} d\nu(k). \tag{5.8}$$

The integral appearing in (5.8) is finite and hence

$$\lim_{|h'|\to 0} \frac{1}{|h'|} v(h,h') = 0.$$

In view of the inequality (5.2), this shows that g_r is F-differentiable and that $Dg_r = g_{r+1}$. □

Let us now consider the abstract statistical model (VI.3.1). We will freely use the notation established in Section VI.3 without any further explanation. Our aim is to prove, under suitable conditions on ξ, g, that the conditional expectation $E_\alpha(g|Qy)$ is infinitely Fréchet differentiable for any orthogonal projection Q. By the Bayes formula, Theorem VI.3.4, for $g \in \mathcal{L}^1(\Omega,\mathcal{A},\Pi)$,

$$E_\alpha(g|Qy) = \frac{\sigma_Q(g,y)}{\sigma_Q(1,y)} = \pi_Q(g,y) \tag{5.9}$$

where

$$\sigma_Q(g,y) := \int g(\omega)\, \exp((y,Q\xi(\omega)) - \tfrac{1}{2}|Q\xi(\omega)|^2)d\Pi(\omega). \tag{5.10}$$

By linearity, it is enough to take $g \geq 0$ such that $\int g d\Pi = 1$. Both $\sigma_Q(g,y)$ and $\sigma_Q(1,y)$ are function of the form:

$$f(h) = \int \exp((h,\eta) - \tfrac{1}{2}|\eta|^2)d\nu(\eta) \tag{5.11}$$

for $\nu \in \mathcal{M}_0(H)$. If we take $\nu = \Pi\circ(Q\xi)^{-1}$, then $f = \sigma_Q(1,\cdot)$ and if $\nu = \Pi'\circ(Q\xi)^{-1}$, $d\Pi' = gd\Pi$, then $f = \sigma_Q(g,y)$. Thus for the time being, let us concentrate on f. As proved in Lemma 5.1, f is r-times F-differentiable for all r, $D^r f = g_r$ which is given by (5.4) and $D^r f \in L^r_{(2)}(H)$. Since $g_r(h)[h_1,\ldots,h_r] \in \mathcal{I}(H)$, it follows that $D^r f$ is a QCM from H into $L^r_{(2)}(H)$.

The next step is to show that $D^k f$ is an accessible random variable if ν satisfies a moment condition.

<u>Lemma 5.2</u>: Let f be given by (5.7). Suppose

$$\int |\eta|^{2p} d\nu(\eta) < \infty. \tag{5.12}$$

Then for $1 \leq k \leq p$,

$$D^k f(y) \in \mathcal{L}^*(E,\mathcal{E},\alpha;\ L^k_{(2)}(H)). \tag{5.13}$$

<u>Proof</u>: Let $\{\phi_i\}$ be a CONS in H. In this proof, j will denote a multiindex $(j_1,j_2,\ldots,j_k)$. The function $D^k f(\cdot)[\phi_{j_1},\ldots,\phi_{j_k}]$ will be denoted by f_j and $\theta_j(\eta)$ will stand for $(\eta,\phi_{j_1})\ldots$

(η, ϕ_{j_k}). Thus

$$f_j(h) = \int \exp((h,\eta) - \tfrac{1}{2}|\eta|^2)\theta_j(\eta)d\nu(\eta). \tag{5.14}$$

Let $P_i \xrightarrow{s} I$, $\{P_i\} \subset \mathscr{P}$ be arbitrary. As seen in the proof of Theorem VI.3.2,

$$R_\alpha([f_j(y)]_{P_i})(\tilde{\omega}) = \int Z_i(\eta,\tilde{\omega})\theta_j(\eta)d\nu(\eta). \tag{5.15}$$

Note that from the definition of lifting for cylinder functions, for any i,r,

$$\begin{aligned} U_{ir}(\tilde{\omega}) &:= \|R_\alpha([D^k f(y)]_{P_i})(\tilde{\omega}) - R_\alpha([D^k f(y)]_{P_r})(\tilde{\omega})\|^2_{k,2} \\ &= \Sigma_j |R_\alpha([f_j(y)]_{P_i})(\tilde{\omega}) - R_\alpha([f_j(y)]_{P_r})(\tilde{\omega})|^2 \\ &\le \Sigma_j [\int |Z_i - Z_r|(\eta,\tilde{\omega})\theta_j(\eta)d\nu(\eta)]^2 \\ &\le \Sigma_j [\int |Z_i - Z_r|(\eta,\tilde{\omega})d\nu(\eta)][\int |Z_i - Z_r|(\eta,\tilde{\omega})\theta_j^2(\eta)d\nu(\eta)] \\ &= [\int |Z_i - Z_r|(\eta,\omega)d\nu(\eta)][\int |Z_i - Z_r|(\eta,\tilde{\omega})|\eta|^{2k}d\nu(\eta)]. \end{aligned} \tag{5.16}$$

In the above we have used Hölder's inequality and $\Sigma_j \theta_j^2(\eta) = |\eta|^{2k}$.

It is proved in the proof of Theorem VI.3.2 that there exists a probability measure Π' on the representation space $(\tilde{\Omega},\tilde{\mathscr{A}})$ such that $\tilde{\Pi} \ll \Pi'$, and

$$\iint |Z_i - Z_r|(\eta,\tilde{\omega})d\mu(\eta)d\Pi'(\tilde{\omega}) \longrightarrow 0 \tag{5.17}$$

as $(i,r) \to \infty$ for any finite measure μ on H. Hence for such a μ,

$$\int |Z_i - Z_r|(\eta, \tilde{\omega}) d\mu(\eta) \longrightarrow 0 \quad \text{in } \tilde{\Pi}\text{-probability}. \tag{5.18}$$

Using (5.18) for $\mu_1 = \nu$ and for μ_2 defined by $d\mu_2(\eta) = |\eta|^{2k} d\nu(\eta)$, it follows from inequality (5.16) that

$$U_{ir} \longrightarrow 0 \text{ in } \tilde{\Pi}\text{-probability as } i,r \to \infty.$$

Note that the assumption (5.12) implies that μ_2 is a finite measure. This proves (5.13). □

For a Banach space B, $q \geq 1$, define $\mathscr{L}^{q*}(E,\mathscr{E},\alpha;B)$ to consist of all $f \in \mathscr{L}(E,\mathscr{E},\alpha;B)$ for which

$$\int |R_\alpha(f_{P_i}) - R_\alpha(f_{P_r})|^q d\tilde{\Pi} \longrightarrow 0$$

as $i,r \to \infty$, for all $P_i \xrightarrow{s} I$, $\{P_i\} \subset \mathscr{P}$.

<u>Lemma 5.3</u>: Suppose ξ is bounded, i.e., $\Pi(|\xi| \leq M) = 1$ and

$$\nu(\eta\colon |\eta| > M) = 0 \qquad \text{for some } M. \tag{5.19}$$

Then for all k, $q \geq 1$,

$$D^k f(h) \in \mathscr{L}^{q*}(E,\mathscr{E},\alpha; L^k_{(e)}(H)). \tag{5.20}$$

<u>Proof</u>: From Lemma 2.1 $D^k f$ exists and belongs to $L^k_{(2)}(H)$ for all $k \geq 0$. We need to prove that

$$\int [U_{ir}]^{q/2} d\tilde{\Pi} \longrightarrow 0 \qquad \text{as } i,r \to \infty. \tag{5.21}$$

In view of (5.16), this would follow if we prove

$$\iint |Z_i - Z_r|^q(\eta,\tilde{\omega})(1 + |\eta|^{2kq})d\nu(\eta)d\tilde{\Pi}(\tilde{\omega}) \longrightarrow 0. \tag{5.22}$$

Here

$$Z_i(\eta,\tilde{\omega}) = \exp\left[(\xi(\omega),\eta) + L_0(P_i\eta)(\omega_0) - \tfrac{1}{2}|\eta|^2\right], \tag{5.23}$$

$\tilde{\omega} = (\omega,\omega_0) \in \tilde{\Omega}$. It was shown in Theorem VI.3.2 that

$$Z_i \text{ converges in } \nu\otimes\tilde{\Pi}\text{-probability to } Z. \tag{5.24}$$

Note that assumption (5.19) and $|\xi| \leq M$ imply that for any $q \geq 1$

$$\begin{aligned}\iint |Z_i(\eta,\tilde{\omega})|^q d\nu(\eta)d\tilde{\Pi}(\tilde{\omega}) &\leq M_1 \iint \exp(qL_0(P_i\eta))d\nu(\eta)d\tilde{\Pi}(\tilde{\omega}) \\ &\leq M_1 \int \exp(q^2|P_i\eta|^2)d\nu(\eta) \leq M_2\end{aligned} \tag{5.25}$$

where M_1, M_2 are constants depending on M, q. Thus $\{|Z_i|^q\}$ is uniformly integrable for all $q \geq 1$. As a consequence

$$\iint |Z_i - Z|^q d\nu d\tilde{\Pi} \longrightarrow 0 \quad \text{as } i \to \infty. \tag{5.26}$$

In view of (5.19), this yields (5.22) and hence (5.21). □

Returning to our abstract statistical model, we have

<u>Theorem 5.4</u>: For all $r \geq 1$, $D^r\sigma_Q(g,y)$ exists and belongs to $L^r_{(2)}(H)$, for all $y \in H$, for all integrable g. If ξ satisfies

$$E_\Pi|\xi|^{2k} < \infty, \tag{5.27}$$

then for all g bounded,

$$D^k\sigma_Q(g,y) \in \mathscr{L}^*(E,\mathscr{E},\alpha;\ L^k_{(2)}(H)). \tag{5.28}$$

Furthermore, if $|\xi|$ is bounded, then for all $q \geq 1$, $k \geq 1$

$$D^k\sigma_Q(g,y) \in \mathscr{L}^{q*}(E,\mathscr{E},\alpha;\ L^k_{(2)}(H)). \tag{5.29}$$

Proof: It has been seen above that $\sigma_Q(g,y) = f(y)$ is given by (5.7) if we take

$$\nu(A) := \int g(\omega)1_A(Q\xi(\omega))d\Pi(\omega). \tag{5.30}$$

For this choice of ν,

$$\int|\eta|^{2k}d\nu(\eta) = \int|Q\xi(\omega)|^{2k}g(\omega)d\Pi(\omega) \leq M\int|\xi(\omega)|^{2k}d\Pi(\omega) < \infty$$

in view of (5.27). Thus the existence of $D^k\sigma_Q(g,y)$ and (5.28) follow from Lemmas 5.1, 5.2 respectively. (5.19) and hence (5.29) follow from Lemma 5.3. □

Our final result on smoothness of the conditional expectation $\pi_Q(g,y)$ can now be proved.

Theorem 5.5: Suppose ξ satisfies (5.27). Then for all g bounded, $\pi_Q(g,y)$ is k-times Fréchet differentiable and

$$D^k\pi_Q(g,y) \in \mathscr{L}^{*}(E,\mathscr{E},\alpha;\ L^k_{(2)}(H)). \tag{5.31}$$

Proof: $\pi_Q(g,y) = f_1(y)/f_2(y)$, where

$$f_1(y) = \sigma_Q(g,y),\quad f_2(y) = \sigma_Q(1,y) \tag{5.32}$$

are as in (5.7). Now, f_1, f_2 are both k-times F-differentiable and $f_2 > 0$. From this it is easy to check that f_1/f_2 is also k-times F-differentiable. It can be shown that $D^k(f_1/f_2)$ can be expressed as

$$D^k(f_1/f_2) = \Lambda_k(f_1,f_2,Df_1,Df_2,\ldots,D^kf_1,D^kf_2,1/f_2) \quad (5.33)$$

where Λ_k is a continuous mapping from

$$\mathbb{R} \times \mathbb{R} \times L^1_{(2)}(H) \times L^1_{(2)}(H) \times \ldots \times L^k_{(2)}(H) \times L^k_{(2)}(H) \times \mathbb{R}$$

into $L^k_{(2)}(H)$. Since $f_1,f_2,1/f_2 \in \mathscr{L}^*(E,\mathscr{E},\alpha;\mathbb{R})$, (see Theorem VI.3.2) and $D^if_1,D^if_2 \in \mathscr{L}^*(E,\mathscr{E},\alpha;\ L^i_{(2)}(H))$, $1 \leq i \leq k$, assertion (5.31) follows from the continuity of Λ_k (appearing in (5.33)) and Theorem V.3.4. □

We have thus shown that if ξ satisfies

$$E_\Pi|\xi|^r < \infty \quad \text{for all } r \geq 1$$

then $\pi_Q(g,y)$ is infinitely F-differentiable for all bounded functions g and for all orthogonal projections Q and all the derivatives are accessible random variables.

As pointed out at the beginning of the section, Theorem 5.5 refers to the abstract statistical model and thus gives F-differentiability of conditional expectations occurring in filtering, prediction and smoothing problems for a finite dimensional signal as well as S-valued signal processes, where the noise itself may be either finite or infinite dimensional.

CHAPTER XI

STATISTICAL APPLICATIONS

In this chapter we discuss some applications of the theory developed thus far. Some of the examples are traditional but involve features that distinguish them from the usual (i.e., countably additive probabilistic) treatment. Others, such as the false alarm problem or filtering of infinite dimensional processes, arise from quality control and neorophysiology.

1. PARAMETER ESTIMATION IN LINEAR MODELS

Consider the usual problem of linear estimation when observation and noise are both infinite dimensional and where the parameter is also allowed to be infinite dimensional. Let H and H_1 be real separable Hilbert spaces and $T: H_1 \to H$ be a continuous linear operator (i.e., $T \in L(H_1, H)$). The unknown parameter θ lies in H_1 and observations on θ are corrupted by additive Gaussian white noise on H, i.e.,

$$y = T\theta + e. \tag{1.1}$$

Let m be the canonical Gauss measure on $(H, \mathscr{C})$ and $n_\theta :=$ $m \circ y^{-1}$ be the measure induced by y when θ is the true parameter. By Theorem VI.2.3 $n_\theta << m$ and

$$\frac{dn_\theta}{dm}(h) = \exp\{(T\theta, h) - \tfrac{1}{2}|T\theta|^2\}, \quad h \in H. \tag{1.2}$$

For a given h, $\rho_\theta(h) := \dfrac{dn_\theta}{dm}(h)$ regarded as a function of θ

is the likelihood function. We can therefore obtain the maximum likelihood estimate (MLE) $\hat{\theta}(y)$ for each $y \in H$. From (1.2), maximizing $\rho_\theta(y)$ with respect to θ is equivalent to minimizing $|T\theta-y|^2$. Let us now assume that

$$T^*T \text{ is invertible,} \tag{1.3}$$

where $T^* \in L(H,H_1)$ is the adjoint of T. It is easy to see from (1.3) that range(T) is closed in H. Denoting by S the orthogonal projector on H with range(S) = range(T) we have

$$|T\theta-y|^2 = |T\theta-Sy|^2 + |y-Sy|^2. \tag{1.4}$$

Let $\hat{\theta}(y)$ be that value of θ (which exists since $Sy \in$ range(T)) such that

$$T\hat{\theta}(y) = Sy. \tag{1.5}$$

$\hat{\theta}(y)$ is uniquely determined because of (1.3). Furthermore, from the definition of S, it is easy to see that $T^*y = T^*Sy$ for all $y \in H$. Hence, from (1.5), $T^*T\hat{\theta}(y) = T^*y$ and we get

$$\hat{\theta}(y) = (T^*T)^{-1}T^*y. \tag{1.6}$$

<u>Properties of $\hat{\theta}(y)$</u>.

(i) $\hat{\theta}$ is a QCM: This is immediate from (1.6).

(ii) For each $h_1 \in H_1$, $(h_1,\hat{\theta}(y))$ is an unbiased estimate of θ.

<u>Proof</u>: $E_{n_\theta}(h_1,\hat{\theta}(y)) = E_{n_\theta}(h_1,(T^*T)^{-1}T^*y)$

$$= E_{n_\theta}(T(T^*T)^{-1}h_1, y) = E_m(T(T^*T)^{-1}h_1,\ T\theta+e)$$

$$= (T(T^*T)^{-1}h_1, T\theta) + E_m(T(T^*T)h_1, e)$$

$$= ((T^*T)^{-1}h_1,\ (T^*T)\theta) = (h_1, \theta).$$

(iii) Under n_θ, $\hat{\theta}(y)$ has a Gaussian cylinder distribution with mean θ and covariance operator $(T^*T)^{-1}$.

<u>Proof</u>: Let $\bar{n}_\theta$ be the distribution of $\hat{\theta}$ under n_θ, i.e.,

$$\bar{n}_\theta := n_\theta \circ [\hat{\theta}(y)]^{-1}.$$

A direct computation of the characteristic functional of $\bar{n}_\theta$ yields the result:

$$\int_{H_1} \exp\{i(h_1, k)\} d\bar{n}_\theta(k) = \int \exp\{i(h_1, \hat{\theta}(y)\} dn_\theta(y)$$

$$= \int \exp\{i(T(T^*T)^{-1}h_1,\ T\theta+e(h))\} dm(h)$$

$$= e^{i(h_1,\theta)} e^{-\frac{1}{2}|T(T^*T)^{-1}h_1|^2} = e^{i(h_1,\theta)-\frac{1}{2}(h_1,(T^*T)^{-1}h_1)}.$$

(iv) For every $h_1 \in H_1$, $(h_1, \hat{\theta}(y))$ is the best continuous linear unbiased estimator of (h_1, θ).

<u>Proof</u>: Fixing $h_1 \in H_1$ and letting $h_2 = T(T^*T)^{-1}h_1$ $(\in H)$, we have

$$(h_1, \hat{\theta}(y)) = (h_2, y). \tag{1.7}$$

Any continuous linear unbiased estimator $\eta(y)$ of (h_1, θ) is of the form $\eta(y) = (h_3, y)$ for some $h_3 \in H$. Hence for all $\theta \in H_1$

$$(h_1, \theta) = E_{n_\theta} \eta(y) = E_m(h_3,\ T\theta + e) = (h_3, T\theta) = (T^* h_3, \theta). \tag{1.8}$$

We thus have

$$T^* h_3 = h_1. \tag{1.9}$$

Using (1.9) it is easily verified that

$$E_{n_\theta}[\eta(y) - (h_1, \theta)]^2 = |h_3|^2 \tag{1.10}$$

and similarly

$$E_{n_\theta}[\hat{\theta}(y) - (h_1, \theta)]^2 = |h_2|^2. \tag{1.11}$$

Now,

$$\begin{aligned}(h_2, h_3 - h_2) &= (T(T^*T)^{-1} h_1,\ h_3 - h_2) \\ &= ((T^*T)^{-1} h_1,\ T^* h_3 - T^* h_2) = 0\end{aligned}$$

since $T^* h_3 = T^* h_2 = h_1$ and so

$$|h_3|^2 = |h_2|^2 + |h_3 - h_2|^2 \geq |h_2|^2.$$

The assertion follows from (1.10) and (1.11).

Consistency and Asymptotic Normality.

For the sake of simplicity assume that $H_1 = H$ and $T = I$. Let $y^N = (y_1, \ldots, y_N)$ be a sample of N independent observations from the model (1.1). Let $L_N(y^N, \theta) = \Pi_{i=1}^N \rho_\theta(y_i)$ be the likelihood function. The MLE of θ based on y^N is $\hat{\theta}_N = \bar{y}_N := \frac{1}{N}\sum_{i=1}^N y_i$. We cannot expect $\hat{\theta}_N$ to be a consistent estimator of θ in the sense that $E|\hat{\theta}_N-\theta|^2 \to 0$ because $|\hat{\theta}_N-\theta|^2$ is not an accessible random variable. However, it is easy to see that $\hat{\theta}_N$ is consistent in a weak sense. For every $h \in H$, $E[(h,\hat{\theta}_N) - (h,\theta)]^2 \to 0$. Also from property (iii) we find that the characteristic functional of $\bar{n}_\theta^N := n_\theta \circ (\hat{\theta}_N)^{-1}$ is given by $\exp[i(h,\theta) - \frac{1}{2N}|h|^2]$. It follows that, for each N, $\xi_N := \sqrt{N}(\hat{\theta}_N-\theta)$ has a cylinder distribution on H whose characteristic functional is $e^{-\frac{1}{2}|h|^2}$.

Let us now compare this situation with what happens in the countably additive case (i.e., the traditional statistical model) corresponding to (1.1)

Suppose that (j,H,B) is an abstract Wiener space with abstract Wiener measure μ on the Borel sets of the Banach space B. The model considered now is

$$\xi = j\theta + \eta \tag{1.12}$$

where η is a B-valued Gaussian random variable on $(\Omega,\mathcal{A},\Pi)$ such that for each $h \in H^*$, $\eta[j^*h]$ is a $N(0,|h|^2)$ r.v. under Π, where $j^*: B^* \hookrightarrow H^*$ is the adjoint of j. Let P_θ be the distribution of ξ when θ is the true parameter ($P_0 = \mu$). Since $\theta \in H$, $P_\theta \ll \mu$ and the R-N derivative $\frac{dP_\theta}{d\mu}(\eta) = \exp\{X_\theta(\eta) - \frac{1}{2}|\theta|^2\}$ where $X_\theta(\eta)$ is $N(0,|\theta|^2)$ under μ. Let $L(\eta)$ = closed linear span$\{\eta[j^*h],\ h \in H^*\} \subset L^2(B,\mathcal{B}(B),\mu)$. Then $\frac{dP_\theta}{d\mu}(\eta) = \exp\{\sum_{j=1}^\infty \theta_j' X_j - \frac{1}{2}\sum_{j=1}^\infty \theta_j'^2\}$ where $X_j \in L(\eta)$ and the X_j's are independent $N(0,1)$ under μ, and $\theta_j' = (\theta, e_j)$ where (e_j) is a CONS in H (H^* is identified with H). Formally, the maximum likelihood estimate (MLE) of θ is given by $\hat{\theta} := \Sigma_{j=1}^{\infty}\hat{\theta}_j' e_j$ where $\hat{\theta}_j' = X_j$ for each j. But $\hat{\theta} \notin H$ since $\Sigma_{i=1}^{\infty} X_j^2 = \infty$ μ-a.s. Hence there is no MLE. This observation is not new. It has been noted in the literature (e.g., by Grenander [79]) and is a motivation for the estimation procedure known as the method of sieves. Its application to the present example is as follows: The likelihood function for a sample $\xi^{(N)} = (\xi_1, \ldots, \xi_N)$ of independent observations on the model (1.12) is

$$L(\xi^{(N)}; \theta) = \exp\left[\sum_{k=1}^{N}\left[\sum_{j=1}^{\infty}\theta_j' X_j^k - \tfrac{1}{2}\sum_{j=1}^{\infty}\theta_j'^2\right]\right]$$

where (X_j^k) corresponds to ξ_k. Let $d_n \uparrow \infty$ with n and let P_n be the orthogonal projection on H with range $P_n = \{h \in H: \Sigma_{j=1}^{d_n} h_j e_j\}$. Let $\theta_n = P_n\theta$, $(\theta \in H)$. Then the MLE $\hat{\theta}_n$ of the d_n-dimensional parameter θ_n is given by $\hat{\theta}_n = \Sigma_{j=1}^{d_n}\overline{X}_j^n e_j$ where $\overline{X}_j^n = \frac{1}{n}\Sigma_{k=1}^{n} X_j^k$. Since $\{\overline{X}_j^n\}$ is a sequence of independent $N(\theta_j', \frac{1}{n})$ r.v.'s under P_θ, we have the following:

$$E_{P_\theta}|\hat{\theta}_n - \theta_n|^2 = \frac{dn}{n} \tag{1.13}$$

and

$$E_{P_\theta}\exp\{i(h, \sqrt{n}(\hat{\theta}_n - \theta_n))\} = e^{-\frac{1}{2}|P_n h|^2}, \quad (h \in H). \tag{1.14}$$

From (1.13), $E_{P_\theta}|\hat{\theta}_n - \theta|^2 = \frac{dn}{n} + \Sigma_{j>d_n}\theta_j'^2 \to 0$ as $n \to \infty$ if we choose d_n s.t., $d_n \to \infty$ and $d_n/n \to 0$, i.e., $\hat{\theta}_n$ is norm consistent in the L^2 sense. It is not possible, in general, to conclude from (1.14) that

$$E_{P_\theta} \exp\{i(h, \sqrt{n}(\hat{\theta}_n - \theta))\} \longrightarrow e^{-\frac{1}{2}|h|^2} \tag{1.15}$$

for all $h \in H$. (1.15) holds, for example, if the parameter set $(\overline{H}) \subsetneq H$ is given by $(\overline{H}) = \{\theta \in H:\ n\Sigma_{j>d_n}^{\infty} \theta_j'^2 \to 0 \text{ as } n \to \infty\}$. Assuming this, and writing $\zeta_n = \sqrt{n}(\hat{\theta}_n - \theta)$ it follows from (1.15) that

$$\lim_{n\to\infty} E_{P_\theta} \exp\{i(j\zeta_n)[k]\} = e^{-\frac{1}{2}\|j^* k\|_B^2} \quad \forall\ k \in B^*. \tag{1.16}$$

(1.16) shows the asymptotic normality of the B-valued sequence $j\sqrt{n}(\hat{\theta}_n - \theta)$.

<u>General Models</u>.

A more general model than (1.1) is the basic filtering model

$$y = \xi + e \tag{1.17}$$

where ξ is an H-valued random variable whose distribution is μ_θ, $\theta \in (\overline{H})$, θ being an unknown parameter. If $(\overline{H}) = H_1$ and μ_θ is the degenerate measure at $\{T\theta\}$, then (1.17) reduces to (1.1). Another example is $\xi := f(X,\theta)$ where X is an S-valued random varible and $f\colon S\times(\overline{H}) \to H$ is a measurable mapping.

Letting n_θ = the distribution of y if ξ is distributed as μ_θ it follows from Theorem VI.3.1 that $n_\theta \ll m$ and

$$\rho_\theta(h) = \frac{dn_\theta}{dm}(h) = \int_H \exp[(\xi,h) - \tfrac{1}{2}|\xi|^2] d\mu_\theta(\xi). \tag{1.18}$$

M.L. estimation of θ can, in principle, be carried out using the likelihood function in (1.18) provided it can be shown in each case *that the MLE is an accessible random variable or a QCM*. The latter can be a difficult problem in some cases.

2. LIKELIHOOD RATIOS AND SIGNAL DETECTION

In this section we will be considering the problems of testing of hypotheses and signal detection. An important role will be played by likelihood ratios in these problems. We being by obtaining a formula for the likelihood ratio.

Once again consider the model

$$y_t = h_t(X_t) + e_t \tag{2.1}$$

of Section VIII.2.

For $t \geq 0$, let $n_t = \alpha\circ(Q_t y)^{-1}$ be the measure induced by $Q_t y$ under α on $H_t = Q_t H$. Let m_t denote the canonical Gauss measure on H_t. We have seen (Theorem VI.2.4) that $n_t \ll m_t$ and that

$$\rho_t(y) := \frac{dn_t}{dn_t}(y) \tag{2.2}$$

$$= E_{\Pi}\exp\left[\int_0^t (h_s(X_s), y_s)_{\mathcal{H}} ds - \tfrac{1}{2}\int_0^t \|h_s(X_s)\|^2_{\mathcal{H}} ds\right].$$

The function $\rho_t(y)$ will be called the likelihood function or likelihood ratio (LR).

We will now obtain a recursive or dynamic formula for $\rho_t(y)$.

<u>Theorem 3.1</u>: Suppose $E_{\Pi}\int_0^T \|h_s(X_s)\|^2_{\mathcal{H}} ds < \infty$. Then

$$\rho_t(y) = \exp\{\int_0^t (\pi_s(h_s), y_s)_{\mathscr{H}} ds - \tfrac{1}{2}\int_0^t \pi_s(\|h_s\|^2_{\mathscr{H}})ds\} \qquad (2.3)$$

where

$$\pi_s(f) \equiv \pi_s(f,y) := \int f(x)dF_t(y)(x), \qquad (2.4)$$

$F_t(y)$ being the conditional distribution of X_t given $Q_t y$.

If we assume that (X_t) is a Markov process, then this result can be deduced from Theorem VIII.2.1. However, the proof given below does not make this assumption.

Proof: Letting

$$c_u(x) = (h_u(x), y_u)_{\mathscr{H}} - \tfrac{1}{2}\|h_u(x)\|^2_{\mathscr{H}}$$

we have $c_u(x) \le \tfrac{1}{2}\|y_u\|^2_{\mathscr{H}}$ and

$$E_{\Pi}\int_0^T |c_u(X_u)|du < \infty. \qquad (2.5)$$

$$\begin{aligned}\rho_t(y) &= E_{\Pi}[1 + \int_0^t c_u(X_u)\ \exp(\int_0^u c_\tau(X_\tau)d\tau)du] \\ &= 1 + \int_0^t E_{\Pi}[c_u(X_u)\ \exp(\int_0^u c_\tau(X_\tau)d\tau)]du \\ &= 1 + \int_0^t \sigma_u(c_u)du,\end{aligned}$$

the use of Fubini theorem being justified by (2.5), and $\sigma_u(f) = \sigma_u(f,y)$ is the unnormalized conditional expectation of $f(X_t)$ given $Q_t y$, i.e., $\sigma_u(f) = \rho_u(y)\pi_u(f)$. Thus $\rho_t(y)$ is an absolutely continuous function of t and

$$\begin{aligned}\frac{d}{dt}\log \rho_t(y) &= \frac{1}{\rho_t(y)}\frac{d}{dt}\rho_t(y) \\ &= \frac{1}{\rho_t(y)}\sigma_t(c_t) \quad \text{for a.e. } t\end{aligned}$$

$$= \pi_t(c_t).$$

Thus

$$\rho_t(y) = \exp(\int_0^u \pi_u(c_u)du).$$

But

$$\pi_u(c_u) = (\pi_u(h_u), y_u)_{\mathcal{H}} - \tfrac{1}{2}\, \pi_u(\|h_u\|_{\mathcal{H}}^2).$$

This completes the proof. □

When $\mathcal{H} = \mathbb{R}$, the expression (2.3) can be recast in another form. For f_1, f_2 such that $E_{\Pi}|f_i(X_t)|^2 < \infty$, $i = 1,2$, write the conditional covariance

$$\operatorname{cov}_t(f_1, f_2) := E_{\alpha}[\{f_1(X_t) - E_{\alpha}(f_1(X_t)|Q_t y)\} \times \{f_2(X_t) - E_{\alpha}(f_2(X_t)|Q_t y)\}|Q_t y]$$

and the conditional variance

$$\operatorname{var}_t(f_1) := \operatorname{cov}_t(f_1, f_1).$$

Using Theorem V.4.6 it is easy to see that

$$\operatorname{cov}_t(f_1, f_2) = E_{\alpha}[f_1(X_t)f_2(X_t)|Q_t y] - E_{\alpha}[f_1(X_t)|Q_t y]\cdot E_{\alpha}[f_2(X_t)|Q_t y].$$

It follows that

$$\pi_t(h_t^2) = \operatorname{var}_t(h_t) + [\pi_t(h_t)]^2.$$

Then the formula (2.3) reduces to

$$\rho_t(y) = \exp\Big[\int_0^t \pi_s(h_s)y_s ds - \tfrac{1}{2}\int_0^t \{\pi_s(h_s)\}^2 ds \qquad (2.6)$$

$$- \tfrac{1}{2}\int_0^t \mathrm{var}_s(h_s)ds\Big].$$

Testing of Hypotheses

The following result will be useful in hypotheses testing and signal detection problems. It can be deduced from results in Section III.5, but a direct and short proof is included here because of its importance.

Lemma 2.2: Let $f \in \mathscr{L}(E,\mathscr{E},\alpha)$ and $A \in \mathscr{B}(\mathbb{R})$ be such that $\alpha\circ f^{-1}(\partial A) = 0$ (where ∂A is the boundary of A). Then

$$g = 1_A(f) \in \mathscr{L}(E,\mathscr{E},\alpha).$$

In particular, for $a \in \mathbb{R}$ such that $\alpha\circ f^{-1}(\{a\}) = 0$,

$$1_{\{f\leq a\}} \in \mathscr{L}(E,\mathscr{E},\alpha). \qquad (2.7)$$

Proof: Let $\{P_k\} \subseteq \mathscr{P}$ be such that $\{P'_k\} \subseteq \mathscr{P}$, $P_k \leq P'_k$ implies

$$R_\alpha(f_{P'_k}) \longrightarrow R_\alpha(f) \quad \text{a.s.}\tilde{\Pi} \qquad (2.8)$$

where R_α is the α-lifting corresponding to a representation $(\rho,L,\tilde{\Pi})$ of α. From the assumption $\alpha\circ f^{-1}(\partial A) = 0$ we have

$$\tilde{\Pi}(R_\alpha(f) \in \partial A) = 0.$$

Hence (2.8) implies

$$1_A(R_\alpha(g_{P'_k})) \longrightarrow 1_A(R_\alpha(f)) \quad \text{a.s. } \tilde{\Pi}. \qquad (2.9)$$

Now $R_\alpha(g_{P'_k}) = R_\alpha(1_A(f_{P'_k})) = 1_A(R_\alpha(f_{P'_k}))$ since $f_{P'_k}$ is a cylinder function. Hence

$$R_\alpha(g_{P'_k}) \longrightarrow 1_A(R_\alpha(f)). \tag{2.10}$$

Since (2.10) holds for all $\{P'_k\} \subseteq \mathscr{P}$ with $P_k \leq P'_k$, we conclude that $g \in \mathscr{L}(E,\mathscr{E},\alpha)$. □

For a set $F \subseteq E$ for which 1_F is an accessible random variable, i.e., $1_F \in \mathscr{L}(E,\mathscr{E},\alpha)$, we will write

$$\int 1_F d\alpha = \alpha(F). \tag{2.11}$$

$\alpha(F)$ was denoted by $\hat{\alpha}(F)$ in Section III.5, but the simpler notation $\alpha(F)$ is preferable here. It is easy to check that for any $f \in \mathscr{L}(E,\mathscr{E},\alpha)$ and $a \in \mathbb{R}$ such that $\alpha \circ f^{-1}(\{a\}) = 0$, we have

$$\alpha\{f \leq a\} = \alpha \circ f^{-1}((-\infty,a]). \tag{2.12}$$

Suppose that in the model (2.1), the function h is not completely known and that there are two alternatives: $h = h^0$ or $h = h^1$. Based on the observations $\{y_s: s \leq t\}$, we can test the null hypothesis H_0: $h = h^0$ against the alternative H_1: $h = h^1$. One can proceed as in the usual statistical practice and use the Neyman-Pearson lemma. Fix t and let n_t^0, n_t^1 be the distribution $\alpha \circ (Q_t y)^{-1}$ of y under H_0 and H_1 respectively. Then

$$\frac{dn_t^1}{dn_t^0}(y) = \left[\frac{dn_t^1}{dm_t}(y)\right] \cdot \left[\frac{dn_t^0}{dm_t}(y)\right]^{-1} \tag{2.13}$$

$$= \rho_t^1(y)/\rho_t^0(y) = f(y)$$

where ρ_t^1, ρ_t^0 are given by (2.2) or (2.3) with $h = h^1$ and $h = h^0$. Then $f(y)$ is the likelihood ratio of H_1 with respect to H_0. For a continuity point a of the distribution of f under n^0 [i.e., $n^0 \circ f^{-1}(\{a\}) = 0$], let $\gamma = n^0(f \geq a)$. Then the following test (called the LR test) is the 'best' test of H_0 versus H_1 at level γ:

Accept H_0 if $f(y) < a$.
Accept H_1 if $f(y) \geq a$.

The region $\{f(y) \geq a\}$ is called the critical region. Here the property of 'best' at level α is to be understood as follows.

Suppose $F \subseteq H$ is such that $1_F \in \mathcal{L}(H, \mathcal{C}, n^0)$. F determines an γ-level test for H_0 against H_1 if $n^0(y \notin F) = \gamma$. The acceptance region for H_0 is $\{y \in F\}$. Then the LR test is best in the sense that it has greater power that F, i.e., $n^1\{f(y) \geq a\} \geq n^1\{y \notin F\}$. Similarly, one could test the hypothesis that the distribution of (X_t) is Π^0 versus the alternative that it is Π^1, the likelihood ratio is $\rho_t^1(y) \cdot [\rho_t^0(y)]^{-1}$, where $\rho_t^i(y)$ is defined by (2.2) or (2.3) with $\Pi = \Pi^i$, $i = 0,1$.

Remark 2.1: For constructing the LR test, the fact that the Radon-Nikodym derivative is a random variable on $(H, \mathcal{C}, n)$ (or $(E, \mathcal{E}, \alpha)$) as opposed to a random variable on the representation space (see Section IV.1) is very important. Here, if the Radon-Nikodym derivative dn^1/dn^0 does not exist, but n^1 is absolutely continuous with respect to n^0 in the ϵ-δ sense, then the best test in the sense described above does not exist.

Signal Detection

An important problem is to detect whether the observation y_t contains a deterministic signal $s(t)$ in the presence of additive white noise e_t or if y_t is just white noise. The two hypotheses can be represented as

$$\begin{aligned} &\text{signal absent:} \quad y_t = e_t \\ &\text{signal present:} \quad y_t = s(t) + e_t \end{aligned}$$

where $\{e_t\}$ is white noise. As a first case, suppose $s(t)$ is known. Then the LR for observations over $u \in [0,t]$ is

$$\rho_t(y) = \exp\left(\int_0^t s(u)y_u du - \tfrac{1}{2}\int_0^t s(u)^2 du\right) \tag{2.14}$$

and a γ-level test for testing if signal is present can be carried out as explained earlier. In this case, the critical region $\{\rho_t(y) \geq a\}$ is equivalently $\{\int_0^t s(u)y_u du \geq a'\}$. The cutoff a' can be determined easily as under the null hypothesis: signal absent, the distribution of $\int_0^t s(u)y_u du$ is Gaussian with mean zero and variance $\int_0^t s(u)^2 du$.

The Bayesian approach to the signal detection problem when the signal is of known form but involves an unknown parameter θ can be handled by using the Bayes formula and the Zakai equation. The use of nonlinear filtering in this connection has been discussed by Davis in his work on gravitational wave data analysis [18]. We give below the white noise version of the solution to the problem.

Let us assume that $\theta \in \mathbb{R}^d$, and taking a Bayesian point of view, suppose that the prior distribution of θ has a density $p_0(\theta)$. Here we have to choose between

signal absent $\quad y_t = e_t$

and

signal present $\quad y_t = s(t,\theta) + e_t$

where $\{e_t\}$ is Gaussian white noise. We can regard θ as a random variable and then the formula (2.2) for the LR yields

$$\rho_t(y) = \int \exp\left[\int_0^t s(u,\theta)y_u du - \tfrac{1}{2}\int_0^t s^2(u,\theta)du\right]p_0(\theta)d\theta \qquad (2.15)$$

and (2.3) gives

$$\rho_t(y) = \exp\left[\int_0^t \pi_u(s(u))du - \tfrac{1}{2}\int_0^t \pi_u(s^2(u))du\right] \qquad (2.16)$$

where

$$\pi_u(s(u)) = \pi_u(s(u,\theta)) = E(s(u,\theta)|Q_u t).$$

In this case, we can obtain the density of the conditional distribution of θ given $Q_u y$ by using the methods of Chapter VII. Take $X_t = \theta$ for $t \geq 0$. Then X_t is a Markov process, whose generator L is given by $Lf = 0$. Thus the Zakai equation for the unnormalized conditional density $p_t(\theta,y)$ (see Section VII.3) is

$$\frac{\partial}{\partial t} p_t(\theta,y) = \{s(t,\theta)y_t - \tfrac{1}{2} s^2(t,\theta)\}p_t(\theta,y) \qquad (2.17)$$

with $p_0(\theta,y) = p_0(\theta)$. Thus

$$p_t(\theta,y) = p_0(\theta)\exp[\int_0^t s(u,\theta)y_u du - \tfrac{1}{2}\int_0^t s^2(u,\theta)du] \quad (2.18)$$

and

$$\pi_t(f) = E[f(\theta)|y_u:u\leq t] = \frac{\int f(\theta)p_t(\theta,y)d\theta}{\int p_t(\theta,y)d\theta}. \quad (2.19)$$

The likelihood ratio $\rho_t(y)$ can then be represented as

$$\rho_t(y) = \int p_t(\theta,y)d\theta. \quad (2.20)$$

We observe that the formulae (2.18), (2.19), (2.20) can be obtained directly from the Bayes formula.

Likelihood Ratios for Random Fields

We now turn our attention to the case where the signal process and noise processes are random fields. For simplicity in notation, consider the case of two index parameters and real valued observations. The general case when observations are $\mathcal{H}$-valued can be treated similarly.

Let $\underline{T} = [0,T_1]\times[0,T_2]$. We will denote a generic element (t_1,t_2) of $\underline{T}$ by $\underline{t}$. Let $H = L^2(\underline{T},\mathbb{R})$ with the usual inner product. Let m be the canonical Gauss measure on $(H,\mathcal{C})$ and let $e = (e_{t_1,t_1})$ be the identity map from $H \to H$. Then for any Borel set $A \subseteq \underline{T}$

$$\iint_A e_{t_1,t_2}dt_1dt_2 = (1_A,e)$$

has Gaussian distribution with mean zero, variance $\lambda(A)$ - the Lebesgue measure of A. Further, if A, B are disjoint, then $(1_A,e)$, $(1_B,e)$ are independent. Thus, $e = (e_{t_1,t_2})$ is Gaussian white noise on $\underline{T}$. Let $\{X_{t_1,t_2}: (t_1,t_2) \in \underline{T}\}$ be an

S-valued random field on $(\Omega,\mathcal{A},\Pi)$. Let $h: \underline{T}\times S \to \mathbb{R}$ be a measurable mapping such that

$$\int_0^{T_2}\int_0^{T_1}|h_{t_1,t_2}(X_{t_1,t_2})|^2 dt_1 dt_2 < \infty \quad \text{a.s. } \Pi. \tag{2.21}$$

Taking $\xi_{\underline{t}}(\omega) = h_{\underline{t}}(X_{\underline{t}}(\omega))$, $\xi = (\xi_{\underline{t}})$ is then an H-valued random variable on $(\Omega,\mathcal{A},\Pi)$. Let $(E,\mathcal{E},\alpha) = (\Omega,\mathcal{A},\Pi) \otimes (H,\mathcal{C},m)$. Consider then the model

$$y_{\underline{t}} = h_{\underline{t}}(X_{\underline{t}}) + e_{\underline{t}}, \qquad \underline{t} \in \underline{T}. \tag{2.22}$$

For $\underline{t} \in \underline{T}$, $\underline{t} = (t_1,t_2)$, define

$$H_{\underline{t}} = \{\eta \in H: \int_0^{T_2}\int_0^{T_1}\eta^2_{s_1,s_2} ds_1 ds_2 = \int_0^{t_2}\int_0^{t_1}\eta^2_{s_1,s_2} ds_1 ds_2\}$$

and let $Q_{\underline{t}}$ be the orthogonal projection onto $H_{\underline{t}}$. Then as in the one parameter case, $Q_{\underline{t}}y$ represents $\{y_{s_1,s_2}: s_1 \le t_1, s_2 \le t_2\}$, the observations over the rectangle $[0,t_1]\times[0,t_2]$.

Let $n_{\underline{t}} = \alpha\circ(Q_{\underline{t}}y)^{-1}$ and $m_{\underline{t}}$, the canonical Gauss measure on $H_{\underline{t}}$. $n_{\underline{t}}$ is the distribution of $Q_{\underline{t}}y$ when the signal ξ is present, i.e., $y = \xi + e$ and $m_{\underline{t}}$ is the distribution of $Q_{\underline{t}}y$ when $y = e$ (i.e., signal ξ is absent). As in the one parameter case, the LR

$$\rho_{\underline{t}}(y) = \frac{dn_{\underline{t}}}{dm_{\underline{t}}}(y)$$

will play a crucial role in testing hypotheses about the signal ξ, and in particular in signal detection problems. Note that by Theorem VI.2.4,

$$\rho_{\underline{t}}(y) = E_{\Pi}\, q_{\underline{t}}(\cdot,y) \tag{2.23}$$

where $q_{\underline{t}}(\omega,y)$ is defined by

$$q_{\underline{t}}(\omega,y) = \exp\left[\int_0^{t_1}\int_0^{t_2}\{h_{t_1,t_2}(X_{t_1,t_2})y_{t_1,t_2} - \tfrac{1}{2}h^2_{t_1,t_2}(X_{t_1,t_2})\}dt_1dt_2\right]. \tag{2.24}$$

We will now obtain a recursive expression for $\rho_{\underline{t}}(y)$ as in the one-parameter case. For a function $f: S \to \mathbb{R}$ such that $E_{\Pi}|f(X_{\underline{t}})| < \infty$, let

$$\pi_{\underline{t}}(f) := E_{\alpha}(f(X_{\underline{t}})|Q_{\underline{t}}y) = \frac{E_{\Pi}[f(X_{\underline{t}})q_{\underline{t}}(\cdot,y)]}{E_{\Pi}q_{\underline{t}}(\cdot,y)} \tag{2.25}$$

by the Bayes formula. Also, for f,g such that $E_{\Pi}f^2(X_{\underline{t}}) < \infty$, $E_{\Pi}g^2(X_{\underline{t}}) < \infty$, let

$$\text{cov}_{\underline{t}}(f,g) := \pi_t(\{f - \pi_t(f)\}\{g - \pi_t(g)\}) \tag{2.26}$$

be the conditional covariance of $f(X_{\underline{t}})$, $g(X_{\underline{t}})$ given $Q_{\underline{t}}y$. Then from Theorem V.4.6,

$$\text{cov}_{\underline{t}}(f,g) = \pi_{\underline{t}}(fg) - \pi_{\underline{t}}(f)\cdot\pi_{\underline{t}}(g). \tag{2.27}$$

<u>Theorem 2.3</u>: Suppose $E_{\Pi}\int_0^{T_2}\int_0^{T_1}|h_{t_1,t_2}(X_{t_1,t_2})|^4dt_1dt_2 < \infty$. Then the likelihood ratio $\rho_{\underline{t}}(y)$ is given by the formula

$$\log \rho_{t_1,t_2}(y) = \int_0^{t_1}\int_0^{t_2}\{\pi_{s_1,s_2}(h_{s_1,s_2})y_{s_1,s_2} - \tfrac{1}{2}\pi_{s_1,s_2}(h^2_{s_1,s_2}) + G(s_1,s_2)\}ds_1ds_2 \tag{2.28}$$

where

$$G(s_1,s_2) = \int_0^{s_1}\int_0^{s_2}\{\tfrac{1}{4}\,\mathrm{cov}_{\underline{s}}(h^2_{s_1,u_2},\ h^2_{u_1,s_2})$$

$$-\ \tfrac{1}{2}\,\mathrm{cov}_{\underline{s}}(h^2_{s_1,u_2},\ h_{u_1,s_2})y_{u_1,s_2}$$

$$-\ \tfrac{1}{2}\,\mathrm{cov}(h_{s_1,u_2},\ h^2_{u_1,s_2})y_{s_1,u_2}$$

$$+\ \tfrac{1}{4}\,\mathrm{cov}(h_{s_1,u_2},h_{u_1,s_2})y_{u_1,s_2}y_{u_2,s_1}\}du_1du_2.$$

Proof: The expression (2.23) - (2.24) for $\rho_{\underline{t}}(y)$ implies that the function $F(t_1,t_2) = \log \rho_{t_1,t_2}(y)$ satisfies

(i) for fixed t_1, $F(t_1,t_2)$ is absolutely continuous in t_2,

(ii) for a.e. t_2 (fixed), $\dfrac{\partial F(t_1,t_2)}{\partial t_2}$ is absolutely continuous in t_1,

(iii) $F(t_1,0) = 0$, $\dfrac{\partial F(0,t_2)}{\partial t_2} = 0$.

From these properties it easily follows that

$$F(t_1,t_2) = \int_0^{t_2}\int_0^{t_1}\left[\frac{\partial^2 F(s_1,s_2)}{\partial s_1\partial s_2}\right]ds_1ds_2.$$

Straightforward computations and the relations (2.25), (2.26) yield the formula

$$\frac{\partial^2 F(s_1,s_2)}{\partial s_1\partial s_2} = \pi_{s_1,s_2}(h_{s_1,s_2})y_{s_1,s_2} - \tfrac{1}{2}\,\pi_{s_1,s_2}(h^2_{s_1,s_2}) + G(s_1,s_2).$$

Thus, $\rho_{\underline{t}}(y)$ can be represented as (2.28). □

Remark 2.1: Bagchi [2] has obtained an alternative representation of $\rho_{\underline{t}}$ by expressing $G(s_1, s_2)$ in terms of 'four-fold integrals' analogous to those introduced by Wong and Zakai so that formula (2.28) then resembles (in the finitely additive theory) the expression for the likelihood ratio for two-parameter processes obtained by Wong and Zakai [74].

3. THE FILTERING PROBLEM FOR COUNTABLE STATE MARKOV PROCESSES WITH APPLICATIONS

We now specialize the results of Chapter VIII to the case where the signal is a countable state Markov process. The measure-valued equation for the optimal filter reduces to an infinite system of ordinary differential equations. An algorithm for obtaining an approximate solution to these equations is given in Theorem 3.1. Later in the section we consider some concrete problems arising in industry (e.g., the disruption problem in industrial production) to which the filtering model is applicable.

In the setup of Section VIII.2, let the state space S of the signal process (X_t) be a countable set. Let us assume that the paths of (X_t) are piecewise constant, i.e., there exist $\tau_k(\omega)$, $0 = \tau_0(\omega) < \tau_1(\omega) < \ldots < \tau_k(\omega) < \ldots$, $\tau_k(\omega) \longrightarrow \infty$ as $k \to \infty$ such that

$$X_t(\omega) = X_{\tau_k}(\omega) \quad \text{for } \tau_k(\omega) \leq t < t_{k+1}(\omega). \tag{3.1}$$

The times τ_i are called jump times of (X_t). Let $P(s,i;\ t,j)$ be the transition probability function:

$$P(s,i;\ t,j) = \Pi(X_{s+t} = j \,|\, X_s = i). \tag{3.2}$$

We make the following assumptions:

$$\frac{P(s,i;t,j) - \delta_{ij}}{t} \longrightarrow a_s(i,j) \text{ as } t\downarrow 0 \text{ uniformly in } (s,i,j). \tag{3.3}$$

$$a_s(i,j) \text{ is right continuous in s for each } i,j. \tag{3.4}$$

$$|a_s(i,j)| \leq K \quad \text{for some } K < \infty. \tag{3.5}$$

$$\Sigma_j a_s(i,j) = 0 \quad \text{for all } s,i. \tag{3.6}$$

Here $\delta_{ij} = 1$ if $i = j$ and $\delta_{ij} = 0$ if $i \neq j$. It is easy to see that $a_s(i,j) \geq 0$ if $i \neq j$ and $a_s(i,i) \leq 0$. Let L be the weak generator associated with $P(s,i;t,j)$ and $\mathcal{D}$ its domain (see Chapter II for the definition of L, $\mathcal{D}$, J_0). It can be shown that under the assumptions made above, J_0 consists of all function $f: \hat{S} = [0,\infty)\times S \to \mathbb{R}$ such that $f(s,i)$ is right continuous in s for all i and $\mathcal{D}$ consists of all functions $f: \hat{S} \to \mathbb{R}$ such that $f(s,i)$ is continuous, the right derivative $\frac{\partial^+ f(s,i)}{\partial s}$ exists, is right continuous for all i and $|\frac{\partial^+ f(s,i)}{\partial s}| \leq K$ for some $K < \infty$. Further, for $f \in \mathcal{D}$,

$$(Lf)(s,i) = \frac{\partial^+ f}{\partial s}(s,i) + \Sigma_j a_s(i,j)f(s,j). \tag{3.7}$$

Starting with $P(s,i;\ t,j)$ satisfying the conditions given above, or more generally, given $a_s(i,j)$ satisfying (3.4)-(3.6), and any initial distribution, one can construct a Markov process (X_t) with paths of the type (3.1), such that the transition probability function is $P(s,i;\ t,j)$ (or satisfies (3.1) if P is not given to begin with). Thus, condition (VII.1.63) imposed on (X_t), $P(s,i;\ t,j)$ is satisfied.

We now consider the filtering model VIII.2.4 discussed

in Section VIII.2. Assume for notational simplicity that $\mathcal{K} = \mathbb{R}^1$. The treatment in the general case is no different. Let Q_t be the orthogonal projection on $H = L^2([0,T])$ defined by (VIII.2.5).

Let $F_t(y)$ ($\Gamma_t(y)$) be the conditional (unnormalized conditional) distribution of X_t given $Q_t y$ and let

$$F_t(y,i) = F_t(y)(\{i\}) = E_\alpha[1_{(X_t=i)}|Q_t y].$$

Similarly, write $\Gamma_t(y,i) = \Gamma_t(y)(\{i\})$. Equation (VIII.2.11) for $\Gamma_t(y)$ then takes the form

$$\Sigma_j f(t,j)\Gamma_t(y,j) = \Sigma_j f(0,j)N_0(j) \tag{3.8}$$

$$+ \int_0^t \Sigma_i \{\frac{\partial^+ f}{\partial s}(s,i) + \Sigma_j a_s(i,j)f(s,j)\}\Gamma_s(y,i)ds$$

$$+ \int_0^t \Sigma_i \{h_s(i)y_s - \tfrac{1}{2}h_s^2(i)\}f(s,i)\Gamma_s(y,i)ds$$

where $N_t(j) = \Pi(X_t=j)$. It can be shown that (3.8) holds for all $f \in \mathcal{D}$ if and only if it holds for $f(s,j) = a_j$ (where a_j is an arbitrary bounded sequence). (This step is similar in spirit to the proof of Lemma VII.3.3). It can then be seen that (3.8) is equivalent to

$$\Gamma_t(y,i) = N_0(i) + \int_0^t \{\Sigma_j a_s(j,i)\Gamma_s(y,j)\}ds \tag{3.9}$$

$$+ \int_0^t \{h_s(i)y_s - \tfrac{1}{2}\, h_s^2(i)\}\Gamma_s(y,i)ds.$$

It follows from Theorem VIII.2.1 that for all $y \in H$, $\{\Gamma_t(y,i)\}$ is the unique solution to (3.9) in the class of $\{K_t(i)\}$ satisfying $|K_t(i)| \leq C_y N_t(i)$, where C_y is a constant. Equation

(3.9) is an infinite system of linear differential equations:

$$\frac{d}{dt}\Gamma_t(y,i) = \Sigma_j\{a_s(j,i) + \delta_{ij}(h_s(i)y_s - \tfrac{1}{2}h_s^2(i))\}\Gamma_t(y,j). \tag{3.10}$$

For the rest of the section, we will concentrate on the case of a finite state space S (identified with $\{1,2,\ldots,d\}$ without loss of generality). Let us assume also that

$$\int_0^T h_s^2(i)ds < \infty \quad \text{for each } i. \tag{3.11}$$

Let $A_s(y)$ be the d×d matrix defined by

$$A_s(y)(i,j) = a_s(j,i) + \delta_{ij}(h_s(i)y_s - \tfrac{1}{2}h_s^2(i)).$$

Then equation (3.10) takes the form

$$\frac{d}{dt}\Gamma_t(y) = A_t(y)\Gamma_t(y) \tag{3.12}$$

where $\Gamma_t(y)$ is considered as a vector in $\mathbb{R}^d$. It is well known that the solution to (3.12) can be expressed as a product or multiplicative integral and the solution can be approximated by Riemann-Stieltjes products. We will directly prove the approximation result.

Theorem 3.1: Fix $y \in H$. For each integer $k \geq 1$, let $\Gamma_t^k(y)$ be inductively defined by

$$\Gamma_0^k(y) = N_0$$

and for $\frac{r}{k} \leq t \leq \frac{r+1}{k}$,

$$\Gamma_t^k(y) = \Gamma_{r/k}^k(y) + \left[\int_{r/k}^{k} A_s(y)ds\right]\cdot\Gamma_{r/k}^k(y). \tag{3.13}$$

Then

$$b_k(t) := \sup_{o\leq s\leq t}\ \max_{1\leq i\leq d} |\Gamma_s^k(y)(i) - \Gamma_s(y)(i)| \longrightarrow 0. \tag{3.14}$$

Proof: Assumptions (3.4) and (3.11) imply that if

$$q(s) = \|A_s(y)\|,$$

then $\int_0^T q(s)ds < \infty$, where $\|A\|$ denotes the (Hilbert-Schmidt) norm of the matrix A. Using (3.13), it can be proved that for $\frac{r}{k} \leq t \leq \frac{r+1}{k}$, $r \geq 1$,

$$b_k(t) \leq b_k(\tfrac{r}{k}) + C\left[\int_{r/k}^{t} q(s)ds\right] b_k(\tfrac{r}{k}) \tag{3.15}$$

where C is a constant depending only on the dimension d. Also, for $t \leq \frac{1}{k}$,

$$b_k(t) \leq C\left[\int_0^t q(s)ds\right]\left[\sum_i N_0(i)\right] \leq C\left[\int_0^{1/k} q(s)ds\right]\left[\sum_i N_0(i)\right]. \tag{3.16}$$

Since $\int_0^T q(s)ds < \infty$ and (3.16) holds for $t = 1/k$, we get

$$b_k(\tfrac{1}{k}) \to 0 \quad \text{as } k \to \infty. \tag{3.17}$$

The inequality (3.15) for $t = \frac{r+1}{k}$ gives

$$b_k(\tfrac{r+1}{k}) \leq \left[1 + C\int_{r/k}^{(r+1)/k} q(s)ds\right] b_k(\tfrac{r}{k})$$

$$\leq \exp\left[C\int_{r/k}^{(r+1)/k} q(s)ds\right] b_k(\tfrac{r}{k})$$

and hence

$$b_k(\tfrac{r+1}{k}) \le \exp\left[C\int_0^{(r+1)/k} q(s)ds\right] b_k(\tfrac{1}{k}). \tag{3.18}$$

Since $b_k(\frac{1}{k}) \to 0$, (3.18) implies $b_k(T) \to 0$. □

The preceding result gives a simple approximation procedure to compute $\Gamma_t(y)$. In practical examples, like the ones considered below, we may fix a small step size $1/k$ and use (3.13) to recursively update $\Gamma_t(y)$.

Example 3.1: Suppose a component in a machine has a (random) life time τ. The distribution G of τ is assumed to be known. Before the failure of the component, let the (true) output of the machine be f_t at time t. After the component has failed, let the output at time t be f'_t. Suppose the output is observed, or measured, in the presence of additive white noise so that the observation y_t can be modelled as

$$y_t = f_t 1_{(\tau>t)} + f'_t 1_{(\tau\le t)} + e_t \tag{3.19}$$

where $\{e_t\}$ is white noise independent of τ. The quantity of interest is the conditional probability of $\tau > t$ given $(y_s:$ $s \le t)$, i.e.,

$$E_\alpha(1_{(\tau>t)} | y_s: s \ge t). \tag{3.20}$$

It is easy to see that (X_t), where $X_t := 1_{(\tau>t)}$, is a Markov process with state space $\{0,1\}$. The transition probability function $P(s,i;\ t,j)$ of (X_t) is given by

$$P(s,0;\ t,0) = 1,$$

and

$$P(s,1;\ t,1) = \Pi(\tau > t+s|\tau > s) = \frac{1 - G(t+s)}{1 - G(s)} .$$

Suppose $G(t)$ admits a density $g_1(t)$ which is right continuous. Then writing $g(s) = \frac{g_1(s)}{1-G(s)}$, we get

$$a_s(1,1) = -g(s),\ a_s(1,0) = g(s),\ a_s(0,0) = a_s(0,1) = 0. \tag{3.21}$$

Suppose $p = \Pi(\tau > 0) = 1-G(0)$. Then $\Pi(X_0=1) = p$, $\Pi(X_0=0) = 1-p$. Thus $N_0(1) = p$, $N_0(0) = 1-p$. The observation model (3.19) can be expressed as

$$y_t = h_t(X_t) + e_t.$$

where $h_t(1) = f_t$ and $h_t(0) = f'_t$. Let us assume that $\int_0^T h_t^2(i)dt < \infty$. Also note that

$$F_t(y,1) = E_\alpha(1_{(X_t=1)}|Q_t y) = E_\alpha(1_{(\tau>t)}|y_s:\ s \le t).$$

The differential equations for $\Gamma_t(y,i)$ are

$$\frac{d\Gamma_t(y,1)}{dt} = -g(t)\Gamma_t(y,1) + \{f_t y_t - \tfrac{1}{2} f_t^2\}\Gamma_t(y,1)$$

and

$$\frac{d\Gamma_t(y,0)}{dt} = g(t)\Gamma_t(y,1) + \{f'_t y_t - \tfrac{1}{2} f'^2_t\}\Gamma_t(y,0)$$

with initial conditions $\Gamma_0(y,1) = p$, $\Gamma_0(y,0) = 1-p$. The solution can be easily verified to be

$$\Gamma_t(y,1) = p \exp(\int_0^t\{f_s y_s - \tfrac{1}{2} f_s^2 - g_s\}ds)$$

$$\Gamma_t(y,0) = (1-p)\ \exp(\int_0^t\{f'_s y_s - \tfrac{1}{2}\ f'^2_s\}ds)$$

$$+\ p\int_0^t g(s)\exp(\int_0^s\{f_u y_u - \tfrac{1}{2}\ f_u^2 - g_u\}du)$$

$$\times\ \exp(\int_s^t\{f'_u y_u - \tfrac{1}{2}\ f'^2_u\}du)ds.$$

From these, $F_t(y,1)$ can be computed.

When the distribution G of τ is exponential with parameter λ, then $p = 1$ and $g(s) \equiv \lambda$. In this case, (X_t) is a time homogeneous Markov process.

This problem is known as the 'order-disorder problem,' 'disruption problem' or 'false alarm problem.'

Example 3.2: Consider a component in an equipment that has N+1 levels of performance, denoted by $\{0,1,..,N\}$, $i = 1,..,N$ denoting the performance of a new component and 0 denoting total failure. Let X_t denote performance at time t. Let $X_0 = N$. Suppose (X_t) is a Markov process and that performance can go down by one step at a time. This amounts to assuming

$$a_s(i,j) = 0 \quad \text{if } j \neq i \text{ and } j \neq i-1,\ i \geq 1.$$

Also $a_s(0,0) = 0$ since once the performance reaches 0, it remains so. Let

$$g_i(s) = a_s(i,i-1)$$

be the instantaneous failure rate at time s given that the performance level is i; Then $a_s(i,i) = -g_i(s)$. Let $h_s(i)$ be the output at time s if performance is i (assume that $\int_0^t h_s^2(i)ds < \infty$) so that output is $h_s(X_s)$ at time s.

Suppose that the observed output is

$$y_t = h_t(X_t) + e_t$$

where (e_t) is white noise independent of (X_t).

The quantities of interest are the posteriori probabilities of $(X_t=i)$ having observed $\{y_s: s \leq t\}$, $i = 0,1,\ldots,N$, i.e.,

$$F_t(y,i) = E_\alpha(1_{(X_t=i)}|Q_t y). \tag{3.22}$$

These can be computed from $\Gamma_t(y,i)$ which satisfy the following equations

$$\frac{d}{dt}\Gamma_t(y,N) = -g_t(N)\Gamma_t(y,N) + (h_t(N)y_t - \tfrac{1}{2}h_t^2(N))\Gamma_t(y,N)$$

and for $0 \leq i \leq N-1$ (writing $g_s(0) = 0$),

$$\frac{d}{dt}\Gamma_t(y,i) = g_t(i+1)\Gamma_t(y,i+1) - g_t(i)\Gamma_t(y,i)$$
$$+ (h_t(i)y_t - \tfrac{1}{2}h_t^2(i))\Gamma_t(y,i).$$

The initial conditions are $\Gamma_0(y,N) = 1$, $\Gamma_0(y,i) = 0$, $i < N$. The quantities $\Gamma_t(y,i)$ can be approximately computed using the recursive algorithm considered in Theorem 3.1. The emplicit solution is given by

$$\Gamma_t(y,N) = \exp(\textstyle\int_0^t\{h_s(N)y_s - \tfrac{1}{2}h_s^2(N) - g_s(N)\}ds).$$

$$\Gamma_t(y,i) = \textstyle\int_0^t g_s(i+1)\Gamma_s(y,i+1)\exp(\int_s^t\{h_u(i)y_u - \tfrac{1}{2}h_u^2(i)$$
$$- g_u(i)\}du)ds.$$

Example 3.3: Consider the problem of estimating a signal θ_t which is transmitted by modulating the amplitude of a sinusoidal wave sin ωt and is observed corrupted by additive white noise (e_t):

$$y_t = h_t(\theta_t) \sin \omega t + e_t$$

where $h_t(x)$ is a known function. Let $\bar{h}_t(x) = h_t(x) \sin \omega t$.

Suppose $\theta_t = Z_j$: $\tau_j \leq t < \tau_{j+1}$ where $\{Z_i\}$ is a sequence of independent identically distributed signals and τ_j are jump times $(\tau_0 = 0)$ independent of the Z_j's.

If the random variables $\{\tau_{j+1}-\tau_j\}$ are indepedent with common exponential distribution with parameter λ, then θ_t is a Markov process. If Z_j (or θ_t) is an analog signal, then its state space $S = \mathbb{R}$ or $[0,1]$. If the signal is digital, then its state space is a finite set $S = \{1,2,..,N\}$. In the latter case, suppose

$$\Pi(Z_j=i) = b_i, \quad 1 \leq i \leq N,$$

where of course $\Sigma b_i = 1$. Then it can be verified that

$$a_s(i,i) = -\lambda(1-b_i), \; a_s(i,j) = \lambda b_j, \quad j \neq i.$$

Again, what we are interested in is $F_t(y)$ or $\Gamma_t(y)$. The latter is the unique solution to

$$\begin{aligned}\frac{d}{dt} \Gamma_t(y,i) &= -\lambda(1-b_i)\Gamma_t(y,i) + \sum_{j\neq i} \lambda b_i \Gamma_t(y,j) \\ &\quad + \{\bar{h}_t(i)y_t - \tfrac{1}{2}\bar{h}_t^2(i)\}\Gamma_t(y,i) \\ &= \lambda b_i \Sigma_j \Gamma_t(y,j) + \{\bar{h}_t(i)y_t - \tfrac{1}{2}\bar{h}_t^2(i) - \lambda\}\Gamma_t(y,i)\end{aligned}$$

with initial condition $\Gamma_0(y,i) = b_i$. The approximate solution can be obtained by using the algorithm given in Theorem 3.1.

An interesting problem is to obtain an estimate $\hat{\theta}_t$ of θ_t. In this problem the natural choice is

$$\hat{\theta}_t = i_0 \qquad \text{if} \qquad F_t(y,i_0) \geq F_t(y,i) \text{ for all } i.$$

Since $F_t(y,i)$ is proportional to $\Gamma_t(y,i)$, $\hat{\theta}_t$ can also be determined by

$$\hat{\theta}_t = i_0 \qquad \text{if} \qquad \Gamma_t(y,i_0) \geq \Gamma_t(y,i) \text{ for all } i.$$

4. FILTERING FOR INFINITE DIMENSIONAL PROCESSES

We will illustrate in this section how a two parameter process can be viewed as an infinite dimensional one parameter process so that the filtering problem for such a process can be solved by using results of Chapter VIII.

The signal process we have in mind is the following: $\{X_{tx}: 0 \leq t \leq T, 0 \leq x \leq b\}$ is a real valued process with continuous paths (on a probability space $(\Omega,\mathcal{A},\Pi)$). Let $B = C([0,b],\mathbb{R})$. Then

$$(X_t(\omega))(x) = X_{tx}(\omega) \tag{4.1}$$

defines a B-valued stochastic process, also denoted by X_t on $(\Omega,\mathcal{A},\Pi)$. We will assume further that (X_t) is a B-valued Markov process.

Here, the parameter t refers to time and x stands for the space variable, assumed for simplicity to vary in the interval $[0,b]$. The Markov property of (X_t) means that at any time instant t, the future evolution $\{X_{ux}: u \geq t, x \in [0,b]\}$ is independent of the past history $\{X_{sx}: s \leq t, x \in [0,b]\}$ given the present $\{X_{tx}: x \in [0,b]\}$. An example of such a process occurs in the study of spatially distributed neurons, where X_{tx} represents the (random) voltage potential at time t at 'site' x of the neuron, the neuron being modelled as a segment [0,b] or a cylinder (see [81], [82]).

Another example is the Ornstein-Uhlenbeck (O-U) process which is the unique solution to the SDE

$$dX_{tx} = -\tfrac{1}{2} X_{tx}dt + dW_{tx}, \tag{4.2}$$

X_{0x} being Gaussian, independent of W and W the space-time or Yeh-Wiener process.

For such a B-valued Markov process (X_t), the filtering problem for the following model of observations has been proposed.

$$Y_t = \int_0^t \int_0^x h_{sv}(X_{sv})dvds + W'_{tx} \tag{4.3}$$

where W' is a Yeh-Wiener process, independent of (X_t), and

$$\int_0^T \int_0^b \hat{h}_{tx}(\eta_{tx})^2 dxdt < \infty \quad \text{for all } \eta \in C([0,T]\times[0,b], \mathbb{R}).$$

We will now consider the white noise model which corresponds to (4.3) in the sense of Section X.3.

Let $\mathcal{H}$ be the reproducing kernel Hilbert space of the one dimensional Wiener process on [0,b],

$$\mathcal{H} := \{f: [0,b] \to \mathbb{R}: \int_0^T (f'(u))^2 du := \|f\|^2 < \infty\}. \tag{4.5}$$

Let $H = L^2([0,T],\mathcal{H})$ be equipped with the natural inner product, as in VIII.2 and let $h_t: B \to \mathcal{H}$ be defined by

$$[h_t(\theta)](x) = \int_0^x \hat{h}_{tv}(\theta_v)dv, \tag{4.6}$$

for $\theta = (\theta_v) \in B$. In view of (4.4),

$$\int_0^T \|h_t(X_t)\|^2_{\mathscr{H}} dt = \int_0^T \int_0^b [\hat{h}_{tx}(X_{tx})]^2 dxdt < \infty. \tag{4.7}$$

Let $\xi(\omega) = (h_t(X_t(\omega)): 0 \le t \le T)$ be the H-valued mapping on Ω, $e = (e_t)$ be the $\mathscr{H}$-valued white noise on $(H, \mathscr{C}, m)$ (see Section VIII.2, replace $\mathscr{K}$ by $\mathscr{H}$). The abstract statistical model (VIII.2.3) (where ξ, e are considered as mappings on $(E, \mathscr{E}, \alpha) = (\Omega, \mathscr{A}, \Pi) \otimes (H, \mathscr{C}, m))$ can be viewed as

$$y_t = h_t(X_t) + e_t \tag{4.8}$$

or

$$y_t(x) = \int_0^x \hat{h}_{tv}(X_{tv})dv + e_t(x). \tag{4.9}$$

That the model (4.8) corresponds to (4.3) in the sense of Section X.3 can be seen by taking $\mathscr{K} = \mathscr{H}$ and checking that the resulting B-valued process W is nothing but the Yeh-Wiener process.

Now if we assume that either

$$\|h_t(x)\|_{\mathscr{H}} \le a(t); \quad \int_0^T a^2(t)dt < \infty \tag{4.10}$$

or that the process (X_t) satisfies (VIII.1.63), then the results in Chapter VIII are applicable and we can conclude that the conditional distribution $F_t(y)$ and the unnormalized conditional distribution $\Gamma_t(y)$ of X_t given $Q_t y$, i.e., $(y_u(x): u \le t, x \in [0,b])$ are unique solutions to the measure valued equations given in VIII.2.

Thus, the technique of converting a two parameter process into an infinite dimensional one parameter process yields results for the filtering problem in a situation where at any time instant t, the enite observations $\{y_t(x): x \in [0,b]\}$ are available. Incidently, these observations, along with the consistency results in Section X.3, give a solution to the filtering problem for (4.3), i.e., for

$$E(f(X_t)|\sigma(Y_{ux}: u \leq t, x \in [0,b])). \tag{4.11}$$

There are other examples of infinite dimensional processes (X_t) which do not arise from a multiparameter process. One such example occurs in the study of neuronal behavior when the geometric shape of the neuron is more complicated. In such models, X_t is an O-U process, or more generally a non-Gaussian diffusion process, whose sample paths lie in $C([0,T],\mathcal{H})$, $\mathcal{H}$ being a separable infinite dimensional Hilbert space. A natural filtering problem is given by the linear model

$$y_t = X_t + e_t \tag{4.12}$$

where $e = (e_t)$ is $\mathcal{H}$-valued noise. This can be treated by the methods of Chapter VIII.

CHAPTER XII

LINEAR AND QUASILINEAR FILTERING THEORY

Section 1 deals with linear filtering. Since the linear (Gaussian) theory can be handled using Hilbert space techniques, the stochastic calculus and white noise approaches are virtually identical. We shall therefore content ourselves with giving a brief derivation of the Kalman filter from the general white noise equation for the nonlinear filtering model.

Section 2 is devoted to the quasilinear filtering problem, i.e., filtering when the initial distribution of the signal process is not Gaussian. A statistical robustness property for the Kalman filter is also established. This result appears to be new and is an almost immediate consequence of the general theorems on statistical robustness for the nonlinear theory proved in Chapter XI.

In Section 3, an explicit formula for the likelihood ratio in the general, non-Markovian case is obtained as a direct consequence of the Bayes formula.

1. LINEAR FILTERING

Linear filtering, equivalently known as the Kalman-Bucy theory, is the most completely developed part of the subject. In this section, we consider the finite dimensional white noise linear filtering model and obtain the solution as an application of the analogue of the FKK equation for $\pi_t(f,y)$ obtained in Section VII.2.

In the conventional approach, the same technique yields the result (see [34, p. 271]). This is based on the well known observation that the conditional distribution is Gaussian. We will show that in the white noise setup too, the conditional distribution is Gaussian, and then obtain equations for the conditional mean and variance.

In the setup of Section VII.2, assume that the signal process (X_t) satisfies the stochastic differential equation

$$dX_t = A_t dW_t + B_t X_t dt \tag{1.1}$$

where A_t, B_t are d×d matrix valued functions of t (only) such that $\int_0^T |A_t^{ij}|^2 dt < \infty$, $\int_0^T |B_t^{ij}| dt < \infty$ for all i, j, $A_t A_t^*$ is positive definite, and W_t is a d-dimensional Wiener process. Suppose X_0 is Gaussian and is independent of (W_t). Then it is known that the SDE (1.1) has a unique strong solution and that (X_t) is a Gaussian process.

The process X_t can be equivalently described as a diffusion process with diffusion and drift coefficients $A_t A_t^*$ and $B_t x$ respectively (A_t^* being the transpose of A_t).

In the filtering model VII.1.10, suppose that

$$h_t(x) = C_t x$$

where C_t is a N×d matrix valued function with $\int_0^T |C_t^{ij}|^2 dt < \infty$. Thus the filtering model is

$$y_t = C_t X_t + e_t \tag{1.2}$$

where (e_t) is an N-dimensional Gaussian white noise, independent of (X_t). We have seen that the conditional distribution $F_t(y)$ of X_t given $\{y_s : 0 \le s \le t\}$ can be represented as

$$F_t(y)(A) = \frac{\int 1_A(X_t)\exp[\int_0^t y_s^* C_s X_s ds - \frac{1}{2}\int_0^t X_s^* C_s^* C_s X_s ds]d\Pi}{\int \exp[\int_0^t y_s^* C_s X_s ds - \frac{1}{2}\int_0^t X_s^* C_s^* C_s X_s ds]d\Pi} . \tag{1.3}$$

Our first observation is that $F_t(y)$ is a Gaussian measure. For this purpose we need the following lemma.

<u>Lemma 1.1</u>: Let $Z_0, Z_1, \ldots, Z_k$ be real valued random variables whose joint distribution is Gaussian. Then for all $a = (a_1, \ldots, a_k)$ and $b = (b_1, \ldots, b_k) \in \mathbb{R}^k$ ($b_j > 0$ for all j),

$$\phi(\lambda;a,b) := \frac{E\ e^{i\lambda Z_0}\exp(\Sigma_{j=1}^k Z_j a_j b_j - \frac{1}{2}\Sigma_{j=1}^k Z_j^2 b_j)}{E\ \exp(\Sigma_{j=1}^k Z_j a_j b_j - \frac{1}{2}\Sigma_{j=1}^k Z_j^2 b_j)} \tag{1.4}$$

is the characteristic function of a Gaussian measure, i.e.,

$$\phi(\lambda;a,b) = \exp\{i\lambda\mu(a,b) - \tfrac{1}{2}\lambda^2\sigma^2(a,b)\} \tag{1.5}$$

for some constants $\mu(a,b)$, $\sigma^2(a,b)$.

<u>Proof</u>: Let $\{W_j\}$ be independent, Gaussian random variables, independent of $\{Z_0, \ldots, Z_k\}$ such that $EW_j = 0$, $EW_j^2 = b_j^{-1}$. Then by an elementary argument (or, see [34, p. 280]) it is easy to see that

$$\phi(\lambda;a,b) = E\left[e^{i\lambda Z_0} | Z_j + W_j = a_j;\ 1 \le j \le k\right],$$

i.e.,

$$\phi(\lambda;(Y_j),b) = E\left[e^{i\lambda Z_0} | \sigma(Y_1, \ldots, Y_k)\right]$$

where $Y_j = Z_j + W_j$, $1 \leq j \leq k$. Since the joint distribution of $(Z_0, Y_1, \ldots, Y_k)$ is Gaussian, it follows from well known properties of Gaussian distributions that the conditional distribution of Z_0 given $(Y_1, \ldots, Y_k)$ is Gaussian. Hence, for almost all $a \in \mathbb{R}^k$, $\phi(\lambda;a,b)$ is of the form (1.5). Since $a \to \phi(\lambda;a,b)$ is continuous it follows that for all a, $\phi(\lambda;a,b)$ is of the form (1.5). □

Theorem 1.2: For all $y \in H$, $0 \leq t \leq T$, $F_t(y)$, defined by (1.3), is a Gaussian measure on $\mathbb{R}^d$.

Proof: Fix $0 \leq t \leq T$, $y \in H$ and let $\phi(\lambda)$ be the characteristic function of $F_t(y)$, i.e., for $\lambda \in \mathbb{R}^d$

$$\phi(\lambda) = \int e^{i\lambda^* x} dF_t(y)(x)$$

$$= \frac{\int \exp[i\lambda^* X_t + \int_0^t y_s^* C_s X_s ds - \frac{1}{2}\int_0^t X_s^* C_s^* C_s X_s ds] d\Pi}{\int \exp[\int_0^t y_s^* C_s X_s ds - \frac{1}{2}\int_0^t X_s^* C_s^* C_s X_s ds] d\Pi} .$$

Let (X_t) be the continuous process given by (1.1). For $k \geq 1$, let

$$X_{k,u} = X_{i/k} \quad \text{for } \frac{i}{k} \leq u < \frac{i+1}{k};\ 0 \leq i \leq kT. \tag{1.6}$$

and

$$\phi_k(\lambda) = \frac{\int \exp[i\lambda^* X_{k,t} + \int_0^t y_s^* C_s X_{k,s} ds - \frac{1}{2}\int_0^t X_{k,s}^* C_s^* C_s X_{k,s} ds] d\Pi}{\int \exp[\int_0^t y_s^* C_s X_{k,s} ds - \frac{1}{2}\int_0^t X_{k,s}^* C_s^* C_s X_{k,s} ds] d\Pi}.$$

Using the fact that $X_{k,u}$ is piecewise constant, we can deduce from Lemma 1.1 that ϕ_k is a Gaussian characteristic function. It is easy to see using (1.6) that for each λ,

$\phi_k(\lambda)$ converges to $\phi(\lambda)$. This implies that $\phi(\lambda)$ is also Gaussian. □

Since $F_t(y)$ is a Gaussian distribution, it is uniquely determined by its mean vector $\hat{X}_t = (\hat{X}_t^i)$ and the variance-covariance matrix $P_t = (P_t^{ij})$.

Let $f_i(x) = x^i$ and $g_{ij}(x) = x^i x^j$ for $x = (x^1, \ldots, x^d) \in \mathbb{R}^d$, $1 \le i,j \le d$. Then

$$\hat{X}_t^i = \pi_t(f_i, y) \tag{1.7}$$

and

$$P_t^{ij} = \pi_t(g_{ij}, y) - \pi_t(f_i, y)\pi_t(f_j, y). \tag{1.8}$$

For simplicity, let us suppress y in $\pi_t(f,y)$ and write it as $\pi_t(f)$. For $f,g\colon \mathbb{R}^d \to \mathbb{R}$, let

$$\text{cov}_t(f,g) := \pi_t(fg) - \pi_t(f)\pi_t(g) \tag{1.9}$$

so $\text{cov}_t(f,g)$ is the covariance between f and g under the measure $F_t(y)$ on $\mathbb{R}^d$. In other words

$$P_t^{ij} = \text{cov}_t(f_i, f_j). \tag{1.10}$$

Also, equation (VII.2.26) for $\pi_t(f,y)$ can be written as

$$\frac{d}{dt}\pi_t(f) = \pi_t(L_t f) + \text{cov}_t(f,\ y_t^* h_t - \tfrac{1}{2} h_t^* h_t) \tag{1.11}$$

for $f \in C^2(\mathbb{R}^d)$. Equation (1.11) as well as (1.14), (1.16), (1.17), (1.20) and (1.21) which appear below hold for a.e. t. It is easy to verify that

$$(L_t f_i)(x) = \sum_{j=1}^{d} B_t^{ij} x^j = \sum_{j=1}^{d} B_t^{ij} f_j(x) \tag{1.12}$$

and

$$(L_t g_{ij})(x) = \tfrac{1}{2}(A_t A_t^*)^{ij} + \Sigma_k B_t^{ik} g_{kj}(x) + \Sigma_k B_t^{jk} g_{ki}(x). \tag{1.13}$$

Thus

$$\frac{d}{dt}\hat{X}_t^i = \frac{d}{dt}\pi_t(f_i) \tag{1.14}$$

$$= \pi_t\Big(\sum_{j=1}^{d} B_t^{ij} f_j\Big) + \mathrm{cov}_t(f_i, y_t^* h_t - \tfrac{1}{2}h_t^* h_t)$$

$$= \sum_{j=1}^{d} B_t^{ij}\pi_t(f_j) + \sum_{k=1}^{d}(C_t^* y_t)^k \mathrm{cov}_t(f_i, f_k)$$

$$- \tfrac{1}{2}\sum_{k=1}^{d}\sum_{j=1}^{d}(C_t^* C_t)^{kj}\mathrm{cov}_t(f_i, g_{kj})$$

since

$$y_t^* h_t(x) = \Sigma_k (y_t^* C_t)^k x^k,$$

$$h_t^* h_t(x) = x^* C_t^* C_t x = \sum_{k,j}(C_t^* C_t)^{kj} g_{kj}(x).$$

Using the fact that $F_t(y)$ is Gaussian with mean vector $\hat{X}_t$, variance-covariance matrix P_t, we have

$$\mathrm{cov}_t(f_i, g_{kj}) = P_t^{ik}\hat{X}_t^j + P_t^{ij}\hat{X}_t^k. \tag{1.15}$$

Hence (1.14) yields

$$\frac{d}{dt}\hat{X}^i_t = \sum_{j=1}^{d} B^{ij}_t\hat{X}^j_t + \sum_{k=1}^{d}(y^*_tC_t)^k P^{ik}_t \tag{1.16}$$

$$- \tfrac{1}{2}\sum_{k=1}^{d}\sum_{j=1}^{d}(C^*_tC_t)^{kj}(P^{ik}_t\hat{X}^j_t + P^{ij}_t\hat{X}^k_t).$$

Writing (1.16) in matrix-vector notation,

$$\frac{d}{dt}\hat{X}_t = B_t\hat{X}_t + P_tC^*_ty_t - P_tC^*_tC_t\hat{X}_t \tag{1.17}$$

$$= [B_t - P_tC^*_tC_t]\hat{X}_t + P_tC^*_ty_t.$$

To obtain the evolution equation for P_t, note that P^{ij}_t is absolutely continuous as $\pi_t(f)$ is and

$$\frac{d}{dt}P^{ij}_t = \frac{d}{dt}[\pi_t(g_{ij}) - \pi_t(f_i)\pi_t(f_j)]$$

$$= \frac{d}{dt}\pi_t(g_{ij}) - \pi_t(f_i)\frac{d}{dt}\pi_t(f_j) - \pi_t(f_j)\frac{d}{dt}\pi_t(f_i).$$

Using (1.11), (1.13), (1.16), we have

$$\frac{d}{dt}P^{ij}_t = \tfrac{1}{2}(A_tA^*_t)^{ij} + \sum_{k=1}^{d}[B^{ik}_t\mathrm{cov}_t(f_k,f_j) + B^{jk}_t\mathrm{cov}_t(f_k,f_i)]$$

$$+ \sum_{k=1}^{d}(C^*_ty_t)^k\Big[\mathrm{cov}_t(f_k,g_{ij}) - \mathrm{cov}_t(f_k,f_j)\pi_t(f_i)$$

$$- \mathrm{cov}_t(f_k,f_i)\pi_t(f_j)\Big]$$

$$- \tfrac{1}{2}\sum_{k=1}^{d}\sum_{j=1}^{d}(C^*_tC_t)^{kr}\Big[\mathrm{cov}_t(g_{kr},g_{ij})$$

$$- \mathrm{cov}_t(g_{kr},f_i)\pi_t(f_j) - \mathrm{cov}_t(g_{kr},f_j)\pi_t(f_i)\Big].$$

Using (1.15), it follows that the coefficient of $(C_t^* y_t)^k$ above vanishes. Also,

$$\operatorname{cov}_t(g_{kr}, g_{ij}) = P_t^{ik} P_t^{jr} + P_t^{ir} P_t^{kj} + \hat{X}_t^i \hat{X}_t^k P_t^{jr} + \hat{X}_t^j \hat{X}_t^r P_t^{ik} + \hat{X}_t^i \hat{X}_t^r P_t^{kj} + \hat{X}_t^j \hat{X}_t^k P_t^{ir}. \tag{1.19}$$

The above expression and (1.18) yield

$$\frac{d}{dt} P_t^{ij} = \tfrac{1}{2}(A_t A_t^*)^{ij} + \sum_{k=1}^{d} [B_t^{ik} P_t^{kj} + P_t^{ik} B_t^{jk}] - \tfrac{1}{2} \sum_{k=1}^{d} \sum_{j=1}^{d} (C_t^* C_t)^{kr} [P_t^{ik} P_t^{rj} + P_t^{ir} P_t^{kj}]. \tag{1.20}$$

In matrix-vector notation, equation (1.20) becomes

$$\frac{d}{dt} P_t = \tfrac{1}{2} A_t A_t^* + B_t P_t + P_t B_t^* - P_t C_t^* C_t P_t. \tag{1.21}$$

It should be noted that as expected, P_t is independent of y_t. We have proved the following result.

<u>Theorem 1.3</u> (Kalman-Bucy filter): For the linear model (1.2), the condition distributional $F_t(y)$ of X_t given $Q_t y$ is a Gaussian measure. Its mean vector $\hat{X}_t$ and variance-covariance matrix P_t are unique solutions (in the class of vector valued and non-negative definite matrix valued continuous functions respectively) to

$$\hat{X}_t(y) = EX_0 + \int_0^t [B_s - P_s C_s^* C_s] \hat{X}_s(y) ds + \int_0^t P_s C_s^* y_s ds \tag{1.22}$$

and

$$P_t = E(X_0-EX_0)(X_0-EX_0)^* \tag{1.23}$$
$$+ \int_0^t[\tfrac{1}{2} A_sA_s^* + B_sP_s + P_sB_s^* - P_sC_s^*C_sP_s]ds.$$

Equations (1.22) and (1.23) are nothing but (1.17) and (1.21) rewritten as integral equations. We have also used the fact that $F_0(y)$ is the (unconditional) distribution of X_0. Using Lemma VIII.1.8, one can conclude that equation (1.22) has a unique solution. Uniqueness of solution to equation (1.23) is proved in [34, p. 253]. If the coefficient functions in (1.1) and (1.2) are continuous, then the differential equation (1.21) which holds for every t is the matrix version of the well known Riccati equation.

The solution to equation (1.22), namely $\hat{X}_t(y)$, can be expressed in terms of B_t, C_t, P_t as follows. Let R_t be the d×d matrix valued function, which is the solution to

$$R_t = I + \int_0^t[B_s - P_sC_s^*C_s]R_sds. \tag{1.24}$$

When d = 1, $R_t = \exp(\int_0^t[B_s - P_sC_s^*C_s]ds)$. It can be verified that

$$\hat{X}_t(y) = R_t[EX_0 + \int_0^t R_s^{-1}P_sC_s^*y_sds]. \tag{1.25}$$

The likelihood ratio for the model (1.2) can be written explicitly now that we have an expression for $\hat{X}_t(y)$. The formula XI.2.3 for $\rho_t(y)$ can be simplified as follows. Note that $h_s(x) = C_sx$ and $\|h_s(x)\|^2 = x^*C_s^*C_sx$,

$$\pi_s(h_s) = C_s\hat{X}_s(y) \tag{1.26}$$

and

$$\pi_s(\|h_s\|^2) = \tilde{\pi}_s(X_s^* C_s^* C_s X_s) = \sum_{ij} (C_s^* C_s)^{ij} \tilde{\pi}_s(X_s^i X_s^j) \tag{1.27}$$

$$= \sum_{ij} (C_s^* C_s)^{ij} [P_s^{ij} + \hat{X}_s^i(y) \hat{X}_s^j(y)]$$

$$= \text{Trace } (C_s^* C_s) P + \hat{X}_s^*(y) C_s^* C_s \hat{X}_s(y)$$

$$= \text{Trace } C_s P C_s^* + \hat{X}_s^*(y) C_s^* C_s \hat{X}_s(y).$$

These computations lead us to the following result.

Theorem 1.4: The likelihood ratio $\rho_t(y)$ for the model (1.2) is

$$\rho_t(y) = \exp\left[\int_0^t \{y_s^* C_s \hat{X}_s(y) - \tfrac{1}{2} \hat{X}_s^*(y) C_s^* C_s \hat{X}_s(y) - \tfrac{1}{2} \text{Trace } C_s P_s C_s^*\} ds\right]. \tag{1.28}$$

2. QUASILINEAR FILTERING

Let the signal process (X_t) be a solution to the SDE (1.1) as in Section 1. If the distribution μ of X_0 is not Gaussian the process (X_t) is no longer Gaussian and the analysis of the previous section fails. Deriving equations for the conditional mean and conditional variance is not going to be enough for these may not identify the conditional distribution. In this case we can use the fact that the conditional distribution of X_t given X_0 is still Gaussian.

For $z \in \mathbb{R}^d$, let X_t^z be the solution to the SDE (1.1) with the initial condition $X_0^z = z$, or equivalently, let X_t^z

be the diffusion process with the same generator as X_t and with $X_0^z = z$. Let $F_t^z(y)$, $\Gamma_t^z(y)$ be the conditional and unnormalized conditional distribution of X_t^z given $Q_t y$ (where the model is (1.2) with X_t^z in place of X_t). Then

$$\Gamma_t(y) = \int \Gamma_t^z(y) d\mu_0(z) \tag{2.1}$$

is the unnormalized conditional distribution of X_t given $Q_t y$. This is a consequence of the following: The conditional distribution of $X_\cdot$ under Π given $X_0 = z$ is the distribution of $X_\cdot^z$ and $\Pi \circ X_0^{-1} = \mu_0$ and hence for any functional G: $C([0,T] \times \mathbb{R}^d) \to \mathbb{R}$,

$$E_\Pi G(X_\cdot) = \int EG(X_\cdot^z) d\mu_0(z). \tag{2.2}$$

It may be noted that the analogue of (2.1) for the normalized conditional distributions $F_t(y)$, $F_t^z(y)$ is not valid as $F_t(y)$ cannot be represented as $E_\Pi G(X_\cdot)$. Since the (unconditional) distribution of X_t^z is Gaussian for $t > 0$, it admits a density with respect to Lebesgue measure and hence by Theorem VII.3, the unnormalized conditional density $p_t^z(x,y)$ exists and

$$p_t(x,y) = \int p_t^z(x,y) d\mu_0(z) \tag{2.3}$$

is the unnormalized conditional density of X_t given $Q_t y$.

Now as seen earlier in Theorem 1.3, $F_t^z(y)$ is Gaussian with mean $\hat{X}_t^z(y)$ and variance $\overline{P}_t$ which in turn are unique solutions to

$$\frac{d\hat{X}_t^z(y)}{dt} = \{B_t - \overline{P}_t C_t^* C_t\}\hat{X}_t^z(y) + \overline{P}_s C^*{}_s y_s \tag{2.4}$$

$$\frac{d\overline{P}_t}{dt} = \tfrac{1}{2} A_t A_t^* + B_t \overline{P}_t - \overline{P}_t B_t - \overline{P}_t C_t^* C_t \overline{P}_t \tag{2.5}$$

and initial conditions $\hat{X}_0^z(y0 = z$ and $\overline{P}_0 = 0$. The formula (1.28) for $\Gamma_t^z(y)(S) = \rho_t^z(y)$ takes the form

$$\rho_t^z(y) = \exp\Big[\int_0^t \{y_s^* C_s \hat{X}_s^z(y) - \tfrac{1}{2}\hat{X}_s^{z*}(y) C_s^* C_s \hat{X}_s^z(y) \tag{2.6}$$

$$- \tfrac{1}{2}\,\text{Trace}\, C_s \overline{P}_s C_s^*\}ds\Big].$$

Since $p_t^z(x,y)$, upto the normalizing constant $\Gamma_t^z(y)(S) = \rho_t^z(y)$ is also the density of the Gaussian distribution $F_t^z(y)$, we get

$$p_t^z(x,y) = \rho_t^z(y)\cdot(2\pi)^{-d/2}\cdot(\det \overline{P}_t)^{-\frac{1}{2}} \tag{1.7}$$

$$\cdot\exp\{-\tfrac{1}{2}[x - \hat{X}_t^z(y)]^*[\overline{P}_t]^{-1}[x - \hat{X}_t^z(y)]\}.$$

Thus we have proved the following result.

<u>Theorem 2.1</u>: The unnormalized conditional density $p_t(x,y)$ of X_t given $Q_t y$ for the model (1.2) is given by

$$p_t(x,y) = \int p_t^z(x,y)d\mu_0(z) \tag{2.8}$$

where $p_t^z(x,y)$, $\hat{X}_t^z(y)$, $\rho_t^z(y)$, $\overline{P}_t$ are in turn given by (2.7), (2.4), (2.5), (2.6).

This result gives an explicit solution to the quasilinear filtering problem. For practical applications, one should first obtain $\overline{P}_t$ by solving (2.5), then obtain $\overline{R}_t$ via

$$\overline{R}_t = I + \int_0^t [B_s - \overline{P}_s C_s^* C_s] \overline{R}_s ds. \tag{2.9}$$

Then, for the observation path (y_s), obtain $\hat{X}_t^z(y)$ by

$$\hat{X}_t^z(y) = \overline{R}_t \{z + \int_0^t \overline{R}_s^{-1} \overline{P}_s C_s^* y_s ds\}.$$

Finally, $\hat{X}_t^z(y)$, $\overline{P}_t$ can now be substituted in (2.6), (2.7), (2.3) to yield the unnormalized conditional density $p_t(x,y)$ explicitly for the quasilinear problem.

<u>Statistical Robustness of the Kalman-Bucy Filter</u>

As a consequence of the robustness result Theorem X.4.3, it follows that the unnormalized conditional distribution $\Gamma_t(y)$ of X_t given $Q_t y$ depends continuously on the distribution μ_0 of X_0. Let μ_0^k be a sequence of possible distributions for X_0 and let $\Gamma_t^k(y)$ be the corresponding conditional distribution. By Theorem X.4.3,

$$\mu_0^k \xrightarrow{d} \mu_0 \quad \text{implies} \quad \Gamma_t^k(y) \xrightarrow{d} \Gamma_t(y). \tag{2.10}$$

We have seen that if $p_0^k(\cdot) = d\mu_0^k/d\lambda$, $p_0(\cdot) = d\mu_0/d\lambda$ exist, then $p_t^{(k)}(\cdot,y) = d\Gamma_t^k(y)/d\lambda$ and $p_0(\cdot,y) = d\Gamma_t(y)/d\lambda$ also exist. By Theorem X.4.5, it follows that if A_t, B_t are Hölder continuous, $\frac{d}{dt}C_t$ exists and is Hölder continuous, and if

$$\sup_{|x| \leq R} |p_0^k(x) - p_0(x)| \longrightarrow 0 \quad \text{for all } R > 0 \tag{2.11}$$

then for all $y \in H$,

$$\sup_{0 \le t \le T} \sup_{|x| \le R} |p_t^{(k)}(x,y) - p_t(x,y)| \longrightarrow 0. \tag{2.12}$$

Since in the quasilinear case, we have an explicit solution for $p_t^{(k)}(x,y)$ and $p_t(x,y)$, we can directly verify a weaker form of (2.12) under less restrictive hypotheses. This is the next result.

<u>Theorem 2.2</u>: Suppose that $\mu_0^k \xrightarrow{d} \mu_0$. Then for all $y \in H$,

$$\sup_{|x| \le M} |p_t^{(k)}(x,y) - p_t(x,y)| \longrightarrow 0 \tag{2.13}$$

for all $t > 0$, $M < \infty$.

<u>Proof</u>: Fix $t > 0$, $y \in H$. Note that $\hat{X}_t^z(y)$ is a continuous function of $z \in \mathbb{R}^d$; indeed it is a linear function of z given by (2.10). Using this expression for $\hat{X}_t^z(y)$ and the assumption that $\int_0^T |C_s^{ij}|^2 ds < \infty$, it can be verified that $\rho_t^z(y)$ expressed by (2.6) is a continuous function of z. Thus $p_t^z(x,y)$ is continuous in z for t,x,y fixed. Moreover,

$$\rho_t^z(y) \le \exp(\tfrac{1}{2}|y|^2) \tag{2.14}$$

and hence

$$p_t^z(x,y) \le \exp(\tfrac{1}{2}|y|^2) \cdot (\det \overline{P}_t)^{-\frac{1}{2}}. \tag{2.15}$$

Therefore $\mu_0^k \xrightarrow{d} \mu_0$ implies

$$\int p_t^z(x,y)d\mu_0^k(z) = p_t^{(k)}(x,y) \tag{2.16}$$

$$\longrightarrow \int p_t^z(x,y)d\mu_0(z) = p_t(x,y)$$

for all $t > 0$, $x \in \mathbb{R}^d$ fixed. It can be verified that for all $t > 0$, $y \in H$ fixed, the family $\{p_t^z(x,y): |x| \leq M\}$ is equicontinuous in z and is uniformly bounded (in view of (2.15)). Hence the convergence in (2.16) is uniform over $\{|x| \leq M\}$. (See [11], p. 17). □

Remark 2.1: Since (2.10) in any case implies $\Gamma_t^k(y)(\mathbb{R}^d) \longrightarrow \Gamma_t(y)(\mathbb{R}^d)$, we also have

$$\int_{\mathbb{R}^d} p_t^{(k)}(x,y)dx \longrightarrow \int_{\mathbb{R}^d} p_t(x,y)dx$$

and thus (2.11) holds even when the unnormalized densities are replaced by the normalized densities.

A practical application of the robustness result is the following. Suppose $\mu_0^k \xrightarrow{d} \mu_0$, where μ_0 is Gaussian. Suppose that the mean vector and variance-covariance matrix of μ_0^k is equal to that of μ_0 for all k. Then the limit of the conditional density corresponding to μ_0^k is the Gaussian density with mean $\hat{X}_t(y)$ and variance-covariance matrix P_t given by (1.22) - (1.23).

Thus, if the distribution of X_0 is approximately Gaussian, the conditional density of X_t given Q_ty can be approximated by the Gaussian density with mean $\hat{X}_t$, variance-covariance matrix P_t still given by the linear filtering equations (1.22), (1.23).

3. LINEAR FILTERING: GENERAL CASE

In this section, we will consider linear filtering model with general Gaussian signal which need not be Markovian, and obtain expressions for the filter and the likelihood ratio. This is achieved by using the well known factorization theorem due to Gohberg and Krein.

Consider the filtering model VII.1.10 with $d = N = 1$, $h_t(x) = x$, on $(E,\mathcal{E},\alpha) = (\Omega,\mathcal{A},\Pi)\odot(H,\mathcal{C},m)$, H being $L^2([0,T])$. The model then can be represented as

$$y_t = X_t + e_t. \tag{3.1}$$

Let the signal process (X_t) (defined on $(\Omega,\mathcal{A},\Pi)$) be a Gaussian process with mean (θ_t) and covariance kernel $R(t,s)$. Assume that the process (X_t) is measurable and that θ_t, $R(t,s)$ are continuous. This implies that the paths $X_{\cdot}(\omega) \in H$ a.s. Π, so that condition (VII.1.6) (with $h_t(x) = x$) holds.

The covariance kernel $R(t,s)$ is a symmetric non-negative definite kernel, which has been assumed to be continuous. Let R denote the integral operator on $H = L^2([0,T])$ by

$$(Rf)(t) = \int_0^T R(t,s)f(s)ds.$$

Then R is a trace class operator and $(I+R)$ is invertible. Let R_a denote the restriction of the kernel R to $[0,a] \times [0,a]$ and R_a be the corresponding integral operator acting on $L^2([0,a])$. Then, as operators on $L^2([0,a])$, $(I+R_a)$ is invertible and the resolvent $K_a := I - (I+R_a)^{-1}$ is again an integral operator on $L^2([0,a])$,

$$(K_a f)(t) = \int_0^a K_a(t,s)f(s)ds \tag{3.2}$$

with kernel $K_a(t,s)$, $0 \leq s,\ t \leq a$.

The following is taken from [26].

<u>Theorem 3.1</u>: $K_a(s,t)$ is jointly continuous in a, s, t. Further, if M is the Volterra integral operator on H corresponding to the kernel

$$\begin{aligned} M(t,s) &= K_t(t,s), \quad 0 \leq s \leq t \leq T \\ &= 0\ , \qquad\quad 0 \leq t < s \leq t, \end{aligned} \tag{3.3}$$

then M is Hilbert-Schmidt. Also

$$(I+R)^{-1} = (I-M^*)(I-M) \tag{3.4}$$

where M^* is the adjoint of M with kernel $M^*(t,s) = M(s,t)$. The decomposition (3.4) is known as Gohberg-Krein factorization of $(I+R)^{-1}$.

Note that continuity of $K_a(t,s)$ in (a,t,s) implies that $M(t,s)$ is continuous in the region $\{0 \leq s \leq t \leq T\}$ and $M^*(t,s)$ is continuous in the region $\{0 \leq t \leq s \leq T\}$.

Fix $a \in [0,T]$. Let $\{\lambda_j\}$ be the eigenvalues of R_a and $\{\phi_j\}$ be the corresponding eigenfunctions. Then ϕ_j are continuous and

$$R_a(t,s) = \sum_{j=1}^{\infty} \lambda_j \phi_j(t)\phi_j(s),$$

the series converging uniformly for $(s,t) \in [0,a]\times[0,a]$. The resolvent kernel is then given by the uniformly convergent series,

$$K_a(t,s) = \sum_{j=1}^{\infty} \left[\frac{\lambda_j}{1+\lambda_j}\right]\phi_j(t)\phi_j(s), \tag{3.6}$$

and is itself a symmetric non-negative definite kernel.

Let $\underline{X}_j(\omega) = \int_0^t X_s(\omega)\phi_j(s)ds$. Then $\{\underline{X}_j\}$ is a sequence of independent random variables, with normal distribution. The mean of $\underline{X}_j$ is $\underline{\theta}_j = \int_0^t \theta_s\phi_j(s)ds$ and variance of $\underline{X}_j$ is λ_j. Further,

$$X_t = \sum_{j=1}^{\infty} \underline{X}_j\phi_j(t) \tag{3.7}$$

a.s. Π and in $L^2(\Omega,\mathcal{A},\Pi)$. Also

$$\int_0^a X_t^2 dt = \sum_{j=1}^{\infty} \underline{X}_j^2, \quad \text{a.s. } \Pi. \tag{3.8}$$

Note that λ_j, ϕ_j, $\underline{\theta}_j$, $\underline{X}_j$ depend on a, though for convenience, this is not made explicit in the notation.

Let $H_a = \{f\cdot 1_{[0,a]},\ f \in H\}$ and Q_a be the orthogonal projection on H with range H_a. The L^2-convergence of the series in (3.7) and Theorem V.4.6 implies

$$E_\alpha(X_t|Q_a y) = \lim_{r\to\infty} E_\alpha\left[\sum_{j=1}^{r} \underline{X}_j\phi_j(t)|Q_a y\right] \tag{3.9}$$

$$= \sum_{j=1}^{\infty} E_\alpha(\underline{X}_j|Q_a y)\phi_j(t)$$

$$E_\alpha(X_t^2 | Q_a y) = \sum_{j=1}^{\infty} \sum_{i=1}^{\infty} E_\alpha(\underline{X}_j \underline{X}_i | Q_a y)\phi_j(t)\phi_i(t) \tag{3.10}$$

The Bayes formula, Theorem VII.1.1, now yields, writing $\underline{y}_j = \int_0^t y_s \phi_j(s)ds$

$$E_\alpha(\underline{X}_j | Q_a y) = \frac{E_\Pi[\underline{X}_j \exp(\int_0^t X_s y_s ds - \frac{1}{2}\int_0^t X_s^2 ds)]}{E_\Pi[\exp(\int_0^t X_s y_s ds - \frac{1}{2}\int_0^t X_s^2 ds)]}$$

$$= \frac{E_\Pi[\underline{X}_j \exp\{\Sigma_{j=1}^{\infty} \underline{X}_j \underline{y}_j - \frac{1}{2}\Sigma_{j=1}^{\infty} \underline{X}_j^2\}]}{E_\Pi[\exp(\Sigma_{j=1}^{\infty} \underline{X}_j \underline{y}_j - \frac{1}{2}\int_{j=1}^{t} \underline{X}_j^2\}]}$$

$$= \frac{E_\Pi \underline{X}_j \exp(\underline{X}_j \underline{y}_j - \frac{1}{2}\underline{X}_j^2)}{E_\Pi \exp(\underline{X}_j \underline{y}_j - \frac{1}{2}\underline{X}_j^2)}$$

using independent of $\underline{X}_j$ and $\{\underline{X}_k : k \neq j\}$. Straightforward calculation with normal distribution yields

$$E_\alpha(\underline{X}_j | Q_a y) = \underline{\theta}_j + \lambda_j(1+\lambda_j)^{-1}(\underline{y}_j - \underline{\theta}_j). \tag{3.11}$$

Similarly it follows that

$$E_\alpha(\underline{X}_j \underline{X}_i | Q_a y) = E_\alpha(\underline{X}_j | Q_a y) E_\alpha(\underline{X}_i | Q_a y) + 1_{\{j=i\}} \lambda_j (1+\lambda_j)^{-1}. \tag{3.12}$$

Hence (3.9), (3.10) imply

$$E_\alpha(X_t | Q_a y) = \sum_{j=1}^{\infty} \underline{\theta}_j \phi_j(t) + \sum_{j=1}^{\infty} \lambda_j (1+\lambda_j)^{-1} (\underline{y}_j - \underline{\theta}_j)\phi_j(t) \tag{3.13}$$

$$= \theta_t + [K_a(y-\theta)](t)$$

$$= \theta_t + \int_0^a K_a(t,s)(y_s - \theta_s)ds$$

and

U

$$E_\alpha(X_t^2|Q_a y) = [E_\alpha(X_t|Q_a y)]^2 + \sum_{j=1}^{\infty} \lambda_j(1+\lambda_j)^{-1}\phi_j^2(t) \tag{3.14}$$

$$= [E_\alpha(X_t|Q_a y)]^2 + K_a(t,t).$$

Substituting $a = t$ in these formulae, we get

$$E_\alpha(X_t|Q_t y) = \theta_t + \int_0^t M(t,s)(y_s-\theta_s)ds \tag{3.15}$$

$$P_t := E_\alpha(X_t^2|Q_t y) - [E_\alpha(X_t|Q_t y)]^2 = M(t,t). \tag{3.16}$$

Since $M(t,s) = 0$ if $s > t$, (3.15) can be rewritten as

$$E_\alpha(X_t|Q_t y) = [\theta + M(y-\theta)](t). \tag{3.17}$$

We have thus proved:

Theorem 3.2: The conditional distribution $F_t(y)$ of X_t given $Q_t y$ is Gaussian with mean $\hat{X}_t$ and variance P_t given by

$$\hat{X}_t = [\theta + M(y-\theta)](t) \tag{3.18}$$

and

$$P_t = M(t,t) \tag{3.19}$$

where M is the Volterra operator appearing in the Gohberg-Krein factorization (3.4) of $(I+R)^{-1}$.

Note that we had used the resolvent K_a of R_a in the computations, but the final result is free of K_a and is expressed only in terms of M.

Likelihood Ratio

The likelihood ratio $\rho_T(y)$ for the model (3.5), i.e., the Radon-Nikodym derivative $\frac{dn}{dm}$, $n = \alpha \circ y^{-1}$, can be expressed as

$$\rho_T(y) = \exp\{\int_0^T \hat{X}_s y_s ds - \tfrac{1}{2}\int_0^T (\hat{X}_s)^2 ds - \tfrac{1}{2}\int_0^T P_s ds\} \tag{3.20}$$

as seen in (XI.2.6). Note that in Theorem XI.2.1, the expression for $\rho_T(y)$ was obtained without assuming that X_t is a Markov process. So here

$$\rho_T(y) = \exp\{(\theta+M(y-\theta),y) - \tfrac{1}{2}|\theta+M(y-\theta)|^2 - \int_0^T M(t,t)dt\}. \tag{3.21}$$

This expression for $\rho_T(y)$ involves the kernel $M(t,s)$. To obtain a form in terms of operators M, M^* and free of the kernel $M(t,s)$, we proceed as follows. Define a kernel

$$\begin{aligned} G(t,s) &= M(t,s), \qquad 0 \leq s \leq t \leq T \\ &= M(s,t), \qquad 0 \leq t < s \leq T. \end{aligned} \tag{3.22}$$

Then $G(t,s) = M(t,s)$ on the triangular region $\{0 \leq s \leq t \leq T\}$ and $G(t,s) = M^*(t,s)$ on the region $\{0 \leq t \leq s \leq T\}$. Thus G is a continuous, symmetric kernel. If G denotes the corresponding operator, then for any $f \in H$,

$$\begin{aligned} (Gf)(t) &= \int_0^T G(t,s)f(s)ds \\ &= \int_0^t G(t,s)f(s)ds + \int_t^T G(t,s)f(s)ds \\ &= \int_0^t M(t,s)f(s)ds + \int_t^T M^*(t,s)f(s)ds \\ &= (Mf)(t) + (M^*f)(t) \end{aligned} \tag{3.23}$$

and hence as operators on H,

$$G = M + M^*. \tag{3.24}$$

Note that $G = M + M^* = [I - (I+R)^{-1}] + M^*M = K_T + M^*M$. As seen earlier, K_T is a non-negative definite operator and hence G is a non-negative definite operator as well. Since the kernel G(t,s) of G is continuous, G is trace class and Trace G = $\int_0^T G(t,t)dt$. Hence

$$\int_0^T M(t,t)dt = \int_0^T G(t,t)dt = \text{Trace } G = \text{Trace}(M+M^*) \quad (3.25)$$

using (3.24). These computations lead us to the following result.

Theorem 3.3: The likelihood ratio $\rho_T(y)$ for the model (3.5) is given by

$$\rho_T(y) = \exp\{(\theta + M(y-\theta),y) - \tfrac{1}{2}|\theta + M(y-\theta)|^2 - \tfrac{1}{2} \text{Trace } (M + M^*)\}. \quad (3.26)$$

Equivalently, it is also given by (3.21) or (3.22).

Remark 3.1: Before we proceed, we want to note that the terms appearing in the exponent in (3.26) are accessible random variables. By arguments similar to those given in Examples III.2.1, III.3.1, it can be proved that if B is any trace class operator on H, then

$$(By,y) \in \mathscr{L}^*(E,\mathscr{E},\alpha). \quad (3.27)$$

Since M is Hilbert-Schmidt, M^*M is trace class and hence $|My|^2 = (M^*My,y)$ is an accessible random variable. Also $(My,y) = \frac{1}{2}\{(My,y) + (M^*y,y)\} = \frac{1}{2}(Gy,y)$ where G is trace class. Thus, $|My|^2$ and (My,y) belong to $\mathscr{L}^*(E,\mathscr{E},\alpha)$. Using this it is easy to see that

$$(\theta + M(y-\theta), y) \in \mathcal{L}^*(E, \mathcal{E}, \alpha) \tag{3.28}$$

$$|\theta + M(y-\theta)|^2 \in \mathcal{L}^*(E, \mathcal{E}, \alpha). \tag{3.29}$$

We have used the Bayes formula in deriving the expression (3.26) for the likelihood ratio which has been obtained for a real valued Gaussian signal process. The likelihood ratio formula can be derived without using the Bayes formula following the steps given below in a more general context.

Let $\mathcal{H}$ be a real separable Hilbert space; $H = L^2([0,T], \mathcal{H})$, Q_t: $0 \leq t \leq T$ be as in Section VIII.2. Let $e = (e_t)$ be $\mathcal{H}$-valued white noise. Let $X = (X_t)$ be a $\mathcal{H}$-valued Gaussian process (on $(\Omega, \mathcal{A}, \Pi)$) with paths in H. Then X is itself a H-valued Gaussian random variable with mean θ and covariance operator R, which is a positive self adjoint, trace class operator on H. For the model

$$y = X + e, \tag{3.30}$$

the likelihood ratio $\rho(y)$ (= dn/dm, where m is the canonical Gauss measure on H, $\alpha = \Pi \Theta m$ and $n = \alpha \circ y^{-1}$) can be shown to be given by

$$\rho(y) = |\det(I+R)|^{-\frac{1}{2}} \cdot \exp\{\tfrac{1}{2}(y, R(I+R)^{-1}y) + (y, (I+R)^{-1}\theta) - \tfrac{1}{2}(\theta, (I+R)^{-1}\theta)\}.$$

where $\det(I+R) = \prod_{j=1}^{\infty} (1+\lambda_j)$, $\{\lambda_j\}$ being the eigenvalues of R. The infinite product converges because R is a trace class operator. The expression (3.30) can be obtained by evaluating the integrals appearing in the expression (VI.2.27) with μ being the distribution of X. Let

$$(I+R)^{-1} = (I-M^*)(I-M) \tag{3.31}$$

be the Gohberg-Krein factorization, where M is a Hilbert-Schmidt integral operator on $L^2([0,T],\mathcal{H})$ with Volterra kernel $M(t,s)$ which vanishes a.e. in the region $\{(t,s): t < s\}$. Then

$$R(I+R)^{-1} = I - (I+R)^{-1} = M^* + M - M^*M$$

and hence

$$\tfrac{1}{2}(y,\ R(I+R)^{-1}y) = (My,y) - \tfrac{1}{2}|My|^2.$$

Also

$$\begin{aligned}(y,\ (I+R)^{-1}\theta) &= (y,\theta) - (y,\ R(I+R)^{-1}\theta)\\ &= (y,\theta) - (y,\ (M^* + M - M^*M)\theta)\end{aligned}$$

and

$$\begin{aligned}(\theta,\ (I+R)^{-1}\theta) &= |\theta|^2 - (\theta,\ R(I+R)^{-1}\theta)\\ &= |\theta|^2 - (\theta,\ (M^* + M - M^*M)\theta).\end{aligned}$$

Using these relations, it is easy to see that

$$\rho(y) = |\det(I+R)|^{-\frac{1}{2}}\cdot\exp\{(\theta+M(y-\theta),y) - \tfrac{1}{2}|\theta+M(y-\theta)|^2\}. \tag{3.32}$$

Note that the resolvent $K = I - (I+R)^{-1}$ has eigenvalues $\{\alpha_j\}$, where $\alpha_j = \lambda_j(1+\lambda_j)^{-1}$. The Gohberg-Krein factorization yields $(I+R)^{-1} = (I-K) = (I-M^*)(I-M)$. The following formula is taken from [43].

$$\prod_{j=1}^{\infty} (1-\alpha_j)\exp(\alpha_j) = \exp\{- \text{Trace } M^*M\}. \tag{3.33}$$

Since $\exp(\Sigma_j \alpha_j) = \exp\{\text{Trace } K\}$ and Trace K = Trace (M^*+M) - Trace M^*M, this gives

$$\prod_{j=1}^{\infty} (1-\alpha_j) = \exp(-\text{ Trace } (M^*+M)). \tag{3.34}$$

Since $(1-\alpha_j) = 1 - \lambda_j(1+\lambda_j)^{-1} = (1+\lambda_j)^{-1}$, (3.34) implies

$$|\det (I+R)| = \prod_{j=1}^{\infty} (1+\lambda_j) = \prod_{j=1}^{\infty} (1-\alpha_j)^{-1}$$

$$= [\exp(-\text{ Trace}(M^*+M)]^{-1}.$$

Hence

$$|\det (I+R)|^{-\frac{1}{2}} = \exp(-\tfrac{1}{2}\text{ Trace } (M^*+M)). \tag{3.35}$$

These computations show that the formula (3.26) for the likelihood ratio is valid for the model (3.30) as well.

APPENDIX

THE STOCHASTIC CALCULUS APPROACH TO FILTERING

Detailed expositions of nonlinear filtering based on Itô stochastic calculus are available in the books and survey articles mentioned in the References. In the brief account presented here, our aim is to acquaint the reader with some of the principal results on the derivation of the stochastic differential equation (SDE) of the optimal filter. We focus our attention on the case of independent signal and noise.

We begin with the general problem. The signal process (X_t) is not observable directly. The observation process is

$$Y_t = \int_0^t h_s ds + W_t, \quad 0 \leq t \leq T, \tag{1.1}$$

where h_s is $\mathcal{F}_s^X$-measurable. The noise (W_t) in (1.1) is modelled by a Wiener process. The processes Y_t, h_t, W_t are taken to be $\mathbb{R}^m$-valued and are defined on a countably additive probability space $(\Omega, \mathcal{A}, \Pi)$. It is further assumed that

$$\int_0^T |h_t|^2 dt < \infty \quad \text{a.s. } \Pi. \tag{1.2}$$

The most general assumptions on the noise and signal process are that for each t, the σ-fields $\sigma\{X_u, W_u, 0 \leq u \leq t\}$ and $\sigma(W_s - W_t, t \leq s)$ are independent.

The problem is to find the conditional distribution $\Pi(X_t \in \cdot \,|\mathcal{F}_t^Y)$ of the state of the system at time t. It is further required that the solution be a recursive estimate,

v

so that it is possible to update the conditional distribution at time $t + \Delta t$ ($\Delta t > 0$) using its value at time t. In other words, one seeks an SDE for the conditional distribution, or equivalently, for the conditional expectation $\hat{\pi}_t(f) := E(f(X_t)|\mathcal{F}_t^Y)$ for a rich class of functions f. When (X_t) is a Markov process, this is possible. The SDE for $\hat{\pi}_t(f)$ is known as FKK equation. For its derivation see [25, 34].

Reference Probability Method

In the model (1.1), suppose that $h_t \equiv h_t(X_t)$ for a suitable function h_t, satisfying

$$\int_0^T |h_t(X_t)|^2 dt < \infty \quad \text{a.s. } \Pi. \tag{1.3}$$

Let

$$R_t = \exp\{\int_0^t \sum_{i=1}^m h_s^i(X_s) dY_s^i - \tfrac{1}{2}\int_0^t |h_s(X_s)|^2 ds\}. \tag{1.4}$$

Define a measure Π_1 on $(\Omega,\mathcal{A})$ by $d\Pi_1 = R_t^{-1} d\Pi$. If h is bounded, or satisfies a suitable integrability condition, Π_1 is a probability measure on $(\Omega,\mathcal{A})$. In the important special case when X and W are independent, it is sufficient to assume that h satisfies (1.3). Then by Girsanov's theorem it follows that (Y_t) is a Wiener process under Π_1 and is a martingale with respect to $\mathcal{G}_t = \mathcal{F}_t^{X,W}$. Then for any $\mathcal{F}_T^X$ measurable integrable function g,

$$\pi_t'(g) := E_{\Pi}[g|\mathcal{F}_t^Y) = \frac{E_{\Pi_1}[gR_t|\mathcal{F}_t^Y]}{E_{\Pi_1}[R_t|\mathcal{F}_t^Y]} . \tag{1.5}$$

A proof of (1.5) is as follows. Noting that $(R_t, \mathcal{G}_t)$ is a Π_1-martingale, it follows that R_t is the Radon-Nikodym derivative of Π restricted to $\mathcal{G}_t$ with respect to Π_1 restricted to $\mathcal{G}_t$. Hence, for any $B \in \mathcal{F}^Y_t$,

$$\int_B g d\Pi = \int_B E_\Pi[g|\mathcal{F}^Y_t]d\Pi = \int_B E_\Pi[g|\mathcal{F}^Y_t]R_t d\Pi_1 \tag{1.6}$$

$$= \int_B E_\Pi[g|\mathcal{F}^Y_t]\cdot E_{\Pi_1}[R_t|\mathcal{F}^Y_t]d\Pi_1.$$

On the other hand

$$\int_B g d\Pi = \int_B g\cdot R_T d\Pi_1 = \int_B E_{\Pi_1}[g\cdot R_T|\mathcal{F}^Y_t]d\Pi_1 \tag{1.7}$$

$$= \int_B E_{\Pi_1}[E_{\Pi_1}[gR_T|\mathcal{G}_t]|\mathcal{F}^Y_t]d\Pi_1$$

$$= \int_B E_{\Pi_1}[g\cdot R_t|\mathcal{F}^Y_t]d\Pi_1.$$

The validity of (1.6), (1.7) for all $B \in \mathcal{F}^Y_t$ implies

$$E_{\Pi_1}[g\cdot R_t|\mathcal{F}^Y_t] = E_\Pi[g|\mathcal{F}^Y_t]\cdot E_{\Pi_1}[R_t|\mathcal{F}^Y_t] \quad \text{a.s. } \Pi_1$$

from which (1.5) follows since $R_t > 0$ a.s. Π_1.

The probability measure Π_1 is called the *reference probability* and the expression (1.5) has been found to be useful in deriving SDE for $\hat{\pi}_t(f)$.

Bayes Formula

Let us now consider the case when the signal process (X_t) and the noise (W_t) are independent. In this case, the formula (1.5) yileds the Bayes formula. This formula was obtained in [44] and has played an important role in filtering theory. The corresponding formula in the finitely additive white noise theory has been obtained in Chapter VI.

So now we are considering the model (1.1) where $h_t = h_t(X_t)$, where h satisfies (1.3) and (X_t), (W_t) are independent. Under these conditions, it is straightforward to verify that

$$(X_t) \text{ and } (Y_t) \text{ are independent under } \Pi_1. \tag{1.8}$$

As noted above, (Y_t) is a Wiener process under Π_1 and hence $E_{\Pi_1}[R_T|\mathcal{F}_T^X] = 1$. Hence for any $\mathcal{F}_T^X$ measurable integrable random variable g, $E_{\Pi}g = E_{\Pi_1}[gR_T] = E_{\Pi_1}[gE_{\Pi_1}[R_T|\mathcal{F}_T^X]] = E_{\Pi_1}g$. Thus X has the same distribution under Π and under Π_1, i.e.,

$$\Pi\circ(X_\bullet)^{-1} = \Pi_1\circ(X_\bullet)^{-1}.$$

To obtain the Bayes formula in a familiar form, we introduce the following setup. Consider the product space $(\Omega\times\Omega,\ \mathcal{A}\otimes\mathcal{A},\ \Pi\otimes\Pi_1)$. Let us denote $\Pi\otimes\Pi_1$ by P. A generic point of $\Omega\times\Omega$ will be denoted by (ω',ω). Given a mapping g on Ω, g will denote the function on $\Omega\times\Omega$ defined by $g(\omega',\omega) := g(\omega)$. Also g' will denote the function on $\Omega\times\Omega$ defined by $g'(\omega',\omega) = g(\omega')$. Thus, $X_t'(\omega',\omega) = X_t(\omega')$ and $X_t(\omega',\omega) = X_t(\omega)$. So under P, X_t' is an independent copy of X_t.

The formula (1.5) now reads

$$\pi_t'(g) = \frac{E_P[gR_t|\mathcal{F}_t^Y]}{E_P[R_t|\mathcal{F}_t^Y]}\ . \tag{1.9}$$

Since (X_t), (Y_t) are independent under P and (X_t') is an independent copy of (X_t), the joint distribution of (X_t),

(Y_t) is the same as that of (X'_t), (Y_t) and hence

$$E_P[gR_t|\mathscr{F}^Y_t] = E_P[g'q_t|\mathscr{F}^Y_t]$$

where $q_t: \Omega\times\Omega \to \mathbb{R}$ is defined by

$$q_t(\omega',\omega) = \exp\Big\{\sum_{i=1}^m \int_0^t h^i_s(X_s(\omega'))dY^i_s(\omega) \tag{1.10}$$

$$- \tfrac{1}{2}\int_0^t |h_s(X_s(\omega'))|^2 ds\Big\}.$$

The conditional expectation $E_P[g'q_t|\mathscr{F}^Y_t]$ can be evaluated by integrating with respect to the first factor on the product space and thus

$$E_P[gR_t|\mathscr{F}^Y_t] = \int g'(\omega',\omega)q_t(\omega',\omega)d\Pi(\omega) \tag{1.11}$$

$$= \int g(\omega')q_t(\omega',\omega)d\Pi(\omega).$$

Thus, we have proved the following.

<u>Theorem 1.1</u> (Bayes formula): Let g be $\mathscr{F}^X_T$-measurable integrable random variable on $(\Omega,\mathscr{A},\Pi)$. Then

$$E_\Pi[g|\mathscr{F}^Y_t] = \sigma'_t(g,Y)/\sigma'_t(1,Y) \tag{1.12}$$

where

$$\sigma'_t(g,Y(\omega)) = \int g(\omega')q_t(\omega',\omega)d\Pi(\omega'). \tag{1.13}$$

The derivation of the Bayes formula given above is different from that given in [44, 34].

The above formula enables us, in principle, to compute the conditional expectation by evaluation of function space integrals given in (1.13). In practice, however, such an evaluation is not always an easy task.

Let us write

$$\hat{\sigma}_t(f,Y) = \sigma'_t(f(X_t),Y) \tag{1.14}$$

for f such that $f(X_t)$ is Π-integrable. So (1.12) gives

$$E_\Pi[f(X_t)|\mathcal{F}^Y_t] = \hat{\pi}_t(f) = \hat{\sigma}_t(f,Y)/\hat{\sigma}_t(1,Y). \tag{1.15}$$

Zakai Equation

We now derive a SDE for $\hat{\sigma}_t(f,Y)$, known as the Zakai equation, from the Bayes formula. The proof is based on a Fubini theorem for stochastic integrals. The idea of the proof is due to Ocone [64].

Assume that the signal process (X_t) is an S-valued Markov process with generator L whose domain is $\mathcal{D}$. (See Chapter II.) Assume that paths of (X_t) are progressively measurable. Let $\mathcal{D}_0$ consist of all $f: S \to \mathbb{R}$ such that f_1 defined by

$$f_1(s,x) = f(x) \tag{1.16}$$

belong to $\mathcal{D}$ and for $f \in \mathcal{D}_0$, define

$$(L_t f)(x) := (Lf_1)(t,x) \tag{1.17}$$

where f_1 is given by (1.16).

Theorem 1.2: Assume that

$$E_\Pi \int_0^T |h_s(X_s)|^2 ds < \infty. \tag{1.18}$$

Then for all $f \in \mathcal{D}_0$, $\hat{\sigma}_t(f,Y)$ satisfies the following SDE

$$d\hat{\sigma}_t(f,Y) = \hat{\sigma}_t(L_t f,Y)dt + \sum_{j=1}^{m} \hat{\sigma}_t(h_t^j f,Y)dY_t^j. \tag{1.19}$$

Proof: Let $g_t = f(X_T) - \int_t^T (L_s f)(X_s)ds$. The definition of $\mathcal{D}_0$ and $\mathcal{D}$ imply that f, $L_s f$ are uniformly bounded. Here $|g_t| \leq C$ for some C. Moreover, from Theorem II.2.2 it follows that

$$E_\Pi[g_t|\mathcal{F}_t^X] = f(X_t), \quad 0 \leq t \leq T. \tag{1.20}$$

Using the fact that for fixed ω, as a function of ω', $q_t(\omega',\omega)$ is a $\mathcal{F}_t^{X'}$-measurable function, (1.20) yields

$$\hat{\sigma}_t(f,Y) = \sigma_t'(g_t,Y). \tag{1.21}$$

A simply application of Itô's formula gives $dg_t' = (L_t f)(X_t')dt$, $dq_t = q_t \cdot \Sigma_{j=1}^m h_t^j(X_t')dY_t^j$ and

$$d(g_t'q_t) = (L_t f)(X_t')q_t dt + g_t'q_t \sum_{j=1}^{m} h_t^j(X_t')dY_t^j.$$

Hence

$$\sigma_t'(g_t,Y) = E_\Pi[g_t'q_t] \tag{1.22}$$

$$= E_\Pi[g_0'q_0 + \int_0^t (L_s f)(X_s')q_s ds + \int_0^t g_s'q_s \sum_{j=1}^{m} h_s^j(X_s')dY_s^j]$$

$$= E_\Pi g_0' + \int_0^t E_\Pi(L_s f)(X_s')q_s ds + A_t$$

$$= E_\Pi g_0' + \int_0^t \hat{\sigma}_s(L_s f,Y)ds + A_t$$

where

$$A_t(\omega) = \int[\int_0^t g_s(\omega')q_s(\omega',\omega) \sum_{j=1}^{m} h_s^j(X_s(\omega'))dY_s^j(\omega)]d\Pi(\omega'). \tag{1.23}$$

In (1.22) above, E_Π stands for integration with respect to ω', i.e., the first factor in the product space $(\Omega\times\Omega,\ \mathcal{A}\times\mathcal{A},\ \Pi\otimes\Pi_1)$ and X_s' stands for $X_s(\omega')$, q_s stands for $q_s(\omega',\omega)$. In the second term, we have used Fubini theorem which is justified as $L_s f$ is bounded and

$$\int\int_0^t q_s(\omega',\omega)dsd\Pi(\omega') < \infty \quad \text{a.s. } \Pi \tag{1.24}$$

because $\Pi << \Pi_1$ and

$$\int\int\int_0^t q_s(\omega',\omega)dsd\Pi(\omega')d\Pi_1(\omega) = 1. \tag{1.25}$$

To complete the proof, it remains to show that

$$A_t = \sum_{j=1}^{m} \int_0^t \hat{\sigma}_s(h_s^j f, Y)dY_s^j. \tag{1.26}$$

The expressions (1.13) and (1.23) can be rewritten as

$$\sigma_t'(g,Y) = E_P[g'q_t|\mathcal{F}_t^Y] \tag{1.27}$$

and

$$A_t = E_P[\int_0^t g_s'q_s \sum_{j=1}^{m} h_s^j(X_s')dY_s^j|\mathcal{F}_t^Y]. \tag{1.28}$$

To prove (1.26), we need to use Fubini theorem for stochastic integrals, which we state below. This is an extension of Theorem 5.14 of [58], due to Liptsar and Shiryayev.

<u>Proposition 1.3</u>: Let $W_t = (W_t^1,\ldots,W_t^m)$ be a Wiener martingale relative to $(\mathcal{F}_t)$, $0 \leq t \leq T$. Let $a_s = (a_s^1,\ldots,a_s^m)$ be $(\mathcal{F}_s)$-

progressively measurable and such that $M_t := \Sigma_{j=1}^m \int_0^t a_s^j dW_s^j$ is an $(\mathcal{F}_t)$-martingale. Assume the following further conditions on a.

(i) $\sum_{j=1}^{m} \int_0^T |a_s^j|^2 ds < \infty$ a.s.,

(ii) $\sum_{j=1}^{m} \int_0^T \{E(|a_s^j| \,|\mathcal{F}_s^W)\}^2 ds < \infty$ a.s.

Then

$$E\left[\sum_{j=1}^{m} \int_0^t a_s^j dW_s^j |\mathcal{F}_t^W\right] = \sum_{j=1}^{m} \int_0^t E(a_s^j|\mathcal{F}_s^W) dW_s^j \quad \text{a.s.}$$

To apply the result in our setup, recall that (Y_t) is a Wiener process on $(\Omega, \mathcal{A}, \Pi_1)$. Hence taking $\mathcal{G}_t = \mathcal{A} \otimes \mathcal{F}_t^Y$, $(Y_t, \mathcal{G}_t)$ is a Wiener martingale on $(\Omega\times\Omega,\ \mathcal{A}\times\mathcal{A},\ P)$.

Take $a_s^j(\omega',\omega) = g_s(\omega')q_s(\omega',\omega)h_s^j(X_s(\omega'))$. We now verify that for this choice of a_s^j, the conditions of Proposition 1.3 are satisfied. Here

$$M_t = \int_0^t \sum_{j=1}^{m} a_s^j dY_s^j = \int_0^T \Sigma g_s' q_s h_s^j(X_s') dY_s^j = \int_0^t g_s' dq_s \tag{1.29}$$

since $q_t = 1 + \int_0^t \Sigma_{j=1}^m h_s^j(X_s') q_s dY_s^j$, which is itself a $\mathcal{G}_t$-martingale. Hence, we have

$$M_t = \int_0^t g_s' dq_s = g_t' q_t - g_0 - \int_0^t q_s dg_s'$$

$$= g_t' q_t - g_0 - \int_0^t q_s (L_s f)(X_s') ds.$$

Hence if K is a bound of g_s, $L_s f$, then

$$|M_t| \leq K(1 + q_t + \int_0^T q_s ds). \tag{1.30}$$

q_t being a martingale is uniformly integrable and $\int_0^T q_s ds$ is an integrable random variable. Therefore (1.30) shows that M_t is a uniformly integrable and from (1.29) easily to be local martingale, and hence a martingale. It remains to verify the conditions (i) and (ii). The first follows upon observing that

$$\sum_{j=1}^{m} \int_0^T (a_s^j)^2 ds \leq K^2 \sup_{0\leq s\leq T} q_s^2 \int_0^T |h_s(X_s')|^2 ds < \infty \quad \text{a.s.}$$

For the second one,

$$\int_0^T \{E_P(|a_s^j| \,|\mathcal{F}_s^Y)\}^2 ds \leq K^2 \int_0^T \{E_P(|h_s^j(X_s')|q_s|\mathcal{F}_s^Y)\}^2 ds. \tag{1.31}$$

Note that by (1.27)

$$\begin{aligned} E_P(|h_s^j(X_s')|q_s|\mathcal{F}_s^Y) &= \hat{\sigma}_s(|h_s^j|,Y) \\ &= \hat{\pi}_s(|h_s^j|,Y)\cdot\hat{\sigma}_s(1,Y) \end{aligned} \tag{1.32}$$

using (1.15). Again using (1.27), we get

$$\hat{\sigma}_s(1,Y) = E_P(q_s|\mathcal{F}_s^Y). \tag{1.33}$$

Since $(q_t,\mathcal{G}_t)$ is a P-martingale and $\mathcal{F}_t^Y \subseteq \mathcal{G}_t$, (1.33) implies that $\hat{\sigma}_s(1.Y)$ is a $\mathcal{F}_t^Y$-martingale and hence $\hat{\sigma}_s(1,Y)$ has continuous paths (or has a continuous version) as every martingale adapted to $\mathcal{F}_t^Y$ can be represented as a stochastic integral with respect to (Y_t). Hence

$$\int_0^T\{\hat{\sigma}_s(|h_s^j|,Y)\}^2 ds \le \{\sup_{0\le s\le T}|\hat{\sigma}_s(1,Y)|^2\}\int_0^T\{\hat{\pi}_s(|h_s^j|,Y)\}^2 ds. \tag{1.34}$$

Moreover,

$$\int_0^T\{\hat{\pi}_s(|h_s^j|,Y)\}^2 ds = \int_0^T\{E_\Pi(|h_s^j(X_s)|\,|\mathcal{F}_s^Y)\}^2 ds \tag{1.35}$$

$$\le \int_0^T E_\Pi(|h_s^j(X_s)|^2|\mathcal{F}_s^Y)ds < \infty \quad \text{a.s.}$$

since $\int_0^T E_\Pi\{E_\Pi(|h_s^j(X_s)|^2|\mathcal{F}_s^Y\}ds = \int_0^T E_\Pi(|h_s^j(X_s)|^2)ds < \infty$ by assumption (1.18). Hence condition (ii) of Proposition 1.3 is verified. Using Proposition 1.3 in (1.28),

$$A_t = \int_0^t E_P\{g_s' q_s \sum_{j=1}^m h_s^j(X_s')|\mathcal{F}_s^Y\}dY_s^j.$$

Using that $\mathcal{F}_t^Y \subseteq \mathcal{G}_t$ and that $E_P\{g_s'|\mathcal{G}_s\} = f(X_s')$, it now follows that

$$A_t = \int_0^t E_P\{q_s f(X_s')\sum_{j=1}^m h_s^j(X_s')|\mathcal{F}_s^Y\}dY_s^j = \int_0^t \sum_{j=1}^m \hat{\sigma}_s(fh_s^j,Y)dY_s^j.$$

This proves (1.26) and hence completes the proof of the Theorem. □

Stochastic Partial Differential Equation
For the Unnormalized Density

Suppose that $\hat{\sigma}_t(f,Y)$ admits a density, i.e.,

$$\hat{\sigma}_t(f,Y) = \int_{\mathbb{R}^d} f(x)p_t(x,Y)dx$$

where $p_t(x,Y)$ is an $\mathcal{F}_t^Y$ adapted process. Here it is assumed that X_t takes values in $\mathbb{R}^d$. From the Bayes formula it follows that

$$E_{\Pi}[f(X_t)|\mathcal{F}_t^Y] = \frac{\int f(x)p_t(x,Y)dx}{\int p_t(x,Y)dx}$$

and hence $p_t(x,Y)$ is called the unnormalized conditional density of X_t given $\mathcal{F}_t^Y$. From the Zakai equation for $\sigma_t(f,Y)$ it is easily seen that $p_t(x,Y)$ formally satisfies the stochastic PDE

$$dp_t(x,Y) = L_t^* p_t(x,Y)dt + \sum_{j=1}^{m} h_t^j p_t(x,Y)dY_t^j \tag{1.36}$$

with initial condition $p_0(x,Y) = \phi(x)$ where ϕ is the initial (unnormalized) density of X_0 assumed to exist. The operator L_t^* in (1.36) is the formal adjoint of L_t.

The problem of identifying the unnormalized conditional density as the unique solution of a stochastic PDE has been investigated by Krylov-Rozovskii [51, 52], Pardoux [65, 66] and, more recently, by Kunita [54, 55]. These authors consider the case when signal and noise satisfy the more general condition of the independence of $\sigma(X_s, W_s, \; s \leq t)$ and $\sigma\{W_v - W_u, \; t \leq u < v \leq T\}$ stated at the beginning of this section.

NOTES

Chapter I.

For the basic results on measure theory and probability theory used in this monograph, we refer to [1, 13, 59]. For the construction and properties of the Wiener process as well as for the facts on martingales needed in later chapters, see [22, 24, 31, 34]. Weak convergence of probability measures has been treated in detail in [11, 67].

Chapter II.

The treatment of Markov processes, associated semigroup, and weak generator discussed here closely follows [21], the difference being that time inhomogeneous processes are considered here.

In Section 3, results from [72] on existence, uniqueness and convergence of diffusion processes are stated.

The versions of the Feynman-Kac formula presented here are more general than those available in the literature, e.g., in [24, 48].

For properties of Hilbert spaces and linear operators on a Hilbert space, including orthogonal projections, see [20].

Chapter III.

In [4, 27, 35, 38, 70], a theory of integration with respect to cylinder probability has been developed. Whereas some of the earlier treatments were situation specific and some dealt only with the Gauss measure m, we have made an attempt to present a complete treatment of the problem of integration with respect to a cylinder probability measure n in the first three sections. This includes a discussion on the class of 'measurable' functions, the integral, lifting as a tool for handling integrals and examples.

The question as to whether the space $\mathcal{L}^1(H,\mathcal{C},n)$ for a cylinder probability n is complete for the $\mathcal{L}^1$-norm has not been settled to the best of our knowledge.

The proof of Theorem 4.1 presented here is from [33]. For a proof of Sazonov's theorem, a good reference is [56], and for the Minlos theorem, see [32, 75].

For a reference to Gaussian white noise measure on $\mathcal{S}[\mathbb{R}^d)^*$, see [29].

In Section 5, the results on the relation between cylinder integral and the general integral with respect to a finitely additive measure are presented here for the first time. Theorem 5.4 is taken from [47]. A proof of Daniell's theorem is given in [63].

Chapter IV.

As mentioned in Section 1, there are three possible definitions of absolute continuity. One is the $\epsilon-\delta$ definition which, incidentally, is the one used in the context of general finitely additive measures [20]. The second is due to Gross [28]. The latter can be described as follows.

Let $\Omega = \mathbb{R}^H$ and $L(h)$, the coordinate mappings on Ω. Consider the mapping $F: H \to \Omega$ defined by

$$F(h_0) = \{(h_0,h): h \in H\}.$$

Then every cylinder probability n induces a finitely additive measure on the field of finite dimensional cylinders on Ω, which has a countably additive extension denoted by $n \circ F^{-1}$. This is essentially the gist of Theorem 3.1. Now $(L, n \circ F^{-1})$ is a representation of n. Gross's definition can be shown to be equivalent to the following:

$$n_1 \ll n_2 \text{ (Gross)} \quad \text{if } n_1 \circ F^{-1} \ll n_2 \circ F^{-1}.$$

Also, Gross defines the R-N derivative of $n_1 \circ F^{-1}$ with respect to $n_2 \circ F^{-1}$, which is a random variable on the representation space $(\Omega, \mathcal{A}, n_2 \circ F^{-1})$ as the R-N derivative of n_1 with respect to n_2. The $\epsilon-\delta$ definition can be shown to be equivalent to Gross's definition.

If $\mathcal{L}^1(H,\mathcal{C},m)$ is not complete, then one can find n such that $n \circ F^{-1} \ll m \circ F^{-1}$ but $\frac{dn}{dm}$ does not exist in the sense of Definition IV.1. Take a Cauchy sequence $[f_k\}$ which does not converge. Let $n_k = \int f_k dm$ where without loss of generality, we may assume n_k to be a probability measure. Then it can be shown that n_k converges to a cylinder probability n in an appropriate sense (e.g., their characteristic functionals converge to that of n). It is easy to see that $n \circ F^{-1} \ll m \circ F^{-1}$ but $\frac{dn}{dm}$ in the sense of Definition IV.1 does not exist, for if it did, then f_k would converge in $\mathscr{L}^1$ to $\frac{dn}{dm}$. The R-N derivative, say X in the sense of Gross, is the $\mathscr{L}^1$-limit of $R_m(f_k)$. The point here is that there is no $f \in \mathscr{L}^1(H,\mathscr{C},m)$ for which $R_m(f) = X$.

A definition of conditional expectation satisfying our requirement was first given in [35]. A systematic treatment of conditional expectation with cylindrical mappings as conditioning variables was given in [40]. The class of cylindrical mappings considered in [40] was found to be rather restrictive and a generalization is given in Section 2.

Chapter V.

The notion of a quasi-cylinder probability (QCP) was introduced in [40] to enable us to define the white noise model on a fixed probability space.

In this chapter, all the notions considered earlier for cylinder probabilities are examined for QCP's.

The conditional expectation $E_\beta(f|\phi)$ defined in Section 4 satisfies the following properties as well.

(i) Suppose f, f_k, $k \geq 1$ are such that $g = E_\beta(f|\phi)$ and $g_k = E_\beta(f_k|\phi)$ exist. Then

$$E_\beta(|f_k - f|) \longrightarrow 0 \quad \text{implies} \quad E_\beta(|g_k - g|) \longrightarrow 0.$$

If further, $E_\beta(f_k^2) < \infty$, $E_\beta(f^2) < \infty$ then $E_\beta(g_k^2) < \infty$, $E_\beta(g^2) < \infty$ and

$$E_\beta(|f_k - f|^2) \longrightarrow 0 \quad \text{implies} \quad E_\beta(|g_k - g|^2) \longrightarrow 0.$$

(ii) Suppose $\phi: (E,\mathcal{E},\beta) \to (H_1,\mathcal{C}_1)$ is a QCM. Let $h_1 \in H_1$ be fixed. Define $\phi_1: E \to H_1$ by $\phi_1(\omega,h) = \phi(\omega,h) + h_1$. Then it is easy to see that ϕ_1 is a QCM and $\mathcal{U}(\phi) = \mathcal{U}(\phi_1)$. Moreover, if $f \in \mathcal{L}^1(E,\mathcal{E},\beta)$ is such that $E_\beta(f|\phi)$ exists, then so does $E_\beta(f|\phi_1)$ and

$$E_\beta(f|\phi_1) = E_\beta(f|\phi).$$

Both these properties can be proved directly. Alternatively, they can be deduced from Theorem 4.5 and the corresponding properties of conditional expectations in countably additive probability theory. For (ii), all one needs to verify is that $\mathcal{D}_\phi = \mathcal{D}_{\phi_1}$, which is an immediate consequence of

$$R_\alpha((\phi_1,h_1')_1) = R_\alpha((\phi,h_1')_1) + (h_1,h_1')_1.$$

Chapter VI.

The main result of this chapter is Theorem 3.4. The Bayes formula was proved in [40] using the definition of conditional expectation given in Chapter V. Earlier, in [35, 36], the Bayes formula was stated and proved using a weaker definition of conditional expectation. The formula is also treated in [5], but again with a weaker notion of conditional expectation.

Chapter VII.

The first three sections have their origin in [35] where most of the results in Sections 1, 2 and some results in Section 3 were derived. The result on uniqueness of the classical solution in the class of generalized solutions to the Cauchy problem, Theorem 3.4, is of independent interest. The results in Section 4 for unbounded h, a, b are from [36, 37, 40]. See also [4, 7]. In [30], filtering equations have been obtained when the signal process is a Lévy process, a diffusion with jumps or a diffusion with boundary.

In the conventional approach to filtering theory, most of the early work was for the case of bounded h. Some results for the case of unbounded h have been obtained in [8, 66, 71].

Chapter VIII.

The results in the first two sections are contained in [38, 41, 46]. The Markov property of Γ_t, F_t on $(H,\mathscr{C},n)$ was proved in [39]. See [67] for a proof of the fact that $\mathscr{M}(S)$ with the topology of weak convergence is a Polish space if S is Polish.

Chapter IX.

The equations for the finite dimensional smoothing problem for bounded coefficients were obtained in [2]. The rest of the material presented here is new. For corresponding results in the countably additive theory, see [66].

Chapter X.

The first consistency result was proved in [35] for the filtering model of Chapter VII with bounded coefficients. This was extended in [36] to cover the case when h and the coefficients are unbounded.

The results of Sections 1 and 2 pertaining to filtering were proved in [40]. An interesting outcome of this approach via Theorem 1.3 is that one has an independent proof of the fact that the unique solution to the robust form of equation (2.26) suitably transformed using (2.29) is the unnormalized conditional density for the filtering problem in the conventional theory. This proof does not use the theory of stochastic differential equations. The result has also been obtained in [8] under different sets of conditions, using stochastic differential equation techniques.

Consistency results for smoothing and prediction as well as the results in the third section on the infinite dimensional case are new.

Some robustness results for the white noise model were obtained in [30]. For robustness results in the countably additive theory, see also [15, 16].

Chapter XI.

A survey of practical applications of signal detection problems to gravitational wave data analysis is given in [18]. This paper also discusses the relevance of nonlinear filering theory to the Bayes approach to signal detection and has motivated the material of Section 2.

See [5] for a treatment of likelihood ratios in the white noise setup. A likelihood ratio formula for random fields with an application to a problem in physical geodesy is to be found in [6].

Early references to statistical problems in Hilbert spaces are [77, 78].

Some two-parameter filtering problems for the finitely additive white noise model are considered in [43].

The order-disorder problem (sometimes also referred to as the disruption or false alarm problem) has been investigated in [5, 30, 58].

Chapter XII.

A treatment of linear filtering for infinite dimensional signals in the white noise framework is given in [5] where a likelihood ratio formula is also given.

The derivation of the Kalman-Bucy filter from the Bayes formula in the white noise setup seems to be new. So is the explicit solution of the quasilinear filtering problem. See [9, 60] for a treatment of the latter using the countably additive approach.

Appendix.

For a comprehensive account of filtering theory from the stochastic calculus point of view, see [22, 34, 58]. Derivations of the SDE for the conditional expectation and for the optimal filter are given in [25, 45, 57, 62, 68, 76].

REFERENCES

1. Ash, R.B. (1972). *Real Analysis and Probability*. Academic Press: New York.

2. Bagchi, A. (1985). Cylindrical measures in Hilbert space and likelihood ratio for two parameter signals in additive white noise. *Control Theory and Advanced Technology*, *1*, 139-153.

3. Bagchi, A. (1986). Nonlinear smoothing algorithms using white noise model. *Stochastics*, *17*, 283-312.

4. Balakrishnan, A.V. (1976). Applied functional analysis. *Applications of Mathematics*, *Vol*. 3. Springer-Verlag: New York.

5. Balakrishnan, A.V. (1977). Likelihood ratios for signals in additive white noise. *Applied Mathematics and Optimization*, 3, 341-356.

6. Balakrishnan, A.V. (1982). A likelihood ratio formula for random fields with application to physical geodesy. *Applied Mathematics and Optimization*, 8, 97-102.

7. Balakrishnan, A.V. (1980). Non-linear white noise theory. *Multivariate Analysis*, 5, (P.R. Krishnaiah, ed.). North-Holland: Amsterdam.

8. Baras, J.S., Blankenship, G.L. and Hopkins, W.E. (1983). Existence, uniqueness and asymptotic behavior of solutions to a class of Zakai equations with unbounded coefficients. *IEEE Trans. Automat. Control*, 28, 203-214.

9. Benes, V.E. and Karatzas, I. (1983). Estimation and control for linear partially observable systems with non-Gaussian initial conditions. *Stochastic Process Appl.*, *14*, 233-248.

10. Besala, P. (1979). Fundamental solution and Cauchy problem for a parabolic system with unbounded coefficients. *J. Differential Equations*, 33, 26-38.

11. Billingsley, P. (1968). *Convergence of Probability Measures*. Wiley: New York.

12. Bodanko, W. (1966). Sur le probleme de Cauchy et les problèmes de Fourier pour les equations paraboliques dans un domaine non borné. *Ann. Polon. Math.*, *18*, 79-94.

13. Breiman, L. (1968). *Probability*. Addison-Wesley Publication Company: Massachusetts.

14. Chaleyat-Maurel, M. (1986). Robustesse du filtre et calcul des variations stochastique. *J. Func. Anal.*, 68, 55-71.

15. Clark, J.M.C. (1978). The design of robust approximations to the stochastic differential equations of nonlinear filtering. In *Communications Systems and Random Process Theory* (J.K. Skwirzynski, ed.). NATO Advanced Study Institute Series, Sijthoff and Noordhoff: Alphen aan den Rijn.

16. Davis, M.H.A. (1979). Pathwise solutions and multiplicative functionals in nonlinear filtering. *18th IEEE Conference on Decision and Control*, Fort Lauderdale, Fla.

17. Davis, M.H.A. (1980). On a multiplicative functional transformation arising in nonlinear filtering theory. *Z. Wahrsch. verw. Gebiete*, 54, 125-139.

18. Davis, M.H.A. (1988). A review of the statistical theory of signal detection. To appear in *Gravitational Data Analysis* (B.F. Schutz, ed.). D. Riedel: Dorecht.

19. Dubins, L. and Savage, L.J. (1965). *How to Gamble If You Must*. McGrew-Hill Book Company:

20. Dunford and Schwartz (1958). *Linear Operators, Vol. 1*. Interscience: New York.

21. Dynkin, E.B. (1964). *Markov Processes, Vol. 1*. Springer-Verlag: New York.

22. Elliott, R.J. (1982). Stochastic calculus and applications. *Applications of Mathematics, Vol. 18*. Springer-Verlag: Berlin.

23. Friedman, A. (1964). *Partial Differential Equations of Parabolic Type*. Prentice-Hall: New York.

24. Friedman, A. (1975). *Stochastic Differential Equations and Applications, Vol. 1*. Academic: New York.

25. Fujisaki, M., Kallianpur, G. and Kunita, H. (1972). Stochastic differential equations for the nonlinear filtering problem. *Osaka J. Math.*, *1*, 19-40.

26. Gohberg, I.C. and Krein, M.G. (1970). Theory and applications of Volterra operators in Hilbert space. *American Mathematical Society, Translations of Mathematical Monographs, Vol. 24*.

27. Gross, L. (1960). Integration and nonlinear transformations in Hilbert space. *Trans. Amer. Math. Soc.*, 94, 404–440.

28. Gross, L. (1962). Measurable functions on Hilbert space. *Trans. Amer. Math. Soc.*, *105*, 372–390.

29. Hida, T. (1979). *Brownian Motion*. Springer-Verlag: New York.

30. Hucke, H. (1985). *Estimation of continuous time Markov processes in a finitely additive white noise model*, Ph. D. thesis, University of North Carolina at Chapel Hill.

31. Ikeda, N. and Watanabe, S. (1981). *Stochastic Differential Equations and Diffusion Processes*. North-Holland: Amsterdam.

32. Itô, K. (1984). *Foundations of Stochastic Differential Equations in Infinite Dimensional Spaces*. Society for Industrial and Applied Mathematics: Philadelphia.

33. Kallianpur, G. (1971). Abstract Wiener processes and their reproducing kernel Hilbert spaces. *Z. Wahrsch. Verw. Gebiete*, *17*, 113–123.

34. Kallianpur, G. (1980). *Stochastic Filtering Theory*. Springer-Verlag: New York.

35. Kallianpur, G. and Karandikar, R.L. (1983a). A finitely additive white noise approach to nonlinear filtering. *Appl. Math. Optim.*, *10*, 159–185.

36. Kallianpur, G. and Karandikar, R.L. (1983b). Some recent developments in nonlinear filtering theory. *Acta Appl. Math.*, *1*, 399–434.

37. Kallianpur, G. and Karandikar, R.L. (1984a). The nonlinear filtering problem for the unbounded case. *Stochastic Process Appl.*, *18*, 57–66.

38. Kallianpur, G. and Karandikar, R.L. (1984b). Measure valued equations for the optimum filter in finitely additive nonlinear filtering theory. *Z. Wahrsch. verw. Gebiete*, 66, 1–17.

39. Kallianpur, G. and Karandikar, R.L. (1984c). Markov property of the filter in the finitely additive white noise approach to nonlinear filtering. *Stochastics*, *13*, 177–198.

40. Kallianpur, G. and Karandikar, R.L. (1985). White noise calculus and nonlinear filtering theory. *Ann. Probab.*, *13*, 1033–1107.

41. Kallianpur, G. and Karandikar, R.L. (1988). The filtering problem for infinite dimensional processes. *Stochastic Differential Systems, Stochastic Control Theory and Applications* (W. Fleming and P.L. Lions, eds.). Springer Verlag: New York.

42. Kallianpur, G. and Korezlioglu, H. (1986). White noise calculus for two parameter filtering. *Stochastic Differential Systems* (Englebert and Schmidt, eds.). Springer-Verlag: New York.

43. Kallianpur, G. and Oodaira, H. (1973). Non-anticipative representations of equivalent Gaussian processes. *Ann. Probab.*, *1*, 104-122.

44. Kallinapur, G. and Striebel, C. (1968). Estimation of stochastic processes: arbitrary system process with additive white noise observation errors. *Ann. Math. Statist.*, 39, 785-801.

45. Kallianpur, G. and Striebel C. (1969). Stochastic differential equations occurring in the estimation of continuous parameter stochastic processes. *Theory of Prob. and its Applications*, *4*, *Vol. 14*, 567-594.

46. Karandikar, R.L. (1987). On the Feynman-Kac formula and its applications to filtering theory. *Appl. Math. Optim.*, *16*, 263-276.

47. Karandikar, R.L. (1988). A general principle for limit theorems in finitely additive probability: the dependent case. To appear in *J. Multivariate Analysis*.

48. Kesten, H. (1986). The influence of Mark Kac on probability theory. *Ann. Probab.*, *14*, 1103-1128.

49. Kolmogorov, A.N. (1950). *Foundations of the Theory of Probability* (English translation). Chelsea Publishing Company: New York.

50. Korezlioglu, H. and Martias, C. (1984). Martingale representation and nonlinear filtering equation for distribution-valued processes. *Filtering and Control of Random Processes, Lecture Notes in Control and Information Sciences*, *Vol. 61*. Springer-Verlag: Berlin.

51. Krylov, N.V. and Rozovskii, B.L. (1978). On the conditional distribution of diffusion processes. *Math. USSR-Izv.*, *12*, 336-356.

52. Krylov, N.V. and Rozovskii, B.L. (1981). Stochastic evolution equations. *J. Soviet Math.*, *16*, 1233-1276.

53. Kunita, H. (1971). Asymptotic behavior of the nonlinear filtering errors of Markov Processes. *J. Multivariate Analysis*, *1*, 365-393.

54. Kunita, H. (1981). Cauchy problem for stochastic partial differential equations arising in nonlinear filtering theory. *Systems and Control Letters*, *1*, 37-41.

55. Kunita, H. (1983). Stochastic partial differential equations connected with nonlinear filtering. *Nonlinear Filtering and Stochastic Control*, *Lecture Notes in Math.* 972 (S.K. Mitter and A. Moro, eds.). Springer-Verlag: New York.

56. Kuo, H.H. (1975). *Gaussian Measures in Banach Spaces*. Springer-Verlag: New York.

57. Kushner, H. (1967). Dynamical equations for optimal nonlinear filtering. *J. Differential Equations*, 3, 179-190.

58. Liptser, R.S. and Shiryaev, A.N. (1977). *Statistics of Random Processes*, *Vol. 1*. Springer-Verlag: New York.

59. Loeve, M. (1977). *Probability Theory*, 4th edition. Springer-Verlag: New York.

60. Makowski, A.M. (1986). Filtering formulae for partially observed linear systems with non-Gaussian initial conditions. *Stochastics*, *16*, 1-24.

61. Meyer, P.A. (1966). *Probability and Potentials*. Blaisdell, Waltham: Massachusetts.

62. Mortenson, R.E. (1966). Optimal control of continuous-time stochastic systems. Report ERL-66-1, Electronics Research Laboratory, College of Engineering, University of California, Berkeley.

63. Neveu, J. (1965). *Mathematical Foundations of the Calculus of Probability* (English translation). Holder-Day: San Francisco.

64. Ocone, D. (1984). Remarks on the finite energy condition in additive white noise filtering. *Systems and Control Letters*, 5, 197-203.

65. Pardoux, E. (1979). Stochastic partial differential equations and filtering of diffusion processes. *Stochastics*, 3, 127-203.

66. Pardoux, E. (1982). Equations du filtrage non lineaire de la prediction et du lissage. *Stochastics*, 6, 193-231.

67. Parthasarathy, K.R. (1967). *Probability Measures on Metric Spaces*. Academic Press: New York.

68. Rozovskii, B.L. (1975). On stochastic partial differential equations. *Math. USSR Sbornik*, 25, 295-322 (English translation).

69. Savage, L.J. (1954). *The Foundations of Statistics*. John Wiley and Sons: New York.

70. Segal, I.E. (1956). Tensor algebras over Hilbert spaces. *Trans. Amer. Math. Soc.*, *81*, 106-134.

71. Sheu, S.J. (1983). Solution of certain parabolic equations with unbounded coefficients and its application to nonlinear filtering. *Stochastics*, *10*, 31-46.

72. Stroock, D.W. and Varadhan, S.R.S. (1979). *Multidimensional Diffusion Processes*. Springer-Verlag: New York.

73. Szpirglas, J. (1978). Sur l'équivalence d'équations differentielles stochastiques a valeurs mesures intervenant dans le filtrage markovien non linéaire. *Ann. Insti. Henri Poincare, Sec. B.*, *XIV*, 33-59.

74. Wong, E. and Zakai, M. (1977). Likelihood ratios and transformation of probability associated with two parameter processes. *Z. Wahrsch. Verw. Gebiete*, *40*, 283-308.

75. Yamasaki, Y. (1985). *Measures on Infinite Dimensional Spaces*. World Scientific: Singapore-Philadelphia.

76. Zakai, M. (1969). On the optimal filtering of diffusion processes. *Z. Wahrsch. verw. Gebiete*, *11*, 230-243.

77. Bensoussan, A. (1969). *L'identification et le Filtrage*. IRIA, cahier 1.

78. Bensoussan, A. (1970). Statistical Problems in Hilbert Spaces. *Kybernetika cislo 4*, 270-271.

79. Grenander, U. (1981). *Abstract Inference*. John Wiley: New York.

80. Hucke, H., Kallianpur, G. and Karandikar, R.L. (1987). Smoothness properties of the conditional expectation in finitely additive probability. Preprint.

81. Kallianpur, G. and Wolpert, R. (1984). Infinite dimensional stochastic differential equation models for spatially distributed neurons. *Appl. Math. Optim.*, *12*, 125-172.

82. Walsh, J.B. (1981). A stochastic model of neural response. *Adv. Appl. Prob.*, *13*, 231-281.

INDEX

SYMBOLS